PROGRESS IN TEMPERATE FRUIT BREEDING

Developments in Plant Breeding

VOLUME 1

Progress in Temperate Fruit Breeding

Proceedings of the Eucarpia Fruit Breeding Section Meeting held at Wädenswil/Einsiedeln, Switzerland from August 30 to September 3, 1993

Edited by

HANNA SCHMIDT
Bundesanstalt für Züchtungsforschung, Ahrensburg, Germany

and

MARKUS KELLERHALS
Swiss Federal Research Station, Wädenswil, Switzerland

Chapters indicated with an asterisk in the table of contents were first published in *Euphytica*, Volume 77: 1–2, 1994

Springer-Science+Business Media, B.V.

Library of Congress Cataloging-in-Publication Data

Progress in temperate fruit breeding / edited by Hanna Schmidt and
Markus Kellerhals.
p. cm. -- (Developments in plant breeding ; v. 1)
Proceedings of an Eucarpia Fruit Section conference held at
Waedenswil/Einsiedeln, Switzerland.
Includes index.

1. Fruit--Breeding--Congresses. 2. Fruit--Genetics--Congresses.
3. Fruit--Biotechnology--Congresses. 4. Fruit--Micropropagation--Congresses. I. Schmidt, Hanna. II. Kellerhals, Markus.
III. Eucarpia. Fruit Section. IV. Series.
SB359.35.P76 1994
634'.0423--dc20 94-21241

ISBN 978-94-010-4209-3 ISBN 978-94-011-0467-8 (eBook)

DOI 10.1007/978-94-011-0467-8

Printed on acid-free paper

Originally published by Kluwer Academic Publishers in 1994
Softcover reprint of the hardcover 1st edition 1994

Contents

Chapter 3: Breeding methods and genetics

Chapter 4: Breeding for specific aims

Chapter 5: Molecular markers

Chapter 6: in vitro methods and biotechnology

Chapter 7: Genetic resources

List of participants

Acatrinei, Florin
Statuinea de Cercetare si Productie Pomicola, BACAU, Calea Romanului 237, 5000 Bacau, Romania

Akça, Yasar
100 Yil Universitesi, Ziraat Fakültesi, Bahçe Bitskleri Bölümü, 65080, Van kileri, Turkey

Aldwinckle, Herb
Cornell University, New York State Agric. Exp. Stn., Geneva, New York 14456, USA

Alston, Frank H.
Horticulture Res. International, East Malling, West Malling, ME19 6BJ Kent, UK

Apostol, Janos
Enterprise for Research and Extension in Fruitgrowing and Ornamentals, P.O. Box 108, H-Budapest 1775, Hungary

Arús, Pere
IRTA Cabrils (Barcelona), Spain

Balan, Viorica
Statiunea de Cercetare si Productie Pomicola, Bulevardul Ion Ionescu de la Brad Nr. 4 Sect. 1, Bucuresti/Romania

Barut, Erdoğan
Uludag University, Fac. of Agriculture, Dept. of Horticulture, 16036 Hürriyet, Bursa, Turkey

Bauer, Annelise
Hechlstrasse 11, D-83254 Breitbrunn, Germany

Bell, Judith
Horticulture Research International, East Malling, West Malling, Kent, ME19 6BJ, UK

Bessho, Hideo
Morioka Branch, Fruit Tree Res. Stn., Shimokuriyagawa, Morioka, 020-01, Japan

Blažek, Jan
Research and Breeding Institute of Pomology, 50751 Holovousy, Czech Republic

Buchter, Helga Anna
Hauptstr. 140, D-67127 Roedersheim, Germany

Chat, Joëlle
INRA, Station de Recherches Fruitières, BP 81, F-33883 Villenave d'Ornon Cedex, France

Chevalier, Michel
Dept. de Biologie I.U.T., BP 2018, 4 Bd Lavoisier F-49016 Angers Cedex, France

Coman, Mihail
Fruit Research Institute, Pitesti-Maracineni, 0300, Romania

Cossio, Ferdinando
Istitutto Sperimentale di Frutticoltura, Provincia di Verona, Via S. Giacomo 25, I-37135 Verona, Italy

da Câmara Machado, Artur
University of Agriculture, Inst. für angew. Mikrobiologie, Nussdorfer Lände 11, A-1190 Wien, Austria

Darbellay Charly
Swiss Federal Research Station, RAC, Centre des Fougères, CH-1964 Conthey, Switzerland

Den Nijs, Ton P.M.
CPRO-DLO, PO Box 16, NL-6700 AA Wageningen, The Netherlands

Deslauriers, Christiane
Agriculture Canada, Research Station, Kentville, Nova Scotia B4N IJ5, Canada

Dirlewanger, Elisabeth
INRA Centre de Recherches de Bordeaux, BP 81, F-33883 Villenave d'Ornon Cedex, France

Djouvinov, Vassiliiy
Breeding Department, Fruitgrowing Research Inst., 12, Ostromila str., 4004 Plovdiv, Bulgaria

Dublin, Pierre
15, rue des Fauvettes, F-21121 Fontaine les Dijon, France

Dunemann, Frank
Bundesanstalt für Züchtungsforschung, Inst. für Zierpflanzenzüchtung, Bornkampsweg 31, D-22926 Ahrensburg, Germany

Evans, Kate
Horticulture Research International, East Malling, West Malling, Kent, ME19 6BJ, UK

Eyssen, Rik
Fruitteeltcentrum, K.U. Leuven, Willem de Croylaan 42, B-3001 Leuven-Heverlee, Belgium

Fischer, Manfred
Genbank Obst, Dorfplatz 2, D-01326 Dresden, Germany

Fischer, Christa
Bundesanstalt für Züchtungsforschung, Institut für Obstzüchtung Pillnitzer Platz 2, D-01326 Dresden, Germany

Gardiner, Susan
Hort. + Research, Batchelar Research Centre, Private Bag 11 030, Palmerston North, New Zealand

Gessler, Cesare
Institut für Pflanzenwissenschaften, Phytomedizin, ETH-Zentrum, CH-8092 Zürich, Switzerland

Glucina, Paul
Hort. + Research, Private Bag 92169, Auckland, New Zealand

Hanke, Viola
Bundesanstalt für Züchtungsforschung, Institut für Obstzüchtung, Pillnitzer Platz 2, D-01326 Dresden, Germany

Hjeltnes, Stein Harald
Ullensvang Research Station, Substation NJØS, N-5840 Hermansverk, Norway

Höfer, Monika
Bundesanstalt für Züchtungsforschung, Institut für Obstzüchtung, Pillnitzer Platz 2, D-01326 Dresden, Germany

Ivaničhka, Július
Fruit Research Breeding Institute, 972 01 Bojnice, Slovakia

James, David
Horticulture Research International East Malling, West Malling, Kent ME19 6BJ, UK

Karnatz, Annemarie
Rudeloffweg 28A, D-14195 Berlin, Germany

Kellerhals, Markus
Swiss Federal Research Station, CH-8820 Wädenswil, Switzerland

Keulemans, Johan
Fruitteeltcentrum K.U. Leuven, Willem de Croylaan 42, B-3001 Leuven-Heverlee, Belgium

Khanizadeh, Shahrokh
Agriculture Canada, Research Station, 430, Boul. Gouin, St-Jean-sur Richelieu, Quebec, Canada, 93B 3E6

King, Graham
Horticulture Research International, Wellesbourne, Warwick CV35 9EF, UK

Koch, Thomas
Institut für Pflanzenwissenschaften, Phytomedizin, ETH-Zentrum, CH-8092 Zürich, Switzerland

Koller, Bernhard
Institut für Pflanzenwissenschaften, Phytomedizin, ETH-Zentrum, CH-8092 Zürich, Switzerland

Kolova, Lilly
Fruitgrowing Research Institute, 12, Ostromila str., 4004 Plovdiv, Bulgaria

Korban, Schuyler S.
Dept. of Horticulture, 310 PABL, University of Illinois, 1201 W. Gregory, Urbana, Illinois 61801, USA

Kozlovskaya, Zoya
Byelorussian Research Institute for Fruit Growing, 2, Kovaleva St., Samokhvalovitchy, 223013 Minsk Region, Byelorussia

Krüger, Jutta
Bundesanstalt für Züchtungsforschung, Institut für Zierpflanzenzüchtung, Bornkampsweg 31, D-22926 Ahrensburg, Germany

Lateur, Marc
Centre de Recherche Agronomiques, Station de Phytopathologie, Chemin de Liroux, B-5030 Gembloux, Belgium

Laurens, François
INRA Centre de Recherche d'Angers, Station d'Amélioration des Espèces, Fruitières et Ornementales, BP 57, F-49071 Beaucouzé Cedex, France

Ledbetter, Craig
USDA, ARS/HCRL/PQ6, 2021 S. Peach Avenue, Fresno, CA 93727-5951, USA

Lespinasse, Yves
INRA Centre de Recherches d'Angers, Station d'Amélioration des Espèces, Fruitières et Ornementales, BP 57, F-49071 Beaucouzé Cedex, France

Maliepaard, Chris
Schaepmanstraat 26, NL-6702 AW Wageningen, The Netherlands

Markussen, Torsten
Bundesanstalt für Züchtungsforschung, Institut für Zierpflanzenzüchtung, Bornkampsweg 31, D-22926 Ahrenburg, Germany

Me, Giovanni
Università degli studi di Torino, Dip. di colture arboree, Via Ormea 99, I-10126 Torino, Italy

Miaja, Maria Luisa
Università degli studi di Torino, Dip. di colture arboree, Via Ormea 99, I-10126 Torino, Italy

Müller, Walter
Swiss Federal Research Station, CH-8820 Wädenswil, Switzerland

Nadosy, Ferenc
University of Horticulture and Food Industry, Dept. of Fruit Production, Villanyi ut 29–43, H-1118 Budapest, Hungary

Nestby, Rolf
Kvithamar Research Station, N-7500 Stjørdal, Norway

Nybom, Hilde
The Swedish University of Agriculture Sciences, Balsgard, Dept. of Hort. Plant Breeding, SLU, Fjälkestadsvägen 123-1, S-29194 Kristianstad, Sweden

Ognjanov, Vladislav
Faculty of Agriculture, Institute of Fruit-Growing, D. Obradovica 8, 21000 Novi Sad, Yugoslavia

Paprštein, František
Research and Breeding Institute of Pomology, 50751 Holovousy, Czech Republic

Parisi, Luciana
INRA Centre de Recherches d'Angers, Station de pathologie végétale, BP 57, F-49071 Beaucouzé Cedex, France

Pascal, Tierry
INRA Station de Recherches Fruitières Méditerranéennes, Domaine Saint Paul, BP 91, F-84143 Montfavet, France

Poessel, Jean-Luc
INRA Station de Recherches Fruitières Méditerranéeenes, Domaine Saint Paul, BP 91, F-84143 Montfavet, France

Popescu, Aurel
Research Institute of Pomology, 0300 Piteşti – Mărăcineni, Romania

Rapillard, Charly
Swiss Federal Research Station, RAC, Centre des Fougères, CH-1964 Conthey, Switzerland

Rati, Joan Viorel
Statuinea de Cercetare si Productie Pomicola BACAU, Calea Romanului 237, 5500 Bacau, Romania

Riggs, Tim
Horticulture Research International, Wellesbourne, Warwick, CV35 9EF, UK

Rudenko, Irina
Botanical Garden, Lesnaya 18, 277001 Kishinev, Moldova, GUS/CIS

Sansavini, Silviero
Università degli studi di Bologna, Dipartimento di Colture arboree, Via Filippo Re 6, I-40126, Bologna, Italy

Scheewe, Petra
Bundesanstalt für Züchtungsforschung, Institut für Zierpflanzenzüchtung, Bornkampsweg 31, D-22926 Ahrensburg, Germany

Schmidt, Hanna
Bundesanstalt für Züchtungsforschung, Institut für Zierpflanzenzüchtung, Bornkampsweg 31, D-22926 Ahrensburg, Germany

Schumacher, Robert
Swiss Federal Research Station, CH-8820 Wädenswil, Switzerland

Sharma, Ram Lok
Dept. of Fruit Breeding, University of Horticulture + Forestry, P.O. Nauni, Solan H.P., Pin 173230, India

Sierotzki, Helge
Institut für Pflanzenwissenschaften, Phytomedizin, ETH-Zentrum, CH-8092 Zurich, Switzerland

Silbereisen, Robert
Franz Beer-Strasse 62, D-88250 Weingarten, Germany

Socias i Company, Rafael
Servicio de Investigacion agraria, Apartado 727, 50080 Zaragoza, Spain

Solar, Anita
Biotechnidal Faculty, Agronomy Department, Jamnikarjeva 101, 61000 Ljubljana, Slovenia

Spiegler, Gertrud
Häberli Obst- und Beerenzentrum AG, 9315 Neukirch-Egnach, Switzerland

Štampar, Franci
Biotechnical Faculty, Agronomy Department, Jamnikarjeva 101, 61000 Ljubljana, Slovenia

Stanisavljević, Miloljub
Agriculture Research Institute Serbia, Fruit and Grape Research Centre, Kralja Petra I 9, 32000 Cacak, Yugoslavia

Theiler, Robert
Swiss Federal Research Station, CH-8820 Wädenswil, Switzerland

Thoss, Hagen
Häberli Obst- und Beerenzentrum AG, 9315 Neukirch-Egnach, Switzerland

Tobutt, Kenneth
Horticulture Research International East Malling, West Malling, Kent ME19 6BJ, UK

Tóth G., Magdolna
University of Horticulture and Food Industry, Dept. of Pomology, Villanyi ut 35–43, H-1118 Budapest/Hungary

Trifonova, Adelina
Inst. of Genetic Engineering, 2232 Kostinbrod-2, Bulgaria

Tschabold, Jean Luc
Service Migros Sano, Case Postale 203, CH-1010 Lausanne, Switzerland

Vallania, Rosalina
Centro Migloramento Genetico, Vite – CNR, Via P. Giuria 15, I-10126 Torino, Italy

Végvari, György
University of Horticulture and Food Industry, Dept. of Fruit Production, Villanyi ut 29–43, H-1118 Budapest/Hungary

Viršč💩ek-Marn, Mojca
Biotechnical Faculty, Agronomy Department, Jamnikarjeva 101, 61000 Ljubljana, Slovenia

Weeden, Norman F.
Cornell University, Dept. of Hort. Sciences,. Sturtevant Hall, Geneva, N.Y. 14456, USA

White, Allan G.
Hort. + Research, Havelock North Research Centre, Private Bag 1401, Havelock North, New Zealand

Yang, Hongyu
Bundesanstalt für Züchtungsforschung, Institut für Zierpflanzenzüchtung, Bornkampsweg 31, D-22926 Ahrensburg, Germany

Yao, Yingmou
Department of Plant Biology, University of Helsinki, 00710 Helsinki, Finland

Li, Zai-long
Department of Horticulture, Zhejiang Agricultural University, Hangzhou, Zhejiang 310029, The People's Republic of China

Zayats, Vasil
Uzhgorod Universiti, 54 Zhovtneva St., 294000 Uzhgorod, Ukraine

Zhukov, Oleg
I.V. Michurin All-Russian Research Institute of Fruit Crops, Genetics and Selection, 393740 Michurinsk-10 Tambov Region, Russia

Zilkah, Shmuel
Institute of Horticulture, ARO The Volcani Center, P.O. Box 6, Bet Dagan 50250, Israel

Züblin, Johann
Migros Aargau/Solothurn, Postfach 182, CH-5034 Suhr, Switzerland

Welcome

Dear colleagues,

It is a great pleasure to welcome you here on behalf of EUCARPIA Fruit Section. As you know we were scheduled to meet two years earlier at Novi Sad, Yugoslavia. Dr. Ognjanov had already organised everything, when we had to cancel the meeting a few weeks before opening. The short notice was due to our inability to foresee and imagine such a cruel, long-lasting and widespread war in the Balkans. We should give our colleague, Dr. Ognjanov, a warm welcome here.

We are also very grateful to Dr. Kellerhals, who offered to host the Fruit Section at Wädenswil/Einsiedeln and put a lot of effort into organising this week's meeting.

EUCARPIA Fruit Section Conferences were always dominated by the pome fruits, especially apples, and by participants from pome fruit producing countries. If you look at the present programme you will see that again nearly 50% of papers/posters will deal with pome fruit, mainly apples. This situation always made it difficult for breeders of other fruit crops to 'feel at home' in the Fruit Section. However, we welcome some 20 contributions on stone fruit and 13 on soft fruit among others. The crop-specific meetings organised by ISHS are often more attractive to non-apple breeders, as these meetings include breeding together with other fields of research. EUCARPIA Fruit Section never wanted to exclude the breeders of other fruit crops. However, either the apple is so important and creates so many problems or EUCARPIA Fruit Section is regarded more or less as an APPLE MEETING, preventing other fruit breeders from participating. Since progress in breeding is based on many other fruit crops we should not exclude the colleagues dealing with these other crops.

High priority is given to *stable resistances*, a rather minor topic at the last few meetings. A quarter of the contributions will deal with this field.

The availability of well-documented *genetic resources* becomes more important than ever, especially in breeding for resistances. However, we should not spend more time in collecting and describing genetic resources than for using them in breeding.

Direct *transfer* of resistance genes will be an important field of future research. We should, however, not forget that after having transferred a gene a long, quite conventional testing procedure in the field will be necessary, not always accelerating breeding quite as much as many colleagues postulate. And we should not forget the consumers' fear of genetically engineered food.

The chances for *marker-aided selection* are good. First results in fruit crops will be presented. A prerequisite for both marking as well as transfer is the identification of genes, a field where we did not work hard enough in the past.

The use of *in vitro techniques* in fruit breeding was neither a topic at the Rome nor at the Hradec Kralove meetings. This week we have 22 contributions on the programme, dealing with regenerating callus, leaves, cotyledons, embryo rescue as well as production of haploids and genetic engineering.

Let me wish everyone an inspiring conference and as much exchange of information as possible, leading us to be aware of the most recent standards in fruit breeding – not to forget personal communication, if possible not *only* on fruit breeding!

Hanna Schmidt
Chairwoman of EUCARPIA
Fruit Section

Opening address

Let me first express to Eucarpia the greetings of the ISHS (International Society for Horticultural Science). The fact that both organizations are represented by their respective fruit sections underscores their shared objectives of furthering scientific research, the exchange of information and cooperation in horticultural genetics and breeding. The activities of Eucarpia are thus rooted in the same terrain as those of the ISHS, which will continue to support, as it has always done in the past, the initiatives of Eucarpia and promote the dissemination of the results of the joint Eucarpia-ISHS Fruit Section meetings.

The world of science looks with increasing interest to genetics in its promise to solve such key problems in agriculture as environmental adaptation of genotypes, resistance to biotic and abiotic stress, yield performance and quality and breeding genetic variability in new cultivars to meet the commercial, nutritional and health demands of the world's population.

This meeting of the Eucarpia Fruit Breeding Section, like its predecessors, has all the makings of success. It can make an especially important contribution to advances in *in vitro* techniques, selection for disease and pest resistance, and the use of molecular gene markers in apple. I would also add that Wädenswil and Einsiedeln are ideal venues that help to make the entire ambience of the meeting most attractive. Let me conclude my remarks by thanking the Wädenswil Station, and Dr. Kellerhals in particular, for their organisational efforts, and Dr. Hanna Schmidt, our chairwoman, for her firm and guiding hand.

S. Sansavini
Chairman of ISHS Fruit Section

Acknowledgements

It is a pleasure to acknowledge the following colleagues who assisted in the review process of the manuscripts:

Frank Alston, HRI, East Malling, England
Cesare Gessler, ETH Zürich, Switzerland
Graham King, HRI, Wellesbourne, England
Robert Theiler, FAW, Wädenswil, Switzerland
Kenneth Tobutt, HRI, East Malling, England
Allan White, Hort+Research, Havelock North, New Zealand

We are very grateful to these reviewers for taking the time to assess the scientific and improve the linguistic standard of the contributions. Special thanks are due to Tim Riggs, HRI, Wellesbourne, England, for checking the English.

We are grateful to Marianne Engeli, Monica Goerre and Hans-Ueli Nebiker, FAW, Wädenswil, Switzerland, for their assistance during the meeting and for collecting and processing the manuscripts. Thanks also to Gilles Jonker from Kluwer Academic Publishers, Dordrecht, The Netherlands, for the pleasant collaboration.

Hanna Schmidt
Markus Kellerhals

CHAPTER 1

Breeding for stable disease and pest resistance

H. Schmidt and M. Kellerhals (Eds.), Progress in Temperate Fruit Breeding, 3–6.

Progress in combining mildew resistance from *Malus robusta* and *Malus zumi* with fruit quality

Hanna Schmidt
Institut für Zierpflanzenzüchtung, Bornkampsweg 31, D-22926 Ahrensburg

Key words: Fruit quality, fruit size, *Malus*, mildew resistance

Summary

Fruit quality was gradually improved in apples derived from mildew resistant species. Three hybrid generations were raised from a *Malus robusta* hybrid since 1977. The 4th generation from a *M. zumi* derivative is now in the greenhouse. The 1993 seed year generation is segregating for mildew, scab and *Nectria* canker resistance from parents combining good fruit quality with resistances.

Introduction

In 1976 a fruit breeding unit joined the former Federal Research Station for Horticultural Plant Breeding at Ahrensburg. Priority was given to breeding for resistance in apples. Sources of mildew resistance were identified in seedlings of *M. robusta* and *M. zumi* at East Malling (Knight & Alston 1968). Initial support was given to our breeding work by supplying us with open pollinated seed of ‘A143/14’ with the *M. zumi* resistance gene ‘Pl2’ as well as scion material of ‘A142/5’ with the *M. robusta* ‘Pl1’ resistance gene (Alston 1976).

Material and methods

We produced a series of backcross generations (BC) with both resistance genes (Figs. 1 and 2). Whereas in the first BC fruit size of the susceptible parents was the main selection criterion, quality and in the last breeding generation resistance to scab and *Nectria* canker were added as criteria for use as parents. Because of the short juvenile period transferred by *M. zumi* four generations of seedlings could be raised (Fig. 3). *M. robusta* seedlings were somewhat slower to fruit, only 3 generations being produced.

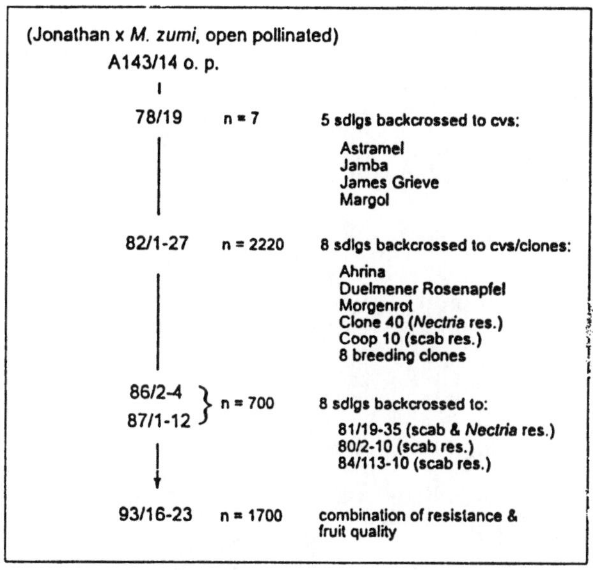

Fig. 1. Transfer of *Malus zumi* mildew resistance

Results

Length of the juvenile period

M. zumi line Progeny 78/19 flowered 3.0 years from seed. Among the 2220 seedlings of the next generation 68% flowered by the 4th year. The third generation had a juvenile period of 3.9 years.

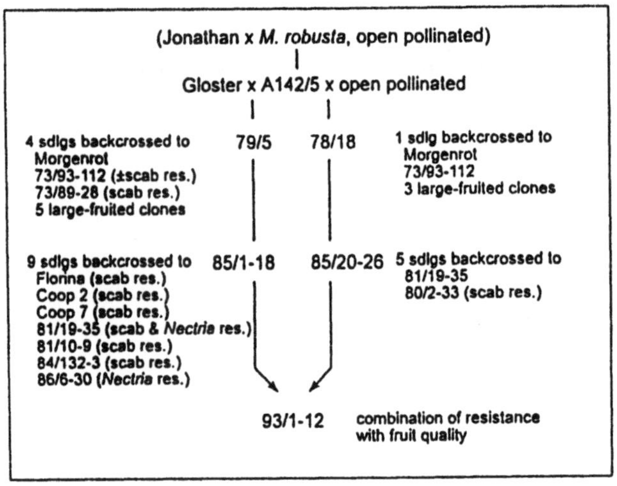

Fig. 2. Transfer of *Malus robusta* mildew resistance

Fig. 3. Increase in fruit size over three generations derived from *M. zumi*

M. robusta line Trees of progeny 79/5 flowered 4,4 years from seed. The progenies from 1985 (85/1–18, 20–26) had a juvenile period of 5.9 years when grafted on to M27, compared with more than 7 years on their own roots, which is partly due to non-optimal cultivation treatment.

Fruit size

M. zumi line Fruit sizes of progeny 78/19 were below 20 mm diameter. The next generation averaged 25 mm with the best selections used in crosses about 40 mm (Fig. 4). The 700 individuals of the 1986/87 progenies (86/2–4, 87/1–12) were up to 51 mm with 11% having at least 60 mm. The best selections chosen for further crosses averaged 55 mm.

M. robusta line Clone 'A142/5' had considerably larger fruit (45 mm) than the *M. zumi* seedlings of 78/19. Thus, the parents selected from 79/5 averaged already more than 50 mm (Fig. 5). The next generation had 60 mm. Selected parents derived from one of the 79/5 seedlings averaged 65 mm and those of 78/18–4 60 mm. Forty percent of their derivatives were at least 60 mm, 14% 65 mm and 5% 70 mm and more.

Taste and sugar/acid ratio

M. zumi line The sugar/acid ratio was estimated on a scale of 1 = very acid to 9 = very sweet, and taste also on a 1–9 scale with 9 = excellent. Within the range of a good sugar/acid ratio from 4 to 7 there is a clear correlation with taste, although in some progenies taste is rated somewhat higher. A high proportion of acid and astringent fruit, often in combination, was found which resulted in low ratings for taste (4.2).

M. robusta line On average progenies showed sugar/acid ratios of 4.7 and taste scores of 4.9. There were marked differences in sugar/acid and taste between the four groups of mildew resistant parents used. Progenies derived from 79/5–19 rated 5.3 for sugar/acid and taste, the progeny groups derived from 79/5–16 and 79/5–21 rated 4,7 and 5.0 and the seedlings derived from 78/18–4 were far more acid with 4.2 for sugar/acid and 4.6 for taste. Among the 79/5–19 seedlings there were only very few with acid fruit, rated 3 and below (4%). Among the 78/18–4 progenies this percentage was up to 30% with only few seedlings rated 6 or 7 (8%).

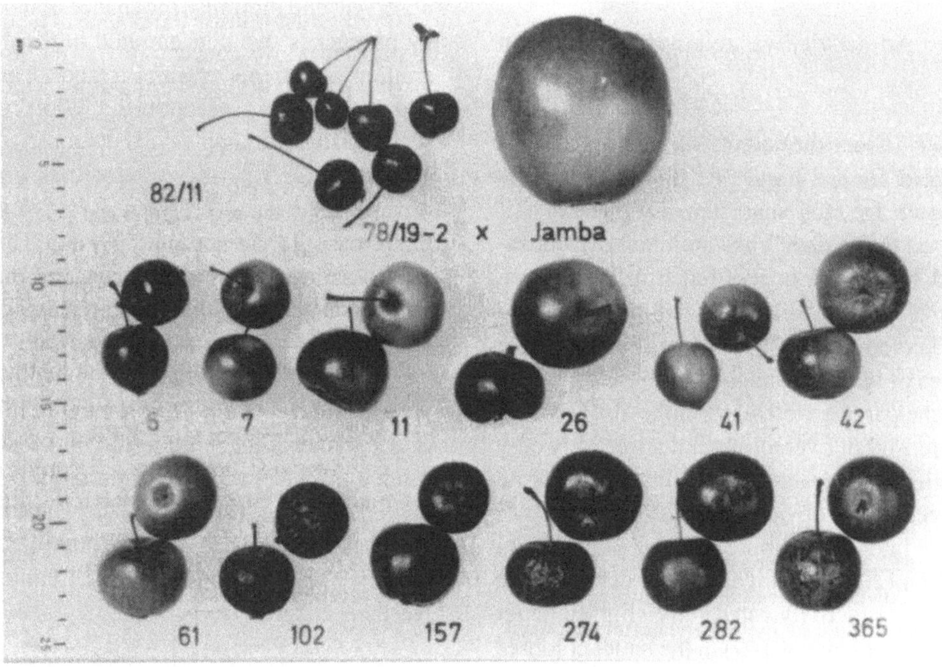

Fig. 4. Fruit size in progeny 82/11

Fig. 5. Fruit size in progeny 79/5

Attractivity

M. zumi line Attractivity was relatively high, 5.7 on a 1–9 scale.

M. robusta line Since the parents 79/5–16 and 78/18–4 carry the gene for red tissue 'Rt' the seedling progenies segregate for this character very clearly 1:1. All fruit of 'red tissue' seedlings are intensely bluish-red and most have more or less red fruit flesh. Even the majority of the green-leafed seedlings produced intensely overcoloured fruit with 2.6 on a 0–3 scale for the derivatives of 79/5–16 and 2.3 for those of 78/18–4. No 'Rt' gene parents were used in the last breeding cycle. For intensity of overcolour and attractivity only those progenies were evaluated that did not segregate for the 'Rt' gene. It is obvious that all four progenies from the more intensely coloured 79/5–21 rated higher in overcolour (2.1) compared to those derived from 79/5–19 (1.3). Attractivity rated only slightly better (5.2 versus 5.0 on a scale of 1–9) in the better coloured progenies.

Discussion

On the basis of first backcross generation mildew resistant material, made available to us by East malling, we continued the transfer of mildew resistance to cultivated apples. Whereas Alston (1976) 'choose cultivars with high susceptibility to mildew as parents, to ensure a 'neutral' background on which the segregation for any major resistance genes could be observed, and minimise the chance of it being obscured by polygenes' we concentrated in the following generations on parents combining good fruit size, quality and low mildew susceptibility. Work on transfer of scab resistance in apples had demonstrated (Rousselle 1974) that the frequency of seedlings with a high level of resistance was decreasing over five generations of backcrossing to susceptible cultivars. This is explained as a loss of modifyers with distance from *M. floribunda*. We can assume that the genetic background does contribute to mildew resistance as well.

Whereas for fruit quality size had highest priority we tried to select large-fruited parents with good eating quality. In each generation the best recombinants for mildew resistance, low scab susceptibility, fruit size and quality were used for the next breeding cycle.

The 1992 and 1993 crosses are made to combine fruit size, quality, and resistance to mildew, scab, and *Nectria* canker.

References

Alston, F.H., 1976. Practical aspects of breeding for mildew (*Podosphaera leucotricha*) resistance in apples. Proc. Eucarpia Meeting on Tree Fruit Breeding, Wageningen, pp. 4–17.

Knight, R.L. & F.H. Alston, 1968. Sources of field immunity to mildew (*Podosphaera leucotricha*) in apple. Canad. J. Genetics Cytology X: 294–298.

Rousselle, G., 1974. Variation of the hypersensitive reaction of the Vf gene for apple scab resistance. Thesis, Rutgers Univ., New Jersey.

H. Schmidt and M. Kellerhals (Eds.), Progress in Temperate Fruit Breeding, 7–12.

Observations on different mildew sources used in apple breeding at Ahrensburg

Jutta Krüger
Institut für Zierpflanzenzüchtung, Bornkampsweg 31, 22926 Ahrensburg, Germany

Key words: Apple, breeding, mildew, *Podosphaera leucotricha*, resistance, *Malus*

Summary

Progenies of *Malus robusta* (A142/5), *M. zumi* (A143/14, A143/24) and Klon 40 (Goldparmäne open-pollinated) as well as those between varieties were evaluated for natural mildew (*Podosphaera leucotricha*) incidence for up to 10 years in the field. In progenies from all resistance sources field resistant plants could be found. *M. robusta* and *M. zumi* transmitted resistance better than Klon 40 and varieties. *M. zumi* had higher numbers of resistant plants in progenies than *M. robusta*. More plants being attacked at least once moderately or severely during the observation time occured in *M. robusta* than in *M. zumi* progenies. Normally in progenies of Klon 40 and of varieties only a small part of the plants was field resistant; the number being higher in Klon 40 than in varieties. The best variety Rote Sternrenette was as good as Klon 40.

Introduction

Apple mildew (*Podosphaera leucotricha* (Ell. and Ev.) Salm.) is one of the most serious fungus diseases of apple, needing high applications of fungicide sprays for its control. Since 1978 resistance breeding against this disease has been carried out at Ahrensburg. The different sources of resistance having been used were *Malus robusta*, *M. zumi*, Klon 40 and different cultivated varieties. The resistance of *M. robusta* and *M. zumi* depends on two dominant genes (Alston, 1983) while the low susceptibility of Klon 40 and cultivated varieties to mildew is polygenic (Schander, 1958; Brown, 1959; Misic, 1969; Korban & Riemer, 1990).

Material and methods

The mildew resistant selections A142/5 (Jonathan × *M. robusta*) and A143/14 and A143/24 (Jonathan × *M. zumi*) and Klon 40 (Goldparmäne open pollinated) were crossed with susceptible varieties or selections; crosses between varieties were made too. Natural mildew incidence was recorded in their progenies.

Young seedlings were evaluated for mildew incidence at the end of the first year in the nursery. Plants infected moderately or severely were eliminated and the remaining trees were planted in unsprayed infection plots. Mildew incidence was recorded each year in June/July and September/October. Four grades were distinguished:

1. healthy;
2. slight: small sporulating spots on single leaves;
3. moderate: numerous small or larger sporulating spots on several leaves and/or single leaves infected completely;
4. severe: many large sporulating spots on many leaves and/or numerous leaves infected completely.

Data from each tree were summarized and classified accordingly:

class 1: never any incidence;

class 2: only slight incidence;

class 3: a moderate or severe incidence on one or two occasions;

class 4: the infection rate was moderate or severe on more than two occasions.

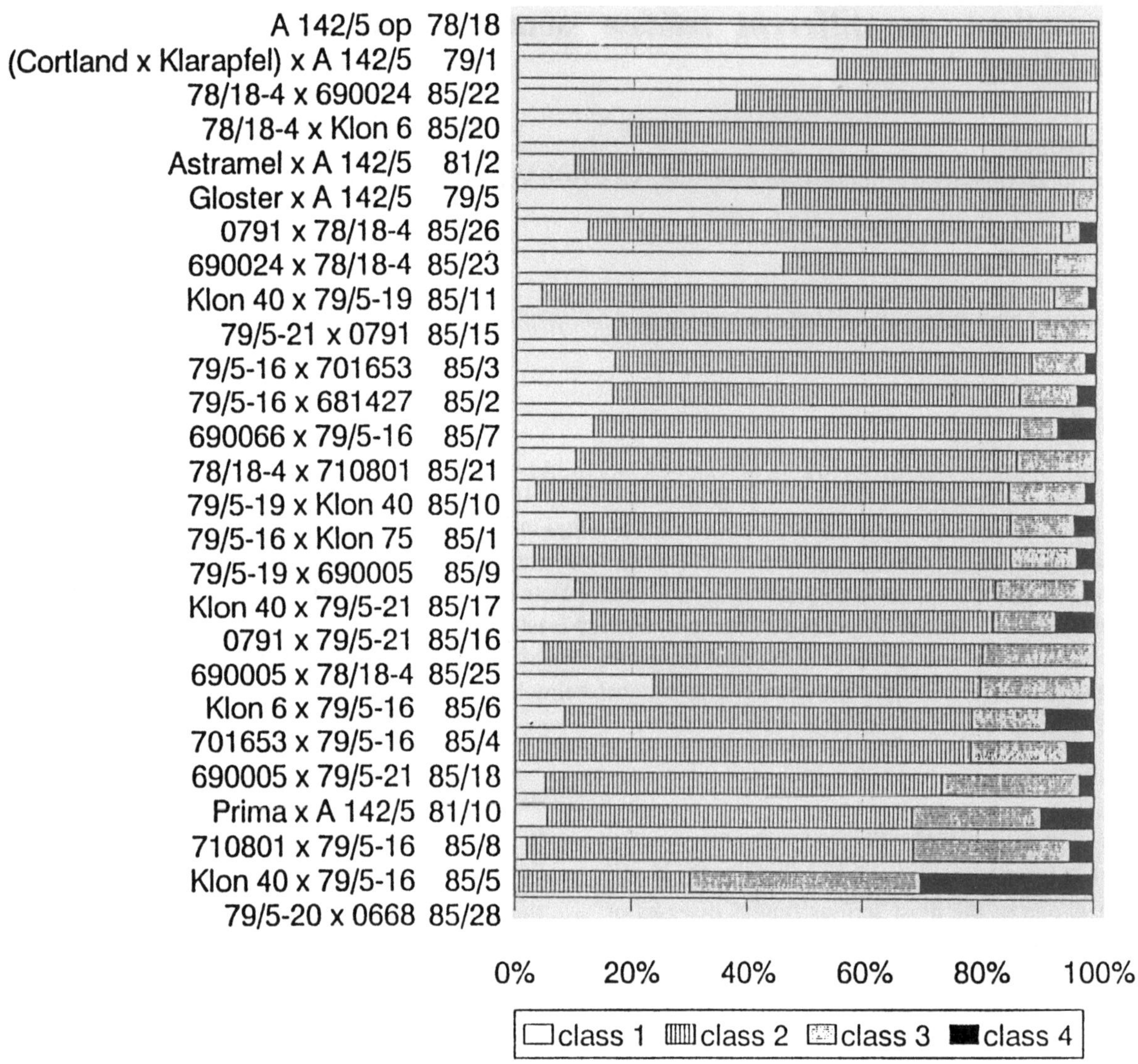

Fig. 1. Natural mildew incidence of cross progenies of *Malus robusta* in the resistance field.

Results

Malus robusta Field resistant plants were found in all progenies from this *Malus* species (Fig. 1). With one exception (85/28) the percentage of seedlings in classes 1 and 2 was greater than 69%, and in class 1 was normally smaller than 50%; two progenies (85/18, 85/28) gave no plants in class 1. In all but one progeny (85/28) less than 10% of the seedlings were in class 4. There were no reciprocal differences.

Malus zumi Figure 2 shows the results of *M. zumi* descendants. Only one progeny (81/3) showed less than 84% in classes 1 and 2; all progenies included plants in class 1 while only one gave plants in class 4 (81/11). Class 3 was represented in only half of the progenies. No reciprocal differences were observed.

Klon 40 Results from progenies with Klon 40 are shown in Fig. 3. Most of the progenies had no plants in class 1 and field resistance (class 2) was not found in three progenies (78/5, 85/125, 86/114). A few pro-

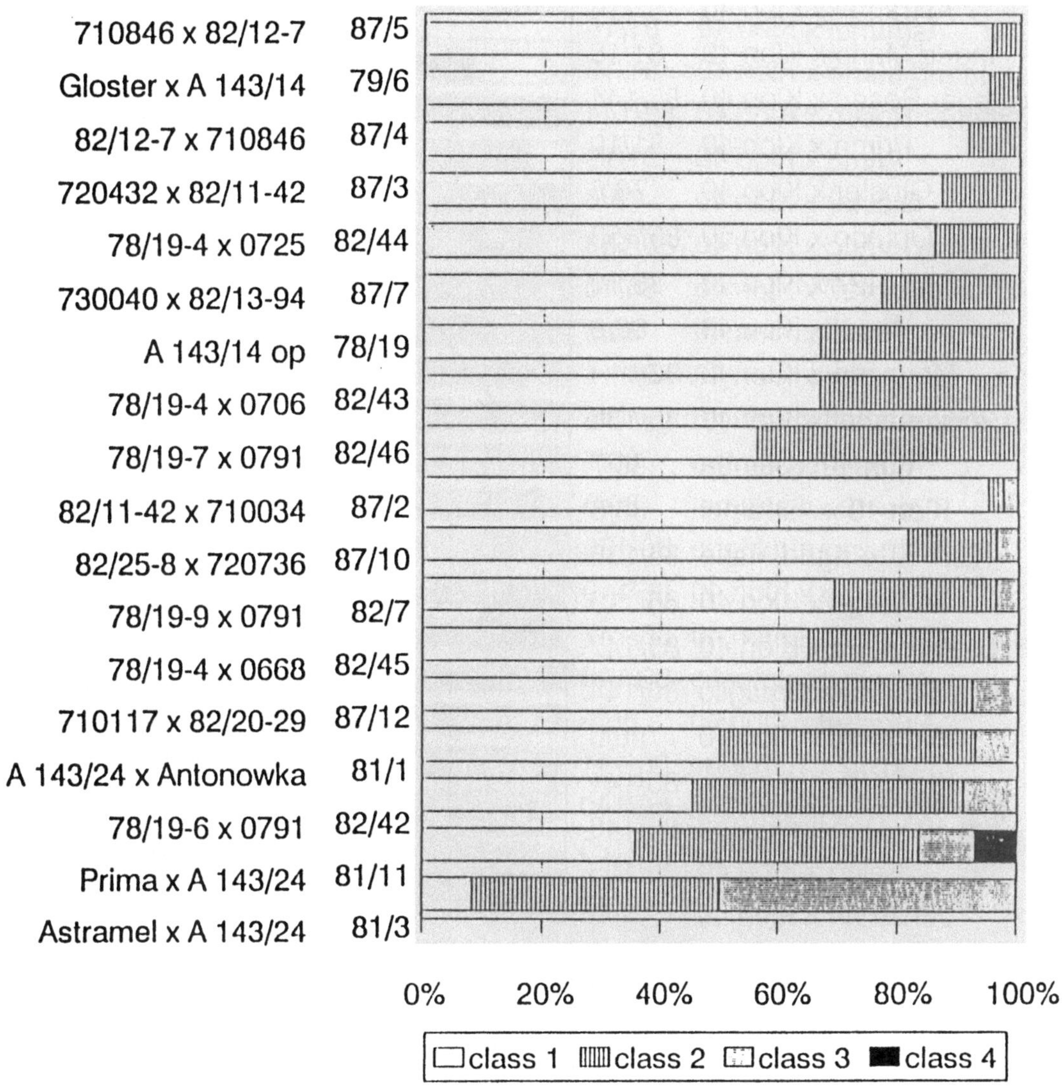

Fig. 2. Natural mildew incidence of cross progenies of *Malus zumi* in the resistance field.

genies had large numbers of plants in class 1 and 2 (81/15, 81/16, 81/19, 85/144). More than 50% of the trees of progenies 86/5, 86/6 and 86/141 belonged to class 4. In four progenies (78/6, 81/16, 85/144, 86/144) no plants were found in class 4. Reciprocal differences were apparent. Klon 40 used as male parent gave a higher proportion of resistant plants than when it was used as a female parent.

Varieties Progenies of 28 crosses between a range of varieties were observed in the field (Fig. 4). Only one progeny (81/17) possessed trees in class 1; nearly half of the progenies gave no plants in the class 2. Crosses with Rote Sternrenette produced the greatest number of field resistant plants. Most of the progenies had more than 50% in class 4; only 81/17 and 85/149 were without any plants in that class.

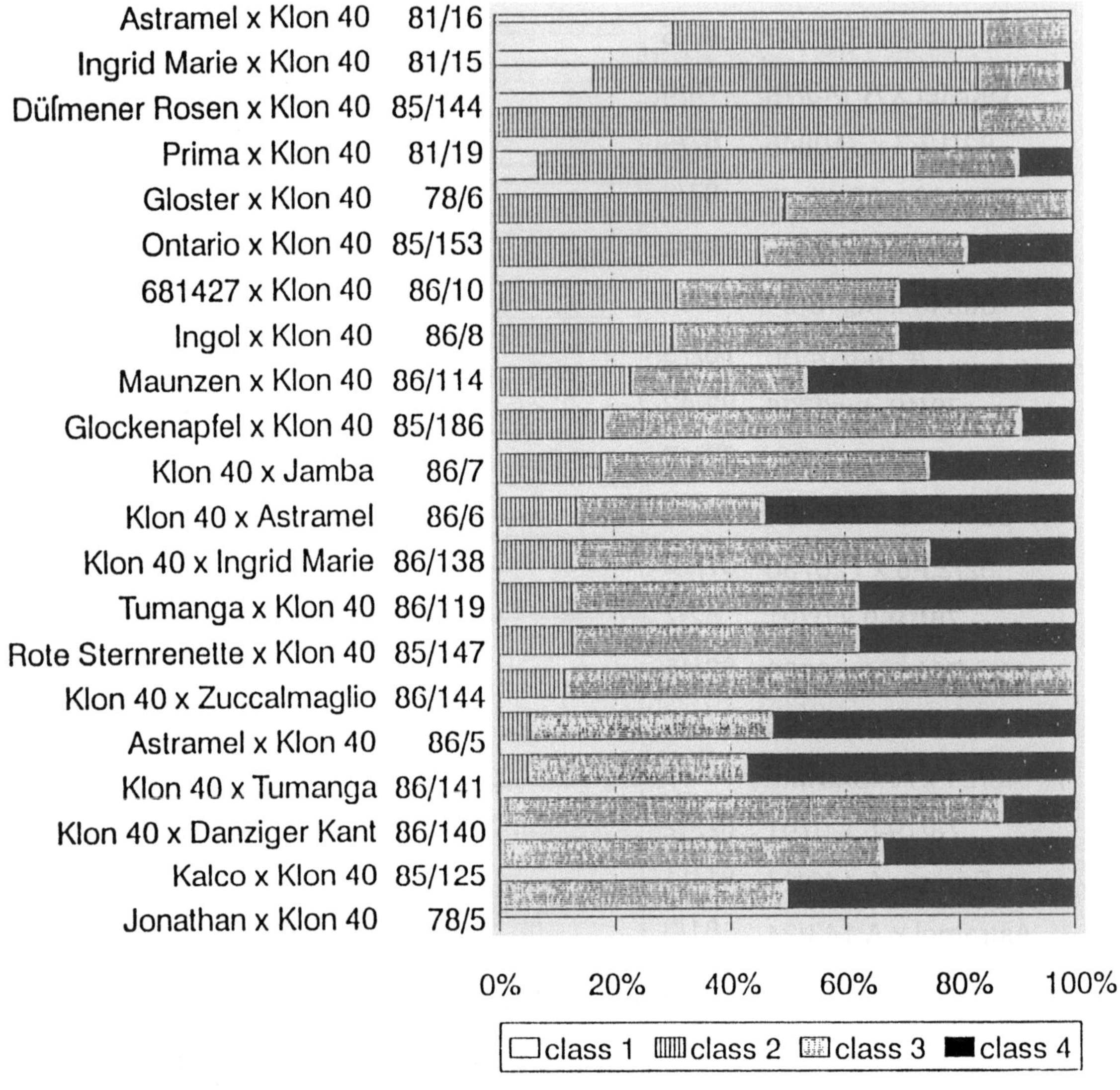

Fig. 3. Natural mildew incidence of cross progenies of Klon 40 in the resistance field.

Discussion

Field resistant plants could be found in all progenies. In *M. robusta* and *M. zumi* progenies most of the seedlings belonged to the classes 1 and 2, some progenies of Klon 40 and of crosses between varieties were without plants in these classes.

M. zumi transmitted resistance to mildew to a higher degree than *M. robusta*; this was evident in each class. No *M. zumi* progeny was without plants in class 1 and only 1 progeny had some susceptible seedlings in class 4. About one third of the *M. robusta* progenies had no seedlings in class 4. The good stability of resistance in descendants of *M. zumi* was described by Lespinasse (1983) and Alston (1983) too. A reduction of resistance through different generations was not observed in *M. zumi* progenies while it seems that resistance has decreased a little from second to third generation in *M. robusta*. Knight & Alston (1968) found 1 : 1 ratios of resistant to susceptible plants in progenies of crosses of *M. zumi* with Jonathan and Edward VII and in those of *M. robusta* with Cox and Jonathan. The

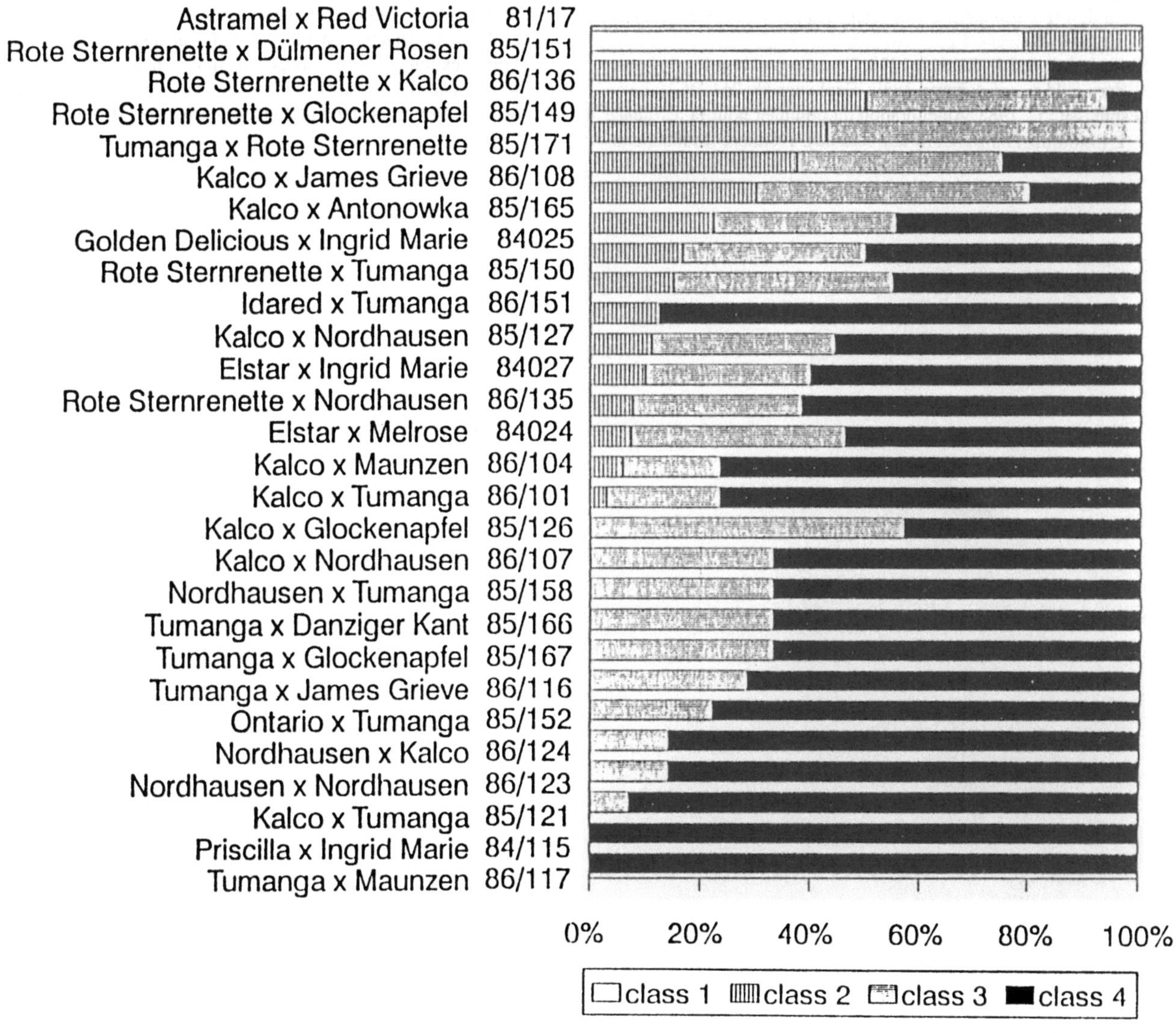

Fig. 4. Natural mildew incidence of cross progenies of varieties in the resistance field.

results presented in this paper based on plants selected in the juvenile stage except progenies of crosses with the sowing numbers 85. In these progenies – with one exception (85/28) – the percentage of field resistant plants was 70% and higher.

Comparing progenies of Klon 40 and of the varieties Klon 40 was better than most varieties except Rote Sternrenette which was as good as Klon 40. Class 1 plants were found very seldom but the number of field resistant seedlings reached considerable values especially in Klon 40 descendants.

Reciprocal crosses brought the same results regarding transmission of mildew resistance; only Klon 40 used as male parent seemed to have a positive influence in this respect.

References

Alston, F.H., 1983. Progress in transferring mildew (*Podosphaera leucotricha*) resistance from *Malus* species to cultivated apple. IOBC (WPRS) Bull. VI: 87–95.

Brown, A.G., 1959. The inheritance of mildew resistance in progenies of the cultivated apple. Euphytica 8: 81–88.

Knight, R.L. & F.H. Alston, 1968. Sources of field immunity to mildew (*Podosphaera leucotricha*) in apple. Can. J. Genet. Cytol. 10: 294–298.

Korban, S.S. & S.E. Riemer, 1990. Genetics and histology of powdery mildew resistance in apple. Euphytica 48: 261–267.

Lespinasse, Y., 1983. Amélioration du pommier pour la résistance à l'oidium (*Podosphaera leucotricha*). Premiers resultats concernant la virulence du champignon. IOBC (WPRS) Bull. VI: 96–110.

Misic, P.D., 1969. An investigation of the inheritance of resistance to apple powdery mildew (*Podosphaera leucotricha* (Ell. and Ev.) Salm.). Hort. Res. 2: 85–92.

Schander, H., 1958. Untersuchungen zur Entwicklung von Frühselektionsmethoden für die Apfelzüchtung. II. Über die Frühselektion auf Resistenz gegen Apfelmehltau (*Podosphaera leucotricha* Salm.), über die Anfälligkeit von Apfelsorten und über die Vererbung der Anfälligkeit. Züchter 28: 105–132.

H. Schmidt and M. Kellerhals (Eds.), Progress in Temperate Fruit Breeding, 13–15.

Early selection for partial resistance to powdery mildew, *Podosphaera leucotricha* (Ell. et Ev.) Salm. in apple progenies

J. Janse, J.J. Verhaegh & A.P.M den Nijs
Department of Vegetable and Fruit Crops, DLO-Centre for Plant Breeding and Reproduction Research CPRO-DLO, P.O. Box 16, 6700 AA Wageningen, The Netherlands

Key words: Apple, powdery mildew, resistance, selection, *Podosphaera leucotricha*, *Malus*

Summary

Selection of apple for partial resistance to powdery mildew in the nursery before planting trees in the orchard can only be relied upon if genotypic differences in expression do not depend on plant age. Five progenies between susceptible cv. Melrose and five partially resistant genotypes (four CPRO-DLO selections from the cross 'Elstar × Prima' and cv. Prima itself) were assessed for mildew resistance following natural infection in the nursery during two successive seasons. Per progeny 60 randomly taken trees were transplanted into the orchard and their mildew resistance levels were assessed for three more years. The mean mildew resistance of the progenies in the nursery was related to that of their resistant parent, but differences in mean resistance levels in the orchard were not consistent with those in the nursery. The individual correlations of resistance between nursery and orchard also varied per progeny. In general the second year's assessment in the nursery gave the best prediction of the orchard performance over three years. In conclusion, nursery selection for partial mildew resistance can be effective, but the efficiency varies between populations.

Introduction

Powdery mildew is a wide-spread disease in apple (*Malus pumila* Mill.), and is caused by the fungus *Podosphaera leucotricha* (Ell. et Ev.) Salm. Many apple breeding programmes around the world aim at the development of resistant varieties. Complete resistance (immunity) to powdery mildew was found in wild and ornamental *Malus*-species (e.g., Gallot et al., 1985; Dayton, 1977; Knight & Alston, 1968) but the use of such species in a breeding programme requires several generations of backcrossing before a high level of fruit quality is attained. Alternatively levels of partial resistance in cultivated varieties can be introduced in high quality varieties. The U.S. variety 'Prima' combines complete resistance to scab (*Venturia inaequalis*) with a high level of partial resistance to mildew. At CPRO-DLO, several selections were developed from the cross 'Elstar × Prima', which have scab resistance and also high to medium levels of resistance to mildew. These selections are now used as parents in the CPRO-DLO resistance breeding programme. However, quantitative assessment of the level of partial resistance is difficult, because infection and further development of the fungus strongly depend on weather conditions. Because of the long juvenile period and large plant size, it is advantageous to select in the nursery before planting the trees into the orchard. The efficiency of such selection strongly depends on the plant stage in which resistance is expressed. Young apple seedlings are known to be more susceptible to mildew than older plants, but it is often assumed that the classification of progenies remains the same (Lespinasse et al., 1976). Mihatsch & Mildenberger (1967) showed that the correlation between observations in the nursery and the orchard stage can vary considerably in a large number of crosses between varieties. In this paper, resistance of five progenies from crosses between a susceptible variety and partially resistant clones derived from 'Prima' are studied at different growth stages. Subsequently, within each population the individual correlation between mildew infection in early and later plant stages is deter-

Table 1. Average mildew infection (and number of trees) of the parental apple selections and two reference varieties in three successive years

Selection/	Average infection (number of trees)			
variety	1990	1991	1992	Average
S8*	1.6	1.7	1.4	1.6 (7)
S18*	1.8	2.0	1.3	1.7 (10)
S26*	2.4	1.9	2.8	2.4 (9)
S27*	1.9	2.5	2.0	2.1 (8)
Prima	1.7	1.6	1.1	1.5 (14)
Elstar	4.1	4.0	4.0	4.0 (12)
Jonathan	5.0	4.6	4.8	4.8 (8)

*: clones derived from the cross 'Elstar × Prima'. For scale of grading see text.

mined. The efficiency of early selection for partial resistance to mildew in breeding programmes is discussed.

Materials and methods

Progenies of five crosses with thc susceptible variety 'Melrose', made in 1987, were studied. Resistant male parents were 'Prima' and four CPRO-DLO selections of the cross 'Elstar × Prima': S8, S18, S26 and S27. Trees from the parental selections are present in a CPRO-DLO orchard, where no fungicides are applied. Every year mildew infection is assessed in this plot, on a 0–5 scale (0 = no visual symptoms, 1 = very slight infection on leaves, 2 = slight infection on leaves or occasionally on shoots, 3 = moderate infection on leaves and some shoots, 4 = heavy infection on leaves and shoots, 5 = very heavy infection on leaves, most shoots badly damaged).

The number of seedlings of the progenies planted in the nursery varied from 107 to 269 per population. (All progenies were screened for scab, and only resistant plants were planted.)

After June of the first year no more fungicides were applied. Mildew infection was assessed at the end of that year and in the summer of the second year, on a 1–9 scale (1 = no visual symptoms, ..., 9 = very severe infection on leaves and shoots).

After two years, 60 trees of each population were dug out at random and planted in an orchard in blocks of 15 trees per population. Mildew infection was assessed on the same 1–9 scale as described above, in the first three years after planting.

Results

The susceptible varieties 'Elstar' and 'Jonathan' were severely infected in three successive years in the orchard, while 'Prima' was only slightly infected (Table 1). S8 and S18 expressed a level of resistance comparable to 'Prima', while the two other selections, S26 and S27, were slightly to moderately infected.

In the nursery the progeny of the cross 'Melrose' with S8 appeared to be most resistant, followed by the crosses with 'Prima' and S18, while S26 and S27 transferred a significantly lower level of resistance to their progenies (Table 2).

In the orchard, however, differences in mean level of resistance were not consistent with the differences found in the nursery. Here, the 'Prima' progeny had significantly the highest level of resistance, followed by the S27 progeny which had the highest level of infection in the nursery. The progenies of S18 and S26 showed a significantly lower level of resistance in the orchard (Table 2).

All but one individual correlations between infection in the nursery and that in the orchard were significant at the 5% confidence level (Table 3), but only few were over 0.6 and are therefore considered meaningful. Within the orchard, the correlation between the first and the third year was lower than that between the first and the second, or that between the second and the third year. This may be due to the initial bad growth of the seedlings, making observations of mildew infection of some trees less reliable. Therefore in Table 3 correlations between nursery infection and orchard infection only for the mean of the second and the third year after planting are presented.

Mildew infection in the first year in the nursery was not highly correlated with the orchard assessments, although differences between populations were present. In all populations, the second year nursery assessments provided a better prediction of infection in the orchard, especially in the population 'Melrose × Prima'.

Combining both assessments made in the nursery did not lead to a better prediction; in 'Melrose × Prima' it even decreased the correlation coefficient.

Table 2. Average mildew infection (N = number of trees) in nursery and orchard of progenies of 'Prima' and 'Elstar × Prima' clones. Nur = nursery; Or = orchard; numbers 1 to 3 refer to the year of testing

Population	N	Average mildew infection						
		Nur 1	Nur 2	mean	Or 1	Or 2	Or 3	mean
Melrose × S8	59	5.68 a	4.69 a	5.19 a	6.54 b	5.42 bc	5.75 bc	5.90 bc
Melrose × S18	59	6.44 b	5.66 b	6.05 b	6.41 b	5.80 c	6.15 c	6.12 c
Melrose × S26	57	6.95 c	6.74 c	6.85 c	6.49 b	5.74 bc	5.88 bc	6.04 c
Melrose × S27	57	6.98 c	6.77 c	6.88 c	6.05 ab	5.19 ab	5.26 b	5.50 b
Melrose × Prima	60	5.95 a	5:63 b	5.79 b	5.63 a	4.62 a	4.28 a	4.84 a
Mean	292	6.39	5.89	6.14	6.22	5.35	5.46	5.68
SED		0.22	0.23	0.20	0.29	0.30	0.32	0.27

SED: standard error of difference
a,b,c: figures followed by the same character (within a column) differ significantly at a 5% level of significance using an LSD test.

Table 3. Correlation coefficients between mildew infection in the nursery and the orchard within each population. Nur = nursery; Or = orchard; numbers 1 to 3 refer to the year of testing

	M × S8	M × S18	M × S26	M × S27	M × P	Mean
Nur1/Nur2	0.52	0.51	0.56	0.57	0.58	0.55
Or1/Or2	0.76	0.63	0.67	0.64	0.78	0.70
Or1/Or3	0.61	0.46	0.65	0.65	0.41	0.56
Or2/Or3	0.75	0.68	0.71	0.76	0.86	0.75
Nur1/Or2,3	0.34	0.46	0.30	0.59	0.24§	0.39
Nur2/Or2,3	0.44	0.57	0.44	0.67	0.62	0.55
Nur1,2/Or2,3	0.44	0.59	0.43	0.71	0.51	0.54

§: correlation not significant at a 5% level of significance.

Discussion

Partial mildew resistance can be selected for both in the nursery and in the orchard. However, important shifts between populations occur as the trees grow older. Therefore, we consider selection based on the population mean more efficient than that based on a fixed threshold. This also reduces any seasonal influences on the severity of selection.

Since these selections have a common genetic background, the large shifts between populations were striking indeed.

First year assessments in the nursery did not provide any additional information and even had an adverse effect. These assessments should therefore be left out. The first year assessments in the orchard did not give any information either, probably due to the bad growth after replanting. In conclusion, selection in the nursery was effective, but the efficiency varied between populations and with the age of the trees.

References

Dayton, D.F., 1977. Genetic immunity to apple mildew incited by *Podosphaera leucotricha*. Hort. Science 12: 225–226.

Gallot, J.C., R.C. Lamb & H.S. Aldwinckle, 1985. Resistance to powdery mildew from some small-fruited *Malus* cultivars. Hort. Science 20: 1085–1087.

Knight, R.L. & F.H. Alston, 1968. Sources of field immunity to mildew (*Podosphaera leucotricha*) in apple. Can. J. Genet. Cytol. 10: 294–298.

Lespinasse, Y., L. Decourtye & A. Renoux, 1976. Selecting minor genes in apple progenies for resistance to mildew (*Podosphaera leucotricha*). Proceedings of the Eucarpia meeting on tree fruit breeding, Wageningen, The Netherlands: 18–32.

Mihatsch, H. & G. Mildenberger, 1967. Beiträge zur Vererbung der Mehltauresistenz bei *Malus*. III. Beobachtungen zum Frühselektionsproblem. Der Züchter 37: 25–32.

H. Schmidt and M. Kellerhals (Eds.), Progress in Temperate Fruit Breeding, 17–20.

Isoenzyme aided selection in the transfer of mildew (*Podosphaera leucotricha*) resistance from *Malus hupehensis* to the cultivated apple

I. Batlle[1] & F.H. Alston
Horticulture Research International, East Malling, West Malling, Kent, ME19 6BJ, Kent, U.K.; [1] present address: IRTA Mas Bové, Apartat 415, 43280 Reus, Spain

Key words: Apple, host resistance, isoenzymes, mildew, plant breeding, *Podosphaera leucotricha*, *Malus*

Summary

The isoenzymic gene *GOT-1* was used as an indicator of ploidy levels and hybridity in the progeny of a 4x × 2x (A878/5, *GOT-1 abdn* × 'Gloster 69', *GOT-1cc*) cross, derived from the apomictic species (*Malus hupehensis*), which segregated for mildew resistance. The distinction between resistant and susceptible plants was sharp. After 3 months in a greenhouse, resistant plants showed no signs of sporulation. Of the hybrid seedlings (diploid and triploid) 78% were mildew resistant. This analysis was facilitated through the use of the *GOT-1c* allele, as a marker of hybridity. It appears that the prospects of transferring high mildew resistance from *M. hupehensis* to the cultivated apple are promising; of the 65 hybrid resistant seedlings 18 were diploid. A similar approach in earlier generations might have facilitated the earlier transfer of this resistance to the cultivated apple.

Introduction

In the search for a source of strong resistance to apple mildew, breeders have investigated the immune species *Malus hupehensis* (Pamp) Rehd. This species is triploid (3x = 51), facultatively apomictic and male sterile (Lespinasse, 1975). Crosses with the cultivated apple gave a high proportion of resistant seedlings (Knight & Alston, 1968); most were apomictic in origin and triploid; although some resistant tetraploid hybrids were produced. These were apomictic also. Thus, the production and identification of resistant diploid (2x = 34) hybrids, necessary for the transfer of resistance, proved difficult. After two generations of backcrossing to the cultivated apple and selecting for mildew resistance, fruit size was still small. A promising resistant tetraploid selection, A878/5, was chosen on the basis of fruit size (*c.* 50 mm). Details of pedigree are summarised (Fig. 1) together with ploidy levels recorded for the parents derived from *M. hupehenis* (Knight & Alston, 1968; Alston & Knight, 1971; Lespinasse et al., 1976; Decourtye et al., 1983; Manganaris, 1989). In this short series of backcrosses to the cultivated apple the ploidy level appeared to alternate with generation. Determination of ploidy levels through chromosome counting is difficult in *Malus*; chromosomes are small and the production of suitable tissue for examination is difficult (Lespinasse et al., 1976).

Isoenzymes offer an alternative approach to the determination of ploidy levels in plants. Polyploids usually give zymograms with more than two distinct allozymic bands, and this approach has been applied to determine ploidy levels in wheat (Hart, 1983) and potato (Martinez-Zapater & Oliver, 1985). In *Malus* Manganaris (1989) examined the *Got-1* genotypes of progenies derived from *M. hupehensis*, including progeny A878, and found that they could be used to distinguish ploidy levels.

In view of the high degree of apomixis present in this breeding material, an isoenzymic marker is needed to facilitate easy recognition of hybrid plants. This paper gives an account of the use of the isoenzymic gene *Got-1* in the recognition of mildew resistant diploid hybrids derived from A878/5 and the cultivated apple.

M. hupehensis (3x) x 'Edward VII' (2x)

|

A37/1 (4x) x 'Lane's Prince Albert' (2x)

|

A434/6 (3x) x 'Lane's Prince Albert' (2x)

|

A878/5 (4x)

Fig. 1. Pedigree of A878/5 (4x).

Table 1. Segregation of E334 [A878/5 (*Got-1 abdn*, mildew immune) × Gloster 69 (*Got-1cc*, mildew susceptible)] for *Got-1* and mildew resistance

Got-1 genotype	Number of seedlings	Mildew	
		Resistant	Susceptible
abc	29	23	6
acd	28	22	6
acn	2	2	0
ab	4	2	2
ac	14	10	4
ad	1	1	0
an	2	0	2
bc	10	8	2
bn	1	0	1
Total	91	68	23

Materials and methods

Apple progeny E334 [A878/5, 4x = 68 (*Got-1 abdn*, mildew resistant) × 'Gloster 69', 2x = 34 (*Got-1cc*, mildew susceptible)] was studied, 'Gloster 69' was chosen as a diploid parent so that *Got-1c* could be used to detect hybrids.

Potted seedlings growing in the greenhouse were dusted with mildew spores by shaking heavily infected shoots over them. The inoculated plants were maintained at 20–25 °C for 1 month, after which the degree of infection on recently expanded leaves was recorded (Batlle, 1993).

Young, actively growing leaves were used for polyacrylamide gel electrophoresis (PAGE) of the parents and seedlings for GOT (glutamate oxaloacetate transaminase). The procedure adopted was similar to that of Manganaris & Alston (1987), except that mercaptoethanol was replaced by dithiothreitol (DTT) in the extraction buffer. Gels were stained for GOT according to Manganaris & Alston (1987).

Results

The growth of the seedlings was variable. Some lacked vigour and died within one month of germination, some made very poor growth, while others grew normally. No clear relationship was found between vigour and ploidy (determined on the basis of isoenzyme constitution) or mildew response.

The reaction of the seedlings to mildew was clear. The distinction between resistant and susceptible plants was clearer than that recorded in greenhouse screening of derivatives of *M. zumi* and *M. robusta*, where resistant plants showed necrotic or chlorotic flecks accompanied by slight sporulation (Alston, 1983). There was a high proportion of resistant plants (75%), showing clean leaves with no sign of sporulation after 3 months in the greenhouse.

All the GOT-I band positions were located and good resolution was also achieved in the GOT-II zone, thus facilitating the clear identification of the d band of GOT-I which overlaps with the GOT-II zone. The *Got-1* genotypes of A878/5 (*abdn*) and 'Gloster 69' (*cc*) recorded by Manganaris & Alston (1989) were confirmed.

Among the 91 seedlings screened for *Got-1*, nine genotypes were identified, three with three alleles, *abc*, *acd* and *acn*, and six with two alleles, *ab*, *ac*, *ad*, *an*, *bc* and *bn* (Table 1). It appears that the female parent (A878/5) provided 33 *ab*, 29 *ad*, 4 *an* and 1 *bn* diploid gametes and 14 *a* and 10 *b* haploid gametes. The *a* and *b* alleles thus appear to be preferentially transmitted. Some expected diploid and triploid combinations were not present and no tetraploid plants were observed. Preferential pairing appears to occur between certain alleles (e.g. *a* and *d* or *b*) which supports the observations of Manganaris (1989).

Of the 91 seedlings recorded for GOT-I, 59 (64.8%) showed three primary bands and were thus considered to be triploid, while the remaining 32 (35.2%) showed two primary bands and were most likely diploid, although some aneuploids may have been produced. A high proportion of plants (91.2%) carried the GOT-I c band and were thus hybrid in origin; these included all the triploids and 24 of the 32 diploids.

Five of the eight maternally derived diploids (seedlings without *Got-1c*) were mildew susceptible. Seedlings of hybrid origin in the two ploidy groups segregated for resistance in a 3 resistant : 1 susceptible fashion; triploids, 47 resistant : 12 susceptible (χ^2 = 0.82, 1 df); diploids, 18 resistant : 6 susceptible (χ^2 = 0.05, 1 df).

Discussion

After three generations of crossing among derivatives of *M. hupehensis*, no dilution of the high level of mildew resistance was apparent. Isoenzyme analysis showed that the high proportion of resistant segregants in the third generation was not a result of apomixis. The use of the allele *Got-1c* as a marker for hybrid genotypes facilitated this analysis. There was no evidence of linkage between *Got-1* and mildew resistance.

Although only triploid and diploid seedlings appeared to be detected, it must be emphasised that this type of analysis cannot distinguish aneuploids carrying *Got-1c* that may arise as a result of the fertilisation of egg cells resulting from irregular cell division in a tetraploid, from triploids. Viable aneuploid seedlings have been reported in *Malus* (Darlington & Moffet, 1930). However, they are weak and often lethal. It seemed that the number of apomictic seedlings was low. Schmidt (1977) observed that the incidence of apomixis, which in *Malus* is inherited as a recessive character, varies according to the ploidy of the species. The presence of viable non-tetraploid seedlings derived from a tetraploid seed parent suggests that regular pairing occurs at meiosis. However, Lespinasse et al. (1976) observed that meiosis in tetraploid apples could be very irregular. The triploids in this study can be assumed to have arisen from the normally reduced 34-chromosome egg cells of the tetraploid parent fertilised by 17-chromosome pollen from the diploid variety 'Gloster 69'. The maternallly derived diploids probably developed from unfertilised 34-chromosome egg cells, while the hybrid diploids developed from 17-chromosome egg cells fertilised by 17-chromosome pollen from 'Gloster 69'.

The reason for the predominance of the *a* allele from A878/5 was not clear, preferential transmission being suggested. Large distortions in the expected proportions of *Got-1* alleles in *Malus* are often explained on the basis of the close linkage between the *Got-1* locus and the *S* incompatibility locus (Manganaris & Alston, 1987). It is fortunate that the *a* allele predominates as it can be traced to the recent recurrent parent 'Lane's Prince Albert', *Got-1aa* (Manganaris & Alston, 1989) of previous generations. No further information on the *Got-1* genotype of the A878/5 pedigree is available, apart from 'Edward VII', *Got-1bd* (Manganaris & Alston, 1989). Assuming a low frequency of crossing over at this stage, the *a* allele might be expected to be accompanied by a sizable portion of the cultivated apple genome. Thus, *ac* diploids might be preferred to *bc* diploids since they are likely to combine a greater proportion of genes from the cultivated apple combined with mildew immunity from *M. hupehensis*.

Chromosome counting in apple is more time-consuming than isoenzyme analysis and not as convenient for the routine analysis of large populations. It was essential to use a stable isoenzyme such as GOT and to select *Got-1* as a marker giving good resolution with PAGE from leaf samples and being highly polymorphic in *Malus*. The use of 'Gloster 69', *Got-1cc* was also critical for the confirmation of hybridity in derivatives of A878/5. However, the dimeric structure of GOT can complicate the interpretation of zymograms through the production of hybrid bands. Ideally, an isoenzymic marker for ploidy should be monomeric, have good resolution, no overlapping of bands, and be highly polymorphic at one locus. *Prx-2* in apple (Manganaris & Alston, 1992) is one such example. However, the alleles of *Prx-2* are not as widely distributed in *Malus* as those of *Got-1*.

Twenty-one mildew resistant diploid seedlings were obtained from a third backcross of *M. hupehensis* to the cultivated apple and 18 were of hybrid origin. In earlier generations diploid derivatives of *M. hupehensis* have been difficult to recognise and were usually found to be susceptible (Alston, 1983). The use of an isoenzymic marker confirmed the hybrid nature of resistant selections when there would otherwise have been some doubt as to their parentage. In addition, it was useful for estimating the contribution of the cultivated apple from a previous generation. Similar recognition and selection of hybrids in the earlier generations might have resulted in a more rapid introduction of this resistance into the cultivated apple. The reason for the similar segregations for mildew resistance within both the triploids and the diploids, and the extent to which they may be due to preferential pairing at meiosis and differential transmission, ist not clear. Subsequent crosses of resistant diploid plants will clarify the genetics of mildew resistance from *M. hupehensis*.

Acknowledgements

The authors are grateful for financial support for this work from the Ministry of Agriculture, Fisheries & Food, the Institut de Recerca i Tecnologia Agroalimentàries (IRTA) and the Commissió Interdepartamental de Recera i Innovació Tecnològica (CIRIT).

References

Alston, F.H., 1983. Progress in transferring mildew (*Podosphaera leucotricha*) resistance from *Malus* species to cultivated apple. IOBC (WPRS) Bulletin: Integrated control in orchards. Angers 6(4): 87–95.

Alston, F.H. & R.L. Knight, 1971. Fruit breeding. Transfer of mildew immunity from wild *Malus* species. Rep. E. Malling Res. Stn. for 1970, pp. 94–95.

Batlle, I., 1993. Linkage of isoenzymic genes with agronomic characters in apple. Ph.D. Thesis, University of London, 181 pp.

Darlington, C.D. & A.A. Moffet, 1930. Primary and secondary chromosome balance in *Pyrus*. J. Genet. 22: 129–151.

Decourtye, L., D. Martin, Y. Lespinasse, & B. Donini, 1983. Looking for solid mutants through irradiation of egg-cells in *Malus hupehensis*. Acta Hort. 140: 87–90.

Hart, G.E., 1983. Hexaploid wheat. In: S.L. Tanksley & T.G. Orton (Eds). Isoenzymes in plant genetics and breeding, part B, pp. 35–36. Elsevier, Amsterdam.

Knight, R.L. & F.H. Alston, 1968. Sources of field immunity to mildew (*Podosphaera leucotricha*) in apple. Can. J. Genet. Cytol. 10: 294–298.

Lespinasse, Y., 1975. Chromosome observations in *Malus* and *Pyrus*, pp. 75–83. Proc. Eucarpia Fruit Sec. Symp. V, Canterbury, 1973.

Lespinasse, Y., F.H. Alston, & R. Watkins, 1976. Cytological techniques for use in apple breeding. Ann. Appl. Biol. 82: 349–353.

Manganaris, A.G., 1989. Isoenzymes as genetic markers in apple breeding. Ph.D. Thesis, University of London, 430 pp.

Manganaris, A.G. & F.H. Alston, 1987. Inheritance and linkage relationships of glutamate oxaloacetate transaminase isoenzymes in apple. 1. The gene *Got-1*, a marker for the *S* incompatibility locus. Theor. Appl. Genet. 74: 154–161.

Manganaris, A.G. & F.H. Alston, 1989. Glutamate oxaloacetate transaminase isoenzymes in apple cultivars and rootstocks. J. Hort. Sci. 64: 9–15.

Manganaris, A.G. & F.H. Alston, 1992. Inheritance and linkage relationships of peroxidase isoenzymes in apple. Theor. Appl. Genet. 83: 392–399.

Martinez-Zapater, J.M. & J.L. Oliver, 1985. Isoenzyme gene duplication in diploid and tetraploid potatoes. Theor. Appl. Genet. 70: 172–177.

Schmidt, H., 1977. Contributions to the breeding of apomictic apple stocks. 4. On the inheritance of apomixis. Z. Pflanzenzücht 78: 3–12.

H. Schmidt and M. Kellerhals (Eds.), Progress in Temperate Fruit Breeding, 21–25.

Breeding apples for scab tolerance at Holovousy

J. Blažek & F. Paprštein
Research and Breeding Institute of Pomology, 507 51 Holovousy, Czech Republic

Key words: Apple-trees, cultivars, breeding, tolerance to scab, resistance

Summary

There are 53 apple cultivars registered for commercial growing in the Czech Republic, at present, from which 4 are resistant and 7 tolerant to scab (*Venturia inaequalis*/Cke/Wint). There seem to be increasing demands especially for these tolerant cultivars in connection with the spread of integrated fruit growing systems. Therefore a majority of cultivars now released from Holovousy and being given to State Variety Trials belongs to this category. In addition to breeding apples with resistance to scab based on a single dominant gene derived from *Malus floribunda* 821, the breeding programme in Holovousy has extensively utilized donors of polygenic resistance to scab for three generations. As donors mostly old and less known cultivars ('Antonovka', 'Emneth Early', 'Hagloe Crab', 'Hájkova reneta', 'Reinette Rouge Étoilée' and 'Spätblühender Taffetapfel') were used but to a lesser extent also some commercially grown cultivars ('Bláhova oranžová', 'Glockenapfel', 'James Grieve', 'Jonathan', 'Mio', 'Mother' and some others). The presented segregation data are from material from three cycles of hybridization comprising 81 crosses and nearly 6,500 seedlings. These seedlings were preselected after early inoculations but the final rating was based on several severe spontaneous scab infections in orchards. No seedling was classified as immune. On average 3% of seedlings were evaluated as field resistant with the highest segregation of 21% in the cross 'Hagloe Crab' × 'Emneth Early'. The group of seedlings rated as 'tolerant' (low susceptibility) segregated in frequencies averaging 10%, with a maximum of 41% for a single cross. From the first generations of the breeding programme at Holovousy 6 new apple cultivars ('Angold', 'Julia', 'Klára', 'Nabella', 'Produkta' and 'Zuzana') with a tolerance to scab have been recently released. Short descriptions are given.

Introduction

In recent years more and more growers have become interested in scab-resistant apple cultivars. There are two main reasons for this: 1) increase of the hectarage of alternate growing systems; 2) an effort of growers to reduce their costs. Despite nearly 50 scab-resistant apple cultivars hitherto released world-wide (Crosby et al. 1992), only four ('Florina', 'Jolana', 'Melodie', and 'Prima') are registered now for commercial growing in the Czech Republic (Blažek et al. 1993). Another seven cultivars ('Akane', 'Auralia', 'Denár', 'Discovery', 'Mother', 'Mio' and 'Wealthy'), from the total of 53 registered for the country, are tolerant to scab. This means that they can be grown successfully under a restricted programme of protective sprays. There seems to be an increasing demand for these tolerant cultivars in connection with the spreading of integrated fruit growing systems in the country. Therefore a majority of apple cultivars now released from Holovousy and being given to the State Variety Trials is from this category.

In this paper attention is focused on segregation of scab-tolerant seedlings from different progenies. Finally, short descriptions of 6 new apple cultivars with tolerance or resistance to scab are also given.

Material and methods

Three cycles of hybridisation were used to create an extensive selection pool for scab resistance. The first crosses were made in 1971, in a framework of an incomplete diallel of the best commercial cul-

Table 1. Percentage of seedlings with low susceptibility to scab from incomplete diallel crosses made in 1971.

Female parent	Male parent												
	Bláhova oranžová[2]	Dukát[1]	Cox's Or.P.[2]	Golden Del.[1]	Idared[1]	James Grieve[1]	Jonathan[2]	Kidd's Orange[1]	McIntosh[1]	Spartan[1]	Starking Del.[1]	Star-krimson D.[1]	Mean
Cox's Or. P.[2]					0							0	0
Starkspur G. D.[1]	5	0			0		0		0			0	1
Idared[1]		0	0						0				0
James Grieve[1]	25	0				18		12	0	6	0	0	6
Jonathan[2]	24	0			0					0			6
Kidd's Orange[2]	0	0		0	4	14			0	14			5
McIntosh[1]	13	10			0								8
Spartan[1]				0	0	0							0
Starkrimson D.[1]		3	13	0	0	12							7
Mean	13	2	6	0	0.5	11	–	–	0	7	–	0	4.2

[1] A highly susceptible cultivar.
[2] A moderately susceptible cultivar.

tivars which were available at that time: 'Bláhova oranžová', 'Dukát', 'Cox's Orange Pippin', 'Golden Delicious', 'Idared', 'James Grieve', 'Jonathan', 'McIntosh', 'Spartan', 'Starkrimson Delicious' and 'Starkspur Golden Delicious'. From this project 41 progenies with a total of 2,609 seedlings were used for the next selection.

Crosses of the second cycle of hybridization were made in 1972–1973. In this cycle different parents were used: scab-resistant cultivars ('Hagloe Crab', 'Hopa Crab', 'Emneth Early'), medium susceptible cultivars ('Glockenapfel', 'James Grieve', 'Ontario', 'King of the Pippins') and highly susceptible ones ('Golden Delicious', 'Spartan', 'Starkrimson Delicious'). From the crosses 17 progenies with a total number of 2,040 seedlings were used for the selection.

Crosses of the third cycle of hybridization were performed in 1974–1976. The majority was made as back-crosses of previously selected scab-resistant seedlings that had been derived from some sources of polygenic resistance (Blažek & Vondráček 1976) to apple cultivars with good fruit quality. The selections were derived from 'Antonovka', 'Hájkova reneta', 'Reinette Rouge Étoileé' and 'Spätblühender Taffetapfel' that had shown the highest level of resistance (Vondráček & Blažek 1973). This cycle consisted of 23 progenies and 1,850 seedlings in total.

All seedlings were assessed for their susceptibility to scab both after artificial inoculation in the plastic house and repeated severe spontaneous infections after natural inoculation in the orchards, where they were grown without any chemical sprays against the disease. For the assessment of the seedlings and their final classification a modified rating scale 1–9 from the IBPGR apple descriptors (Watkins & Smith 1982) was used. The grades of the scale corresponded to the following classes:

- 1–3 to high susceptibility ('Mc Intosh', 'Golden Delicious');
- 4–5 to moderate susceptibility ('Cox's Orange Pippin', 'Jonathan');
- 6–7 to low susceptibility or tolerance ('Reinette Rouge Étoilée', 'Mio');
- 8 to field resistance ('Emneth Early');
- 9 to immunity ('Freedom').

Results and discussion

In the progenies derived from commercial cultivars no seedlings with field resistance /or immunity/ to scab were found. Segregation rates for seedlings with a tolerance to scab in the first cycle of hybridization are given in Table 1. The rating of the parents was taken over from a previous work (Blažek et al. 1977). The mean rate from all progenies was 4.2% tolerant seedlings. In the case of crosses where two highly susceptible parents were combined no scab-resistant seedlings were observed. In the progenies where one parent was highly susceptible whilst the other was moderately suscep-

Table 2. Final rating of seedlings from the second cycle of hybridization (1972–1973) regarding their susceptibility to scab.

Cross	Number of seedlings	Highly susceptible seedlings %	Lowly susceptible seedlings %	Seedlings with field resistance %
Bancroft × Starking Delicious	196	52	2	0
Golden Delicious × Bancroft	84	57	0	0
Golden Delicious × Hopa Crab	288	33	28	10
Starkspur Golden Del. × Glockenap.	62	76	5	0
Hagloe Crab × Emneth Early	122	10	41	21
Hagloe Crab × Glockenapfel	351	23	20	5
Hagloe Crab × Linda	321	0	40	7
James Grieve o.p.	32	52	9	0
Lord Lambourne × Spartan	38	22	11	0
Oldenburg selfing	89	41	8	0
Ontario selfing	30	8	0	0
King of the Pippins o.p.	12	64	18	0
Spartan × Lord Lambourne	11	67	22	0
Starkrimson Del. × Bancroft	110	11	3	0
Starkrimson Del. × Clivia	79	67	0	0
Starkrimson Del. × Linda	71	21	2	0
Starkrimson Del. × Glockenapfel	144	54	4	0
Total	2040	38.7	12.5	3.1

tible, seedlings with tolerance to scab segregated in rates from 0 to 14%. Crosses of two parents with moderate susceptibility gave from 0 to 25 per cent tolerant seedlings.

From the point of view of practical breeding the cross of 'James Grieve' × 'Bláhova oranžová' was remarkable. In the progeny 25% of seedlings with scab-tolerance segregated and most had acceptable fruit quality and other economic characters. On the other hand, in the progeny 'Jonathan' × 'Bláhova oranžová' with 24% of tolerant seedlings, nearly all of them were very susceptible to mildew. Other progenies with some promise were 'Starkrimson Delicious' × 'James Grieve' and 'Starkrimson Delicious' × 'Cox's Orange Pippin' where seedlings with a resistance to mildew and acceptable fruit quality also segregated.

From the progenies of the second hybridization cycle, the cross of two scab-resistant cultivars – 'Hagloe Crab' and 'Emneth Early' was exceptional. In this progeny the proportion of seedlings with field resistance to scab was 21% and of tolerant ones 41% (Table 2). Unfortunately the majority of these seedlings did not have good fruit quality and their storage ability was rather poor. The average of all progenies from this cycle was 3.1 % resistant seedlings and 12.5% tolerant ones.

Progenies of the third cycle gave on average 0.9% of seedlings with field resistance to scab and 10.1% of seedlings with scab-tolerance (Table 3). In this group, all three progenies of the advanced selection A 28/39 ('Antonovka' o.p.) were remarkable. Many of these seedlings had very good fruit quality and even some degree of resistance to mildew. Two new cultivars were selected from these progenies: 'Angold' (A 28/39 × 'Golden Delicious') and 'Produkta' (A 28/39 × 'Golden Spur'). The segregation rates of scab resistance from the 'Antonovka' source correspond to the pattern described by Fischer (1989).

The cross of the early apples 'Quinte' × 'Discovery' proved to be productive, as several promising seedlings with high fruit quality and resistance to mildew segregated . The new cultivar 'Julia' was selected from this progeny.

Table 3. Final rating of seedlings from the third cycle of croses (1974–1976)regarding their susceptibility to scab.

Cross	Number of seedlings	Lowly susceptible seedlings %	Seedlings with field resistance %
A 28/39 (Antonovka o.p.) × Alkmene	86	15	7
A 28/39 (Antonovka o.p.) × Golden Delicious	65	8	5
A 28/39 (Antonovka o.p.) × Golden Spur	30	20	3
Kidd's Orange × B 11/14(Spätblühender T.o.p.)	36	19	3
I/3 18/1 (Späbl. Taff. × Court Pendu P.) × Golden Spur	62	19	3
B 11/14 (Späbl. Taff. o.p.) × Well Spur D.	54	15	2
Kidd's Orange × I/3 11/11 (Court Pendu Plat × Spätblühender Taffetapfel)	178	10	2
Golden Spur × B 11/14	153	9	1
Quinte × Discovery	58	31	0
A23/28 (Reinette Rouge Etoilée × Hajkova reneta) × Melrose	35	31	0
I/3 18/1 × Clivia	34	20	0
A12/74 Späbl. Taff. × Court Pendu P.) × 2 × 57 (Oldenburg × McIntosh	43	16	0
III 12/30 (Jonathan × Ontario) × Rubín	42	14	0
A 12/74 × Šampion	35	14	0
Golden Spur × I/3 17/5 (Spätbl.T. × Court Pendu Plat)	69	12	0
Kidd's Orange × I/3 17/5	111	9	0
Kidd's Orange × A 33/63 (Spätbl.T. × James Grieve)	153	8	0
A 12/74 × Kidd's Orange	77	5	0
A 12/74 × Holiday	75	4	0
A 12/74 × Red Spur Delicious	110	4	0
A 11/1 (Spätbl. Taff. × James Grieve) × Mio	162	2	0
A 16/34 (Mother × J. Grieve) × Atlas Red	182	0	0
Total	1850	10.1	0.9

The conclusion may be drawn from these findings and other experience that sources of polygenic resistance to scab can be successfully used for the development of new cultivars with tolerance to the disease rather than full resistance.

The effectiveness of the method depends to a large extent on the proper choice of parents. However, the procedure of improvement is not easier than the use of monogenic sources of scab resistance.

Short description of new apple cultivars from Holovousy

– 'Angold'

Origin: A 28/39 ('Antonovka' o.p.) × 'Golden Delicious'.

Tree: Moderately vigorous, spreading. Field resistance to scab, moderately resistant to mildew. Early and good cropper.

Fruit: Size intermediate to large, shape globose to globose conical. Green-yellow skin covered mainly by red stripes and red marble. Flesh cream, crisp, juicy. Taste very good, sub-acid to sweet.

Time of ripening: Similar to 'Jonagold', keeping well till February.

– 'Julia'

Origin: 'Quinte' × 'Discovery'.

Tree: Semi-compact growth. Tolerant to both scab and mildew. Very precocious and regular cropper.

Fruit: Medium size, shape flat to globose, very attractive appearance (solid red), good taste. Resistant to bruising.

Time of ripening: Very early like 'Vista Bella'.

– 'Klára'

Origin: 'Reinette Rouge Étoilée' × 'Hájkova reneta'.

Tree: Moderately vigorous, upright, semi-compact. Field resistance to both scab and mildew. Regular and high yields.

Fruit: Size medium, shape intermediate to flat, rectangular to truncate conic. Skin yellow, almost entirely covered by bright red. Flesh sweet sub-acid, slightly performed.

Time of ripening: Similar to 'Idared'.

– 'Nabella'

Origin: 'Mother' × 'Starking Delicious'.

Tree: Moderately vigorous, upright, spurred growth habit. Tolerant both to scab and mildew. Early and regular cropper.

Fruit: Size is intermediate to large, shape globose to globose conical. Skin colour yellow nearly completely covered by marble red and blush. Flesh yellow, firm and crisp. Taste is very good with pleasant flavour.

Time of ripening: Similar to 'Prima', storage ability till December.

– 'Produkta'

Origin: A 28/39 × 'Golden Spur'.

Tree: Moderately vigorous, spreading. Complex field resistance to scab and mildew. Very precocious and high cropper.

Fruit: Size is intermediate to large, shape globose-conical or prolonged. Skin colour green-yellow. Flesh white-cream, texture fine, very juicy. Taste without special flavour.

Time of ripening: Some days before 'Golden Delicious'.

– 'Zuzana'

Origin: 'Glockenapfel' × 'James Grieve'.

Tree: Moderalely vigorous to weak, upright and compact growth, good spurring, tolerant to scab and mildew, moderate but regular cropper, good yield only on dwarf rootstocks.

Fruit: Size intermediate to large, shape globose conical or prolonged, often ribbed. Skin colour yellow with red stripes and blush. Flesh colour white-cream, texture fine, crisp and firm. Taste sub-acid to harmonic (when fully ripped).

Time of ripening: Similar to 'Golden Delicious', keeps well till April or later.

References

Blažek, J., J. Dlouhá, M. Richter, F. Paprštein & J. Blažková, 1993. Cultivars of fruit species recommanded for alternative growing system (in Czech). Ministry of Agriculture, Prague.

Blažek, J., J. Kloutvor & J. Vondráček, 1977. The susceptibility of important apple cultivars to scab (in Czech). Věd. práce ovocn. 6: 61–70.

Blažek, J. & J. Vondráček, 1976. Scab resistance from some less known apple cultivars. Proc. Eucarpia Meeting on Tree Fruit Breeding, Wageningen, 7–9. Sept. 1976: 33–37.

Crosby, J.A., J. Janick, P.C. Pecknold, S.S. Korban, P.A. O'Connor, S.M. Ries, J. Goffreda & A. Voordeckers, 1992. Breeding apples for scab resistance: 1945–1990. Acta horticulturae 317: 43–70.

Fischer, C., 1989. Untersuchungen zur Vererbung der Resistenz gegen Schorf (*Venturia inaequalis* (Cook) Aderh) in Apfel-Nachkommenschaften. Arch. Gartenbau, Berlin 37(1): 23–30.

Vondráček, J. & J. Blažek, 1973. The resistance of apple tree varieties to apple scab (*Venturia inaequalis* Cke. Wint.). Ochrana rostlin (Prague) 9(2): 125–132.

Watkins, R. & R.A. Smith, 1982. Descriptor list for apple (Malus), IBPGR, Rome.

H. Schmidt and M. Kellerhals (Eds.), Progress in Temperate Fruit Breeding, 27–30.

Apple breeding for disease resistance in Hungary *

M. Tóth[1], Zs. D. Rozsnyay[2] & Do X. Quang[1]
[1] *University of Horticulture and Food Industry, Department of Pomology, Villányi út 35–43, H-1118 Budapest, Hungary* [2] *Retired researcher, Research Institute of Plant-Protection, Köbölkút út 13, H-1118 Budapest, Hungary*

Key words: Apple breeding, cultivar, resistance, segregation, *Venturia inaequalis*

Summary

This apple breeding programme started in 1991. It is based partly on earlier results in our department and partly on the breeding methods and results of apple breeders in USA, Canada, France and Germany. Our scientific programme consists of three parts: 1) hybridization; 2) testing *Malus* species as sources of breeding; 3) testing new resistant cultivars origining from France, Germany, USA and Canada. The aim of the crossing programme is to associate good quality, high yield, suitability for Hungarian ecological conditions, good tree shape and suitable maturing time with resistance to scab, powdery mildew, canker and tolerance to aphid and storage damages. This paper deals with hybridization and shows information about the breeding scheme, the parents used and the first results of segregation for scab resistance. The effect of the genes V_f and V_m can be compared.

Introduction

In Hungary, as in West-European countries, American and Canadian states, there is increasing concern about using pesticides and other chemical ingredients. There are efforts to introduce integrated fruit production in orchards and to popularize organic production and 'LISA-principles'. The resistant apple cultivars released abroad are not always suitable for growing in Hungary and sometimes the ripening time and the taste is not exactly convenient for Hungarian consumers. On the other hand consumers need an assortment of resistant varieties as rich as available of non-resistant ones.

In the University of Horticulture and Food Industry at the Department of Pomology 20 years ago Sándor Kovács started the first apple breeding programme. In 1984 G. Tóth started a new programme with the aim of releasing new high yielding cultivars which have good quality, favourable growth habit and are suitable for extensive growing and ecological conditions in Hungary. Progenies of 17 crossing combinations have been selected for growth habit and resistance to powdery mildew and new promising selections are being tested in the second step of evaluation.

In 1991 the newest scientific programme was undertaken. The main objectives of this programme are first to release multiple resistant apple cultivars by hybridization, secondly to search and test new types of *Malus* species as sources of resistance and finally to evaluate resistant cultivars originating from USA, Canada, France and Germany. This paper gives details only of the newest hybridization work undertaken for multiple resistant cultivars.

Breeding policy

Resistant dessert and cider apple cultivars available in the world can be used as parents in breeding for resistance. Among them there are varieties resistant not only to scab, canker or/and mildew but also to fire blight (Lamb et al. 1985, Warner & Potter 1988, Fischer 1991, Sutton & Pope 1990, Lespinasse 1990).

Breeding policy and methods were studied not only from literature but directly in the breeding institutes e.g., in Dresden-Pillnitz (M. Fischer & C. Fischer), in Angers (Y. Lespinasse & M. Le Lezec), in Wädenswil (M. Kellerhals), in East Malling (K. Tobutt K. & F. Alston), too.

* preliminary report.

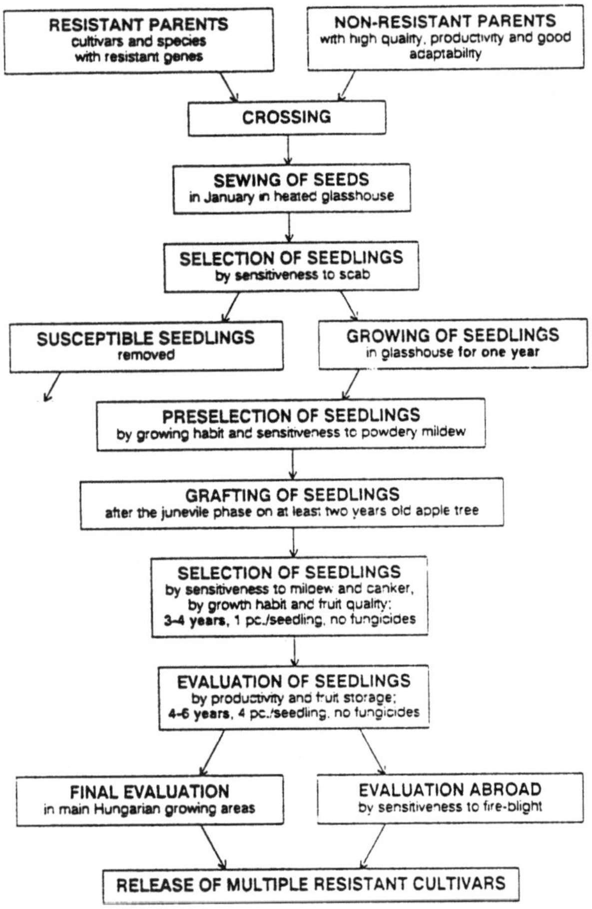

Fig. 1. Scheme of apple breeding.

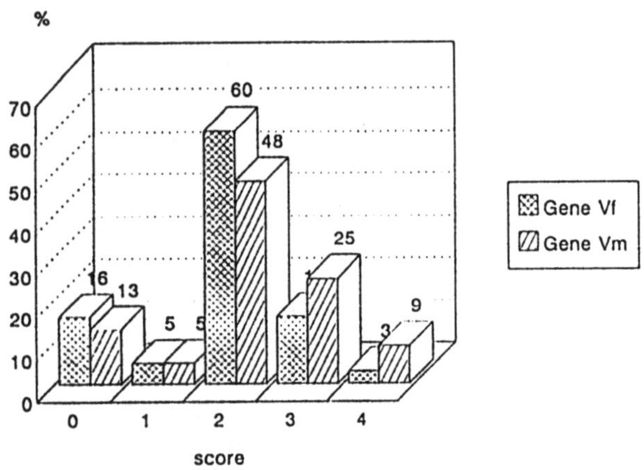

Fig. 2. Percentage of segragation for scab resistance in two V-genes.

Our breeding and selection scheme (Fig. 1) are based on the methods of C. Fischer. This scheme renders an intensive growing and evaluation process. Seeds have been sewed in heated glasshouse for one year. By the end of the first year seedlings have got beyond the juvenile phase. After preselection by growing habit and sensitiveness to mildew seedlings will be grafted on at least two years old apple-tree (e.g., Idared on M26 rootstock). In the following 3–4 years seedlings are selected by growing habit and fruit quality moreover by sensitiveness to mildew and canker. The best ones are being evaluated in two stages. At first productivity and fruit storage of the selections and different foreign resistant cultivars can be compared. Secondly our best selections will be tested in the main Hungarian growing areas on different rootstocks. At the same time sensitiveness to fire-blight can be controlled. Following the scheme Fig. 1 new multiple resistant apple cultivars can be released in 10–12 years after the crossings.

Material and methods

Resistant cultivars with V_f and V_m genes, *Malus* species with supposed V_f, V_b and $V_?$ genes were used as parents in our crosses (Table 1). The non-resistant parents are suitable for growing in Hungarian main apple production areas and favoured by consumers. Kr 5, Kr óriás and 'Kovelit' are promising hybrids of Jonathan and Egri piros (Kovács 1990). All three of them are resistant to powdery mildew.

In order to get the required seeds, only two flowers per cluster have been pollinated. Female flowers have been emasculated.

In 1992 seeds have been sown in a plastic house. Because of hot temperature the inoculation with *Venturia inaequalis* was unaccomplished. In 1993 more than 12 thousand seeds were sown to Jiffy pots in a heated glasshouse in the first decade of January. The seedlings were sprayed twice with the spore suspension of a mixed inoculum of *Venturia inaequalis*. The susceptibility scoring to *Venturia inaequalis* was made three times by the known methods (Shay & Hough in Kellerhals 1989). The scores were: 0 = no symptoms of infection, 1 = very small pin point lesions, 2 = necrotic flecks, 3 = slight sporulation (less than 25% of leaf surface), 4 = severe sporulation. The resistant seedlings were potted to flower pots (10 cm diameter) and to plastic buckets, respectively. Only seedlings with scores 0 and 1 and a few ones with score 2 have been kept.

Table 1. Malus species and cultivars used as parents in apple breeding in Hungary (1991–1993).

Year	Resistant Parents	Resistance	Non-Resistant Parents
1991	Prima	V_f	Idared
	Florina	V_f	Jonathan M41
	Malus floribunda 'Arb-1'	$V_{f?}$	Golden Spur
	Malus baccata var. mandzsurica 'Szcs-2'	$V_{b?}$	Granny Smith
			Kr 5*
			Kr 11*
			Kr orias*
1992	Prima	V_f	Idared
	Richelieu	V_f	Jonathan M41
	Liberty	V_f	All Red Jonathan
	Florina	V_f	Golden Spur
	Priscilla	V_f	Gloster
	Priam	V_f	Kr 5*
	Rouville	V_m	
	Malus floribunda 'Arb-1'	$V_{f?}$	
	Malus × purpurea 'Szcs-1'	$V_?$	
	Malus baccata var. mandzsurica 'Szcs-2'	$V_{b?}$	
1993	Prima	V_f	Idared
	Florina	V_f	Jonathan M41
	Malus floribunda 'Arb-1'	$V_{F?}$	All Red Jonathan
	Malus × purpurea 'Szcs-1'	$V_?$	Golden Delicious
	Malus baccata var. mandzsurica 'Szcs-2'	$V_{b?}$	Gloster
			Granny Smith
			Smoothee
			Imperial Gala

* Hybrids of Jonathan and Egri piros.

? – Supposedly V_f, V_b genes and unknown V gene.

Results and discussion

The first results of the inheritance of scab resistance in 15 progeny groups are shown in Table 2. Some of these progenies gives new information on the inheritance of the resistance to scab because some of these cultivars have not been published as parents earlier (e.g., Richelieu × Idared). By the rate of the highly resistant grades (0 and 1) progenies originated from Prima was the better than the other resistant parents (Table 2). In earlier breeding work Prima also proved to be good parent (Crosby et al. 1992). Among the resistant parents Liberty was not very suitable for getting high proportion of resistant seedlings. Kellerhals (1989) published similar data.

There were differences between progeny groups not only depending on resistant but on non-resistant parents (Table 2). E.g., among progenies originated from Prima the combination of Prima × All Red Jonathan was the best one. From the crossing combination of Richelieu × Kr 5 the only highly resistant (grade 0 and 1) seedling have not been found. In the progenies of Florina and four different non-resistant cultivars Kellerhals (1989) got similar results. The investigation of the inheritance of the scab resistance depending on the non-resistant parents must be continued.

On Fig. 2 the effect of V_f and V_m can be compared. According to the first results the V_f gene looks slightly better than V_m in getting highly resistant young seedlings.

Our intensive growing methods proved to be good, because till the end of the growing season most of the seedlings were more than 2.0 m and have got beyond the juvenile phase.

Table 2. Segregation for scab resistance in apple progenies (1993).

Code	Male parents	Female parents	No. of seedlings	Scab Grade (%) 0	1	2	3	4
71	Prima	Idared	456	7	2	63	23	6
72	Prima	Jonathan M41	298	7	6	55	24	8
74	Prima	Golden Spur	202	37	5	49	7	2
75	Prima	Kr 5	865	20	2	64	11	2
76	Prima	All Red Jonathan	217	51	17	12	17	3
03	Unknown	Prima	3134	14	14	–	–	–
11	Richelieu	Idared	1452	23	3	52	18	3
12	Richelieu	Jonathan M41	127	9	8	60	23	0
15	Richelieu	Kr 5	30	0	0	97	3	0
42	Liberty	Jonathan M41	320	3	3	64	21	9
31	Florina	Idared	396	10	2	74	14	1
51	Priscilla	Idared	918	17	0	65	13	5
61	Priam	Idared	346	6	2	70	17	4
21	Rouville	Idared	1010	22	6	52	16	4
22	Rouville	Jonathan M41	443	5	3	44	33	14

Grading system by Shay & Hough in Kellerhals 1989.

References

Crosby, J.A., J. Janick, P. Pecknold, S.S. Korban, P.A. O'Connor, S.M. Ries, J. Goffreda & A. Voordeckers, 1992. Breeding apples for scab resistance: 1945–1990. Fruit Var. J. 46(3): 145–166.

Fischer, C., 1991. Ergebnisse der Resistenzzüchtung gegenüber Schaderregern beim Apfel. Nachrichtenbl. Deut. Pflanzenschutzd (Braunschweig) 43: 147–150.

Kellerhals, M., 1989. Breeding disease resistant apple cultivars in Switzerland. OILB Working Group Integrated Control of Pome Fruit Diseases. Vol. 2. WPRS Bulletin. 1989/XII/6: 130–136.

Kellerhals, M., 1991. Apfelzüchtung in Wädenswil. Erw. Obstbau. 33: 219–224.

Kovacs, S., 1990. Jonager a new mildew resistant apple variety. Fruit Var. J. 44(6): 66–67.

Lamb, R.C., H.S. Aldwinckle & D.E. Terry, 1985. Freedom, a disease-resistant apple. HortScience 20: 774–775.

Lespinasse, Y., 1990. L'Amélioration Génétique. L'arboriculture fruitière 434: 17–22.

Sutton, T. & P. Pope, 1990. The susceptibility of scab immune cultivars and selections of apple to fire blight and cedar apple rust. Biological and Cultural Test. 5: 4.

Warner, J. & C. Potter, 1988. Performance of scab resistant apple cultivars at the Smiethfield experimental farm. Fruit Var. J. 42(3): 96–102.

H. Schmidt and M. Kellerhals (Eds.), Progress in Temperate Fruit Breeding, 31–34.

Susceptibility of some apple cultivars to scab (*Venturia inaequalis* (Cooke) Aderh.) in Slovenia

M. Virščεk-Marn, F. Štampar, J. Smole & A. Solar
Biotechnical Faculty, Agronomy Department, University of Ljubljana, 61 000 Ljubljana, Slovenia

Key words: Malus, apple, cultivar, resistance, *Venturia inaequalis*, scab, stability

Summary

The susceptibility of 7 apple cultivars to scab was studied in two locations in Slovenia under a minimal spraying programme. Three cultivars with the Vf gene for resistance to apple scab, 'Prima', 'Priscilla' and 'OR 54 T 170', the cvs. 'Alkmene', 'Lord Lambourne' and 'Priols Delicious' showing a low susceptibility to scab in previous tests, and a standard susceptible 'Golden Delicious B' were included in the tests. In the trial orchard 'Pesnica 82' surrounded by modern intensive and old extensive apple orchards the susceptibility of 'Alkmene', 'Lord Lambourne' and 'Priols Delicious' gradually increased from the 4th to the 8th year when scab fungicides were used only 2–4 times per year. In the trial orchard 'Selo 82', with no sources of scab infection in the vicinity and therefore low infection pressure, these cultivars preserved their high level of resistance. The standard 'Golden Delicious B' showed high susceptibility in both orchards. Leaves and fruits of the scab resistant cultivars with the Vf gene occasionally showed few scab lesions in the trial orchard 'Pesnica 82' under extreme infection pressure.

Introduction

Apple scab (*Venturia inaequalis* (Cooke) Aderh.) is the most serious disease of the apple, especially in humid areas and requires numerous fungicide applications for its control (Brown, 1975; Schüepp et al., 1987). The development of integrated pest management has contributed to a considerable reduction of pesticide use, but further reduction seems only possible with the introduction of resistant cultivars (Kellerhals, 1989; Schüepp et al., 1987).

Two types of scab resistance in *Malus* are available for use by the breeder, one polygenic and the other monogenic. The monogenic type of resistance has been found in several wild *Malus* species. So far 6 loci have been identified and 5 physiological races of scab have been recognized which can infect certain of the original sources of resistance. Vf resistance from *M. floribunda*, controlling scab resistance against all known races of scab, is most frequently used in breeding programmes (Brown, 1975; Williams & Kuć, 1969; Fideghelli, 1991).

The low to moderate level of scab resistance exhibited by many commercial cultivars is apparently polygenically conditioned. This form of resistance is not reliable, since a cultivar carrying such resistance can under certain conditions or in certain localities become badly infected by biotypes of the fungus to which it is susceptible. More reliable types of polygenic resistance have been found in some wild *Malus* species and 'Antonovka' (Brown, 1975).

The level of resistance of cultivars is an important factor in reducing the number of sprays needed to control scab. In this respect the most susceptible cultivar in mixed orchards determines the number of treatments. Information on susceptibility of apple cultivars to scab under local conditions are indispensable (van der Scheer, 1989).

Material and methods

In spring 1982 two trial orchards of approx. 0.6 ha were planted with some highly and completely resistant apple cultivars on M 9 on two locations in North-

East Slovenia. The varieties 'Alkmene', 'Lord Lambourne' and 'Priols Delicious' expressing a high level of resistance in previous tests, three cultivars carrying the Vf gene for resistance to apple scab 'Prima', 'Priscilla' and 'OR 54 T 170' and the standard susceptible 'Golden Delicious B' were studied under a highly reduced spraying programme. Numbers of scab fungicide treatments per year and per orchard are presented in Tables 1 and 2 respectively.

Each year an average sample of 100 leaves was examined for each cultivar from mid-June to the end of July. Leaves were classified into 6 categories as follows:

0 = uninfected leaf;
1 = less than 5% of the leaf surface infected;
2 = 5 to 10% of the leaf surface infected;
3 = 10 to 25% of the leaf surface infected;
4 = 25 to 50% of the leaf surface infected;
5 = over 50% of the leaf surface infected.

The incidence of scab was calculated according to the Thownsend Heuberger (Püntener, 1981) equation:

$$\% - \text{infestation} = \frac{\sum_{0}^{i}(n \cdot v)}{i \cdot N} \cdot 100 \qquad (1)$$

in which:
v = value of the category;
i = highest category value;
n = number of plant parts in each category;
N = total number of investigated plant parts.

Results and discussion

In the trial orchard 'Pesnica 82' (Table 1) the susceptibility of 'Alkmene', 'Lord Lambourne' and 'Priols Delicious' gradually increased from the 4th to the 8th year (1985–1989) when scab fungicides were used only 2–4 times a year. In 1989 the incidence of scab on leaves of all three cultivars was as severe as of the susceptible 'Golden Delicious B'. Six applications of scab fungicides in 1990 when the weather conditions were less favourable for the development of scab reduced the incidence of scab to an insignificant amount on all cultivars including 'Golden Delicious B'. In 1991 the number of scab fungicide treatments was further increased. Scab incidence on 'Alkmene', 'Lord Lambourne' and 'Priols Delicious' remained low, whereas the standard cv. 'Golden Delicious B' became badly infected again.

The trial orchard 'Pesnica 82' is located in an apple growing region and is completely surrounded by modern intensive and old extensive apple orchards, therefore the scab inoculum potential in this orchard is very high. We suppose that new pathotypes developed, some of them virulent for 'Alkmene', 'Lord Lambourne' and/or 'Priols Delicious', which multiplied rapidly in the absence of competition. Gessler (1989) concludes that as virulence of scab to apple is controlled by as many as 19 genes, which are mostly independent, redistributed randomly each year and probably widespread, the number of possible races is 2^{16}–2^{19}.

In 1989 the population of scab in the trial orchard 'Pesnica 82' was probably markedly different to that of 1985. In 1990 this population was highly reduced because of the increased number of scab fungicide treatments and relatively unfavourable weather conditions for scab development. The majority of inoculum in spring 1991 originated from neighbouring orchards again as in the first years of observations. Since 1991 was even more unfavourable for scab development than 1990 and the number of scab fungicide treatments was further increased, the incidence of scab on 'Alkmene', 'Lord Lambourne' and 'Priols Delicious' remained as low as it was in 1985 and 1986. Surprisingly the incidence of scab on the susceptible standard 'Golden Delicious B' was also as high as in 1985 in spite of more intensive spraying against scab.

In the trial orchard 'Selo 82' (Table 2) the scab incidence of 'Alkmene', 'Lord Lambourne' and 'Priols Delicious' remained very low, whereas the standard susceptible 'Golden Delicious B' was more or less infected, depending on weather conditions. There are no apple trees in the vicinity of this orchard. We suppose that the development of virulent races for the 3 highly resistant cultivars was hampered due to a limited scab population and low infection pressure.

Our results confirm that the polygenic type of resistance of commercial apple cultivars to scab is unreliable (Brown, 1975). The decrease of the initially high level of resistance of 'Alkmene', 'Lord Lambourne' and 'Priols Delicious' in the trial orchard 'Pesnica 82' was surprisingly fast.

When testing the susceptibility of cultivars currently grown in the Netherlands and of several highly and completely resistant cultivars in an orchard where fungicides were omitted, van der Scheer (1989) found considerable variation in scab incidence among years depending only on the suitability of weather conditions for scab development. None of the cultivars showed

Table 1. Scab incidence on the leaves and number of scab fungicide treatments in the trial orchard 'Pesnica 82'

Cultivar	The incidence of scab in %						
	1985	1986	1987	1988	1989	1990	1991
Golden Del.	68.1	89.8	72.0	80.8	51.2	3.6	53.2
Alkmene	3.8	4.6	4.8	34.0	52.1	0.4	6.6
Lord Lamb.	2.8	12.2	22.2	67.2	49.6	1.4	7.0
Priols Del.	0.05	10.6	16.8	35.2	65.5	2.2	5.8
Prima	0.0	0.4	0.0	0.0	0.0	0.0	1.0
Priscilla	–	–	0.0	0.0	0.0	0.0	–
OR 54 T 170	–	–	0.0	0.0	1.5	0.0	–
No. of scab fungicide treatments	2	4	3	3	2	6	8

Table 2. Scab incidence on the leaves and number of scab fungicide treatments in the trial orchard 'Selo 82'

Cultivar	The incidence of scab in %					
	1985	1986	1987	1988	1989	1990
Golden Del. B	–	–	–	80.0	53.6	18.2
Alkmene	–	–	–	4.8	1.2	1.0
Lord Lambourne	–	–	–	4.8	0.8	0.0
Priols Delicious	–	–	–	0.8	0.0	0.0
Prima	–	–	–	0.0	0.0	0.0
Priscilla	–	–	–	0.0	0.0	0.0
OR 54 T 170	–	–	–	0.0	0.0	0.0
No. of scab fungicide treatments	4	6	1	1	2	3

similar increases in susceptibility as was found in the trial orchard 'Pesnica 82'. 'Alkmene' and 'Priols Delicious' were highly resistant through all 5 years of testing.

Leaves and fruits of scab resistant cultivars carrying the Vf gene occasionally showed a few scab lesions in the trial orchard 'Pesnica 82' (Table 1) with high infection pressure. Similar results were reported by Krüger (1988, 1989), Silbereisen (1989) and van der Scheer (1989).

References

Brown, A.G., 1975. Apples. In: J. Janick & J.N. Moore (Eds) Advances in Fruit Breeding, pp. 3–37. Purdue University Press, West Lafayette.

Fideghelli, C., 1991. Miglioramento genetico del melo per resistenza a *Venturia inaequalis* (Cke.) Wint. L'informatore agrario 24: 53–57.

Gessler, C., 1989. Genetics of the interaction *Venturia inaequalis* – *Malus*: the conflict between theory and reality. In: C. Gessler, D.J. Butt & B. Koller (Eds.), Integrated control of pome fruit diseases II, IOBC-WPRS-Bulletin 12: 168–190.

Kellerhals, M., 1989. Breeding disease resistant apple cultivars in Switzerland. In: C. Gessler, D.J. Butt & B. Koller (Eds.), Integrated control of pome fruit diseases II, IOBC-WPRS-Bulletin 12: 130–136.

Krüger, J., 1988. Beständigkeit der Schorfresistenz aus *Malus floribunda* 821 auf dem Versuchsfeld der Bundesforschungsanstalt für gartenbauliche Pflanzenzüchtung in Ahrensburg. Erwerbsobstbau 52: 254.

Krüger, J., 1989. Scab resistance of apple cultivars, selections and progenies with the Vf gene. In: C. Gessler, D.J. Butt & B. Koller (Eds.), Integrated control of pome fruit diseases II, IOBC-WPRS-Bulletin 12: 161–167.

Püntener, W., 1981. Manual for Field Trials and Plant Protection. pp. 34. Ciba-Geigy Limited, Basle.

Van der Scheer, H.A.Th., 1989. Susceptibility of apple cultivars and selections to scab and powdery mildew in the Netherlands. In: C. Gessler, D.J. Butt & B. Koller (Eds.), Integrated control of pome fruit diseases II, IOBC-WPRS-Bulletin 12: 205–211.

Schüepp, H., W. Siegfried & M. Kellerhals, 1987. Züchtung und Mehltauanfälligkeit schorfresistenter Apfelsorten. Schweiz. Z. für Obst- und Weinbau 123: 91–98.

Silbereisen, R., 1989. Experiences at Bavendorf (FRG) with scab-resistant apple cultivars. In: C. Gessler, D.J. Butt & B. Koller (Eds.), Integrated control of pome fruit diseases II, IOBC-WPRS-Bulletin 12: 148–160.

Williams, E.B. & J. Kuć, 1969. Resistance in *Malus* to *Venturia inaequalis*. Ann. Rev. Phytopath. 7: 223–246.

H. Schmidt and M. Kellerhals (Eds.), Progress in Temperate Fruit Breeding, 35–43.

The behaviour of some apple varieties in the Eastern Subcarpathian area

V.I. Rati, F. Acatrinei & Luminita Rati
Fruit growing research and production station, Bacau, Romania

Key words: Apple varieties, adaptability

Summary

Apple is the main fruit growing species cultivated in the Eastern Subcarpathian region. The existing conditions here necessitated the selection of a range of new valuable varieties for this area, to meet the new requirements of fruit growing (good quality, resistance to disease, suitability for intensive growing). For this purpose 65 apple varieties and hybrids were studied by the Fruit Growing and Research Station in Bacau from 1982 to 1992. The study evaluated the performance and adaption of each variety and hybrid for this area. The most valuable varieties were considered to be those that showed high genetic resistance to diseases and fruit quality, such as: 'Romus 2', 'Romus 3', 'Auriu de Bistrita', 'Florina'. The varieties 'Close', 'Stark Earlist', 'Cardinal', 'Ardelean', 'Radaseni', 'Delicious de Voinesti', 'Mutsu' as well as the hybrids H 3–11–71, BN 13–38–4, V 142–24, NJR 60, NJR 68, also proved to be valuable, merit further attention.

Introduction

The soil and climatic conditions of the Eastern Subcarpathian area of Romania are very favorable for fruit-tree growing. Fruit growing is one of the most efficient forms of agriculture here.

Up to 1960, plum cultivation was widespread, but apple now occupies the largest areas (Cociu 1990). Its suitability for intensive growing makes it the most efficient fruit tree species.

In order to maintain and even to increase the economical efficiency of apple growing, it was necessary to improve the range of available varieties to meet the market requirements (quality, appereance, freedom from toxic residues) to meet the conditions of this area (steady high production resistance to limiting factors, reduced vigour) (Cociu 1981).

Material and methods

The experiment started in 1982 and was grown at the Fruit Growing Research Station located north-east from Bacau, Romania. Apple varieties and hybrids with genetic disease resistance (16), summer-autumn varieties (24), and winter varieties (17) were planted on three experimental fields to which seven more varieties and hybrids resistant to disease were added in 1985 (Cociu 1977) (see Table 1).

The trees were grafted on MM 106 and planted according to a random block design at a distance of 4 × 2,5 m.

A storied fan-shaped espalier with oblique branches was chosen for the crown.

The soil is of an aluvial sandy type having a pH of 8.1–8.6 and organic content of 1.0–1.3. The average annual temperature is 9,1 °C and the mean annual rainfall is 553 mm.

Observations, measurements and tests were made for yield (t/ha), fruit quality (size, dry matter, taste), index of productivity (calculated from the yield (kg/tree) and the trunk cross section area), and scab (*Venturia inaequalis*) on leaves and fruits.

To calculate the Degree of the Attack (DA), the following equation was used:

$$D.A. = (F \times I)/100, \qquad (1)$$

where F% = (percentage of leaves or fruit area attacked); I% = E (i × f)/100, where I = intensity of the attack; E = amount; i = percentage area of leave or fruit attacked (25; 50; 75; 100% – according to the

Table 1. The average annual yield of 65 apple varieties and hybrids (density 1000 trees/ha) (Study 1986–1992).

Variety or hybrid	Average yield t/ha	Index of productivity	Weight fruit (g)	Note of taste (1–9)	Dry matter %
Genetic Disease Resistent Apple Varieties					
Planted in 1982					
1. 3–11–71 P	25.68	0.45	116.4	4.62	13.141
2. Romus 2	23.10	0.32	86.15	4.85	13.16
3. S 25–5–13	17.30	0.27	130.18	4.22	13.05
4. 3–13–71 P	29.83	0.32	103.04	4.50	12.7
5. Priam	27.84	0.41	103.66	5.35	15.1
6. Prima (M)	26.56	0.44	138.32	5.64	13.58
7. DI R57 T120	23.16	0.48	119.25	5.16	13.58
8. DI R57 T195	18.04	0.43	93.28	5.04	14.13
9. 70–80–14 V	9.37	0.06	145.28	4.15	15.24
10. 70–80–11 V	15.10	0.15	130.62	3.22	14.30
11. 70–59–4 V	17.89	0.21	118.81	4.10	14.10
12. 72–33–5 V	15.66	0.18	112.43	3.35	14.10
13. 70–75–7 V	15.14	0.13	237.65	3.94	13.50
14. Sir Prize	22.56	0.27	170.34	5.37	13.74
15. Florina	23.17	0.26	136.65	5.89	15.20
16. Romus 3	17.57	0.35	128.56	6.45	12.22
Mean	20.50				
Planted in 1985					
1. BN 33–39–4	15.50	0.41	164.60	6.12	10.93
2. BN 13–38–4	20.63	0.53	172.65	4.88	12.86
3. Auriu de Bistrita	17.95	0.45	173.66	6.24	18.00
4. Golden Lautswent	17.98	0.45	158.43	5.21	14.6
5. V 142–24	23.45	0.56	176.46	5.77	12.3
6. BN XI–214–7	30.48	0.46	181.22	3.92	13.76
7. Lord Lambourne	22.20	0.56	165.2	4.08	15.7
8. Prima (M)	23.15	0.65	138.32	5.64	13.58
Mean	21.66				
Summer-Autumn Apple Varieties					
Planted in 1982					
1. Close	18.72	0.27	132.37	5.78	12.12
2. Stark Earliest	15.79	0.25	77.37	5.32	12.0
3. Jerseymac	14.33	0.20	128.18	6.36	11.11
4. Quinte	11.64	0.13	133.57	6.36	11.90
5. Cardinal	26.10	0.56	156.37	5.84	12.11
6. James Grieve	19.79	0.45	163.70	5.51	12.44
7. Red Melba	15.96	0.29	106.98	4.83	12.05
8. Ancuta	21.21	0.45	137.38	3.89	13.14
9. Aromat de Vara	13.03	0.14	127.92	5.85	12.65

Table 1. continued

Variety or hybrid	Average yield t/ha	Index of productivity	Weight fruit (g)	Note of taste (1–9)	Dry matter %
10. NJR 49	23.34	0.41	128.57	5.78	12.87
11. NJR 60	20.33	0.30	115.41	6.42	11.74
12. NJR 61	18.83	0.29	114.04	5.21	11.45
13. NJR 63	15.39	0.23	109.71	3.88	12.72
14. NJR 64	18.40	0.41	116.7	4.97	11.74
15. NJR 68	19.57	0.45	172.55	6.12	12.12
16. NJR 71	13.79	0.36	107.42	5.34	12.02
17. NJR 72	13.30	0.18	129.45	4.72	11.97
18. NJR 77	13.37	0.21	107.64	5.04	12.00
19. Mollie's Delicious	13.74	0.18	145.02	5.65	10.50
20. Akane	11.51	0.22	122.34	5.17	13.50
21. S25 R10 101	13.99	0.22	117.45	5.38	11.91
22. Cox Orange Reinette	15.54	0.23	129.6	5.75	13.38
23. Ardelean	22.24	0.55	136.10	4.81	14.82
24. Morspur	22.70	0.48	124.70	4.72	14.28
Mean	17.10				
Winter Apple Varieties					
1. Jonagold	12.04	0.12	143.20	4.81	15.30
2. Kalther de Boemia	14.86	0.33	139.70	5.30	13.70
3. Jonathan	12.36	0.28	129.70	4.50	15.20
4. Starkrimson	19.80	0.45	156.90	5.66	14.40
5. Goldspur	14.54	0.51	134.40	5.26	15.90
6. Delicious de Voinesti	17.14	0.53	115.10	4.80	17.10
7. Mutsu	14.40	0.23	228.40	6.30	12.50
8. Idared	14.48	0.34	139.40	4.53	12.40
9. Wagener Premiat	12.50	0.37	105.60	3.84	14.80
10. Rosu de Cluj	15.06	0.29	88.70	4.80	13.84
11. Feleac	14.18	0.21	168.70	3.95	13.94
12. Radaseni	38.64	1.06	139.70	4.28	13.36
13. Frumos de Voinesti	13.10	0.22	129.00	5.35	15.00
14. Golden Delicious	15.32	0.27	125.80	5.59	15.26
15. Jonathan Heines	12.42	0.18	119.50	4.45	16.
16. Jonathan Watson	11.04	0.16	125.90	4.67	15.82
17. Red Jonathan	11.80	0.19	124.10	4.44	15.16
Mean	10.50				

Table 1a.

The degree of attack of scab (*VENTURIA INAEQUALIS*) on the leaves and fruits of apple disease resistant varieties and hybrids:		
Varieties and hybrids	Leaves (%)	Fruits (%)
1. 3–11–71 P	1.03	0.50
2 Romus 2	1.42	–
3. S 25–5–13	8.23	1.8
4. 3–13–71 P	7.20	1
5. Priam	–	–
6. Prima	–	–
7. D 1 R 57 T 120	1.97	1.8
8. D 1 R 57 T 195	1.13	2.47
9. 70–80–14 V	–	–
10. 70–80–11 V	–	1.80
11. 70–59–4 V	–	1.83
12. 72–33–5 V	0.26	1.22
13. 70–75–7 V	–	1.83
14. Sir Prize	–	1.40
15. Florina	–	–
16. Romus 3	–	0.58
1. BN 33–39–4	1.33	–
2. BN 13–38–4	–	0.35
3. Auriu de Bistrita	0.72	1.50
4. Golden Laustwen	–	1.02
5. V 142–24	–	–
6. BN IX 214–7	4.30	0.97
7. Lord Lambourne	–	0.43
Summer-Autumn Apple Varieties		
1. Close	9.20	1.50
2. Stark Earliest	2.60	3.53
3. Jerseymac	7.57	9.98
4. Quinte	7.57	7.53
5. Cardinal	1.40	1.10
6. James Grieve	1.69	1.20
7. Red Melba	0.23	0.45
8. Ancuta	0.26	1.77
9. Aromat de Vara	1.19	2.40

Table 1a. continued

Varieties and hybrids	Leaves (%)	Fruits (%)
10. NJR 49	1.80	6.93
11. NJR 60	1.56	0.84
12. NJR 61	2.07	3.07
13. NJR 63	0.56	1.83
14. NJR 64	1.19	1.26
15. NJR 68	2.10	1.17
16. NJR 71	3.49	2.63
17. NJR 72	3.76	3.53
18. NJR 77	4.33	1.33
19. Mollies Delicious	1.43	2.37
20. Akane	0.33	–
21. S 25 R 10 101	5.43	3.30
22. Cox Orange Renette	3.23	1.77
23. Ardelean	0.56	–
24. Morspur	0.83	1.24
Winter Apple Varieties		
1. Jonagold	6.03	2.47
2. Kalter de Boiemia	0.92	3
3. Ionathan	5.07	0.71
4. Starkrimson	5.50	2.60
5. Goldspur	5.07	0.61
6. Delicious de Voinesti	1.53	0.96
7. Mutsu	4.97	5.33
8. Idared	8.97	3.97
9. Wagener Premiat	5.67	1.13
10. Rosu de Cluj	1.47	2.15
11. Feleac	0.93	3.67
12. Radaseni	1.44	1.09
13. Frumos de Voinesti	3.33	2.57
14. Golden Delicious	7.47	2.70
15. Jonathan Heines	4.70	1.10
16. Jonathan Watson	4.50	1.57
17. Red Jonathan	3.30	1.30

grades 1, 2, 3, 4); f = number of cases of attack at each level (Cociu & Oprea 1989).

Results

During the study, the flowering period of the varieties and hybrids varied from 5 to 18 days (long flowering periods were recorded in years when the start of blossoming was earlier, and short flowering periods in years when the start of blossoming was later). There were differences of about 40 days between years when flowering was early and those when flowering was late.

As a result, the maturity period varied from 20 to 25 days (see Figs. 1–3).

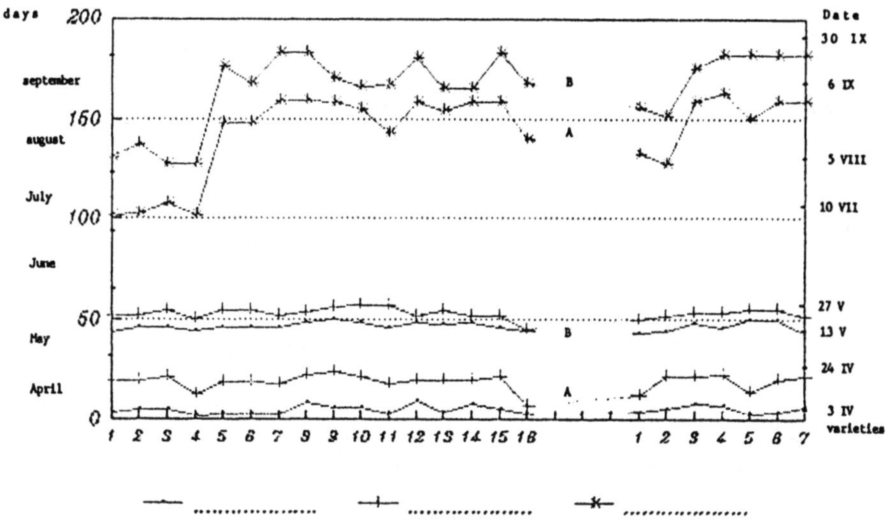

Fig. 1. Phenology of blossom and the period of fruits' maturity on 23 disease resistant apple varieties and hybrids (see Table 1) (Study 1986–1992).

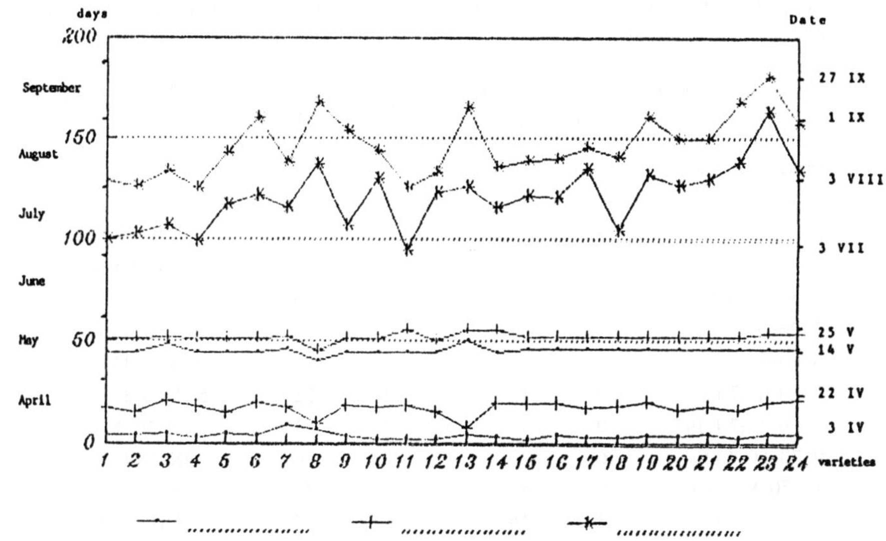

Fig. 2. Phenology of blossom and the period of the fruits' maturity on 24 summer-autumn apple varieties (see Table 1) (Study 1986–1992).

There were no varieties and hybrids whose flowering was outside the flowering period and therefore missed a pollination source.

The index of productivity was calculated as a negative relationship between the two elements: yield (kg/tree) and trunk cross section area (cmp). This value expresses the adaptability of the variety to intensive growing which is in direct proportion to this calculated value. Among the varieties resistant to diseases, those showing the highest values for productivity index were: 'Prima' (0.65 planted 1985 and 0.44 planted in 1982 – the older the tree, the smaller the value because of the size of the trunk); 'Lord Lambourne', 'V 142–24', 'BN 13–38–4', 'Auriu de Bistrita', 'D1 R57 T120', 'H 3.1.71 P', 'Priam', 'Romus 3', 'Romus 2' planted in 1982 (0.32).

Among the summer-autumn varieties the following are mentioned: 'Cardinal' (0.50), 'Ardelean', 'Morspur', 'NJR 68', 'NJR 60' (0.30), and among the winter varieties the following can be mentioned: 'Radaseni' (1.06), 'Delicious de Voinesti', 'Goldspur', 'Starkrimson' (0.45) (see Table 1).

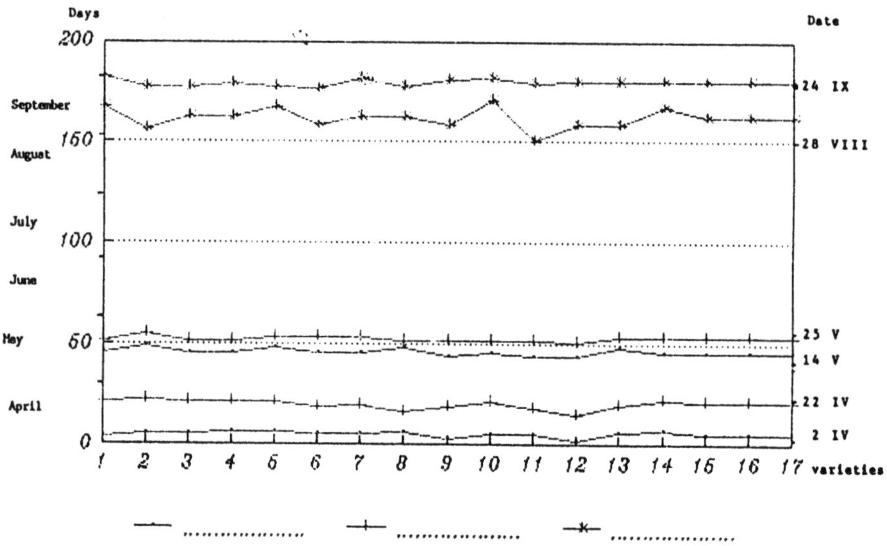

Fig. 3. Phenology of blossom and the period of the fruits' maturity on 17 winter apple varieties (see Table 1) (Study 1986–1992).

Table 2. Phenology of blossom and the period of fruits' maturity on 23 disease resistant apple varieties and hybrids (Study 1986–1992).

Varieties and Hybrids	Beginning Blossom		End of Blossom		Harvest Maturity
	April	May	April	May	
Planted in 1982					
1. 3-11-71 P	4	13	20	22	10 VII–9 VIII
2. Romus 2	5	16	20	22	11 VII–15 VIII
3. S 25-5-13	5	16	21	24	16 VII–5 VIII
4. 3-13-71 P	2	14	13	20	10 VII–5 VIII
5. Priam	3	16	19	24	26 VIII–23 IX
6. Prima	3	16	19	24	26 VIII–15 IX
7. DI R 57 T 120	3	16	18	22	6 IX–30 IX
8. DI R 57 T 195	9	18	22	23	6 IX–30 IX
9. 70-80-14 V	6	20	24	26	5 IX–17 IX
10. 70-80-11 V	6	18	21	27	2 IX–13 IX
11. 70-59-4 V	3	16	18	27	21 VIII–14 IX
12. 72-33-5 V	10	18	20	22	5 IX–27 IX
13. 70-75-7 V	4	17	20	24	2 IX–12 IX
14. Sir Prize	8	18	20	22	5 IX–12 IX
15. Florina	5	16	21	22	5 IX–30 IX
16. Romus 3	3	14	7	15	19 VIII–15 IX
Planted in 1985					
1. BN 33-39-4	4	13	12	20	10 VIII–3 IX
2. BN 13-38-4	5	14	21	22	5 VIII–30 VIII
3. Auriu de Bistrita	8	18	21	23	5 IX–22 IX
4. Golden Laustwen	7	16	22	23	10 IX–29 IX
5. V 142-24	3	20	14	25	28 VIII–29 IX
6. BN XI 214-7	4	20	20	25	5 IX–29 IX
7. Lord Lambourne	6	13	21	22	5 IX–29 IX

Table 3. Phenology of blossom and the period of the fruits' maturity on 24 summer-autumn apple varieties (Study 1986–1992).

Varieties and Hybrids	Beginning Blossom		End of Blossom		Harvest Maturity
	April	May	April	May	
Planted in 1982					
1. Close	4	14	17	21	8 VII–6 VIII
2. Stark Earliest	4	14	15	21	11 VII–4 VIII
3. Jerseymac	5	18	21	22	15 VII–11 VIII
4. Quinte	3	14	18	21	7 VII–3 VIII
5. Cardinal	5	14	15	21	25 VII–20 VIII
6. James Grieve	4	14	20	21	30 VIII–7 IX
7. Red Melba	9	16	18	22	24 VII–15 VIII
8. Ancuta	7	20	10	15	15 VIII–15 IX
9. Aromat de Vara	4	14	19	21	15 VII–30 VIII
10. NJR 49	3	14	18	21	6 VIII–29 VIII
11. NRJ 60	3	14	19	25	3 VII–3 VIII
12. NJR 61	3	14	16	20	1 VIII–10 VIII
13. NJR 63	5	20	8	25	4 VIII–12 IX
14. NJR 64	4	14	20	25	24 VII–13 VIII
15. NJR 68	3	16	20	25	30 VII–16 VIII
16. NJR 71	5	16	20	22	29 VII–17 VIII
17. NJR 72	4	16	18	22	13 VIII–23 VIII
18. NJR 77	4	16	19	22	13 VII–18 VIII
19. Mollie's Delicious	5	16	21	22	10 VIII–7 IX
20. Akane	5	16	17	22	5 VIII–27 VIII
21. S 25 R 10 101	6	16	19	22	8 VIII–27 VIII
22. Cox Orange Reinette	4	16	17	22	16 VIII–15 IX
23. Ardelean	6	16	21	23	1 IX–27 IX
24. Morspur	6	16	22	23	12 VIII–4 IX

Fruits of good quality (size, dry matter, taste score) were given by 'Florina', 'Romus 3', 'Prima', 'Sir Prize', 'Auriu de Bistrita', 'V 142–24', 'Close', 'NJR 68', 'NJR 60', 'Jerseymac', 'Mutsu', 'Starkrimson' and 'Goldspur'.

Of the varieties and hybrids with genetic resistance to diseases the following showed no sign of attack by scab on leaves or fruit, even no fungicide treatment was applied: 'Prima', 'Priam', 'Florina', 'H 70–80', 'V 142–24'. The other varieties and hybrids showed a level of attack lower than 2%.

Under normal fungicide treatment conditions, the summer-autumn and winter varieties showed a different degree of attack. The varieties 'Akane', 'Ardelean', 'Morspur', 'Delicious de Voinesti', 'Radaseni', 'Rosu de Cluj', 'Kalter de Boemia', proved to be the most resistant.

Conclusions

1. The basic range of varieties for this area will consist mainly of varieties showing genetic resistance to diseases.
2. From the varieties studied up to now, the following will be added to the existing range for this area: 'Romus 2', 'Romus 3', as summer varieties; 'Auriu de Bistrita' as an autum variety; 'Florina' as a winter variety.
3. The hybrids 'H3–11–71', 'BN 13–38–4' (summer apple) and 'V 142.24' (autumn-winter apple will keep our attention for further evaluation and possible multiplication.
4. The following apple varieties and hybrids showing no genetic resistance to scab will also keep our attention: 'Close', 'Stark Earliest', 'NJR 60', 'NJR 68', 'Cardinal', 'Ardelean', 'Delicios de Voinesti', 'Radaseni', 'Mutsu', 'Idared', 'Goldspur', 'Starkrimson'.

Table 4. Phenology of blossom and the period of the fruits' maturity on 17 winter apple varieties (Study 1986–1992).

Varieties and Hybrids	Beginning Blossom		End of Blossom		Harvest Maturity
	April	May	April	May	
1. Jonagold	4	16	21	22	15 IX–30 IX
2. Kalter de Boemia	5	19	22	25	3 IX–25 IX
3. Jonathan	5	16	21	22	10 IX–26 IX
4. Starkrimson	6	16	21	22	10 IX–26 IX
5. Goldspur	6	18	21	23	15 IX–25 IX
6. Delicious de Voinesti	5	16	19	23	5 IX–24 IX
7. Mutsu	5	16	20	23	10 IX–29 IX
8. Idared	6	18	16	22	10 IX–25 IX
9. Wagener Premiat	3	14	19	22	5 IX–28 IX
10. Rosu de Cluj	5	16	21	22	18 IX–30 IX
11. Feleac	5	14	18	22	28 VIII–26 IX
12. Radaseni	2	14	15	21	5 IX–27 IX
13. Frumos de Voinesti	6	18	20	23	5 IX–27 IX
14. Golden Delicious	7	16	22	23	15 IX–27 IX
15. Jonathan Heines	5	16	21	23	10 IX–27 IX
16. Jonathan Watson	5	16	21	23	10 IX–27 IX
17. Red Jonathan	5	16	21	23	10 IX–27 IX

References

Cociu, V., 1977. The Results of Research-Work Concerning the Obtaining of New Varieties in Romania. Ceres, Bucharest.

Cociu, V., 1981. Some aspects on fruit and small fruit varieties location in Romania. Scientific papers I.C.P.P. Pitesti IX: 9–23.

Cociu, V. & St. Oprea, 1989. Methods of Research in Improvement of Fruit Breeding. Dacia, Cluj/Napoca.

Cociu, V., 1990. New Varieties, Progress Factor in Fruit Growing. Ceres, Bucharest.

H. Schmidt and M. Kellerhals (Eds.), Progress in Temperate Fruit Breeding, 43–48.

Breeding apple cultivars with multiple resistance

Christa Fischer
Bundesanstalt für Züchtungsforschung an Kulturpflanzen, Institut für Obstzüchtung Dresden-Pillnitz, Pillnitzer Platz 2, D-01326 Dresden, Germany

Key words: Apple, breeding, scab, powdery mildew, fire blight, multiple resistance

Summary

The long-term apple breeding programme in Dresden-Pillnitz is based on the combination of resistances to the economically important diseases scab, mildew and fire blight and a high fruit quality. The multiple resistance programme in apples began against scab and mildew and was followed by breeding against fire blight. As gene sources for scab resistance we used *Malus floribunda*, *M. pumila*, *M. micromalus*, *M. atrosanguinea* and local varieties, for mildew resistance *M. zumi 'Calocarpa'*, *M. robusta 'Persicifolia'*, Dülmener Rosenapfel, Remo, Rewena, Resi and Retina, and for fire blight resistance *M. floribunda*. Double and triple resistant progenies were obtained from crossings between the resistant varieties Remo, Rewena and Pi-AS-44,14. Other resistant varieties are screened in order to test their inheritance of double and triple resistance. Some promising multiple resistant progenies and cultivars with good fruit quality in different mature periods have been selected for consumption as dessert fruit or for processing. They are suitable as an assortment for integrated fruit production in order to reduce the fungicide sprays by 80% and to minimize the risk of fire blight infection.

Introduction

When planning apple breeding programmes the control of economically important pests and diseases and the need to decrease the use of chemical sprays must be carefully considered. Plant breeders should be aiming to raise, in a systematic way, the level of pest and disease resistance in apple cultivars. This will not only reduce the cost of chemical sprays but it will significantly reduce the risk of environmental pollution. It will also greatly contribute to stabilizing yield and quality in orchards. Research at Dresden-Pillnitz is concentrating on obtaining combined resistance to:
scab (*Venturia inaequalis*);
powdery mildew (*Podosphaera leucotricha*);
fire blight (*Erwinia amylovora*);
red spider mite (*Panonychus ulmi*);
bacterial canker(*Pseudomonas syringae*).

Multiple resistance has been obtained by crossing resistant parents, by back crossing and by using special crossing patterns such as diallels. Multiple-resistant types are best obtained by using donors with high levels of durable resistance.

Material and methods

Selections with resistance to scab, mildew and fire blight have been obtained by crossing dessert cultivars with resistant hybrids – including types carrying resistance genes from various *Malus* species and cultivars.

Progenies have been screened for scab, mildew and fire blight in greenhouse and field using standard methods (Schmidt 1938, Shay & Hough 1952, Fischer 1979).

Results have been assessed at the 5% probability level with different methods (variance analysis, χ^2-test).

Table 1. Segregation of scab resistant and susceptible seedlings in apple progenies with resistance gene sources from apple species and varieties (1977 to 1983).

Resistance gene sources	Number of seedlings	Segregation 'resistant to susceptible'	
		% of seedlings with scab	ratio of segregation (χ^2)
M. floribunda	21,540	53	1:1
M. micromalus	1,863	42	1:1
M. atrosanguinea	1,136	52	1:1
M. pumila	134	38	3:1
Antonovka	7,030	69	–

Table 2. Segregation of scab resistant seedlings in progenies with two resistant parents.

Resistance gene sources	Number of seedlings	Number of resistant seedlings	Percentage resistant seedlings
M. floribunda × *M. floribunda*	2,584	1,862	72
M. floribunda × *Antonovka*	256	190	74
M. pumila × *Antonovka*	68	54	79
number of progenies: 30	2,908	2,106	72

Results

Scab (V. inaequalis)

Screening for scab was the first step in multiple-resistance evaluation. Ratios of resistant to susceptible seedlings are shown in Table 1. Genes were used in crossing progenies from the species and hybrid species *Malus floribunda* (probably a hybrid between *M. sieboldii* × *M. baccata*), *M.* × *micromalus* (probably *M. spectabilis* × *M. baccata*) (Rehder 1951) and *M.* × *atrosanguinea* (*M. halliana* × *M. sieboldii*) (Rehder 1951). All these parents segregated with 1:1 ratios. Progenies with *M.* × *pumila* (Paradise Apple-selection) gave ratios of 3(resistant):1(susceptible). In progenies with polygenic genes derived from old local varieties, such as Antonovka, between 5–30% of seedlings were resistant (Fischer 1989).

Progenies involving resistance alleles, with genes derived from both parents (various combinations of *M.* × *floribunda.*, *M.* × *pumila* and/or Antonovka) gave an average of 72% resistant seedlings with a range from 72–79% (Table 2).

Powdery mildew (P. leucotricha)

Mildew resistance can be obtained from both oligogenic and polygenic sources. Oligogenic sources include the dominant allele of the major gene Pl_1 derived from *M.* × *robusta* (*M. baccata*{*M. baccata* × *M. prunifolia*} Rehder 1951) '*Persicifolia*', and the dominant allele of the major gene Pl_2 derived from *M.* × *zumi* (*M. baccata var. mandshurica* × *M. sieboldii*) (Rehder 1951) '*Calocarpa*' (Knight & Alston 1968). Polygenic sources include old local varieties and some scab resistant hybrids. An average ratio of 1(resistant):1(susceptible) was obtained from *M.* × *zumi* and 1:3 from *M.* × *robusta* populations in our screened breeding material.

Additional sources of mildew resistance were found in scab resistant breeding material. Some scab resistant varieties have good levels of resistance to mildew and transmit this character to some of their progeny. In the work at Dresden this type of resistance to mildew was found to be inherited polygenically. In such progenies it has been possible to make selections combining scab and mildew resistance. Double-

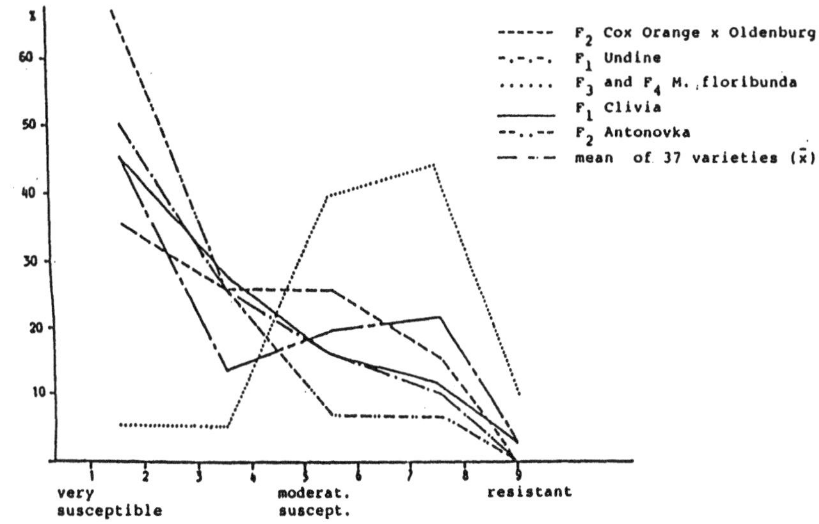

Fig. 1. Resistance to fire blight in apple progenies.

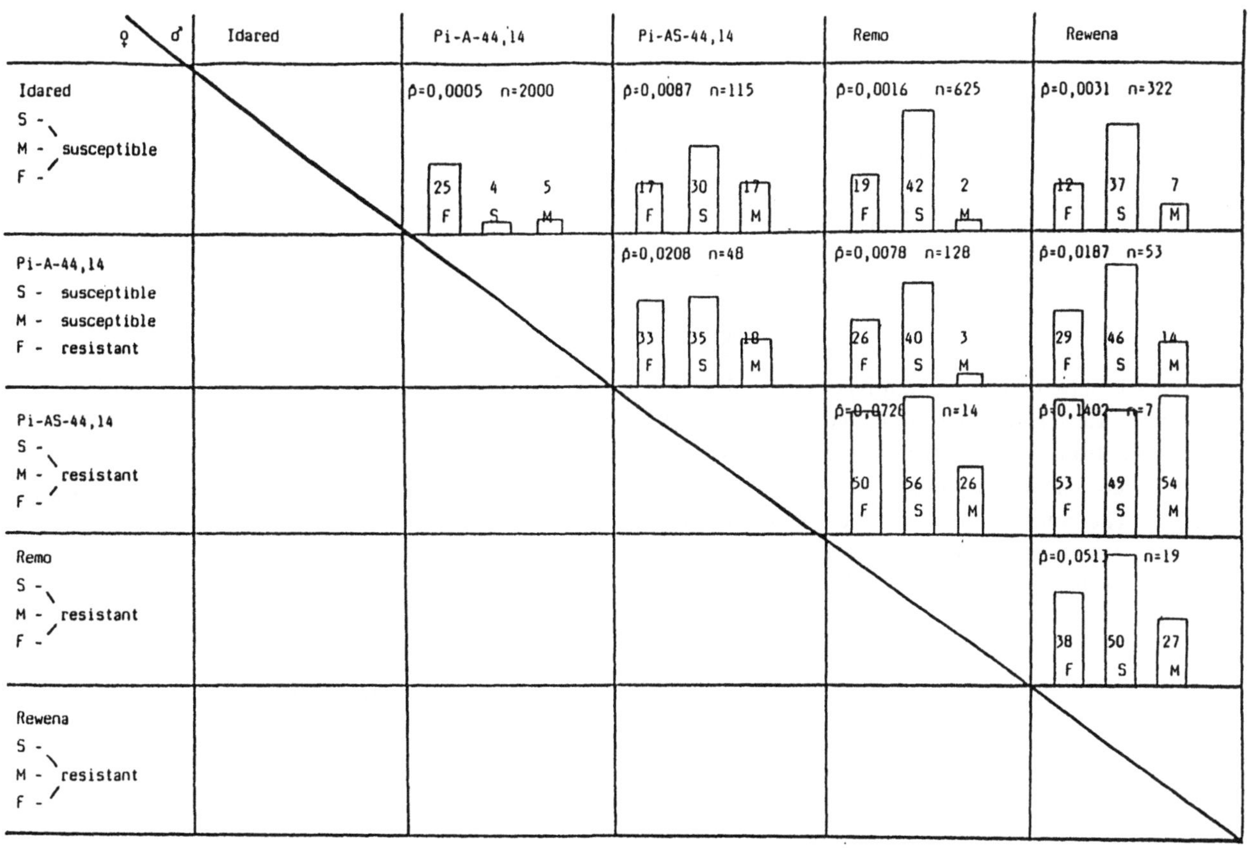

Fig. 2. Diallel – % of seedlings without infection to fire blight (F) and scab (S) and without or with only slight infection to mildew (M), p̂ = estimated probability (Fischer 1989).

Table 3. Segregation of mildew resistant and susceptible seedlings in apple progenies with resistant parents (1 – without mildew, 9 – very susceptible), rate in %.

Progenies	Mildew, rate in %								
	1	2	3	4	5	6	7	8	9
Remo × Rewena	1	17	54	22	6	0	0	0	0
Releika × Rewena	4	20	34	29	6	3	4	0	0
Resi × Releika	12	18	24	31	6	7	2	0	0
Resi × Pi-AS-5,157	4	21	38	23	6	4	3	1	0
Resi × Pi-AS-5,167	0	6	39	33	14	5	0	3	0
Resi × Piros	1	12	33	15	17	11	6	4	1
Retina × Rewena	0	5	16	33	29	14	3	0	0
Retina × Dülmener Rosenapfel	0	19	33	30	3	3	12	0	0
Pi-A-3,163 × Dülmener Rosenapfel	11	16	21	18	17	7	10	0	0
Remo × Dülmener Rosenapfel	13	37	19	18	5	5	3	0	0

Table 4. Segregation of mildew resistance in the progenies of resistant parents, screenings in the field over 5 years (scab 1 – without, mildew 1 – without, 2 – very slow development, 3 – slow development).

Progenies	Number of seedlings	Scab resistance, rate in %	Mildew resistance, rate in %		
		1	1	2 + 3	1–3
Helios × Remo	69	100	26	28	54
Helios × Rewena	35	100	54	31	85
Remo × Ant. 34,15	16	100	56	31	87
Alkmene × Remo	10	100	43	43	86
Rewena × Pi-AS-44,14	39	100	36	46	82
Rewena × Remo	107	100	34	52	86

resistance is inherited by the cultivars Remo, Rewena, Retina, Resi and some numbered clones (Table 3). Results from 5-year-screenings of progenies in the field were similar to those from greenhouse tests (Table 4). The cultivars Dülmener Rosenapfel, Helios and Alkmene gave progenies which included selections with high levels of resistance to mildew.

The percentage of double-resistant seedlings in progenies varied as shown below depending on the nature of the parents:

			Percentages of double-resistant seedlings
Scab susceptible and mildew susceptible	×	scab resistant and mildew susceptible	20
scab susceptible and mildew susceptible	×	scab resistant and mildew resistant	41
scab resistant and mildew resistant	×	scab resistant and mildew resistant	67

Table 5. Combinations of the multiple resistance to important diseases and damages in the new Pillnitz apple varieties (x – high resistant, (x) – moderately resistant, o – moderately susceptible, – – susceptible) (Christa Fischer 1991, Institut für Obstforschung Dresden-Pillnitz).

New apple varieties		Resistance combinations to diseases (biotic) and damages (abiotic)							
Name	Sign of clone	Scab	Scab gene source	Mildew	Fire blight	Bacterial canker	Red spider mite	Spring frost	Winter frost
Remo	Pi-AS-8,5	x	Vf	x	x	o	o	x	x
Reglindis	Pi-AS-8,60	x	V Ant.	(x)	(x)	o	x		x
Retina	Pi-AS-23,3	x	Vf	(x)	o	o	(x)	x	–
Rewena	Pi-AS-8,28	x	Vf	x	x	x	o	x	o
Rene	Pi-AS-4,14	x	Vf	o	x	(x)	–	x	o
Reka	Pi-AS-7,50	x	Vr	(x)	o	x	–	–	(x)
Reanda	Pi-AS-33,59	x	Vf	x	x	o	–	x	o
Relinda	Pi-AS-27,25	x	Vf	(x)	o	x	–		
Releika	Pi-AS-21,73	x	Vf	o	(x)	x	x		
Resi	Pi-AS-33,144	x	Vf	x	o	x	–		
Renora	Pi-AS-21,77	x	Vf	(x)	o	o	o	x	(x)
Regine	Pi-AS-24,132	x	Vf	o	o		x		
Realka	Pi-AS-3,68	x	Vr	–	x	o	o	–	o
Releta	Pi-AS-4,51	x	Vr	–	o	x		o	o
Remura	Pi-AS-7,5	x	Vr	x	o	o		o	x
Regunde	Pi-AS-8,20	x	Vf	x	o	x	–		
	Pi-AS-3,60	x	Vr	o	x	x		–	o
	Pi-AS-22,17	x	Vf	x	–		–		
	Pi-AS-10,124	x	Vr	x	o		o		

V Ant. = polygen Antonovka.

The increase in the percentage of double-resistant seedlings is due to the accumulation of desirable combinations of alleles of genes from the parents which both contribute resistance to scab and mildew (Fischer 1990).

Fire blight (E. amylovora)

Fire blight is very important because there are no efficient bactericides registered for use in apple orchards. Methods available for protection against fire blight are both complicated and expensive. The best prospects for obtaining healthy plants are through breeding resistant selections. Donors of alleles for resistance have been found in progenies involving *M.* × *floribunda* and some cultivars (Fig. 1), e.a. progenies of Clivia, Golden Delicious, Alkmene and Pi-AS-44,14 produce a high percentage of seedlings with a high level of resistance.

Multiple-resistance

A diallel crossing programme (involving parents with various degrees of resistance to scab, mildew and fire blight) was designed to obtain triple-resistant selections (Fig. 2). The percentage of selections which are resistant to the three pathogens differs greatly. Parents with triple-resistance transmit a high degree of resistance to their progenies. The best progeny (Pi-AS-44,14 × Rewena), on average ($\hat{p} = 0{,}14$), produced one triple-resistant plant for every seven seedlings tested (Fischer 1990).

Table 5 shows a range of valuable parents with various multiple-resistant combinations to: scab, mildew, fire blight, bacterial canker, red spider mite, spring frost and winter frost.

Commercial trials (carried out over 12 years) with reduced or no use of fungicidal sprays have demonstrated, that the resistant cultivars (Realka, Reanda, Regine, Reglindis, Regunde, Reka, Releika, Releta, Relinda, Remo, Remura, Rene, Renora, Retina, Rewena and Resi) have durable resistance to scab and good levels of resistance to mildew. Fungicides can be reduced by at least 80% for these cultivars. It was very encouraging to note that diseases caused by other fungi were also absent in these trials, e.a. there was no European canker (*N. galligena*) in these trials – even under irrigated conditions.

Discussion and outlook

The Pillnitz apple resistance breeding programme has produced a large number of desirable 'quality selections' with resistance to economically important diseases. Selection efficiency is high when both parents have high levels of resistance combined with good fruit quality. The best Pillnitz resistant clones and cultivars have been tested in commercial trials in several European countries – thus subjecting them to a range of environmental conditions. They have demonstrated their ability to maintain their resistance and provide dessert fruit and/or for processing. The cultivars Reglindis, Remo, Retina and Rewena are now available for commercial production.

The durable resistance to scab and mildew of the Pillnitz cultivars has been obtained by using combinations of alleles of genes from several sources. It is vital to select varieties where both parents contribute genes for resistance to each disease, and not to depend on backcrossing single major genes into susceptible types.

Trials on the durability of the resistance obtained in the Pillnitz breeding programme need to be continually assessed under a range of climatic conditions over a number of years. Results so far show that very significant reductions in the use of fungicidal sprays are readily and satisfactorily achieved without risk from scab and mildew attack. In years of high risk from the two diseases, sprays can be reduced to 2–3 treatments at the beginning of the vegetative period; whilst in years of low risk, fungicidal sprays can be eliminated completely (Fischer 1993).

Under unsprayed conditions the risk of fire blight is greatly reduced by using resistant cultivars. The use of such types is particularly important in the absence of approved bactericides against fire blight (Fischer & Schaefer 1990).

Resistance breeding contributes to pest and disease to providing fruit growers with cultivars suitable for integrated control programmes. Such production schemes lead to better and healthier fruit and a safer and less hazardous environment.

References

Fischer, C., 1979. Resistenzzüchtung gegen Schorf (*Venturia inaequalis*/Cooke/Aderh.) und Mehltau (*Podosphaera leucotricha*/Ell et Ev./Salm.) Tagungsbericht 174: 97–104.

Fischer, C., 1989. Untersuchungen zur Vererbung der Resistenz gegen Schorf (*Venturia inaequalis* (Cooke) Aderh.) in Apfelnachkommenschaften. Arch. Gartenbau 37: 23–30.

Fischer, C., 1990. Ergebnisse der Resistenzzüchtung beim Apfel in der DDR. Tagungsbericht 292: 25–30.

Fischer, C., 1993. Results in the resistance breeding of apple in Dresden-Pillnitz. Acta Horticulturae 347: 163–168.

Fischer, C. & H.-J. Schaefer, 1990. Vergleichende Untersuchungen der Resistenz von Apfelsorten gegenüber Feuerbrand im Gewächshaus und im Freiland. Gartenbau 37: 299–300.

Knight, R.L. & F.H. Alston, 1968. Sources of field immunity to mildew (*Podosphaera*) in apple. Can J. Genet. Cytol. 10: 294–298.

Rehder, A., 1951. Manual of Cultivated Trees and Shrubs. The Macmillan Company, New York.

Schmidt, M., 1938. *Venturia inaequalis* (Cooke) Aderhold. VIII. Weitere Untersuchungen zur Züchtung schorfwiderstandsfähiger Apfelsorten. Züchter 10: 280–291.

Shay, J.R. & L.F. Hough, 1952. Evaluation of apple scab resistance in selections of *Malus*. Amer. J. Botany 39: 288–297.

H. Schmidt and M. Kellerhals (Eds.), Progress in Temperate Fruit Breeding, 49–52.

Screening methods of apple trees for resistance to *Valsa* canker

H. Bessho,[1] S. Tsuchiya[2] & J. Soejima[1]
[1] *MAFF, Morioka Branch, Fruit Tree Research Station, Shimokuriyagawa, Morioka, 020–01, Japan;* [2] *MAFF, Fruit Tree Research Station, Fujimoto, Tsukuba, 305, Japan*

Key words: Malus, Valsa canker, disease resistance, phloridzin

Summary

Valsa canker is the most serious wood rot disease of apple in Japan. Dormant, excised twig assay in vitro was used to determine resistance to *Valsa* canker. 'Fuji', 'Starking Delicious', 'M.26 EMLA', and 'MM.106' were considered to be highly susceptible, while *Malus sieboldii* showed moderate resistance. Several cultivars belonging to *M. prunifolia, M. baccata, and M.* × *domestica* showed comparative resistance. The phloridzin content in the bark of *Malus* species was studied in relation to *Valsa* canker resistance. Phloridzin content of *M. sieboldii* was lower than in *M.* × *domestica*. Comparatively resistant cultivars had similar or higher phloridzin content compared to susceptible cultivars. No clear correlations were found between resistance and phloridzin contents in the bark of *Malus* species.

Introduction

Valsa canker caused by *Valsa ceratosperma* (Tode ex Fr.) Maire is one of the most important diseases of apple in Japan (Sakuma, 1990). The disease decays the bark of the tree in both scion and rootstock. Young cankers are often first observed at fruit scars, twig stubs, and burrknots. When branches are girdled by the canker, they are killed. No curative fungicides are available for controlling the disease. No commercial apple cultivars grown in Japan are known to be resistant to *Valsa* canker. Therefore it is very important to develop resistant cultivars.

It is reported that the metabolized products of phloridzin in the bark are associated with the symptom development of *Valsa ceratosperma* (Koganezawa & Sakuma, 1982).

The purpose of the present study was to select resistant materials to *Valsa* canker in *Malus* germplasm collections, and to investigate an association between resistance and phloridzin content in the bark tissue of *Malus* species.

Materials and methods

An excised twig assay in vitro (Jeffers et al., 1981) was used to determine resistance to *Valsa* canker. An isolate of *Valsa ceratosperma* (VC88PS) was used in the following experiments.

In experiment 1, 9 genotypes were examined in 1990. Dormant shoots, removed from trees in March in Morioka Branch, Fruit Tree Research Station, were wrapped and stored in a refrigerator at 4 °C for 64 days until they were investigated. Stored shoots were used in order to accelerate the symptom development of *Valsa* canker. They were cut into 75 mm lengths. The basal end of each twig segment was pared by making tangential cuts 10 mm long and 1 mm deep on opposite sides for infection by fungus mycelium. Twenty twig segments per cultivar were inserted vertically into an agar on which fungus mycelium had been cultivated. After 11 days incubation, the twig segments were removed from the agar and the length of necrosis was measured.

In experiment 2, 22 genotypes were examined in 1991. The experiment was similar to Experiment 1 but the storing period was 108 days. Ten twig segments per cultivar were tested. After 10 days incubation, the length of necrosis was measured.

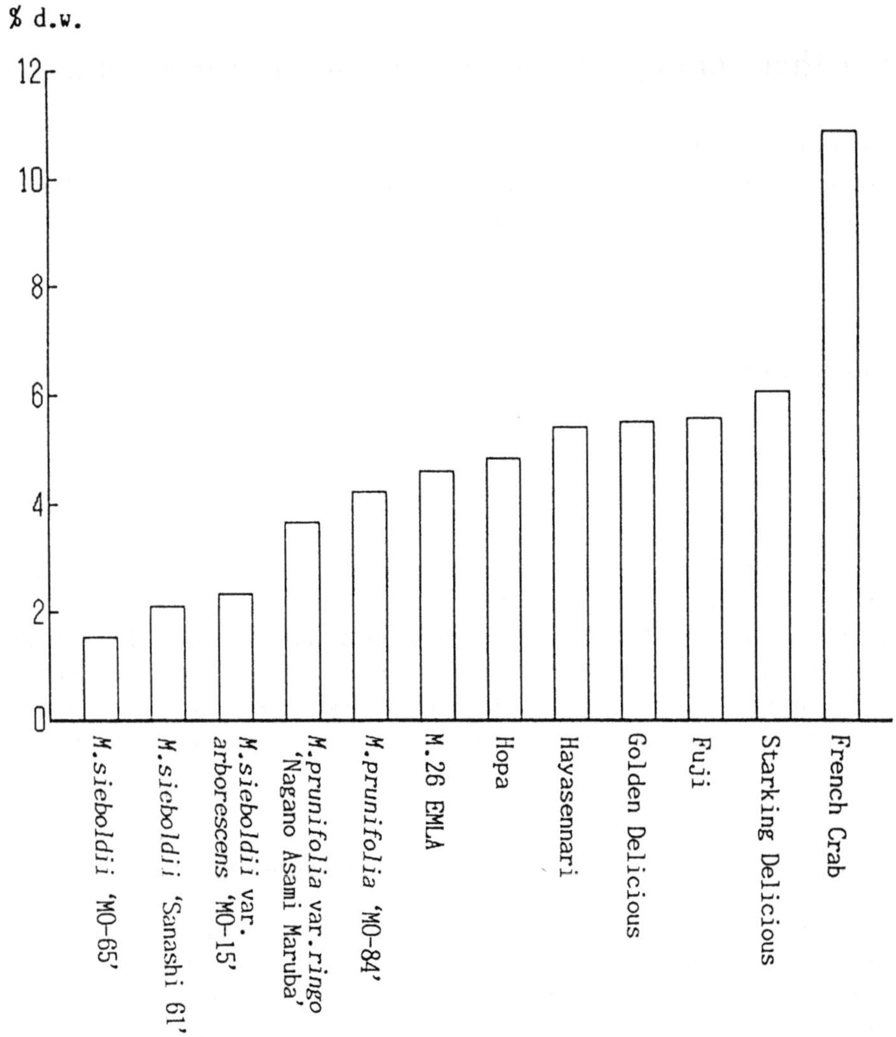

Fig. 1. Phloridzin contents in the bark of several *Malus* species.

Analysis for phloridzin was made according to the method of Martin & Williams (1964). The bark was freeze-dried and ground to pass a 40 mesh screen. Phloridzin was extracted by soaking the powder in 80% ethanol for 4 h at 90 °C and filtering. The extracts were spotted on Whatman's No. 1 filter paper, developed overnight in isopropanol : ammonia: water (4 : 1 : 2 v/v). After drying, the paper spot of phloridzin was observed under UV irradiation. The eluate was made by boiling a cut spot in water, and the absorbance at 283 nm of the eluate was measured.

Results and discussion

Table 1 shows the resistance of *Malus* species to *Valsa* canker. In experiment 1, 'Fuji', 'Starking Delicious', 'M.26 EMLA', and 'MM.106' were considered to be highly susceptible to *Valsa* canker, while *M. sieboldii* showed moderate resistance. In Experiment 2, 9 genotypes, *M. sieboldii* 'Sanashi 61', *M. prunifolia* 'Nagano No.1', *M. baccata* var. *nikkoensis*, *M.* × *domestica* 'Hayasennari', *M. prunifolia* var. *ringo* 'Nagano Asami Maruba', *M.* × *adstringens* 'Hopa', *M.* × *domestica* Unknown, *M. sieboldii* 'Sanashi 60', and *M. sieboldii* var. *arborescens* 'MO–15' were less susceptible than *M.* × *robusta* 'Shirobana Robusta', 'Hyslop', *M.* × *purpurea* 'Purple Lemoine', and *M. pumila*.

Table 1. Length of necrosis on excised, dormant twigs of various species inoculated with *Valsa ceratosperma*

Genotype	Lesion length (mm)	
	Exp. 1	Exp. 2
M. sieboldii 'MO–65'	8.2 c*	
M. pumila var. *paradisiaca* 'M.9 EMLA'	26.9 b	
M. prunifolia 'MO–84'	30.2 b	
M. × *domestica* 'Jonagold'	30.8 b	
M. pumila var. *paradisiaca* 'M.27'	41.4 b	
M. × *domestica* 'Fuji'	45.0 ba	
M. pumila var. *paradisiaca* 'MM.106'	45.4 ba	
M. × *domestica* 'Starking Delicious'	47.6 ba	
M. pumila var. *paradisiaca* 'M.26 EMLA'	63.7 a	
M. sieboldii 'Sanashi 61'		0 z*
M. prunifolia 'Nagano No. 1'		0.1 z
M. baccata var. *nikkoensis*		0.3 z
M. × *domestica* 'Hayasennari'		0.3 z
M. prunifolia var. *ringo* 'Nagano Asami Maruba'		0.5 z
M. × *adstringens* 'Hopa'		0.6 z
M. × *domestica* Unknown		0.6 z
M. sieboldii 'Sanashi 60'		0.8 z
M. sieboldii var. *arborescens* 'MO–15'		0.9 z
French crab		2.1 zy
Flontish		2.8 zy
NY 58–22		4.5 zy
M. prunifolia var. *ringo* 'Seishi'		4.7 zy
M. × *robusta*		5.1 zy
Geneva		6.1 zy
Spy 227		7.5 zy
M. × *domestica* 'Lord Lambourne'		8.8 zy
R12740–7A		9.9 zyx
M. × *robusta* 'Shirobana Robusta'		10.0 yx
Hyslop		11.2 yx
M. × *purpurea* 'Purple Lemoine'		18.8 xw
M. pumila		24.2 w

* Means followed by the same letter do not differ significantly. (P = 0.05)

These results obtained by in vitro assay coincided with field observations. Liu et al. (1990) also reported that *M. sieboldii* and several other wild species were resistant to *Valsa* canker.

Figure 1 shows phloridzin content in the bark of *Malus* species. The contents of phloridzin in the bark were low in *M. sieboldii* and *M. sieboldii* var. *arborescens* 'MO–15', medium in *M. prunifolia* and *M. prunifolia* var. *ringo*, and high in most of cultivars belonging *M.* × *domestica* and crab apples. 'French Crab' showed the highest phloridzin content among all tested genotypes.

Phloridzin content of *M. sieboldii* which showed moderate resistance was less than 50% of susceptible cultivars such as 'Fuji' and 'Starking Delicious'. But the phloridzin contents of cultivars such as *M.* × *adstringens* 'Hopa', and 'Hayasennari' which showed resistance were almost the same as susceptible cultivars. *M.* × *domestica* cultivars thought to be susceptible in field observations were evaluated as comparatively resistant in this experiment. While phloridzin has been investigated in connection with quite a series of diseases resistance (Tschen & Fuchs, 1969; Bondarenko et al., 1977; Alt & Schmidle, 1980), the correlation between phloridzin content in the bark of *Malus* species and resistance to *Valsa* canker is not clear at present. Further experiments are needed to confirm the relationship.

Recently many studies have been conducted on the role of specific substances in disease resistance of plants. *M. sargentii* was resistant to crown rot, and had higher total phenol and flavonol contents than susceptible genotypes (Alt, 1979). Gutmann et al. (1990) also reported that *M. sargentii*, exceeded *M.* × *domestica* cultivars in flavan-3-ol production. It is important to find the specific substances in bark tissue which are concerned with the resistance to *Valsa ceratosperma*.

Acknowledgements

Contribution C-17Ø of the Fruit Tree Research Station.

References

Alt, D., 1979. Investigations on the possible bases for resistance to *Phytophthora cactorum* in apples. Mittelungen aus der biologischen Bundesanstalt für Land- und Forstwirtschaft Berlin-Dahlem 191: 290.

Alt, D. & A. Schmidle, 1980. Investigations on potential resistance factors of apple to *Phytophthora cactorum* (Leb. et Cohn) Schroet. Angew. Botanik. 54: 139–156.

Bondarenko, A.A., N.A.Lapteva & E.S.Artamonova, 1977. Biological action of phloridzin and phloretin on *Fusicladium dendriticum* (Wallr.) Fuck. *in vivo* and *in vitro*. Referativnyi Zhurnal, Biologiya 3: 192.

Gutmann, M., D. Treutter & W. Feucht, 1990. Analysis of phenols of cultivars and wild species from apple by use of HPLC-CRD techniques. Mitt. Klosterneuburg 40: 80–83.

Jeffers, S.N., H.S. Aldwinckle & P.A. Arneson, 1981. Excised twig assay for the study of apple tree crown rot pathogens *in vitro*. Plant Dis. 65: 823–825.

Koganezawa, H. & T. Sakuma, 1982. Possible role of breakdown products of phloridzin in symptoms development by *Valsa ceratosperma*. Ann. Phytopath. Soc. Japan 48: 521–528.

Liu, H.Z., Q.M. Ren & L.J. Liu, 1990. Evaluation of resistance to *Valsa mali* in *Malus* germplasm resources. J. Fruit Sci. 7: 65–70.

Martin, C.M. & M.W. Williams, 1964. Quantitative Phloridzin Analysis. Anal. Biochem. 8: 257–260.

Sakuma, T., 1990. *Valsa* canker. In: A.L. Jones & H.S. Aldwinckle (Eds) Compendium of Apple and Pear Diseases, pp. 39–40. APS Press, St. Paul, U.S.A.

Tschen, J. & W.H. Fuchs, 1969. The effects of phloridzin on infection of *P. vulgaris* by *U. appendiculatus*. Naturwissenschaften 56: 643.

H. Schmidt and M. Kellerhals (Eds.), Progress in Temperate Fruit Breeding, 53–55.

Results of testcrossing in order to get disease resistant sour cherry varieties

János Apostol & Klára Véghelyi
Enterprise for Extension and Research in Fruit Growing and Ornamentals, H-1223 Budapest, Park u. 2.- Hungary

Key words: Breeding, leaf spot, resistance, sour cherry

Summary

Diseases of sour cherries cause considerable losses by leaf spot (*Blumeriella jaapi*). Numerous chemical sprays are required annually to prevent early defoliation from this fungi. Fortunately, disease resistant sour cherry germplasm does exist in Hungary. In 1991 a cooperative breeding programme was started by Michigan State University (U.S.A.) and Enterprise for Research and Extension in Fruit-Growing and Ornamentals (Hungary). The results of this programme were:
As sources of disease resistance the following varieties were found: 'Csengödi' and 'Akasztói'.
'Csengödi' and 'Akasztói' have high heritability when used es male parents.
'Pándy' × 'Csengödi' or 'Akasztói' population gives 19,6% leaf spot resistant items, combined with good fruit quality.
'Érdi bötermö' ('Pándy' × 'Nagy angol') × Csengödi' or 'Akasztói' population gives 41,6% leaf spot resistant items combined with good fruit quality.

Introduction

Pathogen fungi can seriously damage sour cherry growing. This creates a major problem for the producers.

The most dangerous pathogens of sour cherry in Hungary are the leaf spot (caused by *Blumeriella jaapi*) and brown rot (caused by *Monilia laxa*). Numerous chemical sprays are required annually to prevent fruit rot and early defoliation by these fungal pathogens. However the rising growing costs and the decreased yields result in loss of income of sour cherry.

Epidedemic infection of cherry leaf spot causes premature defoliation which may reduce vigour, winter hardiness of the tree and fruit sizes and influences adversely the flower bud development.

On the other hand *Monilia laxa* infection causes flower and fruit rot and decaying of branches. In addition to fruit brown rot can be caused in Europe by *Monilia fructigena* and in the USA by *Monilia fructicola*.

The total yield loss due to the two pathogens can be as much as 20–30% or more, but in case of an epidemic infection the loss can reach 80–90% in some places.

As result several kilograms per hectare of chemicals are put on sour cherry orchards to control these fungus diseases. To avoid this loss a joint breeding programme was started by Enterprise for Extension and Research in Fruit Growing and Ornamentals, Hungary and Michigan State University, U.S.A. (Gergely et al., 1993). Hungary is located in the secondary gene centre of *Prunus* where genetic diversity of cultivated sour cherries were developed, and have been coexisting with pathogenic fungi for centuries.

This is why resistance to *Monilia spp.* and *Blumeriella sp.* fungi has developed there by natural selection.

The first step of our joint effort was to look for and to collect germplasm which was potentially resistant to brown rot and cherry leaf spot.

We have managed to select some resistant or tolerant local plants. The first selected variety was 'Csengödi', the second was 'Akasztói' (Apostol, 1990; Apostol & Iezzoni, 1992). Their level of resistance

Table 1. Distribution of seedlings according to infection level by *Blumeriella jaapi*, Érd, 1992

	Infection levels											Total number of seedlings
	0	1	2	3	4	5	6	7	8	9	10	
Pándy × Csengödi/Akasztói	12	19	9	9	4	4	1	2	1	–	–	61
Érdi bötermö × Csengödi/ Akasztói	22	1	1	–	–	–	–	–	–	–	–	24

0 = non infected; 10 = seriously infected; infection level of Pándy = 4,
Érdy bötermö = 5, Csengödi and Akasztói = 0.

Table 2. Some important fruit characteristics of the resistant seedlings (infection level of seedlings: 0)

Combinations	Fruit size			skin colour			Firm			Total
	19	20–21	22	C_5	C_{5-7}	C_{8-10}	1–2	3	4–5	
Pándy cl 279 ×			2	2					2	
Csengödi/			4		4				4	
Akastói		2			1	1		2		
	4			1	2	1		4		
	4	2	6	3	7	2		6	6	12
Érdi Bötermö ×			3		3				3	
			7			7			7	
Csengödi/Akasztói			4	1	3			4		
		5		2	2	1	2	1	2	
	3					3		3		
	3	5	14	3	8	11	2	8	12	22

C_1–10 = Carmin colour scale.

was checked in vivo and in vitro (Apostol & Véghelyi, 1992).

The second step in our work was the immediate use of these selections as potential parents in the breeding programme in Hungary and to find out the inheritance of this resistance in Michigan (Gergely et al., 1993).

We made test crossings in order to evaluate the heritability of this resistance using 'Csengödi' and 'Akasztói' as male parents in different combinations.

The third step will be the crossing of several hundreds from the best combinations for a selection of best varieties and/or to proceed with the crossing.

Method and results

We used two varieties: 'Pándy' (Köröser) (clon 279) and 'Érdi bötermö' as female parents. 'Pándy' had been used as a female parent for all crossings in the first breeding programme in Hungary from 1950 to 1954, and 'Érdi bötermö' is a 'Pándy' × 'Nagy angol' offspring. As a donor of resistance to *Monilia spp.* and *Blumeriella sp.* 'Csengödi' and 'Akasztói' varieties were used. Those two varieties belong to the local varietal group 'Bosnyák' originating from Bosznia at the end of the last century.

Having evaluated the hybrids we could not find any difference of heritability of resistance between 'Csengödi' and 'Akasztói', therefore their data are shown together in Table 1 and Table 2.

The 'Pándy' cl 279 × 'Csengödi/Akasztói' population has 61 seedlings; among them 12 seedlings (19,6%) showed 0-level infection to leaf spot, the other 49 seedlings showed various level of infection (1–8) (Infection levels: 0 = non infected, 10 = infection seriously, the whole tree defoliated).

The 'Érdi bötermö' × 'Csengödi/Akasztói' population has 24 seedlings, among them 22 seedlings, (91,6%) showed 0-level infection to leaf spot and only two were infected on 1 and 2 level.

As you can see from the data in Table 1:

- the varieties 'Csengödi' and 'Akasztói' can be used as donors for leaf spot resistance,
- 'Érdi bötermö' used as female parent provides an extremely high proportion of seedlings resistant to leaf spot (91,6%),
  'Pándy cl.' 279 used as female parent provides only 19,6% of resistant seedlings,
- Among the 'Érdi bötermö' × 'Csengödi/Akasztói' combination the majority of the leaf spot resistant ones (54,5%) produced large fruits (22 mm diameter or larger) with dark red or black skin-colour, good firmness and a high yield. 41,6% of the whole population showed resistance and good characteristics.

'Pándy cl. 279' × 'Csengödi/Akasztói' combination provided the same good characteristics with only 11,8% of the whole population. There was no serious *Monilia* infection in our experimental plots during the years of testing; that is why we could not evaluate the resistance to *Monilinia* yet.

Conclusions

The conclusions of these test crossings are the following:

1. The varieties 'Csengödi' and 'Akasztói' are good donor sources of leaf spot resistance.
2. The 'Érdi bötermö' × 'Csengödi' or 'Akasztói' combination is suitable for the development of leaf spot resistant hybrids.

References

Apostol, J., 1990. Biomeggy. Az elsö ellenálló meggyfajta. Kertészet és Szölészet. 39 (17): 3.

Apostol, J. & A. Iezzoni, 1992. Sour cherry breeding and production in Hungary. Fruit Varieties Journal 46 (1): 11–15.

Apostol, J. & K. Véghelyi, 1992. Meggymonilia. Igéretesen ellenálló fajták. Kertészet és Szölészet. 41 (20): 8–9.

Gergely, I., J. Apostol, K. Véghelyi, A. Iezzoni & A. Jones, 1993. Breeding disease resistant sour cherry cultivars: A cooperative program between Hungary and Michigan. Hung. Agr. Res. Vol 2 (3): 33–35.

H. Schmidt and M. Kellerhals (Eds.), Progress in Temperate Fruit Breeding, 57–61.

Evaluation of two screening methods for resistance of apricot, plum and peach to *Monilinia laxa*

T. Pascal,[1] A. Levigneron,[1] J. Kervella[1] & C. Nguyen-The[2]
[1] *Station de Recherches Fruitières méditerranéennes, INRA, Domaine St Paul, F-84143 Montfavet cedex, France;*
[2] *Station de Technologie des Produits végétaux, INRA, Domaine St Paul, F-84143 Montfavet cedex, France*

Key words: Apricot, plum, peach, *Monilinia laxa*, resistance, screening tests, brown rot, *Prunus*

Monilinia laxa causes important pre- and post-harvest losses in stonefruit. In order to initiate a breeding programme for increased resistance to *M. laxa*, two screening tests were used. In the "uninjured fruit inoculation" test, 30 mature fruits of each variety were inoculated on their surface by depositing 20 μl of conidial suspension. In the "artificially injured fruit inoculation" test, 10 mature fruits were inoculated on both sides by injecting 20 μl of conidial suspension. Genotypic differences were found in both tests, within the three species studied. Marked differences were observed in the uninjured fruit inoculation test. Differences between genotypes were slighter in the artificially injured fruit inoculation test. Within each species, the rankings of the genotypes according to the two tests were not correlated. Both testing procedures would be usefully applied in a breeding programme to obtain genotypes with combined resistance to *Monilinia laxa*.

Introduction

Brown rot, caused by *Monilinia laxa* (Aderh. and Ruhl.) Honey, is the main storage fungal disease of peaches in Europe (Nguyen-The et al, 1989). It is also responsible for apricot and plum post-harvest fruit losses. These post-harvest losses are serious, especially for prolonged storage of fruit. They may reach 25% and chemical applications at this stage are not recommended. Fruits become increasingly susceptible to *M. laxa* as they ripen. This encourages early harvest, which is detrimental to fruit quality (Arnoux, 1986). Peaches, apricots and plums may also be infected by *M. laxa* in the orchard, in wet conditions.

Feliciano et al. (1987) proposed screening techniques in breeding peaches resistant to *Monilinia fructicola* (Wint.) Honey, which causes similar brown rot damages in Southern Brazil. To start a breeding programme for increasing resistance to *M. laxa*, we tested 7 apricot, 7 plum and 12 peach genotypes with two fruit inoculation techniques. We used higher spore concentrations than Feliciano et al. Spore concentrations were similar to those noted as optimum by Fourie & Holtz (1985) in artificial inoculation studies with *M. laxa*.

Material and methods

Testing procedure

Inoculum. The same strain of *M. laxa*, isolated from a decaying fruit and maintained on Potato Dextrose Agar-PDA (Biomérieux) was used in all tests. This strain was cultured on fruit to provide conidia. Just before inoculation, a conidial suspension was prepared with sterile distilled water and adjusted to 10^6 conidia per millimeter with a hemacytometer.

Two inoculation methods were used. In the "uninjured fruit inoculation" test, 30 uninjured fruit are inoculated by depositing 20 μl on the fruit surface. These fruit were then placed in a hermetic box to ensure high relative humidity and avoid the evaporation of the suspension. Fruit with a rot diameter exceeding 10mm are considered as infected. The number of infected fruit is recorded every 10 to 14 h for 10 days.

In the "injured fruit inoculation" test, 10 fruit were inoculated on both sides by injecting 20 μl of conidial suspension with a Pipetman to a depth of 4 mm. The diameter of rot was recorded every 10 to 14 h for 5

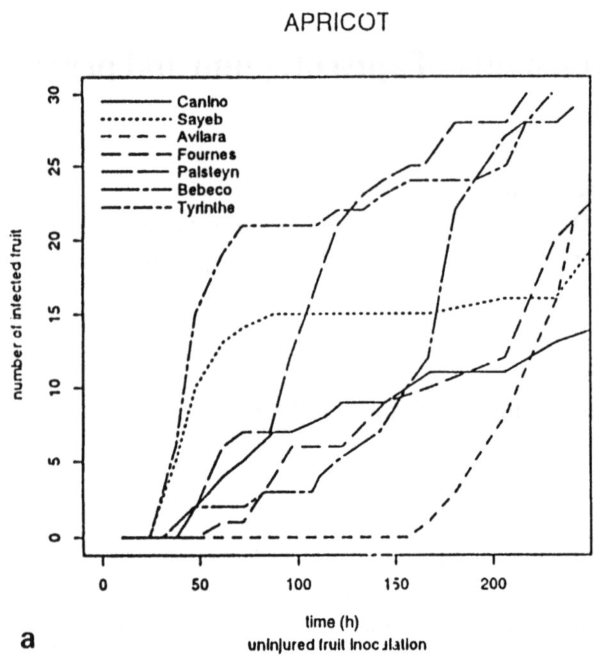

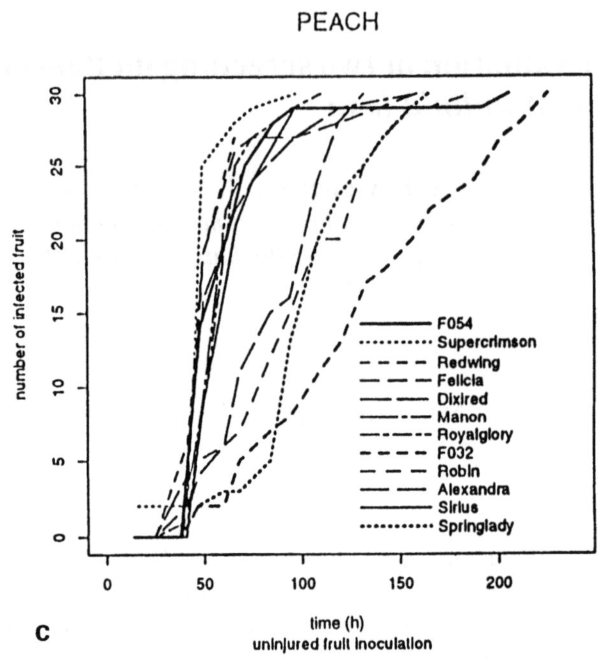

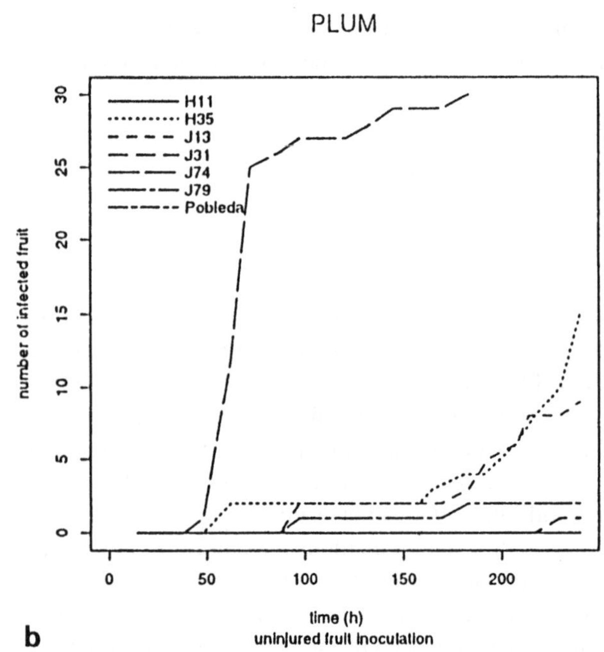

Fig. 1. Uninjured fruit inoculation test. Number of infected fruit out of 30 infected fruit for apricot, plum and peach genotypes.

In the "injured fruit inoculation" test, 10 fruit were inoculated on both sides by injecting 20 μl of conidial suspension with a Pipetman to a depth of 4 mm. The diameter of rot was recorded every 10 to 14 h for 5 days. Fruit were kept at 23 °C after inoculation for both tests.

Fruit material

The fruit were "commercially" mature. They came from experimental orchards and had not received fungicide applications for three weeks before harvest. Among the *Prunus persica* genotypes studied, 3 (F032, F054 and 'Supercrimson') were nectarines, the oth-

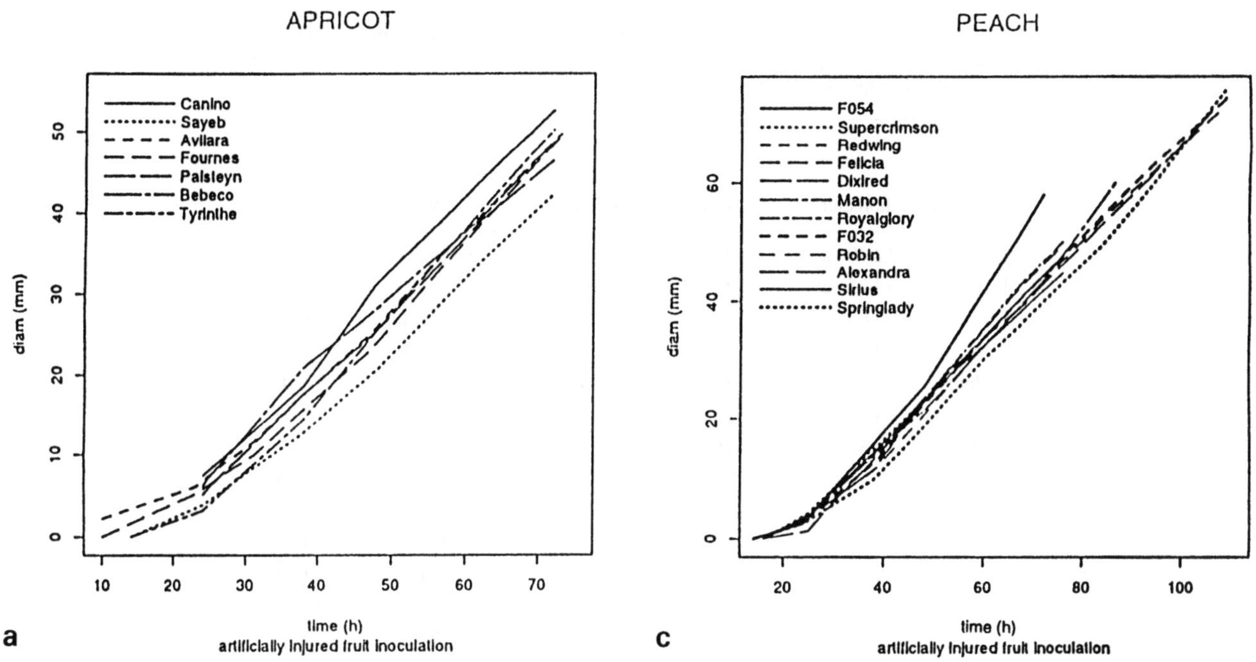

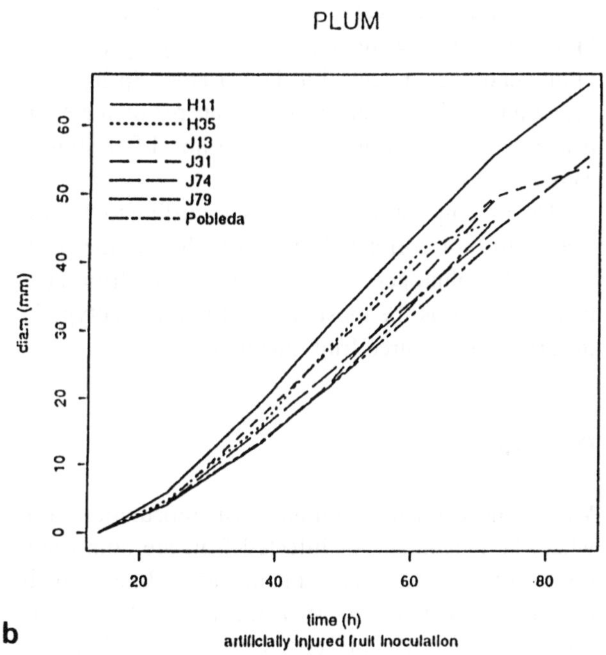

Fig. 2. Artificially injured fruit inoculation test. Mean diameter of rot for 10 fruit for apricot, plum and peach genotypes.

ers ('Alexandra', 'Dixired', 'Felicia', 'Springlady', 'Manon', 'Redwing', 'Robin', 'Royalglory' and 'Sirius') were freestone peaches. Diploid plum genotypes were either *P. cerasifera*, *P. salicina* or interspecific hybrids between those two species (H11, H35, 'Pobieda').

Statistical analysis

Differences between genotypes in the performance at the uninjured fruit inoculation test were assessed by comparing the distributions of time between inoculation and infection with a chi-square test.

Table 1. Mean rot diameter (mm) in the artifically injured fruit inoculation test, 38 hr after inoculation. For each species, values followed by the same letter do not differ significantly according to the Newman-Keuls test, P = 0.05.

Apricot		Plum		Peach	
Genotype	diameter	Genotype	diameter	genotype	diameter
Sayeb	12.70 a	J13	13.30 a	Springlady	9.70 a
Tyrinthe	14.35 ab	Pobieda	13.33 a	Robin	11.15 ab
Fournes	15.65 bc	J74	13.70 a	Alexandra	11.35 ab
Palsteyn	17.50 cd	J31	15.40 ab	Manon	12.45 ab
Avilara	17.55 cd	H11	16.05 b	Dixired	12.76 ab
Canino	18.50 d	J85	17.40 b	F032	13.70 ab
Bebeco	20.85 e	H35	19.45 c	Royalglory	14.12 ab
				Felicia	14.26 ab
				Sirius	14.26 ab
				Redwing	14.35 ab
				Supercrimson	14.39 ab
				F054	15.75 b

Comparisons for the artificially injured fruit inoculation test were done by an analysis of variance of the rot diameter after 24, 38, 48 and 72 hours of incubation.

Results

Marked differences were observed among diploid plum and apricot genotypes in the "uninjured fruit inoculation" test. For apricot genotypes (Fig. 1a), the percentage of infected fruit ranged from 0 to 80% 150 h after inoculation, and from 40 to 100% at the end of the experiment. The infection patterns varied greatly. For some genotypes such as 'Tyrinthe' and 'Fournès', the number of infected fruit increased steadily. For 'Bebeco', it increased rapidly for the first 3 days and remained almost constant afterwards.

All the fruit from *P. cerasifera* clone J74 were rotten after 160 h, whereas none of the fruit from *P. salicina* × *P. cerasifera* hybrids H11 and 'Pobieda' were infected after 250 h (Fig. 1b).

The infection patterns of most peach genotypes were similar (Fig. 1c). About 75% of the fruit were generally infected after 3 days. The infection developed more slowly only on 'Alexandra', 'Robin', 'Springlady' and 'Manon' fruit.

Differences between genotypes were slighter in the "artificially injured fruit inoculation" test (Fig. 2). Brown rot developed on all injured fruit. Significant differences appeared however between genotypes (Table 1). They were important among apricot and plum genotypes. Except for the extreme 'Springlady' peach and F054 nectarine, the mean rot diameter of peach and nectarine genotypes did not differ significantly.

The rankings of the genotypes according to the two tests were not correlated. J74, which was the most susceptible plum genotype in the uninjured fruit inoculation test, was the most resistant plum genotype in the artificially injured fruit inoculation test.

Discussion

We assume that in the "uninjured fruit inoculation" test, when the percentage of infected fruit are compared, the genotypes are ranked mainly according to their epidermal resistance to *M. laxa*. In the "artificially injured fruit inoculation" test, the resistance of flesh to *M. laxa* is the important feature.

The lack of marked resistance of peaches and nectarines in the uninjured fruit test might be due to the microcracks which appear on maturing peach and nectarines (Nguyen-The, 1991; Fogle & Faust, 1976, 1975). High infection levels of artificially inoculated uninjured peaches had already been reported by Fourie & Holz (1985). These authors found low infection levels on uninjured plums. This was the general case in our experiment but one of the plum genotypes, which is known to be very sensitive to *M. laxa* in the orchard, showed an infection pattern similar to those observed for peaches.

Variations were found for both tests for apricot, plums and peaches but the rankings of the genotypes differed according to the test used in the three species. Resistance of epidermis and of flesh to *M. laxa* do not appear to be linked. This has important consequences in a breeding programm. If we aim at selecting resistant varieties from fruiting seedlings, the test allowing better prediction of post-harvest losses should be chosen. The more efficient test might not be the same for plums, which generally are not infected on uninjured fruit, and peaches, which are easily infected in those conditions. In a long-term breeding programme, it would be desirable to combine both types of resistances. To obtain plums with improved resistance to *M. laxa*, for instance, J74 with enhanced resistance to injuring inoculation would might be usefully crossed to H11 or 'Pobieda'. In such a programme, a joint testing procedure might be usefully applied.

References

Arnoux, M., 1986. Qualité des pêches. Maîtrise des différents facteurs: Interférence entre technique et socio-économie. Arboric. Fruit. 384: 46–49.

Feliciano, A., A.J. Feliciano & J.M. Ogawa, 1987. *Monilinia fructicola* resistance in the peach cultivar Bolinha. Phytopathology 77: 776–780.

Fogle, H.W. & M. Faust, 1976. Fruit growth and cracking in nectarines. J. Am. Soc. Hortic. Sci. 101: 434–438.

Fogle, H.W. & M. Faust, 1975. Ultrastructure of nectarine fruit surface. J. Am. Soc. Hortic. Sci. 100: 74–77.

Fourie, J.F. & G. Holz, 1985. Artificial inoculation of stone fruit with *Botrytis cinerea*, *Monilia laxa* and *Rhizopus stolonifer*. Phytophylactica 17: 179–181.

Nguyen-The, C., 1991. Structure of epidermis wall, cuticle and cuticular microcracks in nectarine fruit. Agronomie 11: 909–920.

Nguyen-The, C., R. Hugueney & M. Arnoux, 1989. Contribution à l'étude des voies de pénétration de parasites fongiques des nectarines *Monilinia laxa* (Ascomycète-Discomycète) et *Rhizopus stolonifer* (Zygomycète-Mucorale). Agronomie 9: 271–276.

H. Schmidt and M. Kellerhals (Eds.), Progress in Temperate Fruit Breeding, 63–66.

Breeding for resistance to fungal diseases in strawberry

D.W. Simpson, J.A. Bell & D.C. Harris
Horticulture Research International, East Malling, West Malling, Kent. ME19 6BJ, UK

Key words: Phytophthora cactorum, crown rot, resistance, *Sphaerotheca macularis*, powdery mildew, strawberry, *Verticillium dahliae*, wilt, *Fragaria* × *ananassa*

Summary

Resistance to fungal diseases is given a high priority in the strawberry breeding programme at HRI, East Malling although most primary screening is on the basis of fruit quality.

Wilt, caused by *Verticillium dahliae*, is an important soil-borne disease in the UK. Resistant genotypes are used extensively as parents and advanced selections are tested for resistance by planting in a heavily infested field. Pegasus, released in 1990, has a high level of field resistance to *V. dahliae*. The inheritance of resistance is being studied using seedlings in both the field and glasshouse.

Work on resistance to *Phytophthora cactorum* (crown rot) began in 1992. Inoculation of potted plants with a zoospore suspension is an effective test but symptom development is slow and a large amount of replication is required for reliable results.

Powdery mildew (*Sphaerotheca macularis*) is a serious problem on everbearing strawberries. Each year the seedling population is screened by allowing a natural epidemic to develop in the field. Resistance is partly non-additive, so small test crosses are evaluated and those with good specific combining ability are repeated on a large scale.

There is no screening for resistance to *Phytophthora fragariae* at HRI but potential new cultivars are sent to the Scottish Crop Research Institute for testing their susceptibility to the common races of the disease.

The relative resistance of recently released HRI cultivars to each of the above diseases is shown.

Introduction

The strawberry breeding programme at Horticulture Research International (HRI) East Malling aims to produce new, improved strawberry cultivars for all sectors of the UK industry. The most important attributes of a cultivar for most growers are fruit quality, productivity, agronomic characteristics related to ease of husbandry and resistance to diseases and pests. Of these it is the fruit quality traits which form the basis of most primary screening in the seedling population, since without good fruit quality no cultivar can ever become commercially successful. In most cases screening for resistance is done after the seedlings have been assessed for agronomic and fruit quality characteristics and the inferior ones eliminated.

Nevertheless, resistance to fungal diseases is highly desirable in any cultivar, as it leads to a reduction in pesticide use and thus lower production costs. Consequently resistance breeding and inheritance studies are given a high priority in the breeding programme and we are currently concentrating on two soil-borne fungi, *Phytophthora cactorum* and *Verticillium dahliae*, which cause crown rot and wilt respectively, and the air-borne fungus *Sphaerotheca macularis*, the cause of powdery mildew.

Verticillium dahliae

Verticillium dahliae is an important pathogen of strawberry in the UK and it is particularly widespread in south-east England, the main growing region. The only practical control measure for this disease is chemical soil disinfestation and it is likely that the most widely used fumigation chemical, methyl bromide, will not be available to growers in the near future due to concerns over depletion of the ozone layer. Elsanta, the most widely grown cultivar in the UK, is susceptible to *V. dahliae*, and a large proportion of growers would not be able to use this cultivar if soil fumigation were unavailable.

Resistance to *V. dahliae* has been extensively studied in the USA but little work has been done using European germplasm, particularly during the last 25 years. Bringhurst et al. (1961) studied the reaction of 18 clones of Californian origin to six different isolates and found varying levels of resistance but, even when plants appeared symptomless, it was usually possible to re-isolate the pathogen. Except for the most susceptible clones, the reaction to inoculation varied according to which isolate was used. However, there were no clone × isolate interactions and the range of reactions was due to differences in the pathogenicity of isolates. A similar result was obtained by Maas et al. (1989) when studying the reaction of 36 cultivars from Eastern USA and Canada to four isolates of *V. dahliae*. Bringhurst et al. (1968) proposed that resistance was controlled by more than one locus with partial dominance, if any, over susceptibility. Bowen et al. (1968) reached a similar conclusion but neither group proposed a model based on a particular number of genes.

A number of genotypes which have demonstrated resistance in the field are held in the HRI germplasm collection and used extensively in the crossing programme. Many of these lines are derived from Redgauntlet, which generally shows good field resistance, although Kronenberg & van der Giessen (1968) found this cultivar to be susceptible when inoculated in the glasshouse. However, Jordan (1973) found that progeny from Redgauntlet did show resistance when tested in the glasshouse.

The breeding strategy used at HRI to incorporate resistance is to plant the main seedling population in disease-free land and select initially on the basis of agronomic and fruit quality characteristics. At a later stage, advanced selections are tested for resistance by planting in a heavily infested field where they are assessed for symptoms of *Verticillium* wilt, in comparison with resistant and susceptible standards, over two seasons. This approach has proved effective and Pegasus, released in 1990, has a high level of field resistance to *V. dahliae*. Three other recent releases, Pandora, Calypso and Tango, also have useful levels of resistance.

Since 1991 the inheritance of resistance has been studied by raising seedlings from specific crosses and inoculating half at age eight weeks by immersing the roots in a conidial suspension for five minutes before potting and growing the plants in a glasshouse. Symptom development is observed over the next 12 weeks. The other half of the seedlings are planted in a heavily infested field plot where they are observed for one season. Results from the glasshouse experiments thus far have shown that crosses between two susceptible parents rarely produce any resistant individuals while resistant × resistant and resistant × susceptible combinations may result in a bimodal distribution with a small proportion of seedlings showing an intermediate response. These findings broadly support the conclusions of Bringhurst et al. (1968) and Bowen et al. (1968) and suggest possible major gene involvement with resistance showing some dominance over susceptibility. The field screening has yet to produce any useful data since the weather conditions during 1992 were not conducive to good symptom expression for *Verticillium* wilt.

Phytophthora cactorum

Phytophthora cactorum has recently become a serious problem to growers in the UK, largely because the popular cultivar Elsanta is susceptible to the disease. A complicating factor is that cultivars such as Elsanta can carry a latent infection and appear symptomless in the nursery but collapse the following spring in the fruiting plantation. This can lead to the disease being spread from one farm to another on apparently healthy plants. Once the soil becomes contaminated it is not feasible to eliminate the disease although some soil fumigation treatments and/or fungicides help to reduce the problem.

Work on resistance to *P. cactorum* began in 1992 and has so far concentrated on methods of testing the susceptibility of mature plants. Seemüller (1977) tested 18 cultivars by introducing a piece of mycelium into the crown and obtained good infection levels using fresh plants lifted from the nursery during summer and autumn. Pitrat & Risser (1977) used a similar method to test the susceptibility of 30 cultivars and with nine of these they also used the alternative method of pouring a zoospore suspension on to the crown of the plant. The former method was the more effective. They found

Table 1. Disease resistance in strawberry cultivars from Horticulture Research International

Cultivar	Year of release	*Verticillium wilt*	*Sphaerotheca macularis*	*Phytophthora cactorum*	*Phytophthora fragariae*
Pandora	1988	Moderately resistant	Resistant	Moderately resistant	Susceptible
Pegasus	1990	Resistant	Moderately resistant	Moderately resistant	Resistant to races 1, 2 & 3
Calypso	1991	Moderately resistant	Moderately susceptible on foliage	Susceptible	Susceptible
Tango	1994	Moderately resistant	Moderately susceptible on foliage	Not tested	Moderately resistant to races 1, 2, 11a & 11c

that the susceptibility of the plants varied during the year and that they were at their least susceptible during autumn and winter.

Experience of disease incidence in the UK suggested that plants were most likely to collapse in the spring, during the flowering and fruiting period, and that plants which had been cold stored were more vulnerable than those planted fresh. It was therefore decided to inoculate plants by spraying a zoospore suspension on to ex-cold stored plants which had been potted and then grown on in the glasshouse. Zoospore infection was preferred to introducing mycelium directly into the crown as it was considered closer to the natural infection process. In a preliminary experiment in 1992 this procedure was found to be effective for determining the level of resistance in ten cultivars but symptom developments was slow and some plants, particularly of cultivar Honeoye, were still collapsing 12 weeks after inoculation. Furthermore, it was found that even on the most susceptible cultivar, Tamella, a small proportion of plants did not collapse and it appears that a high level of replication is required to reliably classify genotypes for their resistance. Variation for resistance between the ten cultivars was of a continuous nature, a finding which concurred with those of Seemüller (1977) and Pitrat & Risser (1977) using different germplasm. Of the cultivars examined Tamella was the most susceptible and Pegasus the most resistant.

For the above experiment, a single, highly pathogenic isolate was used but additionally a range of *P. cactorum* isolates from throughout Europe were tested on six cultivars. Large differences in pathogenicity were detected but there was no evidence of any isolate × cultivar interaction.

Although the procedure described above for evaluating the susceptibility of strawberry clones to *P. cactorum* gives generally good correspondence with known field susceptibility, it is time consuming and requires large areas in the glasshouse for adequate replication. Several *in vitro* testing methods are currently under investigation as possibly quicker and simpler alternatives to glasshouse screening.

The feasibility of studying inheritance of resistance by screening young seedlings is under investigation. Crosses have been made using parental genotypes showing a wide range of phenotypic variation for resistance. The continuous nature of the variation for resistance observed between cultivars suggests that control may be polygenic and this hypothesis will be tested.

Sphaerotheca macularis

Environmental conditions in the UK are usually conducive to epidemics of powdery mildew (*Sphaerotheca macularis*) in the field from late July to late September. Consequently the disease presents a serious problem for all late season production, whether this involves using everbearing cultivars or sequentially planted cold stored runners of short-day types. In everbearers the problem is more acute as they flower and fruit over a long period, thus making it difficult to achieve chemical control due to the problems of coordinating a spraying programme with regular harvesting.

Earlier work on *S. macularis* in everbearing and day-neutral germplasm suggested that resistance was quantitatively inherited and with a significant non-additive genetic component (Simpson, 1987). The breeding strategy, therefore, is to use resistant genotypes as parents whenever possible but to do a large number of small scale test crosses each year. Those combinations which show good specific combining ability are then repeated on a large scale. Resistance to powdery mildew is a high priority, since it is considered that to be successful an everbearing or day-neutral cultivar must have a useful level of resistance. Consequently it is treated as a primary selection criterion and the entire seedling population is screened by allowing a natural epidemic to develop each year in the field.

Phytophthora fragariae

Phytophthora fragariae is easier to control with fungicides than *Verticillium dahliae* but is still a serious disease in the UK, particularly in areas of high rainfall and on poorly drained soils. No resistance screening is done at HRI but resistant genotypes from other breeding programmes are often used as parents. Potential new cultivars are sent to the Scottish Crop Research Institute to be tested for their susceptibility to the common races of the pathogen before commercial release. Since 1988 four new cultivars from HRI East Malling have been named and their relative levels of disease resistance are shown in Table 1.

Acknowledgements

This work was carried out as part of a project commissioned by the Ministry of Agriculture, Fisheries and Food (Project number HH0113JFV) except for the studies on *P. cactorum* which were funded by the Fruit Culture Council.

References

Bowen, H.H., L.F. Hough & E.H. Varney, 1968. Breeding studies of *Verticillium* wilt resistance in *Fragaria ananassa* Duch. Proceedings of the American Society for Horticultural Science 93: 340–351.

Bringhurst, R.S., S. Wilhelm & V. Voth, 1961. Pathogen variability and breeding *Verticillium* wilt resistant strawberries. Phytopathology 51: 786–794.

Bringhurst, R.S., P.E. Hansche & V. Voth, 1968. Inheritance of *Verticillium* wilt resistance and the correlation of resistance with performance traits of the strawberry. Proceedings of the American Society for Horticultural Science 92: 369–375.

Jordan, V.W.L., 1973. A procedure for rapid screening of strawberry seedlings for resistance to *Verticillium* wilt. Euphytica 22: 367–372.

Kronenberg, H.G. & A.C. van der Giessen, 1968. Testing strawberries for resistance to *Verticillium* wilt. Euphytica 17: 141–152.

Maas, J.L., G.J. Galletta & A.D. Draper, 1989. Resistance in strawberry to races of *Phytophthora fragariae* and to isolates of *Verticillium* from North America. Acta Horticulturae 265: 521–526.

Pitrat, M. & G. Risser, 1977. Étude de la sensibilité du fraisier à *Phytophthora cactorum* après contamination provoquée. Annales de l'Amélioration des Plantes 27 (1): 49–60.

Seemüller von, E, 1977. Resistenzverhalten von Erdbeersorten gegen den Erreger der Rhizomfäule, *Phytophthora cactorum*. Nachrichtenblatt des Deutschen Pflanzenschutzdienstes 29: 124–126.

Simpson, D.W., 1987. The inheritance of mildew resistance in everbearing and day-neutral strawberry seedlings. Journal of Horticultural Science 62 (3): 329–334.

H. Schmidt and M. Kellerhals (Eds.), Progress in Temperate Fruit Breeding, 67–71.

Identification of pathogenic races of *Phytophthora fragariae* Hickman in Germany

Petra Scheewe
Federal Centre for Breeding Research on Cultivated Plants, Institute for Breeding of Ornamental Species, Bornkampsweg 31, D-22926 Ahrensburg, FRG

Key words: differential set, *Fragaria* × *ananassa*, physiologic races, *Phytophthora fragariae*, red stele disease, strawberry

Summary

The fungus *Phytophthora fragariae* Hickman is the causal agent of red stele disease in strawberry (*Fragaria* × *ananassa*), this being a major disease in many areas with cool and moist conditions. Success of resistance breeding can be nullified by the appearence of specific races of the fungus capable of overcoming the introduced resistance. In some countries (USA, UK, Japan and Canada) races were identified by using a differential set of strawberry cultivars. The absence of an international standard differential set and the use of different test methods make a comparison of the identified races difficult or even lead to contradictory results for one variety/race combination. The aim of this study was to obtain information about the spectrum of pathogenic races in Germany as a basic contribution to research on resistance breeding against the fungus. The susceptibility of different strawberry cultivars to German isolates of *P. fragariae* was evaluated. The inoculation was done by modifying a method described by Milholland et al. (1989). Rooted *in vitro* plants, four weeks after transferring them to the soil, were used for the investigation. Inoculation was done with a zoospore suspension of defined concentration. Up to now three German isolates, G-1, G-2, and G-3, can be separated by their ability to infect and produce oospores in the roots of the strawberry cultivars 'Senga Sengana', either 'Saladin', 'Redgauntlet' or 'Climax' and *Fragaria chiloensis* clone 'YaquinaB'. These results confirm the existence of pathogenic races of *P. fragariae* in Germany and should be taken into consideration for resistance breeding against the fungus.

Introduction

The fungus *Phytophthora fragariae* Hickmann, recently named *P. fragariae* var. *fragariae* (Duncan & Kennedy, unpublished; Laun, 1992), is the causal agent of red stele disease in strawberry (*Fragaria* × *ananassa*), this being a major disease in many areas with cool and moist conditions. It was first dicovered in 1921 in Scotland. In 1935 serious damage in strawberry plantings due to the red stele disease were reported from the United States. The causal agent of the disease was described by Hickman (1940) as a new species *P. fragariae*. In Germany Zinkernagel (1970) could not isolate the fungus from strawberry plants. But by 1978, first losses in southern Bavaria due to red stele disease were reported (Seemüller & Riedel, 1980). The fungus is present in most strawberry production areas in Germany though at the time there are no reports from the former GDR about occurence of the disease (R. Büttner, personal communication).

Symptoms revealed above ground are young leaves blue/green in colour while older leaves show a red/orange or yellow discoloration. Infected plants remain stunted and are generally weak growing. The fruit and runner production is reduced. With warm temperatures, wilting symptoms can occur, and frequently the plants are killed. The root system is reduced altogether, and due to dying of the lateral rootlets, the main roots look like 'rat tails'. When such roots are cut longitudinally the characteristic red stele can be seen.

Healthy roots of strawberry plants are infected by zoospores of the fungus attracted through root exudates. The fungus grows through the stelar root tissue, usually at the tip, into the stele. Zoosporangia are formed at the root surface and oogonia, antheridia and oospores alongside of the stele. While hyphae, sporangia and zoospores are not viable in soil, oospores may persist up to 15 years.

Resistance breeding has been practiced since 1933 in Britain and since 1938 in the United States, and since about 1970 in the Netherlands (Scott et al., 1984). Sources of resistance are the American cultivar 'Aberdeen' (parentage unknown), the British cultivar 'Frith' (parentage unknown), which does not exist anymore, *Fragaria chiloensis* clones, and *F. virginiana.*

However, practical resistance breeding is complicated by two facton: the regional adaptation of cultivars and the presence of physiologic races of the fungus.

In the United States ten physiologic races have been identified (Converse, 1970); in Canada six races (McKeen, 1958); in Japan six races (Morita, 1975) and in Britain eleven races (Montgomerie, 1966). All races where identified by using a differential set of strawberry varieties.

One aim of this study was to find out about races of the fungus in Germany.

Material and methods

Plant sources

Stock plants of the strawberry genotypes used in these experiments were obtained from four sources. The cultivars Redgauntlet, Saladin, and Guardian came from the Federal Registration Office, 'Climax' from the CPRO-DLO, Department of Vegetable and Fruit Crops, Wageningen, 'Senga Sengana' from the firm Hummel, Stuttgart as *in vitro* plantlets, and the *Fragaria chiloensis* clone 'YaquinaB' from the Department of Horticultural Plant Breeding, Balsgard, Sweden, also as *in vitro* plantlets. The cultivars Redgauntlet, Saladin, Guardian, and Climax were established and multiplied *in vitro* using the medium of Boxus (1974) slightly modified, as saccharose (40 mg/l) instead of glucose was used. The plantlets were rooted on the Boxus (1974) medium, again using saccharose instead of glucose. Rooted *in vitro* plants were removed from the medium, transferred into multiplates filled with a soil and sand mixture 1:1 by volume, and moved in the greenhouse for acclimatisation.

Isolates of P. fragariae

Three cultures of *P. fragariae* were obtained from the two following sources: L1 from the Federal Biological Research Center for Agriculture anf Forestry, Institute for Plant Protection in Horticultural Crops, Braunschweig (isolated in 1983 from the cultivar Tenira), and 45 (isolated in 1979 from the cultivar Senga Sengana) and 47 (isolated in 1980, probably from the cultivar Gorella) from the Federal Biological Research Center for Agriculture and Forestry, Institute for Plant Protection in Fruit Crops, Dossenheim. The three isolates were chosen because they were isolated in different parts of Germany and from different cultivars. Single spore isolates of all three cultures were produced by zoospores. Stock cultures were maintained in the dark on red kidney bean agar at 20° C.

Zoospore production

Ten 10 mm diameter disks of mycelium were removed from three to four week old cultures on red kidney bean agar and placed in 50 ml of autoclaved 4%Granini-vegetable juice in 100 ml Erlmeyer-flasks. The cultures were incubated for 18 to 21 days on a shaker at 15° C in darkness. A soil leachate was prepared as described by George & Milholland (1986) using a top soil from a field. After filtering, the soil leachate was autoclaved. For zoospore production, the liquid medium from the cultures was decanted and 75 ml of the sterile soil leachate was added. They were incubated at 15° C in darkness. After 24 hours the soil leachate was changed. Zoospore production reached a maximum after two or three days.

Inoculation procedure

Two or three plants of each genotype were inoculated with each of the three isolates. The experiment was repeated once with the isolate 47/1 and twice with the other isolates. For inoculation the method described by Milholland et al. (1989) was modifed. Plants were removed from the multiplates 4 weeks after planting, and the roots were carefully rinsed in cool tap water. The plants were placed on a wet paper towel and then inoculated by spraying a zoospore suspension on the roots, about 2 ml per plant, using a chromatography sprayer. Although the conditions for production of zoospores were the same for all experiments, there were differences in the maximum number of zoospores produced. Therefore the concentrations used for inoc-

Table 1. Interaction between selected cultivars of strawberry and particular isolates of *Phytophthora fragariae* 2 weeks after inoculation: a. mean percentage roots with oospores in 2 mm long root segments, in brackets number of investigated roots, b. mean number of oospores per root segment c. percentage of plants with oospores in 2 mm long root segments[1)]

Cultivars	Isolates 47/1			L1/2			45/11		
	a	b	c	a	b	c	a	b	c
Senga Sengana	62% (68)	5,2	100%	45% (58)	8,3	100%	38% (58)	4,8	71%
Saladin	62% (55)	5,8	100%	30% (63)	3,2	71%	0% (78)	0	0%
Redgauntlet	59% (68)	4,5	100%	63% (70)	10,2	100%	0% (97)	0	0%
Climax	58% (38)	6,1	100%	45% (47)	8,4	100%	0% (85)	0	0%
YaquinaB	55% (44)	4,7	100%	5% (42)	0,04	14%	0% (58)	0	0%
Guardian	6% (103)	0,1	50%	1% (71)	0,03	14%	15% (91)	0,8	57%

[1)] Means were calculated from 6–8 plants of each cultivar/isolate combination with three replications for isolates 45/11 and L1/2 and two replications for isolate 47/1, each replication with two or three plants.

ulation were between 1.2–2.3x10^4 zoospores per ml. According to Milholland et al. (1989) there were no differences in degree of infection between concentrations of 1–3x10^4 zoospores per ml. The plants were incubated for 48 hours, wrapped in the paper towel at 15° C. Subsequently they were planted in plastic pots (diameter 8 cm) filled with water saturated soil. The plants were kept in closed plastic boxes with each pot placed in a petridish filled with water, 11 pots per plastic box and grown in a phytotron with a 12 hour photoperiod (20.000 Lux) at 15° C.

Scoring was done two weeks after inoculation by counting the number of oospores in 2 mm long root tip segments of all main roots under a light microscope. With strongly diseased plants the root tips were often already completely rotted; oospores in the remaining ends of the roots were counted. The mean number of oospores per root and the percentage of roots with oospores was estimated. Additionally the plants were investigated visually for a red stele.

Results

Table 1 gives the percentage of plants with oospores in 2 mm root tip segments, the percentage of roots with oospores and the mean number of oospores per root segment. With the cultivar/isolate combination 'Senga Sengana' and 45/11 there were two plants with no oospores in the root tips but oospores could be found in lateral roots. With the combination 'Saladin' and L1/2 there were plants showing a red stele but no detectable oospores. With the cultivar Guardian with all three isolates there were plants showing symptoms as a red stele with or without oospores and plants with no symptoms at all. 'Saladin', 'Redgauntlet', 'Climax', and 'YaquinaB' were resistant to isolate 45/11, as the plants did not show any symptoms. With compatible cultivar/isolate combinations the percentages of roots with oospores were between 30% and 66%. With 'YaquinaB' and isolate L1/2 and 'Guardian' with all three isolates only very few roots with oospores could be found. The mean numbers of oospores per root segment was between 3.2 for the combination Saladin/L1/2 and 10.2 for the combination 'Redgauntlet'/L1/2. In the roots of 'Saladin', 'Redgauntlet', 'Climax', and 'YaquinaB' inoculated with isolate 45/11 no oospores could be found.

Discussion

Since the first reports of physiologic races of *P. fragariae* in 1950 (Scott et al.) various systems using many strawberry varieties and different methods for naming the isolates have been used (McKeen, 1958; Converse & Scott, 1962, 1970; Hickman, 1962; Montgomerie, 1967). For example, infested soil, mycelia slurry, or zoospore suspensions, sometimes with defined concentrations were used as inoculum for infection stud-

Table 2. Interaction type of three proposed German races of *Phytophthora fragariae* on a set of six strawberry cultivars. Original isolate numbers are given in brackets

Cultivars	German races		
	G-1 (47/1)	G-2 (L1/2)	G-3(45/11)
Senga Sengana	S	S	S
Saladin	S	S	R
Red gauntlet	S	S	R
Climax	S	S	R
YaquinaB	S	(R)	R

S = susceptible,
R = resistant,
(R) = few roots containing oospores.

ies. Plant material was usually young runner plants but often with different root ages. Conditions and time of incubation and systems for disease assessment were also different. Results for resistance or susceptibility of one cultivar/race combination are often contradictory. Thus Milholland et al. (1989) found 'Surecrop' susceptible to the American races A-2 and A-8, while Maas et al. (1989) found it resistant. At present there is no existing international standard differential set, though the need for more controlled screening procedures for identifying races has been emphasized by several authors (Montgomerie, 1966; Kennedy et al., 1986; Milholland et al., 1989).

Using a differential set of seven strawberry hosts Milholland et al. (1989) could differentiate seven of nine tested isolates representing the American races A-1 to A-10 except A-5 of *P. fragariae.* They found the cultivar/race interaction for A-1 and A-3 as well as A-2 and A-7 the same and they propose a new race system. Their rating system was based on the presence or absence of oospores in root tips and not just on the visual assessment of the amount of root rotting and/or the presence of a red stele. Using this method, partly modified, we could detect different cultivar/isolate reactions of three German isolates of *P. fragariae* tested on six strawberry hosts. 'Senga Sengana' was susceptible to all isolates. 'Saladin', 'Redgauntlet', and 'Climax' were susceptible to 47/1 and L1/2 but resistant to 45/11. The identical response of 'Redgauntlet' and 'Climax' is in accordance with the result of other authors (Kennedy et al., 1986; Milholland et al., 1989). 'YaquinaB' was susceptible to 47/1 and resistant to 45/11. With L1/2, few oospores were found in few roots. A comparable reaction could be observed for 'Guardian' with all three isolates. 'Guardian' is an American cultivar with a high level of resistance. Maas et al. (1989) found it resistant to seven of nine tested American races of *P. fragariae* using a visual scoring system with random samples examined for oospores. Law & Milholland (1992) found oospores in the roots of this cultivar with three American races though with two of these the percentage of roots with oospores and the number of oospores was very low, comparable to our data. The basis of resistance is not well known. Results of Van de Weg et al. (1989) indicate a monogenic dominant and race-specific resistance for some cultivar/race combinations. But other genes may be involved as well. Small physiologic differences in the roots may be responsible for oospore development in roots. Our own data (P. Scheewe, unpublished) and results of Milholland et al. (1989) indicate an influence of the root age on disease development by *P. fragariae*, as the percentage of roots containing oospores decreased with increasing age of roots.

'YaquinaB' was included in the differential set because although it had a compatible reaction with L1/2 the symptoms were only slight compared to those caused by 47/1. The first three German races, with proposed designations G-1, G-2, and G-3, could be identified using the diffentials 'Senga Sengana', 'YaquinaB', and either of the cultivars 'Saladin', 'Redgauntlet' or 'Climax' (Table 2).

References

Boxus, Ph., 1974. The production of strawberry plants by in vitro micropropagation. J. Hortic. Sci. 49: 209–210.

Converse, R.H., 1970. Occurence of *Phytophthora fragariae* race A-10 in California. Plant Disease Reporter 54: 969–971.

Converse, R.H. & D.H. Scott, 1962. Physiologic specialization in *Phytophthora fragariae*. Phytopathology 52: 802–807.

George, S.W. & R.D. Milholland, 1986. Inoculation and evaluation of strawberry plants with *Phytophthora fragariae*. Plant Disease 70: 371–375.

Hickman, C.J., 1940. The red core root disease of the strawberry caused by *Phytophthora fragariae* N. sp. J. Pomol. and Hort. Sci. 18: 89–118.

Hickman, C.J., 1962. Physiologic races of *Phytophthora fragariae*. Ann. Appl. Biol. 50: 95–103.

Kennedy, Diana M., J.M. Duncan, P.I. Dugard & P.H. Topham, 1986. Virulence and aggressiveness of single-zoospore isolates of *Phytophthora fragariae*. Plant Pathology 35: 344–354.

Laun, N., 1992. Untersuchungen zur Wurzelfäule der Himbeere (Erreger: *Phytophthora fragariae* var. *rubi*) und zum Resistenzverhalten von Sorten und Kreuzungsnachkommen. Dissertation am Lehrstuhl für Phytopathologie der Technischen Universität München.

Law, T.F. & R.D. Milholland, 1992. Susceptibility of strawberry genotypes to infection and colonization by races of *Phytophthora fragariae* and the growth responses of inoculated genotypes. Plant Disease 76: 335–339.

Maas, J.L., G.J. Galletta, & A.D. Draper, 1989. Resistance in strawberry to races of *Phytophthora fragariae* and to isolates of *Verticillium* from North America. Acta Horticulturae 265: 521–526.

McKeen, W.E., 1958. Red stele root disease of the loganberry and strawberry caused by *Phytophthora fragariae*. Phytopathology 48: 129–132.

Milholland, R.D., W.O. Cline, & M.E. Daykin, 1989. Criteria for identifying pathogenic races of *Phytophthora fragariae* on selected strawberry genotypes. Phytopathology 79: 535–538.

Montgomerie, Isabel G., 1966. Pathogenicity of British isolates of *Phytophthora fragariae* and their relationship with American and Canadian races. Trans. Br. mycol. Soc. 50: 57–67.

Morita, H., 1975. Studies on red stele root disease of strawberry. Shizuoka Agr. Expt. Sta. Special. Bul. Bo. 10.

Scott, D.H., A.D. Draper, & G.J. Galletta, 1984. Breeding strawberries for red stele resistance. In: J. Janic (Ed.). Plant Breeding Reviews, Vol. 2, pp. 195–213. AVI Publishers, Westport.

Scott, D.H., W.F. Jeffers, G.M. Darrow, & D.P. Ink 1950. Occurence of strains of the strawberry red stele fungus, *Phytophthora fragariae* Hickman, as shown by differential response. Phytopathology 40: 194–198.

Seemüller E. & M. Riedel, 1980. Zum Auftreten der Roten Wurzelfäule (*Phytopthora fragariae*) in Süddeutschland. Nachrichtenbl. Deut. Pflanzenschutzd. 32: 81–85.

Van de Weg, W.E., L.M. Wassenaar, C.P.J. Lindeloof, 1989. Inheritance of resistance to *Phytophthora fragariae* Hickman in strawberry. Euphytica 42: 25–30.

Zinkernagel, V., 1970. Bodenbürtige Krankheiten in Erdbeerkulturen Norddeutschlands und ihre Ursachen. II. Wurzel- und Rhizomfäulen der Erdbeere. Z. Pflanzenkrankh. Pflanzensch. 77: 65–75.

H. Schmidt and M. Kellerhals (Eds.), Progress in Temperate Fruit Breeding, 73–76.

Breeding for resistance to *Phytophthora* root rot in red raspberries

G. Spiegler & H. Thoss
Häberli Obst- und Beerenzentrum AG, 9315 Neukirch-Egnach, Switzerland

Key words: breeding, resistance, *Phytophthora fragariae var. rubi*, *Phytophthora* root rot, *Rubus idaeus*, red raspberry

Phytophthora root rot, caused by *Phytophthora fragariae var. rubi*, has become a major problem in raspberry growing also in Switzerland. Häberli Obst- und Beerenzentrum AG, a private nursery company, with its own breeding programmes in small fruits since 1983, has begun a programme with special respect to Phytophthora resistance in raspberries in 1989. In cooperation with the Institutes of Fruit-Growing and Phytopathology of the Technical University Munich at Freising-Weihenstephan a susceptibility test of the seedlings was set up. Over 3 years 22 different reciprocal crossings were made. The varieties 'Latham', 'Winklers Sämling' and 'Autumn Bliss' were used as donors of resistance. These resistant varieties were crossed with several varieties with good fruit size, fruit quality, yield, but susceptiblity to Phytophthora root rot. In the 1989 test, where 'Latham' and 'Winkler's Sämling' were used as donors of resistance, 106 of 601 seedlings survived the test (17.6%), 18 of which died after planting out. The total survival rate of the 1989 crossings was 14.2%. In the 1991 test with 'Autumn Bliss' as donor of resistance 15 of 429 seedlings survived, the rate was 3.5%. All the surviving seedlings will be evaluated for their crop potential during the next years.

Introduction

The disease in Switzerland, control and breeding

Raspberry root rot or raspberry decline has been known in Switzerland since the 1960 (Lutz et al. 1980). Later this root rot, which was found to be *Phytophthora fragariae var. rubi*, became a major problem in raspberry growing, occuring mainly in the eastern and central parts of the country where heavier soils are prevalent.

One way to control the disease is by growing resistant cultivars. Within the summer fruiting types no high yielding variety with good quality and resistance is available. The primocane variety 'Autumn Bliss' has shown resistance (Daubeny pers. comm., Laun 1989).

Häberli Obst- und Beerenzentrum AG, a private nursery company with its own breeding programmes in small fruit since 1983, established a programme with special respect to *Phytophthora* resistance in red raspberries in 1989.

Material and methods

Breeding goal, donors of resistance, resistance test

The breeding goal is the combination of root rot resistance and a high yield of good quality fruit in new selections.

The cultivars 'Latham' (Barritt et al. 1979), 'Winklers Sämling' (Lutz et al. 1980) and 'Autumn Bliss' are used as donors of resistance. To introduce their resistance they are crossed with large fruited, high yielding cultivars with good fruit quality.

The resistance of the seedlings is assessed in pots by the method described by Laun (1989, 1992).

Crossings 1989–1993

In 1989 'Latham' and 'Winklers Sämling' were used as resistant parents. They were crossed with the cultivars 'Rutrago/Rucanta', 'Rumiloba', 'Rusilva', 'Malling Delight', 'Malling Exploit' for high yield and good

Table 1. Results of the resistance test of crossings 1989.

Seed bearing parent	Pollen parent	Number of seedlings in test	Number of surviving seedlings	Percentage of surviving seedlings
Latham	Rutrago	8	2	25
Rutrago	Latham	17	2	11.8
Latham	Rumiloba	28	3	10.7
Rumiloba	Latham	84	13	15.5
Latham	Rusilva	69	14	20.3
Rusilva	Latham	32	14	43.8
Latham	M. Delight	60	7	11.7
M. Delight	Latham	2	0	0
Latham	M. Exploit	13	1	7.7
Winklers S.	Rutrago	35	5	14.3
Rutrago	Winklers S.	6	1	16.7
Winklers S.	Rumiloba	25	1	4.0
Rumiloba	Winklers S.	1	0	0
Winklers S.	Rusilva	119	20	16.8
Rusilva	Winklers S.	1	0	0
Winklers S.	M. Delight	42	15	35.7
Winklers S.	M. Exploit	27	5	18.5
M. Exploit	Winklers S.	32	3	9.4
Zeva Herbst.	Aut. Bliss	102	4	3.9
	Total	703	110	14.2 (mean)

Table 2. Results of the resistance test of crossings 1991.

Seed bearing parent	Pollen parent	Number of seedlings in test	Number of surviving seedlings	Percencentage of surviving seedlings
Aut. Bliss	Glen Moy	54	1	1.9
Glen Moy	Aut. Bliss	73	1	1.4
Aut. Bliss	Framita	78	2	2.6
Framita	Aut. Bliss	0	0	0
Aut. Bliss	Rusilva	66	0	0
Rusilva	Aut. Bliss	22	0	0
Aut. Bliss	Zeva 2	7	0	0
Zeva 2	Aut. Bliss	2	0	0
Aut. Bliss	Willamette	0	0	0
Willamette	Aut. Bliss	86	10	11.6
unknown	unknown	41	1	2.4
	Total	429	15	3.5 (mean)

fruit quality. The test for resistance was carried out in summer 1991.

In 1990 'Autumn Bliss' was used as donor of resistance. It was crossed with the early ripening summer fruiting varieties 'Glen Moy', 'Framita/Ambition', 'Rusilva', 'Zeva 2' and 'Willamette' to obtain autumn fruiting genotypes with earlier ripening than 'Autumn Bliss'. The resistance test was carried out in winter 1992/93.

In 1992 'Latham', 'Winklers Sämling' and 'Autumn Bliss' were used as donors of resistance. They were crossed with 'Ruby/Watson', 'Zeva Herbsternte' and 'Glen Prosen' to produce resistant autumn fruiting genotypes later than 'Autumn Bliss'. The test was carried out in summer 1993.

In spring 1993 'Winklers Sämling' was crossed with the especially large fruited varieties 'Rumiloba', 'Tulameen', 'Malling Exploit' and 'Himbo Queen/Rafzeter'.

Crosses between all the 3 resistant cultivars were produced to combine their factors of resistance.

Table 3. Resistance test, crossings 1992.

Seed bearing parent	Pollen parent	Number of seedlings in test
Ruby	Latham	5
Latham	Ruby	253
Zeva Herbst.	Latham	44
Latham	Zeva Herbst.	274
Ruby	Winklers S.	20
Winklers S.	Ruby	130
Zeva Herbst.	Winklers S.	58
Winklers S.	Zeva Herbst.	179
Aut. Bliss	Zeva Herbst.	406
Zeva Herbst.	Aut. Bliss	131
Aut. Bliss	Glen Prosen	464
Glen Prosen	Aut. Bliss	481
	Total	2445

Results

Resistant genotypes

Table 1 shows the number and percentage of surviving = resistant genotypes of the 1989 crossings. This summer the resistant genotypes will be assessed for yield and fruit quality.

Table 2 shows the number of resistant genotypes of 1991. They are planted in the field for further assessment.

Table 3 shows the number of seedlings of the 1992 crosses in test.

Use of the different donors of resistance

The three donors of resistance were used in reciprocal crosses.

Table 4 shows the survival rates of the progenies of the different donors of resistance.

Discussion

By using the resistant cultivars 'Latham', 'Winkler's Sämling' and 'Autumn Bliss' as donors of resistance to *Phytophthora* root rot caused by *Phytophthora fragariae var. rubi*, it was possible to select resistant genotypes by testing their resistance in contaminated compost. Resistant genotypes could be selected from all the progenies derived from resistant parents.

In future they will be tested according to the 'testing scheme for the selection of productive and resistant *Rubus* clones' (Laun 1992).

'Winkler's Sämling' and 'Latham' appear to be better donors of resistance than 'Autumn Bliss'. This should be verified in future tests.

There seems to be no difference in the use of the donor of resistance as seed bearing or pollen parent.

Several breeders are working to develop resistant genotypes. Within the next years there will be resistant cultivars with high cropping potential of good quality fruit at the growers' disposal.

Acknowledgements

We acknowledge the support of H. Schimmelpfeng, Institute of Fruit-Growing, Technical University Munich-Weihenstephan, Germany for providing the cultivar 'Latham', permanent readiness for discussion and further hints and support; N. Laun, formerly Institute of Phytopathology, Technical University Munich-Weihenstephan, Germany for instructions of the test method and discussions; and H.P. Lauber, Swiss Federal Research Station Wädenswil for multiplicating the inoculum.

Table 4. Progenies of the donors of resistance.

Variety	Crossing year	Used as	Number of seedlings in test	Number of surviving seedlings	Percentage of surviving seedlings
Latham	1989	total	313	56	17.9
		s.b.p.	178	27	15.2
		p.p.	135	29	21.5
Winklers Sämling	1989	total	288	50	17.4
		s.b.p.	248	46	18.5
		p.p.	40	4	10.0
Autumn Bliss	1989	total	102	4	3.9
		s.b.p.	0	0	0
		p.p.	102	4	3.9
Autumn Bliss	1991	total	388	14	3.6
		s.b.p.	205	13	6.3
		p.p.	183	1	0.6

References

Barritt, B.H., P.C. Crandall & P.R. Bristow, 1979. Breeding for root rot resistance in Red Raspberry. J. Americ. Soc. Hort. Sci. 104(1): 92–94.

Laun, N., 1989. Untersuchungen zur unterschiedlichen Anfälligkeit von Kultursorten und Sämlingspopulationen der Himbeere (*Rubus idaeus L.*) gegenüber *Phytophthora erythroseptica Pethybr. var. erythroseptica*. Weihenstephan.

Laun, N., 1992. Untersuchungen zur Wurzelfäule der Himbeere (Erreger: *Phytophthora fragariae var. rubi*) und zum Resistenzverhalten von Sorten und Kreuzungsnachkommen. Weihenstephan.

Lutz, M., H.P. Lauber & W. Stocker, 1980. Das Wurzelsterben der Himbeere. Schw. Z. f. Obst- u. Weinbau 116: 239–245.

Matthies, A. & C. Lankes, 1993. Das Wurzelsterben der Himbeere. Erwcrbsobstbau 3(35): 63–70.

SCRI, 1992/93. Leaflet 'Raspberry root rot'.

CHAPTER 2

Durability of scab resistance in apple: problems and possible solutions

H. Schmidt and M. Kellerhals (Eds.), Progress in Temperate Fruit Breeding, 79.

A new race of *Venturia inaequalis* virulent to apples with resistance due to the Vf gene

L. Parisi,[1] Y. Lespinasse,[2] J. Guillaumès[1] & J. Krüger[3]
[1] *INRA, Station de Pathologie Végétale et Phytobactériologie, rue Georges Morel, B.P. 57, 49071 Beaucouzé Cedex, France;* [2] *INRA, Station d'Amélioration des Espèces Fruitières et Ornementales, rue Georges Morel, B.P. 57, 49071 Beaucouzé Cedex, France;* [3] *Bundesanstalt für Züchtungsforschung im Wein- und Gartenbau, Institut für gartenbauliche Pflanzenzüchtung, Bornkampsweg 31, 22926 Ahrensburg, Germany*

Summary

The gene Vf from *Malus floribunda* 821 for resistance to scab (*Venturia inaequalis*), has been used successfully for 50 years in apple breeding programs. Since 1984, scab symptoms have been observed in the field at Ahrensburg, Germany, on seedlings of apple cv. Prima that have been selected as resistant in the greenhouse. In 1988, small scab lesions were found on some Vf selections in the same orchard. The inoculum from Ahrensburg was compared with the inoculum currently used at Angers, France, for selecting apple seedlings for resistance to *V. inaequalis*. All Vf-gene cvs. or selections tested were susceptible to the Ahrensburg inoculum, whereas *M. floribunda* 821 itself and the ornamental crabapple Evereste were resistant. The progeny from a cross between a resistant (Vf) and a susceptible cv. segregate into the 5 expected classes to Angers inoculum, but were completely susceptible to Ahrensburg inoculum. These results indicate the urgency of diversifing the sources of resistance to *V. inaequalis* in new breeding strategies. The distinction should be made between the resistance of *M. floribunda* 821, that is resistant to Ahrensburg inoculum, and that of the named cv. and selections, that are susceptible. The new race of *V. inaequalis* is named race 6.

H. Schmidt and M. Kellerhals (Eds.), Progress in Temperate Fruit Breeding, 81–85.

Results on the stability of scab resistance in apple breeding

Christa Fischer[1], Antonina Bondarenko[2] & Eleonore Artamonova[2]
[1] *Bundesanstalt für Züchtungsforschung an Kulturpflanzen, Institut für Obstzüchtung, Pillnitzer Platz 2 01326 Dresden-Pillnitz, Germany;* [2] *Kishinev, Moldova*

Key words: apple, breeding, *Venturia inaequalis,* resistance, durability, scab, *Malus*

Summary

The preservation of durable resistance to scab is important for the practical cultivation of new scab resistant varieties. New results were observed in different years and in a range of locations in connection with the breakdown of scab resistance derived from *Malus floribunda* (Vf). Several scab resistant selections were attacked by scab infections, but preserve their resistance as 'field resistance'. They showed only weakly susceptible reactions (classification 3b) in one or two years without influence on yield and fruit quality. The most durable resistance was observed with a combination of genes from several resistance sources (Vf × Vr, Vf × VA, Vr × VA). The best planting strategy is to mix varieties with resistance from different genetic sources.

Introduction

In the apple resistance breeding programme the most important aim is the introduction of durable resistance to economically important diseases, especially scab (*Venturia inaequalis*). The use of donors with scab resistance of a monogenic type can threaten the long-term stability of scab resistance. The monogenic type of scab resistance determined by the Vf-gene from *Malus floribunda* is included in most of the new scab resistant varieties (Table 1). Only a few scab resistant cultivars derive resistance from other sources. The risk of resistance being overcome by a new race of scab can be very high, particularly when resistant varieties are cultivated as monocultures.

In our apple resistance breeding programme we have included investigations on the durability of scab resistance. The trials were carried out in cooperation between the Institutes of Fruit Research and Breeding at Dresden-Pillnitz (Germany) and Kishinev (Moldova).

Material and methods

The investigations on durable resistance began in 1977 in the Moldavian Institute at Kishinev. The varieties 'Clivia', 'Helios', 'Alkmene', 'Golden Delicious' susceptible to scab were screened together with the resistant variety 'Prima' (Vf), and other scab resistant parents and selections with Vf- (*Malus floribunda*), VA- (Antonovka polygenic) and Vr- (*Malus pumila*) resistance (Table 2).

The inoculations were carried out artificially with different biotypes of scab in the orchard under natural conditions. The climatic conditions in Kishinev are very favourable for the development of scab nearly every year.

The breeding material was inoculated with scab pathotypes, isolated from various cultivars of 'Golden Delicious', 'Prima' and 'Simirenko Renette' and a natural scab population from the mixture of varieties, all of which were cultivated in the orchards at Kishinev.

The evaluation of symptoms was carried out by the classification according to Shay & Hough (1952), changed according to Chevalier et al. (1991) in class 3 (Table 2).

Table 1. Scab resistant varieties (with scab resistance sources).

USA		Canada		France		Great Britain		Germany		Romania		Brazil	
Prima (1970)	Vf	Macfree (1974)	Vf	Priam (1974)	Vf	Gavin (1977)	Vf	Remo (1978)	Vf	Generos (1983)	Vf + MK	Primicia (1988)	Vf
Priscilla (1972)	Vf	Nova Easygro (1975)	Vr	Querina (1977)	Vf			Reglindis (1978)	VA	Pionier (1983)	Vf		
Sir Prize (1975)	Vf	Novamac (1978)	Vf	Judeline (1986)	Vf			Reka (1978)	Vr	Romus 1 (1984)	Vf		
Liberty (1978)	Vf	Moira (1978)	Vf	Judaine (1986)	Vf			Realka (1978)	Vr	Romus 3 (1984)	Vf		
Jonafree (1979)	Vf	Trent (1979)	Vf					Releta (1978)	Vr	Voinea (1986)	Vf+ MK		
Redfree (1980)	Vf	Britegold (1980)	Vf					Remura (1978)	Vr	Gloria (1987)	?		
MacShay (1981)	Vf	Murray (1980)	Vm					Rewena (1978)	Vf				
Freedom (1983)	Vf + VA	Rouville (1983)	Vm					Rene (1978)	Vf				
Dayton (1988)	Vf	Richelieu (1983)	Vf					Relinda (1980)	Vf				
William's Pride (1988)	Vf							Retina (1982)	Vf				
								Reanda (1982)	Vf				
								Renora (1982)	Vf				
								Resi (1984)	Vf				
								Releika (1984)	Vf				
								Regine (1988)	Vf				

Vf – Malus floribunda
Vr – Malus pumila
Vm – Malus micromalus
VA – V-Antonovka polygenic resistance.
MK – Malus × kaido.

Table 2. Reaction of apple varieties, resistant parents and resistant selections after inoculation with scab biotypes in Kishinev.

Cultivar/selection donor	Pathotype	Years/Classification of Symptoms 1977	1979	1980	1981	1982	1983	1984	1985
Simirenko	1	4	4	4	4	4	4	4	4
	2	4	4	4	4	4	4	4	4
	3							4	4
	4	4	4	4	4	4	4	4	4
Clivia	1	3b	4	4	4	0	3b	3b	
	2		3b		2				
	3							3b	
Helios	1	3b	0	3b	0	0	3b	4	
	2		2		2				
	3							2	
Alkmene	1	3b	0	3b	1	0	0	0	
	2		2		2				
	3					2		3b	
Golden Delicious	1	3b	3b	3b		3b	4	4	4
	3								
	4					3b		4	4
Prima Vf	1					3b	3b	4	4
	3					2		3b	
	4								4
Pi-34,10 VA	1	0	0	0	0	0	0	0	
	2		1		2				
	3							2	
Pi-34,15 VA	1	0	0	0	0	0	0	0	
	2		2		2			3b	
	3							3b	
R2T107 VA	1	0	0	0	0	0	3b	3b	
	2		0		2			3b	
	3							3b	
R2T110 VA	1	0	0	0	0	0	3b	0	
	2		2		2				
	3					2		3b	
Dg23T53 Vf	1	0	0	0		0	3b	3b	4
	4								0
OR32T41 Vf	1	0	0	0	0	0	0	0	0
	3							3b	
	4								3b
OR40T43 Vr	1	0	0	0	0	0	0	0	0
	2				2				
	3							1	
	4								3b
Pi-AS-8,60 VA	1	0	0	3b	0	0	0	0	
	3							3b	
Pi-AS-5,20 Vf	1						0	0	0
	3							3b	

Table 2. continued

Cultivar/selection donor	Pathotype type	Years/Classification of Symptoms							
		1977	1979	1980	1981	1982	1983	1984	1985
Pi-AS-8,20 Vf	1	0	0	3b			0	0	2
	3							3b	
Pi-AS-3,60 Vr	1		0	0	0	0	0	0	2
	2				2				
	4								2
Pi-AS-3,68 Vr	1						0	0	
	3							3b	
Pi-AS-4,51 Vr	1	0	0	0		0	0	0	
	2		0						
	3					0		3b	

Annotations.
Classification of symptoms according to Shay & Hough (1952) and Chevalier et al. (1991):
0 = without scab; 1 = hypersensitivity; 2 = resistance; 3a = weak resistance;
3b = weak susceptibility; 4 = susceptibility.
Inoculation with pathotypes:
1 = natural scab population in Kishinev; 2 = isolated pathotype from Golden Delicious in Kishinev;
3 = isolated pathotype from Prima in Kishinev; 4 = isolated pathotype from Simirenko Renette in Kishinev.
VA – Antonovka polygenic source of scab resistance.

Results

The results are shown in Table 2 for the investigation period from 1977 to 1985. Every year the inoculum of the biotypes 1, 2, 3 and 4 was tested on the highly susceptible apple variety 'Simirenko'. The pathogenity of the biotypes was very high on the leaves of 'Simirenko'. All the leaves showed intensive symptoms of scab (classification 4) in all years with all biotypes. The reaction of other susceptible varieties was more variable. The susceptible varieties 'Clivia', 'Alkmene', 'Helios' and 'Golden Delicious' did not develop scab symptoms in every testing year, depending on the weather conditions. In the strongly epiphytotic years 1977, 1980, 1982, 1984 and 1985 the susceptible varieties developed severe scab infections.

The resistant variety 'Prima' with Vf-resistance from *Malus floribunda* 821 showed scab symptoms in the class 3b (weakly susceptible) and 4 (susceptible) with 3 different biotypes in 4 years. The scab inoculation on the variety was successful every year from 1982 to 1985.

The scab resistant parents differ in the source of scab resistance. The scab resistant parents Pi-34,15, R2T107 and R2T110 with polygenic scab resistance from 'Antonovka' (VA) developed only in a few years, namely in 1983 and 1984 a weak susceptibility to scab. Only one resistant parent Pi-34,10 with polygenic scab resistance never showed scab symptoms.

The first attack of scab we observed on resistant selections in 1980. The new scab resistant selection Pi-AS-8,60 (polygenic resistance of Antonovka – VA) could be infected only in 2 years through an inoculum of the natural scab population in 1980 and through an inoculum of scab leaves from the resistant variety 'Prima' in 1984. This selection Pi-AS-8,60 showed only weak symptoms with a very slight sporulation, without influence on yield and fruit quality – that means 'field resistance'.

The resistant parents Dg23T53 and OR32T41 with resistance from *Malus floribunda* 821 were attacked by scab in 3 years 1983–1985 showing strong scab symptoms (classes 3b and 4). The Vf-resistant selections Pi-AS-5,20 and Pi-AS-8,20 had only in 2 years slight scab symptoms.

No symptoms were observed on the resistant parent with Vr-gene source OR40T43. The same results were obtained with the new resistant selection Pi-AS-3,60 with Vr-resistance. The Vr-resistant selections Pi-AS-3,68 and Pi-AS-4,51 were infected only in 1 year, 1984, with a scab pathotype from 'Prima'.

In trials on different locations in East Germany with the same resistant selections scab symptoms were not observed under natural infection conditions without fungicide sprayings in all the years. Only in 1991 we found scab symptoms on the Vf-resistant variety 'Liberty' under natural conditions at Wurzen (near Leipzig). The development of scab symptoms is very much dependent on the ecological conditions and the virulence of scab populations under these conditions.

Other investigations of new resistant selections derived from several combinations of alleles of resistant genes – Vf × Vr, Vf × VA, Vf × Vm, Vr × VA – demonstrated in the last 10 years, that these combined genotypes were not infected by scab.

Discussion and conclusions

The overcoming of scab resistance, especially with the Vf-genetic basis of the variety 'Prima' and 2 other resistant parents may occur in different years on special locations with very good conditions for scab infections. A very high risk is the one-sided genetic basis with monogenic resistance of *Malus floribunda*. The first breakdown of Vf-resistance was described on the variety 'Prima' (Fischer et al., 1983). In the Moldavian Institute Bondarenko & Artamonova (1984) investigated the resistant variety 'Prima' and isolated scab biotypes from this variety. These biotypes were able to infect Vf-resistant parents and clones. Further results concerning a scab attack on the varieties 'Prima' and COOP 2 about were published in 1988 (Krüger). In the same time at Ahrensburg scab symptoms were observed on the Vf-resistant numbers COOP 7, COOP 9 and COOP 10 (Krüger 1988). At Ahrensburg a new scab race was isolated, which infected different varieties carrying Vf resistance: 'Baujade', COOP 28, 'Florina', 'Liberty', 'Priscilla' (Parisi et al. 1993). The durability of scab resistance from *Malus floribunda* is threatened on different locations and in different years.

Although other resistant parents and selections with resistance genes from *Malus pumila* (Vr) and 'Antonovka' (VA polygenic) were attacked by scab, they did not develop scab symptoms regularly every year. That means that such genotypes demonstrated 'field resistance' without influence on yield and fruit quality. The 'field resistance' can be a basis for the stability of natural scab populations in various regions.

These results emphasize the necessity of defining and using new strategies in resistance breeding programmes and in fruit production:

1. Combination of a number of distinct genes for scab resistance in one variety. The durable resistance of a number of new Pillnitz selections results from combinations of alleles from several sources (Vf × Vr, Vf × VA, Vr × VA). It is vital to select varieties with both parents contributing genes for resistance to scab, and not to depend on backcrossing single major genes into susceptible types (Fischer 1993).
2. The cultivation strategy should include a mixed assortment with different resistance genes, not a monoculture of Vf-resistant varieties. The strategy for the breeder and the grower is the prevention of changes of the make up of the natural scab populations and of its virulence. The breeder and the grower must preserve the variation of natural scab populations and the interactions between host (the resistant apple varieties) and pathogen (scab populations). The breeding and selection of cultivars with 'field resistance' to scab can be an effective method. With these strategies it is possible to preserve the balance between newly bred resistant apple varieties and the scab pathogens.

References

Bondarenko, A. & E. Artamonova, 1984. Der Einfluß der Resistenzquellen beim Apfel gegen Mehltau und Schorf und auf virulente Rassen des Erregers beim Apfelschorf. In: Genetik der Immunität und Züchtung landwirtschaftlicher Kulturarten in Molda vien, 108–120. Verlag 'Schtiinza', Kischinjow (in Russian).

Chevalier, M., Y. Lespinasse & S. Renaudin, 1991. A microscopic study of the different classes of symptoms coded by the Vf gene in apple for resistance to scab (*Venturia inaequalis*). Plant Pathol. 40: 249–256.

Fischer, C., 1994. Breeding apple cultivars with multiple resistance. In: H. Schmidt and M. Kellerhals (eds). Progress in Temperate Fruit Breeding. pp. 45–50. Kluwer Academic Publishers Dordrecht.

Fischer, C., V. Bukartschuk, A. Bondarenko & E. Artamonova, 1983. Erste Ergebnisse zur Stabilität der Schorfresistenz beim Apfel unter verschiedenen ökologischen Bedingungen in der UdSSR und DDR – vorläufige Mitteilung. Arch. Gartenbau 31(6): 263–264.

Krüger, J., 1988. Beständigkeit der Schorfresistenz aus *Malus floribunda* 821 auf dem Versuchsfeld der Bundesforschungsanstalt für gartenbauliche Pflanzenzüchtung in Ahrensburg. Erwerbsobstbau 52(2): 52.

Parisi, L., Y. Lespinasse, J. Guillaumes & J. Krüger, 1993. A new race of *Venturia inaequalis* virulent to apple with resistance due to the Vf gene. Phytopathology 83(5): 533–537.

Shay, J.R. & L.F. Hough, 1952. Evaluation of apple scab resistance in selections of *Malus*. Ann. J. Bot. 39: 288–297.

H. Schmidt and M. Kellerhals (Eds.), Progress in Temperate Fruit Breeding, 87–92.

Histological and cytological studies of the interaction between apple selections carrying the resistance Vm and compatible and incompatible *Venturia inaequalis* strains

M. Chevalier & Y. Lespinasse
Institut National de la Recherche Agronomique, Station d'Amélioration des Espèces Fruitières et Ornementales, B.P. 57, 49071 Beaucouzé Cédex, France

Key words: Cytology, *Malus*, overcome, resistance, susceptibility, *Venturia inaequalis*, apple, scab

Summary

Resistance overcome by a new race of pathogen is today one of the greatest preoccupations of breeders. The resistance of Vm gene was overcome and the Vf gene was also broken down under certain conditions. The resistance of Vm gene, present in a *Malus micromalus* clone, was overcome by race 5 of *Venturia inaequalis*; this gene constitutes outstanding material for studying host-parasite interactions. Consecutive inoculation of resistant and susceptible hybrids with race 1 and 5 of *Venturia inaequalis* enables rigorous histological and cytological studies. When the resistance is overcome, susceptible and resistant symptoms coexist. At symptom expression, the conidiogenesis is restricted in comparison to the susceptible control. Host reactions are limited to the upper epidermis with abundant subcuticular stroma and very deep erosion of epidermal walls. Epidermal cells collapse irregularly and wall or organelle modifications vary from one cell to another. The resistance's mechanisms still operate. It is possible to envisage that *Venturia inaequalis* has acquired sites compatible with those of the receiving host.

Introduction

Scab is a harmful disease in the apple orchard. It is caused by a biotrophic ascomycete fungus *Venturia inaequalis* (Cke.) Wint. The majority of the cultivated varieties are susceptible to scab. In the 1940s, Hough (1944), then Hough & Shay (1949) discovered genes conferring complete resistance in *Malus* wild species. Six dominant and independent genes were identified: Vf, Va, Vb, Vbj, Vr, Vm (HOUGH et al. 1970). These genes were introduced in the cultivated varieties by classical hybridization methods followed by selection. Numerous resistant varieties, mainly originating from *Malus floribunda* 821 (Vf gene), are now recommended to fruit growers (Crosby et al. 1992).

Nevertheless, although disease resistance constitutes a very attractive solution, new strains of the pathogen can overcome the resistant genes. Both Vm (Williams & Brown 1968) and Vf (PARISI et al. 1993) have broken down. It is important to investigate the processes involved in resistance breakdown.

Among the different sources of resistance which are available, that coded by the Vm gene, present in *Malus micromalus* 245–38, constitutes outstanding material for studying host-parasite interactions. The selection 9AR2T196 and its seedlings Vm/vm are resistant to race 1, but susceptible to race 5. Consecutive inoculation of plants in the glasshouse with race 1 and 5 on the same host or on seedlings with the same genetic background allow rigorous comparative studies.

Histological and cytological studies of resistance breakdown have not been reported. In this paper leaf tissue reactions, of resistant hybrids overcome by race 5 of *Venturia inaequalis*, are described.

Materials and methods

The host

Hybrids which were used carry the Vm gene and originate from crossing Golden Delicious (genotype vm/vm, scab susceptible cultivated variety) with 9AR2T196 (genotype Vm/vm, differential host of race 5 of *Venturia inaequalis*). Susceptible hybrids vm/vm from the same cross were used as controls. Young apple trees were cultivated in pots under greenhouse conditions as described by Olivier & Lespinasse (1981).

The fungus

Tests required the use of two strains of *Venturia inaequalis* obtained from monospore isolation:

- strain 104 belongs to the wild race (race 1) compatible to genome vm/vm, incompatible to Vm/vm, isolated from Golden Delicious in 1978;

- strain 147 belongs to race 5 compatible to Vm/vm and vm/vm, isolated from differential host 9AR2T196 in 1985.

Conidia of all strains were collected in distilled water and each suspension was adjusted to 5.10^5 conidia ml^{-1}.

Inoculations and samples

Before each inoculation, the second fully expanded leaf from the apex was marked. The conidial suspension was sprayed onto plants in sufficient quantities to form small droplets on the leaf surface without a drip forming; they were then incubated at 20 °C and 99% r.h. for 48 h. Vm/vm and vm/vm hybrids were first inoculated with strain 104 (race 1). After symptom expression shoots were cut back, and 4 weeks later strain 147 (race 5) was sprayed onto the young actively growing new shoots. The control hybrid vm/vm was sprayed with distilled water (referred to as non inoculated control).

Conidia germination and fungi penetration were studied on samples collected 72 h after inoculation. Histological and cytological studies were performed on samples collected at symptom expression: either 4 days after inoculation for resistant hybrids, or 12 days for susceptible hybrids.

Surface observations

Leaf specimens were discoloured in ethanol-acetic acid (V:V/1:1) for 24 h. The leaf tissues were made transparent in lactophenol after 24 h; they were then stained with lactic blue.

Histological and ultrastructural observations

Specimens were fixed in 4% glutaraldehyde in 0,1 M Soerensen buffer (pH 7,4) for 2 h at 4 °C, under vacuum and post-fixed in 2% osmium tetroxide in the same buffer for 1 h 30 at room temperature. Specimens were dehydrated in graded ethanol series and propylene oxide and were embedded in Epon.

Semithin sections were cut at 0.5–1 μm with an ultramicrotome (Reichert-Jung), mounted on glass slides, stained by double coloration P.A.S.-toluidine blue, re-mounted after dehydration in a synthetic resin and examined under a Polyvar Reichert Jung microscope.

Ultrathin sections were obtained with a diamond knife with an ultramicrotome (Reichert-Jung), mounted on 100 mesh copper grids, contrasted with uranyl acetate and lead citrate and examined with a Jeol 100B electron microscope.

Results

Macroscopic aspects

After 72 h, the Vm/vm hybrid leaves inoculated with race 1 had many rust coloured spots; the small pits or 'pin-points' appeared later on. These same hybrids, inoculated with race 5 did not reveal any symptom before 10 to 12 days when the first sporulating lesions became differentiated. The control vm/vm hybrids inoculated with race 1 or race 5 gave the same response; they were susceptible.

Surface observations with optical microscope

At the surface of actively growing leaves of Vm/vm hybrids inoculated with race 1, the conidia gave rise to a long germ tube with numerous ramifications, differentiating one to six appressoria (Table 1). When these hybrids were inoculated with race 5 the number of appressoria per conidia was lower. The results obtained with vm/vm susceptible hosts showed that conidia gave rise to only one short germ tube, almost

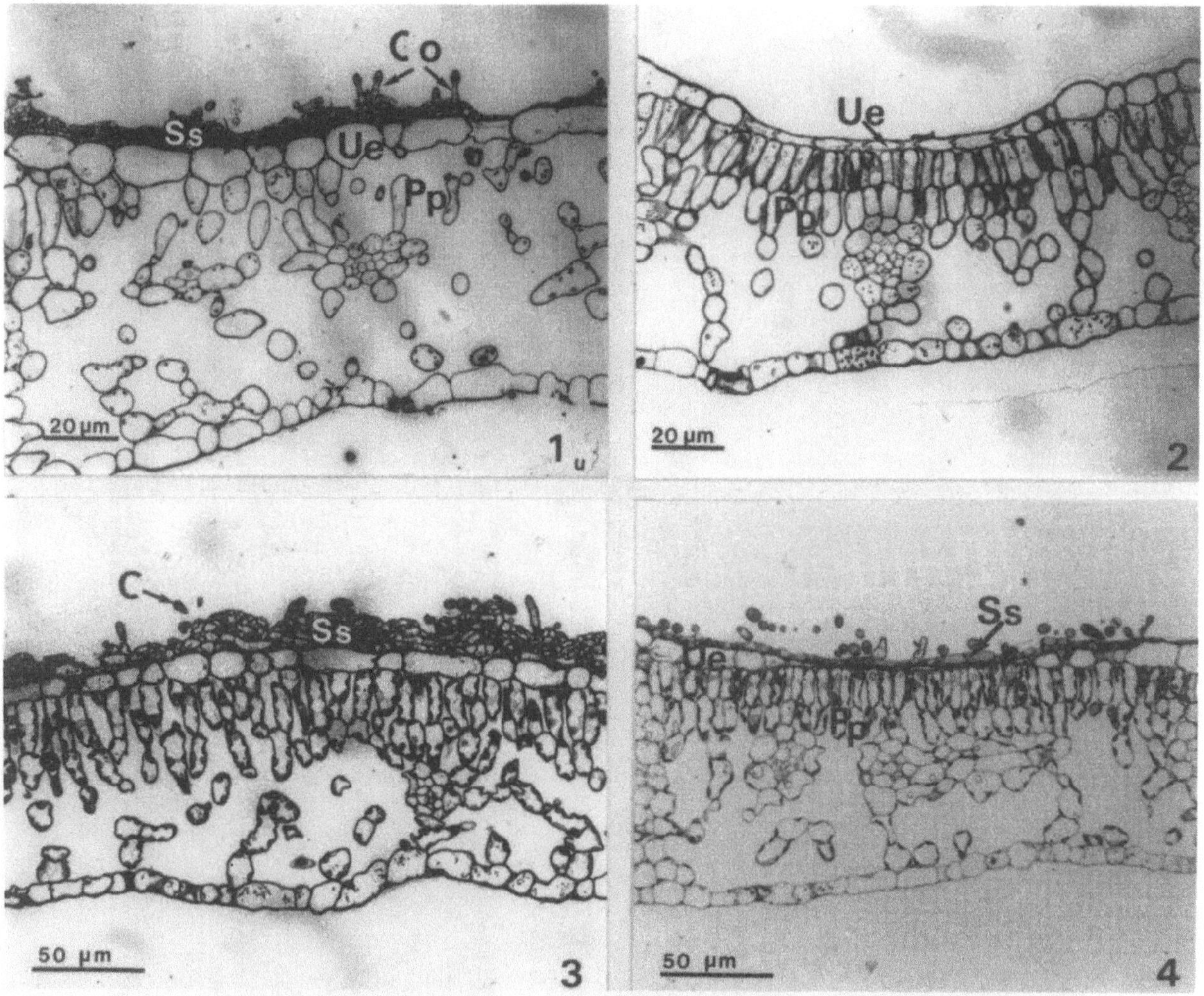

Fig. 1–4. Histological study (1) Symptom of susceptibility: the subcuticular stroma is abundant (Ss) and gives rise to numerous conidiophores (Co). The leaf tissues are slightly modified. (2) Symptom of resistance to race 1 (pin-point): the upper epidermis collapses. The palisade tissue (Pp) is strongly modified. (3) Resistance overcome by race 5. The subcuticular stroma is very abundant and sporulation very low. (4) Resistance overcome by race 5. The upper epidermis gives hypersensitive symptom but the subcuticular stroma is abundant. Symptoms of susceptibility and resistance symptoms coexist. C – conidium; Co – conidiophore; Pp – palisade parenchyma; Ss – subcuticular stroma, Ue – upper epidermis.

Table 1. Number of appressoria per 100 germinated conidia, 72 h after inoculation, 20 °C.

	Number of appressoria/conidia					
	1	2	3	4	5	6
Vm/vm inoculated with race 1	38	37	17	4	3	1
Vm/vm inoculated with race 5	74	22	4	–	–	–
vm/vm (race 1 or 5)	98	2	–	–	–	–

always terminated by one appressorium, whatever the strain used.

Histological study

In susceptible hybrids Vm/vm inoculated with race 1 or 5, the subcuticular stroma, which was often discontinuous (Fig. 1), gave rise to conidiophores with abundant sporulation. The leaf tissues did not undergo any important modifications compared with the non inoculated control.

The leaf tissue response was homogeneous when the Vm/vm hybrids were inoculated with race 1 of *Venturia inaequalis*: many of the upper epidermal cells collapsed and the palisade tissue underwent substantial

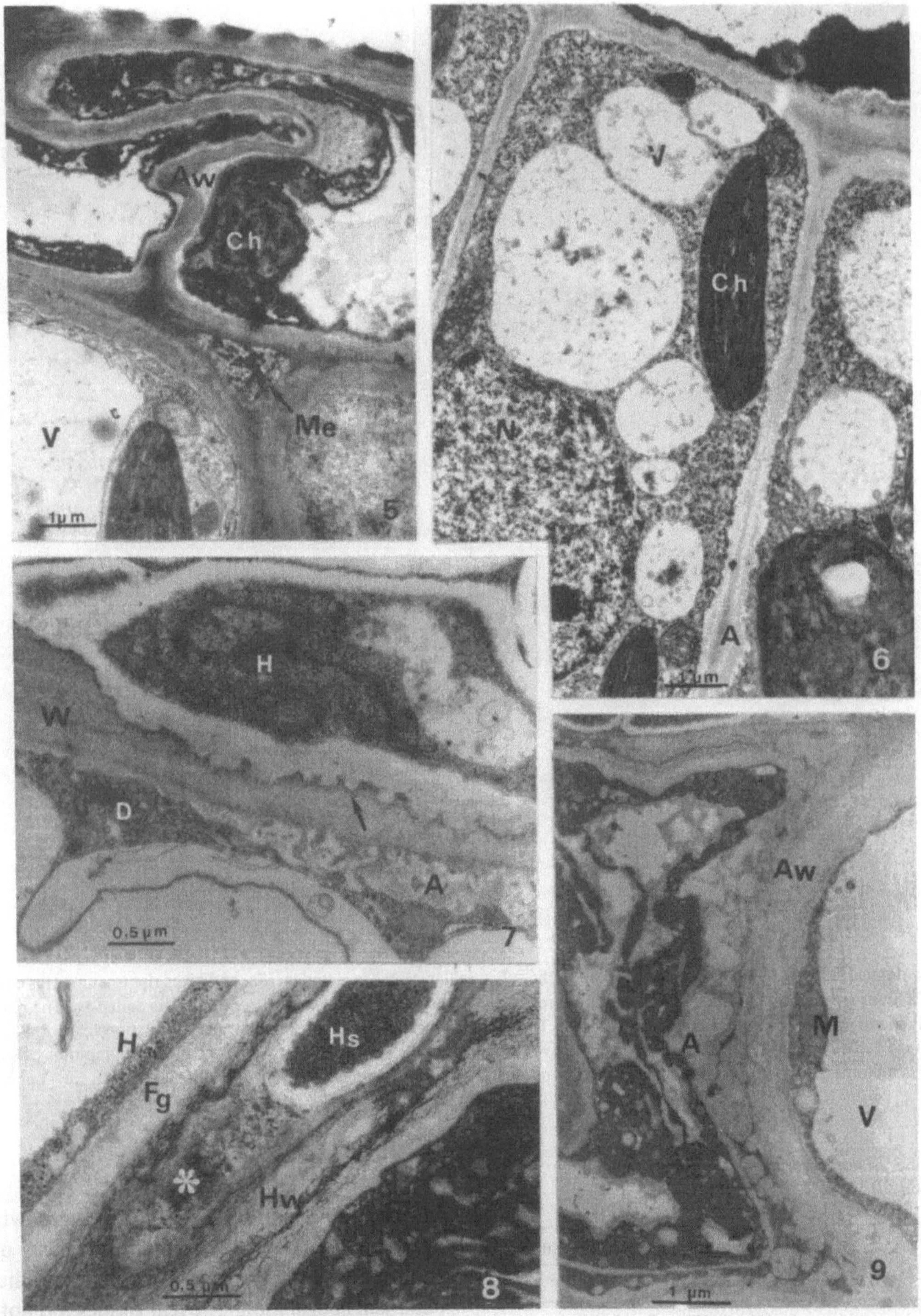

Fig. 5–9. Ultrastructural study Symptom of resistance to race 1. (5) In epidermal cells: The anticlinal wall collapses (Aw), the cytoplasm is highly necrotic and the cellular spaces (Me) are filled with a heterogeneous material. (6) In palisadic cells: Small vacuoles (V) contain precipitates, chloroplasts (Ch) are very dense and wall appositions (A) are numerous. Resistance overcome by race 5. (7) A very deep erosion (arrow) of epidermal wall (W); below the attack there are apposition walls. (8) A small hyphae (Hs) progresses within the wall preceded by an important degradation (*). (9) In some epidermal cells, apposition walls are numerous and the cytoplasm is highly necrotic. A – apposition wall; Aw – anticlinal wall; Ch – chloroplast; D – dictyosome; Fg – fungal wall; Hw – host wall; H – hyphae; Me – cellular space; N – nucleus; V – vacuole.

modifications (Fig. 2). Cell size was reduced, small vacuoles showing polymorphic granules were abundant in the cytoplasm.

Results were different when the leaves of these hybrids were inoculated with race 5. At symptom expression, the sections showed various reactions:

- a very abundant stromatic mycelium with numerous small hyphae and low sporulation (Fig. 3);
- resistant symptoms with abundant stroma (Fig. 4);
- on numerous sections the epidermal cells collapsed discontinuously and the fungus grew normally.

Ultrastructural study

An ultrastructural study of the resistance response after race 1 inoculation, showed:

- in cells of the upper epidermis, collapse of the anticlinal wall and highly necrotic cytoplasm (Fig. 5);
- in cellular spaces, heterogeneous material;
- in the subjacent palisadic cells, numerous necrotic structures, very dense chloroplasts, the vacuole split up into small vacuoles containing precipitates and with many very heterogeneous appositions along the walls (Fig. 6).

Fungi profiles were very rare and always highly necrotic.

After inoculation with race 5, the contact zone between the host and the parasite fungus showed various aspects: a very deep erosion of epidermal walls (Fig. 7), numerous hyphae of small diameter develop within the wall (Fig. 8), considerable degradation at the cell's junction and sometimes very deep progression between anticlinal walls of some epidermal cells.

Epidermal cell organelles underwent modifications which varied from one cell to another:

- in collapsed epidermal cells the wall appositions were absent ; the cytoplasm was highly necrotic;
- in numerous cells, the ultrastructure was severely disturbed : the wall appositions had very heterogeneous appearances (Fig. 9);
- in epidermal cells which are still turgescent, wall appositions were numerous but their size and their distribution varied greatly from one cell to another. The cellular organelles appeared with few modifications.

Epidermal cells collapsed; *subjacent palisade* cells underwent modifications varying from one cell to another. The intercellular spaces were usually empty.

Wall appositions were reduced and always limited to the contact zone with epidermal cells.

The appearance of palisade cells below non collapsed epidermal cells was similar to those in the non inoculated control.

Discussion

Macroscopic observations showed that browning preceding the appearance of pin-point symptoms characterized the resistance reaction. The browning could correspond to phenolic oxidation. This browning reaction does not occur when race 5 overcomes resistance and macroscopic symptoms, although not so widespread, are similar to those found on susceptible vm/vm hybrids.

Differences observed at conidia germination indicate that penetration in susceptible hybrids is faster. The elongation of the germ tube and appressoria multiplication in resistant hybrids show that establishment of the pathogen is more difficult. Sherwood & Vance (1980) think that the development of numerous appressoria characterizes resistance. Johnson (1977) studying cereal mildew suggests that preformed metabolites are exuded at intercellular junctions and would be responsible for appressoria multiplication. When resistance is overcome, these factors still exist; they delay pathogen establishment.

Histological studies are in agreement with the results previously reported for resistant and susceptible reactions (Chevalier & Lespinasse 1991). The susceptible vm/vm hybrids inoculated with race 1 or 5 react in a similar way and sporulate abundantly; they are characterized by only a slight modification of leaf tissues. Resistant hybrids Vm/vm inoculated with race 1 are characterized by a restricted and necrotic stroma and very strong modifications of upper epidermis and palisade cells. When the resistance is overcome, Vm/vm hybrids inoculated with race 5 express reactions limited to the upper epidermis; symptoms of susceptibility and resistance coexist; they significantly restrict the conidiogenesis in comparison to the susceptible control. It seems that in this case *V. inaequalis* has acquired compatible sites in a resistant host.

References

Chevalier, M. & Y. Lespinasse, 1991. La résistance du pommier (*Malus* × *domestica* Bork.) à *Venturia inaequalis* Cke. Wint. Étude histologique et ultrastructurale du symptôme de résistance en piqûre d'épingle. C.R. Acad. Sci., Ser. III, 312: 117–124.

Crosby, J.A., J. Janick, P.C. Pecknold, S.S. Korban, P.A. O'Connor, S.M. Ries, J. Goffreda & A. Voordeckers, 1992. Breeding apples for Scab Resistance: 1945–1990. Fruit Var. J. 46(3): 145–166.

Hough, L.F., 1944. A survey of the scab resistance of the foliage on seedlings in selected apple progenies. Proc. Am. Soc. Hortic. Sci. 44: 260–272.

Hough, L.F. & J.R. Shay, 1949. Breeding for scab resistant apples. Phytopathology 39: 10 (Abstract).

Hough, L.F., E.B. Williams, D.F. Dayton, J.R. Shay, C.H. Bailey, J.B. Mowry, J.J. Janick & F.H. Emerson, 1970. Progress and problems in breeding apples for scab resistance. Proc. EUCARPIA Fruit Section, Sept. 14–18, 1970: 217–230.

Johnson, L.E.B., 1977. Resistance mechanisms to powdery mildew fungi (Erysiphaceae) in non-host and in appropriate host plants. M.S. Thesis Univ. Minnesota, St. Paul, 180 pp.

Olivier, J.M. & Y. Lespinasse, 1981. Évolution des recherches sur la résistance du pommier à la tavelure. II. Étude du parasite et stratégies de lutte. 1er Colloque sur les Recherches Fruitières – Bordeaux: 145–156.

Parisi, L., Y. Lespinasse, J. Guillaumes & J. Krüger, 1993. A new race of *Venturia inaequalis* virulent to apples with resistance due to the Vf Gene. Phytopathology 83: 533–537.

Sherwood, R.T. & C.P. Vance, 1980. Resistance to fungal penetration in graminae. Phytopathology 70: 273–279.

Williams, E.B. & A.G. Brown, 1968. A new physiologic race of *Venturia inaequalis*, incitant of apple scab. Plant Dis. Rep. 52: 799-801.

H. Schmidt and M. Kellerhals (Eds.), Progress in Temperate Fruit Breeding, 93–97.

Approaches for breeding apples with durable disease resistance

Markus Kellerhals & Beat Furrer
Swiss Federal Research Station for Fruit-Growing, Viticulture and Horticulture, CH-8820 Wädenswil, Switzerland

Key words: apple, breeding, durable resistance, scab

Summary

In most apple breeding programmes resistance to diseases has a high priority. The main emphasis is put on scab and mildew resistance. Recent evidence about the risk of breakdown of the Vf scab resistance demonstrates the need to search for new approaches in disease resistance breeding. The following options might be considered:
discover and exploit new sources of resistance to enlarge the genetic basis of resistance;
accumulate different functional resistances in one genotype;
develop orchard designs in order to increase diversity in the host-pathogen interactions and establish a resistance management system.

To implement these strategies the genetic structure of the resistance must be examined by using modern methods of molecular biology. Resistance donors and advanced selections should be examined under high natural inoculum pressure at several locations around the world. We examined the possibility of identifying polygenic or partial scab resistance in apple. Nineteen apple varieties and selections, among them the differential hosts for scab races 1 to 5, were artificially inoculated in the glasshouse with scab conidia. The leaf symptoms were analysed macroscopically and microscopically There were significant differences which are actually exploited in our breeding programme. Conclusions are drawn towards new approaches for breeding and growing apples with durable disease resistance.

Introduction

Breeding for disease resistance has a high priority in most apple breeding programmes throughout the world. In the case of the Vf scab resistance, recent evidence has shown that resistance breakdown is possible (Fischer et al., 1983; Krüger, 1988, 1989, 1991; Van der Scheer, 1989; Silbereisen, 1989; Parisi et al., 1993). Resistance breakdown has been observed with the Vm resistance originating from *Malus micromalus* 245–38 and *M. atrosanguinea* 804 coding for the pit type reaction to scab. This resistance was attacked by race 5 of the fungus (Williams & Brown, 1968; Williams & Kuç, 1969). A new orientation in disease resistance breeding is required.

New sources of resistance should be discovered and exploited to enlarge the genetic basis of resistance. Besides the Vf resistance very few resistances have been incorporated in breeding programmes till now. Other sources such as Vr, Va, Vb, Vbj or the polygenic Antonovka resistance have been used only sporadically. With Vr, Va, Vb, and Vbj there is the same potential risk of breakdown as with Vf and Vm. However, a combination of several resistances in one genotype might be a solution as it is the case, for example in the variety Freedom, which carries Vf and the polygenic Antonovka resistance. Polygenic or partial resistance or low susceptibility of old and modern varieties has always been considered difficult to deal with. A quantitative assessment of the level of this type of resistance is difficult. However, Blazek & Paprstein (1994) found that sources of polygenic resistance to scab can be successfully used to develop new cultivars tolerant to the disease. We examined the possibility of identifying polygenic or partial scab resistance in apple. Nineteen different apple varieties and selections were artificially inoculated in the glasshouse with scab conidia. The leaf symptoms were analysed macroscopically and microscopically.

The idea to develop completely resistant, immune varieties with the inherent risk of resistance breakdown should probably be abandoned in favour of an integrated and sustainable approach to reach the goal of minimum pesticide application in apple orchards. Ongoing research at Wädenswil and Zürich (Gessler & Blaise, 1994) should help to identify strain-specific scab resistances in different apple varieties. These resistances might be accumulated in future breeding efforts. This type of resistance as well as all types of partial resistance could be combined and accumulated with so-called monogenic or oligogenic types of resistance such as Vf, Vr, Va, Vb and Vbj. However, to achieve this goal it would be necessary to exactly recognize the resistances present in each seedling genotype. Modern methods of molecular biology should help to examine the genetic structure of the resistances present in a genotype. The greater the variation in the resistance types the more durable the resistance should be. Orchard designs which include variety mixtures are supposed to further strengthen the durability of genetically determined disease resistances (Blaise & Gessler, 1994).

Materials and methods

From nineteen apple varieties and selections six trees each were grafted on M25 rootstock and grown in pots of 17 cm diameter in the glasshouse. Table 1 shows the varieties, their resistance and origin. The inoculum consisted of *Venturia inaequalis* from different orchards in Switzerland not treated with fungicides and then multiplied on the seedling progenies of the breeding programme. The inoculum suspension contained 250,000 conidia per ml. The inoculation was made in two different ways:

1. spray inoculation in order to create small droplets on the whole surface of the leaves;
2. pipette inoculation in order to create 4 drops of 7 μl on the two youngest fully developed leaves according to Gessler & Stumm (1984).

The trees were incubated under plastic at 20 °C and high relative humidity for 48 hours. Humidity was kept at a high level in case 1 with a humidifier and in case 2 with humid peat. Next the humidity was reduced to 70% and a few days before scoring the leaf symptoms increased again in order to promote sporulation of the fungus. The scoring was made between 14 and 21 days after infection depending on the development of the symptoms. The inoculation with scab conidia was repeated in the case of the spray inoculation when the pruned trees grew new shoots.

To characterise the leaf symptoms the following grading system was used:

1. macroscopically
   - 0 = no visible symptoms;
   - 1 = crinckled leaves, no sporulation;
   - 2 = slight chlorosis or necrosis;
   - 3 = strong chlorosis or necrosis or strongly crinckled leaves;
   - 4 = lesions with slight sporulation;
   - 5 = abundant sporulation.
2. microscopically
   - 1 = germinated conidia with visible appressorium;
   - 2 = slightly developed stroma visible;
   - 3 = strongly developed stroma visible;
   - 4 = few conidiophores visible;
   - 5 = many conidiophores visible;
   - 6 = slight sporulation;
   - 7 = abundant sporulation.

The preparation of the leaves for the microscopic examination was made according to Stadler (1988) with slight modifications. For the microscopic study only leaves from the pipette drop inoculation were used.

Results

Figure 1 shows the average macroscopic scab incidence on the nineteen varieties and selections examined. The varieties tested covered the whole range from highly susceptible to completely resistant. The commercial varieties such as Elstar and Golden Delicious were highly susceptible. Old varieties such as Orléans Reinette, Oberrieder Glanzreinette and Fraurotacher as well as the modern variety Delbard Jubilé and the hybrid FAW 5878 showed a moderate level of resistance. Angold and HL 75–27 both carry a polygenic Antonovka resistance. They showed a good level of resistance. The same level was also found in Discovery which is known for its polygenic resistance (Kellerhals et al, 1993). As expected, the highest level of resistance was carried in the Vf resistant cultivar Florina and the diffential hosts for the scab races 2, 3, 4 and 5. Among the differental hosts, 9AR2T196 was the least resistant. 9AR2T196 is the differental host for race 5 and carries Vm resistance.

On the microscopic level (Figure 2) a similar sequence with slight modifications was found. The

Table 1. Varieties included in the glasshouse trial 1993

Variety	Parentage	Resistance	Origin
Golden Delicious	chance seedling	–	USA
Elstar	Golden Delicious × Ingrid Marie	–	Netherlands, 1972
Discovery	Worcester Pearmain × Beauty of Bath ?	–	England, 1964
Delbard Jubilé	Golden Delicious × Lundbytrop	–	France, 1963
Orléans Reinette	unknown	–	France, 1621
Fraurotacher	unknown	–	Switzerland, 1745
Oberrieder Glanzreinette	unknown	–	Switzerland, 1820
Florina	multiple cross	Vf	France, 1977
FAW 2151	James Grieve × Lobo	–	Switzerland
Marina	Kidds Orange × Idared	–	Switzerland, 1993
Iduna	Golden Delicious × Glockenapfel	–	Switzerland, 1993
FAW 5878	(Idared × Maigold) × Elstar	–	Switzerland
Pinova	Clivia × Golden Delicious	–	Germany, 1986
Angold	Antonovka o.p. × Golden Delicious	Antonovka polygenic	Czech Republic
HL 75–27	Goldspur × Antonovka	Antonovka polygenic	Czech Republic
TSR 34T15	McIntosh × (R # 12740-7A × Red Delicious)	Vr	USA
TSR 8T24	Multiple cross	multiple	USA
TSR 33T239	Multiple cross	multiple	USA
9AR 2T196	McIntosh × (Wolf River × M. micromalus)	Vm	USA

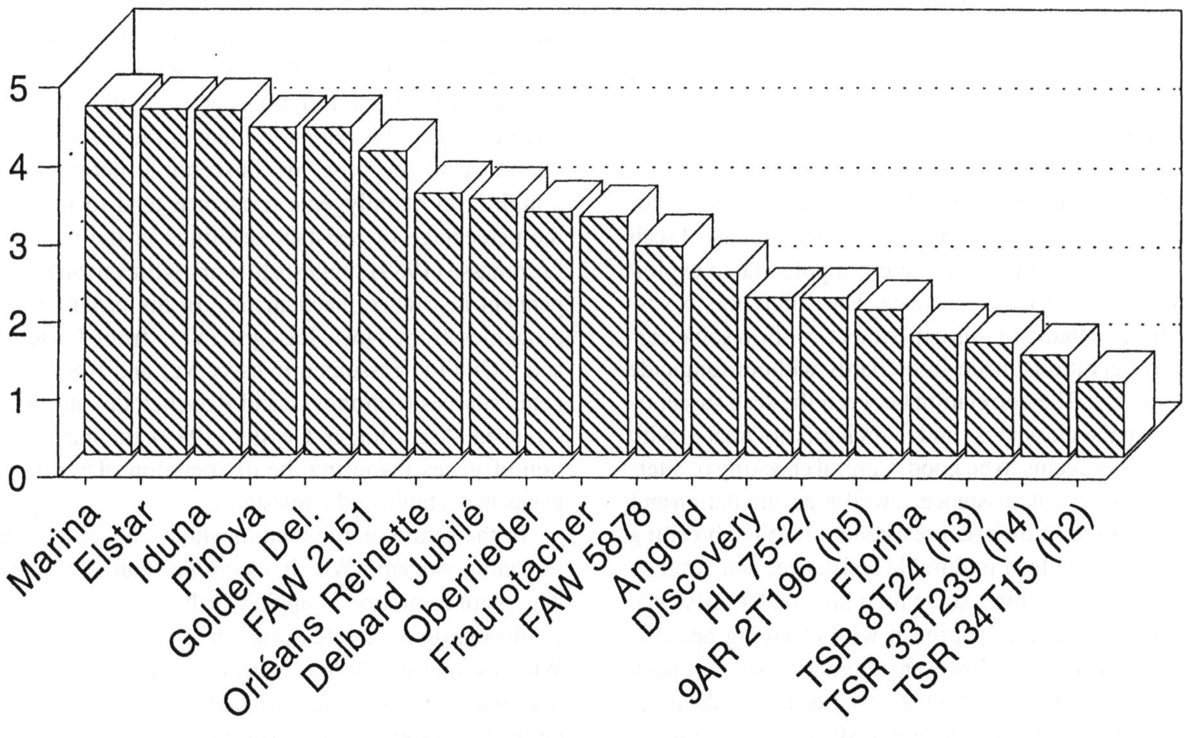

Fig. 1. Macroscopic scab development after artificial inoculation in the glasshouse. Scab development: 0 = no symptoms, 5 = abundant sporulation.

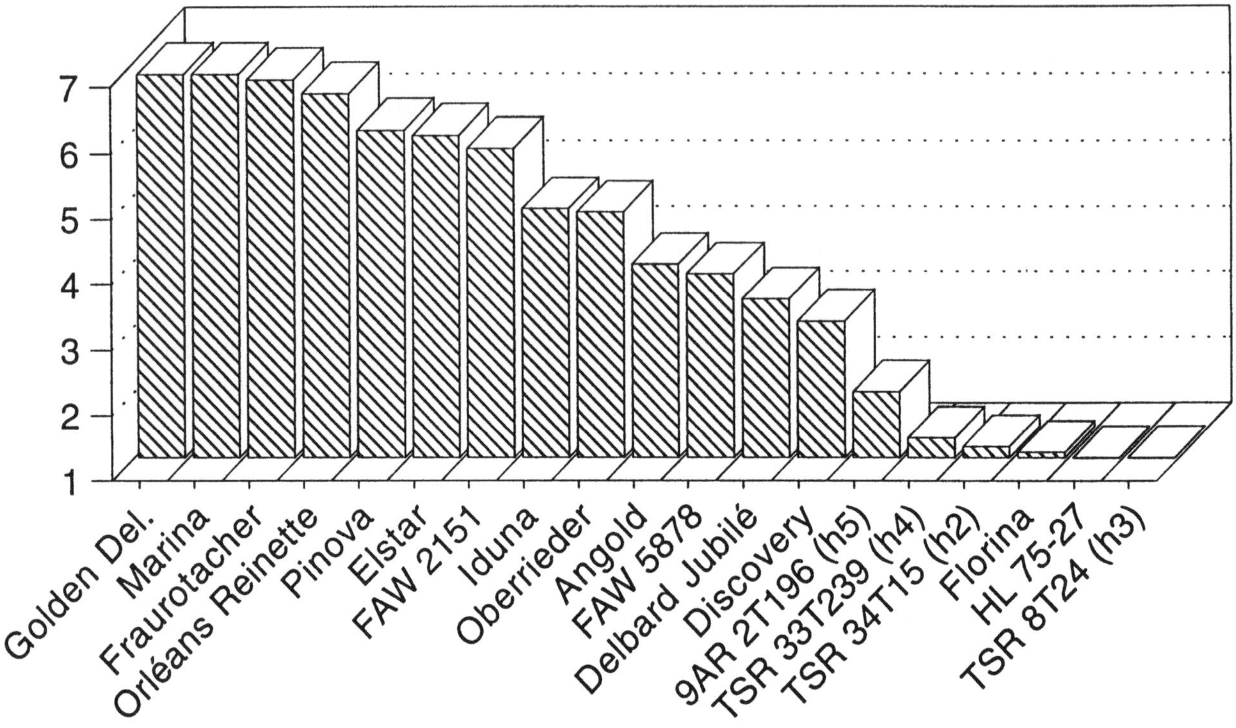

Fig. 2. Microscopic scab development after artificial inoculation in the glasshouse. Scab development: 1 = no stroma, 7 = abundant sporulation.

microscopic study confirmed the macroscopic evidence.

Discussion

Our glasshouse experiment with various apple genotypes carrying a wide range of different sources of scab resistance illustrates that there is a potential to exploit new sources of scab resistance. The differences in susceptibility found in this experiment confirm some earlier observations made in orchards without fungicide treatments. The correlation between the scab susceptibility in the glasshouse and in orchards not treated with fungicides seems to be good. Careful checking of alternative sources of resistance as well as accumulation and combination of different resistances through breeding efforts should lead to durable resistance. Our strategy should change from a static focussing on Vf resistance towards a dynamic approach with changing and differing resistances. However, alternative sources such as Antonovka or other local varieties may reduce the immediate output of resistant seedlings in the progeny (Fischer, 1989). Moreover, Brown (1975) and Virscek-Marn et al. (1994), consider the polygenic type of scab resistance unreliable. A decrease of the initially high level of resistance was observed under orchard conditions. In cross combinations with the Vf resistant cultivar Florina, Kellerhals et al. (1993), showed, that the susceptible parent influences the level of scab resistance in the progeny. Susceptible parents with a high level of polygenic resistance tend to increase the percentage of resistant seedlings in the progeny. Lespinasse (1989) suggests to work with various independent resistance mechanisms. In his programmes, progenies from Nova Easygro (Vr), from CCR1T8 and CCR1T11 (assumed to carry Va) and from polygenic sources (Antonovka and Rouchetode) are being screened. According to Gessler (1989), the incorporation of genes favouring the dissociation of virulence genes may stabilize the resistance.

In the future more attention should be paid to the screening system. Which inoculum source is used, which concentration is applied, which environmental conditions do we create during the screening process? Which symptoms do we consider as susceptible reaction and which ones as resistant? A fixed threshold might be unsatisfactory, especially when alternative sources of resistance are included.

To fully exploit the suggested strategies, identification of resistances through markers and not only phenotypically, will be nesessary. Current efforts to develop scab resistance gene markers could be a tremendous help to select the desired genotypes and to characterise precisely the type of resistance present in a specific genotype (Gianfranceschi et al., 1994). Once these techniques are routinely applicable it could be interesting to start again backcross series with wild species such as *Malus floribunda* 821, *Malus baccata*, and others. In each backcross generation the genetics of the inherited resistance should be carefully analysed to avoid erosion of the resistance. In the case of other diseases such as mildew, canker and fireblight a lower level of resistance compared to scab might be tolerated. Therefore it should be possible to rely on polygenic or oligogenic resistances. Molecular genetics may offer possibilities to improve our understanding of these resistances as well.

Genetically determined disease resistance should also be considered as one component of the orchard ecosystem. Durability of resistance is greatly influenced by its management and one of the most convincing systems is based on the concept of mixtures (Gessler & Blaise, 1994; Fischer et al., 1994). The host-pathogen interaction is not a static one. Therefore it is necessary to gradually develop the breeding approaches for durable disease resistance. To check the quality of resistance it is important to test resistance donors and advanced selections under high natural inoculum pressure at several locations around the world.

References

Blaise, Ph. & C. Gessler, 1994. Cultivar mixtures in apple orchards as a mean to control apple scab? Norwegian Journal of Agricultural Sciences (in press).

Blazek J. & F. Paprstein, 1994. Breeding apples for scab tolerance at Holovousy. pp 21–25 (this volume).

Brown, A.G., 1975. Apples. In: J. Janick & J.N. Moore (Eds). Advances in Fruit Breeding, pp. 3–37. Purdue University Press, West Lafayette.

Fischer, C., 1989. Untersuchungen zur Vererbung der Resistenz gegen Schorf (*Venturia inaequalis* (Cooke) Aderh.) in Apfelnachkommenschaften. Arch Gartenbau 37: 23–30.

Fischer, C., V.F. Bukartschuk, A.A. Bondarenko & E.S. Artamanova, 1983. Erste Ergebnisse zur Stabilität der Schorfresistenz beim Apfel unter verschiedenen ökologischen Bedingungen in der UdSSR und DDR. Arch Gartenbau 31: 263–264.

Fischer, C., A. Bondarenko & E. Artamonova, 1994. Results of the stability of scab resistance in the apple breeding. pp. 83–87 (this volume).

Gessler, C., 1989. Genetics of the interaction *Venturia inaequalis – Malus*: The conflict between theory and reality. OILB Working Group "Integrated Control of Pome Fruit Diseases" Vol. II, WPRS Bulletin XII/6: 168–190.

Gessler, C. & Ph. Blaise, 1994. Differential resistance in apple against scab and its use in breeding and in orchard planting strategies to control the disease. pp. 101–106 (this volume).

Gessler, C. & D. Stumm, 1984. Infection and stroma formation by *Venturia inaequalis* on apple leaves with different degree of susceptibility to scab. Phytopath. Z. 110: 119–126.

Gianfranceschi L., J.M. McDermott, N. Seglias, B. Koller, M. Kellerhals & C. Gessler, 1994. Towards a marker assisted breeding for resistance against apple scab. pp. 293–296 (this volume).

Kellerhals, M., A. Fouillet & Y. Lespinasse, 1993. Effect of scab and the susceptible parent on resistance to apple scab (*Venturia inaequalis)* in cross progenies with the scab resistant cv. 'Florina'. Agronomie 13: 631–636.

Krüger, J., 1988. Beständigkeit der Schorfresistenz aus *Malus floribunda* 821 auf dem Versuchsfeld der Bundesforschungsanstalt für gartenbauliche Pflanzenzüchtung in Ahrensburg. Erwerbsobstbau 30: 52.

Krüger, J., 1989. Scab resistance of apple cultivars, selections and progenies with the Vf gene. OILB Working Group "Integrated Control of Pome Fruit Diseases" Vol. II, WPRS Bulletin XII/6: 161–167.

Krüger, J., 1991. Schorfbefall von Nachkommen aus Kreuzungen mit der schorfresistenten Apfelsorte 'Prima'. J. Plant Dis. Protect. 98: 73–76.

Lespinasse, Y., 1989. Breeding pome fruits with stable resistance to diseases. 3. Genes, resistance mechanisms, present work and prospects. OILB Working Group "Integrated Control of Pome Fruit Diseases" Vol. II, WPRS Bulletin XII/6: 100–115.

Parisi, L., Y. Lespinasse, J. Guillaumes & J. Krüger, 1993. A new race of *Venturia inaequalis* virulent to apples with resistance due to the Vf gene. Phytopathology 83: 533–537.

Silbereisen, R., 1989. Experiences at Bavendorf (FRG) with scab resistant apple cultivars. OILB Working Group "Integrated Control of Pome Fruit Diseases" Vol. II, WPRS Bulletin XII/6: 148–160.

Stadler, B., 1988. Quantitative Untersuchungen der varietalen und ontogenetischen Resistenz der Apfelblätter von Golden Delicious, Liberty und *Malus floribunda* gegenüber *Venturia inaequalis*. Diss. ETH Nr. 8685.

Van der Scheer, H.A.Th., 1989. Susceptibility of apple cultivars and selections to scab and powdery mildew in the Netherlands. OILB Working Group "Integrated Control of Pome Fruit Diseases" Vol. II, WPRS Bulletin XII/6: 205–211.

Virscek-Marn, M., F. Stampar, J. Smole & A. Solar, 1994. Susceptibility of some apple cultivars to apple scab (*Venturia inaequalis/Cooke/Aderh.)* in Slovenia. pp. 31–34 (this volume).

Williams, E.B. & A.G. Brown, 1968. A new physiological race of *Venturia inaequalis*, incitant of apple scab. Plant Dis. Rep. 52: 799–801.

Williams, E.B. & J. Kuç, 1969. Resistance in *Malus* to *Venturia inaequalis*. Ann. Rev. Phytopathol. 7: 223–246.

H. Schmidt and M. Kellerhals (Eds.), Progress in Temperate Fruit Breeding, 99–104.

Differential resistance in apple against scab and its use in breeding and in orchard planting strategies to control the disease

Cesare Gessler & Philippe Blaise
Federal Institut of Technology, Inst. of Plant Sciences, Phytomedicine, Pathology, 8092 ETH-Zürich, Switzerland

Key words: Ephemeral resistance, apple scab, breeding, cultivar mixture, *Venturia inaequalis*, *Malus*

Summary

Early and recent studies demonstrate in many commercial cultivars the presence of functional different resistances overcome by specific races of the scab fungus. It is postulated therefore that the composition of the pathogen population is due to selection by host resistance genes. Those ephemeral resistances are of no effect on the scab populations in homogeneous orchards since they are composed of cultivar adapted races. On the other hand, this information could be used to develop new cultivars which carry several functionally different resistances. The consequences for breeding strategies and for orchard planting to minimise fungicide input are discussed as well as the use of genomic markers to help in adopting the proposed strategies. Secondly, strategies to control apple scab by planting orchards with a mixture of selected cultivars carrying different functional resistances can be developed. To study the effect of cultivar mixtures in apple orchards, results of a simulation model are presented and discussed.

Introduction

Selecting plants best adapted to survive stress situations and to produce the desired crop has been the long term motor of agricultural development. Without new selections, progress in cultural, harvest and storage practices, and changes in consumer habits could not have taken place. Inadvertent selection of plants resistant to pests and diseases took place as these lowered yield and quality of susceptible individuals. This led to varieties relatively resistant or tolerant to diseases and pests under the agricultural systems used. Changes in agricultural systems and in market requirements have often led to increases in losses due to pests and disease, because this selection was too slow to compensate for the losses due to the cultural changes. Rapid changes in the phytosanitary situation could not be compensated by selection of more robust varieties. Dramatic examples of rapid changes are the introductions of diseases unknown to an area. Late blight (*Phytophthora infestans*) on potato and powdery mildew (*Uncinula necator*) and downy mildew (*Plasmopara viticola*) on grape caused great damage in Europe which could not be mitigated by the choice of more tolerant or resistant varieties. The solution was found in fungicides (sulphur and copper). With the availability of fungicides with their excellent efficacy, pathologic stress situations could be avoided even for susceptible varieties. Disease resistance became a secondary argument for the choice of a specific variety and selection and breeding could concentrate on other goals. This was specially true for crops of high value and with specific quality requirements. Few breeding programs incorporated early selection procedures for disease resistance or at least tried to avoid release of highly susceptible cultivars. With the increasing problems posed by the pesticides, e. g. resistance of the target organism against the pesticide, upsurge of secondary pests to primary, reduction of the biodiversity, possible health hazard to consumers and applicant, breeding for disease resistance became a primary goal again (Alston 1989, Kellerhals 1989a,b), even without any of the particular problems being acute. Today, due to consumer and environmental lobby concerns, even short term oriented policy makers turn their attention towards replacement of pesticides by the introduction of environmental safer and sustainable technologies. These often are more costly and bear more risks, and therefore can only be adopted in countries with high living standards. One exception is the use, if available, of resistant cultivars. On the other hand resistance breeding itself can be costly, retard success of the breeders in other goals and, even if obvious, the benefit is not considered in its true extent when resources have to be allocated. A major problem in breeding resistant cultivars of perennial plants is the requirement that the resistance is durable. Johnston (1981) defines 'durable' disease resistance as resistance that lasts at least the period during which the resistant cultivar is widely cultivated and successful on the market (Hogenboom 1993). With apple this period, as the example of Golden Delicious demonstrates, can be extremely long. On the other hand the introduction of resistance should be accomplished in one or a few generations, should be simply inherited, easily selected for and not lost in backcrossing. Genetic control of resistance in plants is thought to be of two basic kinds. Resistance can be displayed against all isolates of a particular pathogen. Often this resistance is not complete, as it allows some pathogen development and is assumed to be controlled by many genes, but should be stable and operating under all conditions (horizontal or general resistance). The second type of resistance in most cases will confer a strong resistance against some isolates of the pathogen, but none toward others (vertical or differential resistance). This type of resistance has been found to be controlled by a single gene (often called 'major gene' or R-gene). Correspondingly the pathogen carries a single gene which determines whether or not it is virulent to a particular host (gene-for-gene-theory (Flor 1942)). A host population uniformly carrying such a vertical resistance will inevitably lead to the selection of a pathogen population carrying the corresponding virulence, and therefore the effect of the resistance gene will disappear (ephemeral resistance). Genes for avirulence have been cloned from some plant pathogenic bacteria (Gabriel et al. 1986, Kobayashi et al. 1989), but in fungi it is still disputed whether genes code for virulence or for avirulence in spite of the increase of papers in favour of avirulence genes (e.g., Silué et al. 1993).

Resistances against the major disease apple scab were found in wild *Malus* sources. Unfortunately the main resistance Vf, on which many pathologists and breeders hope relayed, recently was shown to be overcome by particular isolates of *Venturia inaequalis* (Parisi et al. 1993, Roberts et al. 1994), therefore to be ephemeral. Ephemeral resistances are assumed to be present in old apple varieties (Sierotzki et al. 1994). Also the data presented in older literature suggests a gene-for-gene-relationship, with many functional different resistance genes (Gessler 1994). This raises the question whether the use of ephemeral resistance genes is not a Sisyphus type of effort, with the risk of the 'Vertifolia' effect (Vanderplank 1968, 1982) in which a variety with a low general, ephemeral (vertical) resistance is selected, which is excessively susceptible to a race able to overcome the vertical resistance. Moreover the manifestation of 'per se ephemeral' resistances are related directly to the current status of the pathogen population at the test sites and cannot represent expected performance against the whole range of pathogen variability at present or in the future (Wolfe & Gessler 1992).

Durability of resistance on the other hand is greatly influenced by its management (Wolfe 1983) and may also include resistances that have already been overcome (Hogenboom 1993). Thus, by combining breeding strategies and orchard planning, the use of the many ephemeral resistances known in *Malus* can still be advantageous.

Ephemeral resistances in commercial cultivars

Since the introduction of resistances from wild *Malus* species poses several problems such as the need of several backcrosses to eliminate undesired characteristics, the use of established cultivars is desirable. On the other hand, except for the new scab resistant cultivars developed (Crosby et al. 1992) mostly carrying the Vf-resistance, no commercial cultivars are used for that purpose as the ephemeral resistances are not recognised. Recently we identified functional different resistances in Boskoop, Spartan and James Grieve (Sierotzki et al. 1994, Gessler et al. 1993). These resistances are functional against several scab isolates obtained from the other cultivars. This leads to the hypothesis that the three cultivars each has a different resistance which impedes the normal development of a scab epidemic by scab strains not carrying the virulence corresponding to the resistance of that particular cultivar. More recently further cultivars were tested. Conidias were collected from 10 individual lesions per tree on trees of 6 cultivars. With an inoculum consisting of equal parts of the 10 lesions, potted trees of the same cultivars where infected. The conidias from the resulting lesions were then inoculated on 3 trees of all 6 cultivars, with 4 inoculations per tree. After incubation the resulting interactions were classified after the type of lesions formed from 0 (no visible symptom)

Table 1. Symptoms[1] caused on a set of apple cultivars by inoculation of *V. inaequalis* conidia collected from different cultivars. Each inoculum originated from 10 lesions per cultivar and multiplied once on that particular cultivar.

Origin of inoculum	Inoculated host cultivar					
	Maigold	Golden Delicious	Glockenapfel	Spartan	Boskoop	Idared
Maigold	4	4	1	3	1	4
Golden Delicious	4	4	1	4	1	1
Glockenapfel	4	4–(1)	4	2	1	3
Spartan	3	4	1	3	1	4
Boskoop	3	3	1	2	4	4
Idared	3	3	1	3	2	4

[1] Symptoms: 4, obvious and strong sporulating lesions; 3, well sporulating but weak chlorotic lesions; 2, small, chlorotic or weak sporulating flecks; 1, small, chlorotic flecks; 0, no visible symptoms. Deviations were observed and indicated, in parenthesis deviation by a single lesion.

to 4 (enlarging heavily sporulating lesions). The cultivar Maigold was susceptible to all inoculum sources used. Cultivar Golden Delicious was susceptible to all tested sources of inoculum, except that in one out of three trials the source Glockenapfel induced only small necrotic flecks, suggesting that a part of the inoculum was avirulent to Golden Delicious. Further trial with monosporic isolates will furnish the answers. The other cultivars showed a differential reaction to the various inoculum sources (Table 1). Simplifying the evaluation into compatible (+) and incompatible (−) (Table 2) results into Glockenapfel, Spartan, Boskoop and Idared each having different resistances. Simultaneously, the various sources of inoculum show specific virulences/avirulences. The Boskoop source is just virulent on Boskoop (it misses therefore the gene for Avirulence or has the specific gene for Virulence which overcomes the resistance of Boskoop, depending on the hypothesis followed) and evidently on Maigold and Golden Delicious. The source Maigold is virulent to Golden Delicious, Spartan and Idared, while avirulent to Glockenapfel and Boskoop and so forth (Table 2). Because some of the original 10 isolates from each source carry more than just the corresponding virulence, they can be regarded as complex races. Further research with monoconidial isolates as well as population studies will clarify the frequency of complex races and the number of virulences/avirulences present. Anyhow, these results clearly indicate that functional different resistances are present not only in Boskoop, Spartan and Gravenstein but also in Idared and Glockenapfel. Careful screening of the most popular cultivars used for breeding may reveal more different resistances.

The use of ephemeral resistance in breeding

Breeders have two principal options in resistance breeding. The first is to breed a set of cultivars each having a functionally different resistance. These unrelated cultivars, each with one or more resistances, planted in an appropriate mixture, may differ so much that adaptation of the fungus is difficult if not impossible. The recognition of functional different resistances during screening of the progeny is possible only through the use of differential races. Such a selection scheme would be very cumbersome. An alternative would be to transfer the genome segment carrying the resistance from one cultivar into the target cultivar by genetic engineering. The target cultivar could be a popular commercial cultivar. By incorporating functionally different resistances, the concept of near-isogenic lines (NIL) could be adopted. Even though this cannot be done yet because the resistance genes are not yet isolated and many obstacles are still present (Newbury 1992), it may be possible in the foreseeable future. However, it is not clear whether this will be acceptable to consumers and moreover, by incorporating the resistances into a single genetic background, we may simplify the problem for the pathogen to overcome the resistances (Wolfe 1993), and thus select a fit superrace (a race able to overcome all mixture resistance components) even in a resistance mixture concept.

Table 2. Interaction between cultivars and pathogen assuming a gene for gene relation (+ = compatible, − = incompatible). Putative plant genes are indicated as R = resistance, r = susceptible, pathogen genes A = avirulence and a = virulence.

Origin of inoculum and putative avirulence genes	Inoculated host cultivar and putative resistance genes					
	Maigold $r_g, r_{ga}, r_s, r_b, r_i$	Golden Delicious $R_g(?), r_{ga}, r_s, r_b, r_i$	Glockenapfel $r_g, R_{ga}, r_s, r_b, r_i$	Spartan $r_g, r_{ga}, R_s, r_b, r_i$	Boskoop $r_g, r_{ga}, r_s, R_b, R_i$	Idared $r_g, r_{ga}, r_s, r_b, R_i$
Maigold (A_{ga}, a_s, A_b, a_i)	+	+	−	+	−	+
Golden Delicious (A_{ga}, a_s, A_b, A_i)	+	+	−	+	−	−
Glockenapfel (a_{ga}, A_s, A_b, a_i)	+	+(−)	+	−	−	+
Spartan (A_{ga}, a_s, A_b, a_i)	+	+	−	+	−	+
Boskoop (A_{ga}, A_s, a_b, a_i)	+	+	−	−	+	+
Idared (A_{ga}, a_s, A_b, a_i)	+	+	−	+	−	+

The second option is to breed varieties that contain the greatest possible number of resistance genes (pyramiding). The presence or absence of each gene could be recognised by serial testing of the progenies of carefully selected parents with inoculum of races carrying the appropriate virulences. Alternatively, the races could be mixed in the inoculum to reach the same conclusions in fewer tests except that genes giving incomplete resistance may be missed. This system could again be used only for resistances known to be ephemeral, i.e., if we have the corresponding virulent races which would allow identification of a second (or third) resistance. However, selection for super-races may be stronger than in the NIL concept, particularly for a parasite such as *V. inaequalis* which reproduces sexually each year.

There has been much discussion on the advantages and disadvantages of either strategy (Wolfe 1993), mostly by experts on annual plants. The choice of strategy of multi-cultivar mixture, NIL or pyramiding resistance genes may not be relevant to the apple system because the constraints on success may be of a completely different nature.

Screening for functional different resistances

Screening for scab resistance is mostly done on seedlings whilst in trays in the glasshouse. Seedlings showing a desired type of reaction to an inoculation by a standard conidial suspension are selected. Later the resistance is verified in the field. However, such a system does not allow the recognition of vertical resistances to the inoculum used, unless the corresponding virulence is present at a reasonable frequency. On the other hand if the corresponding virulence is not present it will hide any other resistance. It is not possible to select individuals with more than one functional resistance in such a screening system. As mentioned above a system of serial testing with inoculum of defined virulences could be used. If the corresponding virulences are not available, progeny analysis would have to be made of the individuals suspected to carry more than one resistance.

A more elegant approach is to identify or mark the genome segment carrying the resistance information and to select progeny by the presence of such DNA-markers. The success of such a method is based on the assumption that resistances at different loci are also functionally different, which most probably is the case. Marker-assisted breeding is thought to have a high feasibility and cost-benefit ratio.

Markers can be developed for several vertical resistance genes such as the ephemeral resistance genes of the commercial cultivars (Williams et al. 1990). Efforts to develop scab resistance gene markers in apple are underway (Gianfranceschi et al. 1994), Weeden (USA), King (GB), Gardiner (NZ). All at Eucarpia meeting.

It would be possible to provide each old and new cultivar with a key denoting its resistances which not only facilitate the choice of parents in breeding, but also the choice of optimum partners for a cultivar mixture at a particular site. This information is fundamental for the use of the concept of apple cultivar mixtures as proposed by Blaise & Gessler (1994).

The use of cultivars with ephemeral resistances

Durability of resistance is greatly influenced by its management and one of the most convincing systems is based on the concept of apple cultivar mixtures and may also include resistances that have already been overcome (Wolfe 1985, 1993). Blaise (Blaise & Gessler 1994) developed a simulation model to test the effect of cultivar mixtures in reducing scab epidemics. Under the assumption that no complex races are present, that the initial inoculum is composed of randomly distributed ascospores, and that the various frequencies are proportional to the frequencies of the cultivars, the model shows for example that an orchard system with three cultivars carrying different resistances reduces the numbers of lesion by 65% after six generations when planted in alternate rows, and by 79% in a within-row mixture. Even if simulations never can account for all factors present in an orchard, this still indicates that ephemeral resistances can be employed by carefully choosing the appropriate cultivars and planting strategies.

Conclusions

The recognition of the presence of ephemeral resistances, which are of no practical value in monoculture, in popular commercial cultivars suggests that by combining them into one cultivar or in several cultivars, together with appropriate planting strategies, resistance breeding can contribute significantly to the reduction of the infection pressure by apple scab. Screening techniques for scab resistance would have to be changed accordingly. Resistance gene markers would facilitate such a breeding strategy or in some cases make it feasible. It may not replace the fungicide use under today's quality requirements, but contribute significantly to the requirements of Integrated Production, which postulates the use of all practical means to reduce pesticide use (El Titi et al. 1993).

Acknowledgements

Plant material was kindly provided by the Swiss Federal Research Station Wädenswil (M. Kellerhals). The first author acknowledges the support by the Swiss National Foundation for Scientific Research grants 31–29928.90 and 31–36271.92, grant NF-Biotechnology Module 6 5002–034614, Swiss federal office for education and science grant no. A6010.

References

Alston, F.H., 1989. Breeding pome fruits with stable resistance to diseases. In: C. Gessler, D. Butt & B. Koller (Eds) Integrated Control of Pome Fruit Diseases II. OILB-WPRS Bulletin XII/6: 90–99.

Blaise, Ph. & C. Gessler, 1994. Cultivar mixtures in apple orchards as a mean to control apple scab? Norwegian Journal of Agricultural Sciences (in press).

Crosby, J.A., J. Janick, P.C. Pecknold, S.S. Korban, P.A. O'Connor, S.M. Ries, J. Goffreda & A. Voordeckers, 1992. Breeding apples for scab resistance: 1945–1990. Fruit Varieties Journal 46: 145–166.

El Titi, A, E.F. Boller & J.P. Gendrier, 1993. Integrated production. Principles and technical guidelines. IOBC/WPRS Bulletin 16: 97 pp.

Flor, H., 1942. Inheritance of pathogenicity in *Melampsora lini*. Phytopathology 32: 653–669.

Gabriel, D.W., A. Burges & G.R. Lazo, 1986. Gene for gene interaction of five cloned avirulence genes from *Xanthomonas campestris* pv. *malvacearum* with specific resistance genes in cotton. Proceedings of the National Academy of Science 83: 6415–6419.

Gessler, C., 1994. Biology and biotechnology in strategies to control apple scab. Norwegian Journal of Agricultural Sciences (in press).

Gessler, C., M. Eggenschwiler & H. Sierotzki, 1993. Vertikale Resistanz gegen Schorf in anfälligen Apfelsorten. Schweiz. Landw. Fo. 32: 401–410.

Gessler, C., M. Eggenschwiler, H. Sierotzki & B. Koller, 1993. La resistenza di varietà di meli ritenuti suscettibili contro la ticchiolatura. Frutticoltura, Bologna 9: 49–54.

Gianfranceschi, L., J.M. McDermott, N. Deglias, B. Koller, M. Kellerhals & C. Gessler, 1994. Toward a marker assisted breeding for resistance against apple scab. pp. 293–296 (this volume).

Hogenboom, N.G., 1993. Economic importance of breeding for disease resistance. In: Th. Jacobs & J.E. Parleviet (Eds) Durability of Disease Resistance, pp. 5–10. Kluwer Academic Publishers, Dordrecht

Johnston, R., 1981. Durable resistance: definition of, genetic control, and attainment in plant breeding. Phytopathology 71: 567–568.

Kellerhals, M., 1989a. Breeding of pome fruits with stable resistance to diseases. In: C. Gessler, D. Butt & B. Koller (Eds) Integrated Control of Pome Fruit Diseases II. OILB-WPRS Bulletin XII/6: 116–129.

Kellerhals, M., 1989b. Breeding disease resistant apple cultivars in Switzerland. In: C. Gessler, D. Butt & B. Koller (Eds) Integrated Control of Pome Fruit Diseases II. OILB-WPRS Bulletin XII/6: 130–136.

Kobayashi, D.Y., S.J. Tamaki & N.T. Keen, 1989. Cloned avirulence genes from the tomato pathogen *Pseudomonas syringae* pv tomato confer cultivar specificity on soybeans. Proceedings of the National Academy of Science 86: 8467–8471.

Newbury, H.J. 1992. Fungal resistance: The isolation of a plant R gene by transposon tagging. In: A.M.R. Gatehouse, V.A. Hilder & D. Boulter (Eds.): pages 109–134. CAB Biotechnology in agriculture no 7, 266 pages.

Parisi, L., Y. Lespinasse, J. Guillaumes & J. Krüger, 1993. A new race of *Venturia inaequalis* virulent to apples with resistance due to the Vf gene. Phytopathology 83: 533–537.

Roberts, A.L., F.H. Alston & I.R. Crute, 1994. Apple scab resistance from *Malus floribunda* 821 (Vf) is rendered ineffective by isolates of Venturia inaequalis from *Malus floribunda*. Norwegian Journal of Agricultural Sciences (in press).

Sierotzki, H., M. Eggenschwiler, J. McDermott & C. Gessler, 1994. Specific virulence of isolates of *Venturia inaequalis* on 'susceptible' apple cultivars. Norwegian Journal of Agricultural Sciences (in press).

Silué, D., D. Tharreau & J.L. Notteghem, 1993. Identification of *Magnaporthe grisea* avirulence genes to seven rice cultivars. Abstracts 6th International Congress of Plant Pathology, 1993, Montreal: 198.

Vanderplank, J.E., 1968. Disease Resistance in Plants. Academic Press, New York.

Vanderplank, J.E., 1982. Host-Pathogen Interaction in Plant Disease. Academic Press, New York.

Williams, E.B. & J. Kuc, 1969. Resistance in Malus to *Venturia inaequalis*. Ann. Rev. Phytopathology 7: 223–246.

Williams, J.G.K., A.R. Kubelik, K.J. Livak, J.M. Rafalski & S.V. Tingey, 1990. DNA polymorphisms amplified by arbitrary primers are useful as genetic markers. Nucleic Acid Research 18: 6531–6535.

Wolfe, M.S., 1983. Genetic strategies and their value in disease control. In: T. Kommedahl & Ph. Williams (Eds) Challenging Problems in Plant Health. pp. 461–473. The Americ. Phytopathol. Soc., St. Paul.

Wolfe, M.S., 1985. The current status and prospects of multiline cultivars and variety mixtures for disease resistance. Ann. Rev. Phytopathology 23: 251–273.

Wolfe, M.S., 1993. Can the strategic use of disease resistant hosts protect their inherent durability? In: Th. Jacobs & J.E. Parlevliet (Eds) Durability of Disease Resistance. pp. 83–96. Kluwer Academic Publishers, Dordrecht.

Wolfe, M.S. & C. Gessler, 1992. Resistance genes in breeding: epidemiology. In: T. Boller & F. Meins (Eds) Plant Gene Research. Genes Involved in Plant Defence, pp. 3–23. Springer Verlag, Wien, New York.

H. Schmidt and M. Kellerhals (Eds.), Progress in Temperate Fruit Breeding, 105–106.

Apple scab resistance and durability. New races and strategies for the future

Yves Lespinasse
Institut National de la Recherche Agronomique, Station d'Amélioration des Espèces Fruitières et Ornementales B.P.57, F-49071 Beaucouze Cedex, France

Several scab resistant apple varieties were released these last years and their number (48 according to Crosby et al., 1992) will increase in the coming years. Thirty seven carry Vf resistance from *M. floribunda* 821. There now exists a wide range of genotypes containing Vf resistance ranging in maturity from 75 days to 180 days or longer after flowering, with large fruits, crisp flesh, good storage behaviour, and a wide range in flavor and skin color.

The possibility of development of races of the fungus that would overcome an individual source of resistance was recognized very early in the programme. Efforts were undertaken to find and characterize additional sources of resistance (Table 1). Five races of *V. inaequalis* with particular virulences to chosen differential *Malus* species and selections have been identified; race 5 overcome Vm (Table 2).

Recently, in a cooperative work with J. Krüger (Ahrensburg, Germany) we identified at Angers a new race named race 6 which overcomes the Vf resistance carried by the recent selections (Parisi et al., 1993). More recently T. Roberts found a scab strain which is able to overcome the Vf resistance in the original *M. floribunda* 821 (Roberts et al., 1994).

It is therefore very urgent to discuss these recent results to set up experiments to examine the genetics, to draw conclusions for the durability of resistances in the orchard and to propose new breeding strategies.

Those results were obtained in growth chambers from strains found in the field. At least, there is a big question about the ability of these strains to compete with the standard strains representing race 1. This is another work to do involving epidemiology; this point is not very well known. As far as I know, the people who reported scab on Vf derivatives are:

Table 1. Apple species and varieties used as sources of resistance to scab

I. Monogenic		II. Polygenic
Malus floribunda 821	V_f	*Malus baccata*
Antonovka P.I. 172 623	V_a	*Malus sargenti*
Malus pumila 127 40-7A	V_r	*Malus sieboldii*
Hansen's baccata 2	V_b	*Malus toringo*
Malus baccata jackii	V_{bj}	Old European varieties
Malus micromalus 245.38		
	V_m	race 5
Malus atrosanguinea 804		

- C. Fischer (1983) in a joint work with Bukartschuk & Bondarenko in Dresden & Kichinev; C. Fischer found also scab on 'Liberty' after artificial inoculation in the greenhouse as early as 1981.
- Van der Scheer (1989) in Holland also reported scab on fruits of the variety 'Gavin'; Blommers (1983) on some scab resistant hybrids.
- Silbereisen (1989) in Germany reported scab on leaves of certain varieties such as 'Prima', 'Florina' and 'Jonafree'.
- Gessler (personal communication) reported scab symptoms on 'Novamac' under favourable conditions for scab in southern Switzerland, but not consistent from year to year.
- Bergamini (Trento, Italy) found scab on 'Liberty' and 'Coop17' at Forli.

During the roundtable no other report of scab occurrence on resistant hybrids was outlined. The debate put emphasis on the 3 following points:

Table 2. Differential *Malus* selections for the named races of *Venturia inaequalis*

Pathotype	Pathodeme						
	Dolgo	Geneva	R127-40-7A		*M. micro.*	Florina	Golden
			S_i	S'_i	V_M	V_F	
1	r	r	r	r	r	r	S
2	S	S(*)	S	r	r	r	S
3	r	S	r	r	r	r	S
4	r	r	r	S	r	r	S
5	r	r	r	r	S	r	S
6	r	r	r	r	r	S	S

r = resistant;
S = susceptible;
S(*) = Necrotic lesions in which fungus sporulates abundantly in moist atmosphere;
Si = certain segregates of R127-40-7A;
S'i = other segregates of R127-40-7A.

1. It is very important to survey the places where the virulent strains were observed: Ahrensburg & East Malling are good candidates; there are unique situations for epidemiology works.
2. We have to address some recommendations to the growers which grow Vf resistant varieties. The best might be to spray when inoculum is very high to limit the spread of virulent strains.
3. The breeders have to promote new strategies involving independent gene associations but also, more than in the past, the partial resistance (Jacobs & Parlevliet, 1993).

References

Blommers, L., 1983. Apple scab in mixed stands: varietal susceptibility and field resistance. Bulletin OILB/SROP-VI/4: 67–76.

Fischer, C., V. Bukartschuk, A. Bondarenko & E. Artamonova, 1983. Erste Ergebnisse zur Stabilität der Schorfresistenz beim Apfel unter verschiedenen ökologischen Bedingungen in der UdSSR und DDR-vorläufige Mitteilung. Archiv Gartenbau, 31: 236–264.

Silbereisen, R., 1989. Experiences at Bavendorf (FRG) with scab resistant apple cultivars. Bulletin OILB/SROP-XII/6: 148–160.

Van der Scheer, H.A.Th., 1989. Susceptibility of apple cultivars and selections to scab and powdery mildew in the Netherlands. Bulletin OILB/SROP-XII/6: 205–211.

Crosby, J.A., J. Janick, P.C. Pecknold, S.S. Korban, P.A. O'Connor, S.M. Ries, J. Goffreda & A. Voordeckers, 1992. Breeding apples for scab resistance: 1945–1990. Fruit Var. J., 46(3): 145–166.

Jacobs, Th. & J.E. Parlevliet, 1993. Durability of disease resistance. Kluwer Academic Publishers, 375 pp.

Parisi, L., Y. Lespinasse, J. Guillaumes & J. Krüger, 1993. A new race of *Venturia inaequalis* virulent to apples with resistance due to the Vf gene. Phytopathology 83: 533–537.

Roberts, A.L., I.R. Crute & F.H. Alston, 1993. Apple scab resistance from *Malus floribunda* 821 (Vf) is rendered ineffective by isolates of *Venturia inaequalis* from *M. floribunda*. Norwegian Journal of Agriculture Sciences (in press)

CHAPTER 3

Breeding methods and genetics

H. Schmidt and M. Kellerhals (Eds.), Progress in Temperate Fruit Breeding, 109–116.

The Apple Breeding Program at the University of Bologna

S. Sansavini & M. Ventura
Dipartimento di Colture Arboree (DCA), Universitá di Bologna, Bologna, Italia

Key words: Apple, scab resistance, mildew resistance, spur-compact growth habit inheritance, early selection, *Venturia inaequalis*, *Podosphaera leucotricha*, *Phytophthora cactorum*

Summary

The Bologna apple breeding program was started in 1981 to develop dessert varieties with scab resistance, low mildew susceptibility, compact and spur habit, high fruit quality of enhanced shelf-life and, to a limited extent, the development of compact crab-apple pollinators. The last 12 years have produced 23,000 seedlings from 248 crosses and part of the progeny is in stage-2 selection. While the Vf scab resistance gene has thus far shown an inheritance of about 50%, as expected, the most notable finding is the higher than expected resistance registered by crossing parents carrying both the Vf and Va genes. These polygenic offspring provide greater assurance against new virulent races of *Venturia inaequalis* and the breakdown of the Vf gene. The spur and compact traits have often proved to be associated while the dwarf character seems to be of more complex inheritance. In addition, *in vitro* methods using both cultural filtrate and partially purified phytotoxins from *Phytophthora cactorum* for early screening of resistant rootstocks are showing encouraging results.

Introduction

The University of Bologna's 'Apple Breeding Program' began in 1976 and its scab-resistant screening project in 1981. These projects were undertaken at a time when Italy's apple industry, among Europe's most important in terms of yield, realised that it faced a loss of competitiveness in the marketplace because of its over-reliance on cultivars of US provenance, the sole exception being the country's Alpine orchards whose varietal and market interests were oriented more towards central and northern Europe.

The apple industry thus redirected its breeding efforts on the one hand to meet national consumer demands by concentrating on uniformly coloured apples of large size and fine-textured flesh that were soft, juicy, sweet and aromatic, and on the other by growing standard cultivars of firm, crisp-textured and juicy flesh with a sugar/acid ratio that was either well-balanced or somewhat more acidic for export. The former include such cultivars as 'Red Delicious', 'Golden Delicious', 'Rome Beauty' and 'Stayman Red' and the latter 'Jonagold', 'Granny Smith', 'Jonathan', 'Gloster 69', and the more acidic 'Elstar', 'Braeburn' and 'Boskoop', which have performed well only in some of northern Italy's growing areas. To these can be added the local varieties of value produced in the orchards of several regions, especially the old cv. 'Annurca' in the southern Campania.

The other key factor stimulating breeding was the need to develop scab-resistant varieties with fruit of high quality. This has led to the import and testing of numerous foreign genotypes, mainly from the US, New Zealand, France, Canada, Japan, the Netherlands, the UK and Germany, for their adaptability, cropping and quality traits (Sansavini 1993). This selecting has been accompanied by the identification and retrieval of about 400 clones of old, presumably native cultivars that over the years had been phased out of the country's orchards.

Crosses and parents

The University of Bologna's 'Apple Breeding Program' is designed to achieve the following objectives:

Table 1. Cross combinations and propagated seedlings, 1976–1992.

	Cross No.	Seedling Total No.	Seedling Still under Indexing No.
A – Scab-resistant (since 1981)			
A1 SR × SS/StH	16	2056	1642
A2 SR × SS/CH	7	638	548
A3 SR × G/CH or G/LSS	14	3583	2908
A4 SR × SR	11	2193	1564
A5 SR o.p.	28	634	528
A6 SR × LMS/StH	2	2554	775
A7 SR × LMS/CH	7	3019	1959
Total A	85	14677	9924
B – Scab-susceptible (since 1976)			
B1 SS/StH × SS/CH	46	1835	1468
B2 SS/CH × SS/CH	8	106	92
B3 SS/SH o.p.	20	355	43
B4 SS/CH o.p.	20	1034	722
B5 SS/StH × SS/StH	19	1507	1387
Total B	113	4837	3712
C – Pollinators (since 1976)			
C1 CA × SS/StH	10	617	373
C2 CA × SR	1	59	59
C3 CA × SS/CH	24	1122	520
C4 CA × LMS/CH	3	73	72
Total C	38	1871	1024
D – Low Mildew Susceptible (since 1981)			
D1 LMS/CH × LMS/CH	2	210	208
D2 LMS/CH × SS/StH	6	960	897
D3 LMS/CH × SS/CH	3	1034	979
D4 MR × SS/CH	1	26	26
Total D	12	2230	2110
Total LMS + MR (A6 + A7 + C4 + D)	24	7876	4916
Total A + B + C + D	248	23615	16770

SR – Scab Resistance;
SS – Scab Susceptibility;
StH – Standard Habit;
CH – Compact or Spur Habit;
G – Germplasm.
LSS – Low Scab Susceptibility;
CA – Pollinator (Crab Apple);
LMS – Low Mildew Susceptibility;
MR – Mildew Resistance.

(A) scab resistance, mainly using Vf genotypes (PRI, Crosby et al. 1992), with selection addressed to meet flavour demands of Italian consumers; (B) low mildew susceptibility derived from trait-specific cultivars; (C) trees of compact or spur habit with high-quality fruit (mainly late ripeners); (D) retrieval via old germplasm of such traits as flesh firmness, resistance to environmental stresses, enhanced shelf-life and specific flavour characters; (E) pollinators, mainly via crab apple, for compact habit and extensive bloom; and (F)

Table 2. Seedlings from some scab-resistant × scab-susceptible (and viceversa) crosses screened for scab resistance and segregation.

Cross	Seedling total No.	Infection a) green-house A[1] %	b) field B[2] %	Total infected A + B %	Expected segregation resistant: suscetible	Chi-Square[3]
Perleberg 3 × Florina*	1303	33.2	15.5	48.7	1:1	NS
Red Chief × Florina*	84	29.8	15.5	45.3	1:1	NS
Royal Gala × Florina*	82	31.7	13.4	45.1	1:1	NS
Sel.PRI 2750/1* × Gelata	456	43.6	3.3	46.9	1:1	NS
Sel.PSE R13 T50* × Granny Smith	73	46.6	14.0	60.6	1:1	NS
Spezzano albanese × Florina*	1024	6.0	41.0	47.0	1:1	NS

* Scab-resistant clone (Vf gene). Avg 3-year monitoring: 1 in greenhouse and 2 later in field.
[1] Artificial inoculation in greenhouse and elimination of infected plants.
[2] Natural field infection (1993).
[3] NS – not significant at 5% level.

Table 3. Progeny screening from scab-resistant × scab-resistant crosses.

Parents and SR Gene	Seedling total No.	Infected[1] %	Expected segregation resistant: susceptible	Chi-Square[2]
Florina (Vf) × Antonovka (Va)	52	25.0	3:1	NS
Freedom (Vf + Va) × Antonovka	173	2.3	7:1	**
Freedom × Florina	234	4.3	7:1	**
Sel.PRI 2750/1 (Vf) × Florina	717	33.2	3:1	**
Sel.PRI COOP 11 (Vf) × Florina	179	27.9	3:1	NS
Sel.PRI COOP 31 (Vf) × Florina	751	27.2	3:1	NS

[1] Average of artificial and field inoculation.
[2] NS – not significant; ** – significant at 1% level.

somaclonal variation, regeneration and selection by in vitro biochemical and physical screening of toxin-resistant (e.g., *Phytophthora cactorum*) rootstocks.

The 80 parents used over the 17 years of crosses to date have been selected for the following traits (Table 1): (1) SR: scab resistance, mainly Vf gene genotypes; (2) LSS: low scab susceptibility, mainly old germplasm cultivars; (3) SS/CH: scab susceptibility with compact or spur habit; (4) G/CH: old germplasm carrying compact or spur habit and limited scab susceptibility; (5) SS/StH: scab susceptibility with standard growth habit; (6) LMS: low mildew susceptibility; (7): CA: crab apple (ornamentals) hybrids as pollen donors.

The program has so far resulted in over 23,000 seedlings, of which about 17,000 are still under observation and 100 have been selected, both resistant and not, for *stage-2 screening*, which started in 1990 and will cover mainly phenotipic evaluation of trees for field performance and fruit quality.

Selection

Traditional field selection methods for *scab resistance* were performed up to 1987, given the favourable environment. The seedlings found susceptible to both scab and mildew were eliminated. Since 1988, because

Table 4. Field mildew susceptibility: seedling screening from some crosses.

	Cross No.	Seedling No.	Low Mildew[1] suscept %	High Mildew[2] suscept %
a) Parents: low suscept. × resistant				
Starkrimson × *M.* × *robusta*[3]	1	26	96.2	3.8
Total a	1	26	96.2	3.8
b) Parents: low suscept. × low suscept.				
Red Chief × Fior d'acacia	1	123	43.1	56.9
Red Chief × Sel.B2R6T2	1	85	57.6	42.3
Starkrimson × Crimson Gold	1	18	61.1	38.9
Starkrimson × Evereste	1	15	68.4	31.6
Starkrimson × Golden B	1	20	10.0	90.0
Starkrimson × Jurelle	1	39	74.4	25.6
Total b	6	300	52.4	47.6
c) Parents: low suscept. × highly suscept.				
Belgolden × Florina	1	295	3.7	96.3
Golden SAFA 20 × Florina	1	480	2.1	97.9
Perleberg 3 × Florina, Gelata	3	1639	10.1	89.9
Red Chief × *	6	1206	27.7	72.3
Sel.516/1 × Florina	1	214	7.5	92.5
Sel.B1R7T1 × *	4	522	23.4	76.6
Sel.B2R6T2 × Runsè	1	234	4.3	95.7
Total c	17	4590	11.2	88.8
Total a + b + c	24	4916		

* Highly susceptible parents: Florina, Gelata, Granny Smith, Royal Gala.

[1] Low mildew susceptibility: none or few leaves and shoots with symptoms.

[2] High mildew susceptibility: medium and high intensity of attack.

[3] *M.* × *robusta* clone is supposed to be screened as mildew resistant.

Indexing conducted in 1993 at DCA: Ozzano experimental field.

of the high cost of the discard operation, the crosses have mainly been targeted via artificial inoculation to scab resistance (SR) or low susceptibility (LSS) (the seedling population has increased to 3,000–5,000 plants yearly); thus destructive selection was employed in the first and second year in green- and screenhouse.

The potted seedlings, kept in plastic-covered tunnels at very high relative humidity, were inoculated by spraying a conidial suspension in May-June and the survivors in late September. The *Fusicladium* pathotypes for inoculum were found in the local growing area, isolated and grown in laboratory at the University of Bologna's Plant Protection Department (Dr R. Fiaccadori). Our experience so far indicates that proper evaluation of field infection requires several more years, thus adding to its cost.

The selection process for cropping and fruit quality was completed in the field (up to year 9–10). Screening was in part accelerated by grafting plants to M9 so as to enhance bearing (Table 6).

Table 2 lists the main resistant (Vf) × susceptible crosses and Table 3 the resistant crosses using both Vf and Va genes. Note that in the former (Table 2) the heterozygote character of resistance in the cultivar and selections carrying the Vf gene is found in the progeny at percentages approaching those expected (50%), as shown by the Chi-square test. Only the highly scab-susceptible 'Granny Smith' appears to influence the inheritance of scab resistance, raising the rate of infect-

Table 5. Segregation of compact, dwarf and spur traits in scab-susceptible progenies.

Cross	Seedling No.	Internode Length[1] Standard %	Compact %	Dwarf %	Spur Habit (2) %
a) Standard × Compact/Spur*					
Royal Gala × Gelata*	408	100.0	0.0	0.0	16.8
Royal Gala × Red Chief*	290	65.4	34.6	0.0	16.7
Granny Smith × Gelata*	55	94.0	6.0	0.0	18.0
Red Chief* × Granny Smith	176	95.1	4.9	0.0	29.4
b) Compact/Spur* × Compact/Spur*					
Gelata* × Red Chief*	339	93.0	7.0	0.0	3.0
c) Compact/Spur* × Dwarf+					
Gelata* × Sel.B1R7T1+	93	100.0	0.0	0.0	28.2
Red Chief* × Sel.B2R6T2+	87	35.8	46.9	17.3	32.1
d) Standard × Dwarf+					
Granny Smith × Sel.B1R7T1+	29	84.2	15.8	0.0	0.0
Royal Gala × Sel.B1R7T1+	208	82.5	15.5	2.0	11.3
Total	1685				

* Gelata & Red Chief are spur cvs. with compact-type internodes.
\+ – dwarf parent.
[1] Standard: 1.50 ± 0.32 cm.; Compact: 1.11 ± 0.06 cm.;
Dwarf: apical internode avg. 0.5 cm, shoot length < 20 cm.
[2] Spur: > 30 spurs per m of young branch.

Table 6. Influence of M9 on early bloom and cropping.

Parents	Trees[1] Total No.	Total in Bloom No.	Total in Bloom %	Cropping Tree No.	% Total graft	% Total bloom
Belgolden × Florina	34	12	35.3	9	26.5	75.0
Gelata × Florina	50	29	58.0	14	28.0	48.3
Golden SAFA 20 × Florina	36	19	52.8	6	16.7	31.6
Perleberg 3 × Florina	42	20	47.6	9	21.4	45.0
Red Chief × Florina	12	2	16.6	1	8.3	50.0
Sel.2750/1 × Florina	36	16	44.4	14	38.9	87.5
Sel.516/1 × Florina	32	7	21.9	1	3.1	14.3
Sel.B1 R7 T1 o.p.	49	33	67.3	4	8.2	12.1
Sel.PRI COOP 11 × Florina	38	12	31.6	5	13.1	41.7
Sel.PRI COOP 31 × Florina	36	17	47.2	14	38.9	82.3
Starkrimson × Golden B	16	10	62.5	5	31.2	50.0
Total	381	177	44.1	82	21.3	48.9

[1] The trees were grafted in 1991 (using scions from two-year-old seedlings), established in 1992 and the data collected in 1993.
NB: None of the 381 original seedlings on their own roots, transplanted at the same time, bloomed in 1993.

Table 7. Performance rating of some selections at screening stage 2.

Parents	Fruit Traits			Ripening Date
	Skin Color	Size	Flavor	+/– Golden D. (16/9) No. Weeks
a) Scab-resistant				
81.411.005+ Prima × Crimson Gold	Red	Medium-Large	Fair	+ 2
81.412.005 Prima × Florina	Red	Large	Fair	+ 0,5
81.414.007 Prima × Jerseymac	Yellow-Red	Medium-Large	Fair	– 4
81.414.031 Prima × Jerseymac	Red	Medium-Large	Good	– 1
81.417.002 Prima × Summerred	Red	Medium-Large	Fair	– 6
81.418.002 Prima × Sel.2059/101	Red	Large	Good	+ 0
81.418.003 Prima × Sel.2059/101	Red	Medium	Good	+ 0,5
b) Scab-susceptible				
77.404.001+ Sel.10-C-18-33 S.P.	Red	Medium	Fair	+ 2
77.418.031 Ed Gould Golden O.P.	Yellow	Medium	Fair	– 2,5
77.441.101 Magnolia Gold O.P.	Red	Medium	Fair	+ 2,5
78.415.004 Unknown	Red	Medium-Large	Fair	+ 1
81.413.036 Prima × Golden D.	Red	Medium	Fair	+ 0
81.414.023 Prima × Jerseymac	Yellow-Red	Large	Good	– 4
81.424.010 Starkrimson × Golden	Yellow	Medium	Good	– 2,5
81.425.014 Starkrimson × Golden	Yellow	Medium	Good	+ 1

+ spur habit.

ed progeny (60%), as reported for other very susceptible varieties by Lamb & Hamilton (1969).

The resistant × resistant apple crosses (Table 3) show, despite the low number of seedlings involved, a number of infected progeny among Vf parentals that is very close to the expected ratio (3:1), the variation range being 25–33%, while the 'Freedom' and 'Antonovka' crosses diminished the rate of infected seedlings to less than 5%, well below the expected 12.5% (7:1 ratio). This may be due to the artificial inoculum conditions (carried out only one year), although it also suggests either a change in the expected segregation because of the polygenic nature of one parent (Lamb et al. 1985) or the interaction of the major and minor genes in re-inforcing the resistance trait. This needs to be confirmed in the coming years.

The screening for *mildew low susceptibility*, which involves 4,916 seedlings (Table 4), is somewhat limited by the fact that only one resistance source, *Malus × robusta* (Alston 1983; Gallott et al. 1985), was available for testing. The findings for the crosses of LMS parents show that only 47.6% of the resulting progeny exhibited a high level of infection, whereas the findings for the crosses of higly susceptible parents show a high infection level in nearly 90% of the seedlings. Note that the progeny of 'Golden Delicious' and its clones, although thought to be parents of low susceptibility, exhibited very high infection in over 90% of derivatives. By contrast, the progeny of 'Red Delicious' ('Starkrimson') in the one cross with *M. × robusta* (resistant) showed only 3.8% of seedlings to be highly infected. These findings suggest that the breeding of apples with low susceptibility to mildew can produce acceptable results before achieving complete resistance in a very time-consuming program.

The *compact*, *spur-type*, and *dwarf* growth habit have been screened to date in only 1,685 seedlings (Table 5). As reported by Lapins (1974), since the spur trait is polygenic and, hence, gives rise to a phenotypic variation both continuous and scattered throughout the progeny, we arbitrarily designated as 'spur' the trees with more than 30 spurs per meter of 2–3 yr branch (Buscaroli & Sansavini 1991). The spur trait is often associated with the compact habit (i.e., internode shorter than 1.2 cm), whereas the dwarf character (i.e., very short shoots and apical internodes less than 0.5 cm) may or may not be independent of both. We employed two dwarf types (Faust & Zagaya 1984), one of which ('B2R6T2') segregates this trait to a certain extent (17.3%), whereas both can transmit to some extent the compact and/or spur traits. The dwarf Sel. 'BIR7T1' in some crosses does not transmit this trait.

Other significant data concern the compact habit, although the findings for spur density are still too few because of tree juvenility: 'Red Chief', which has both spur and compact traits, has a significant transmission rate for the latter habit (34.6% of the progeny with 'Royal Gala' and 46.9% with the dwarf parent 'B2R6T2'). Note too that 'Red Chief', when used as the female parent, seems to transmit the spur trait to a greater number of seedlings. A possible cytoplasmatic inheritance should be investigated in this connection. Yet it still appears too early to draw any sound conclusions about the inheritance of the compact-spur-dwarfing traits.

Early cropping is currently being screened in only 381 seedlings grafted to M9 in 1991, when they were two years old (Table 6). The results to date show that 44% of these seedlings bloomed in their second orchard year, although flowering increased to over 60% in the progeny of 'Starkrimson' × 'Golden Delicious' cross. This good rate of bloom was matched in 1993 by an excellent cropping percentage, a fact that will make it possible markedly to accelerate selection.

Several considerations are in order as regards *stage-2 selection*, which involves 100 seedlings selected in stage 1 and budded to M9. Of these trees, only 15 are being re-selected, seven of which are scab-resistant and two with spur habit (Table 7). Red is the dominant colour and four of these, derived from the parent 'Prima', are rated as good in taste; the selections 81.424.010 and 81.425.014, both derived from the 'Starkrimson' × 'Golden Delicious' cross, exhibit the most promising fruit appearance and flavour (for the Italian market), as well as good yellow blush coloring and medium-large size.

Pollinators

Cultivated varieties were employed here in 38 crosses with crab and ornamental species. Of the several hundred seedlings bred to date, fifteen have been selected with compact or erect habit and extensive flowering, and a few with semi-dwarfing habit. The aptitude for pollen production and its fertility is being assayed as added selection criteria.

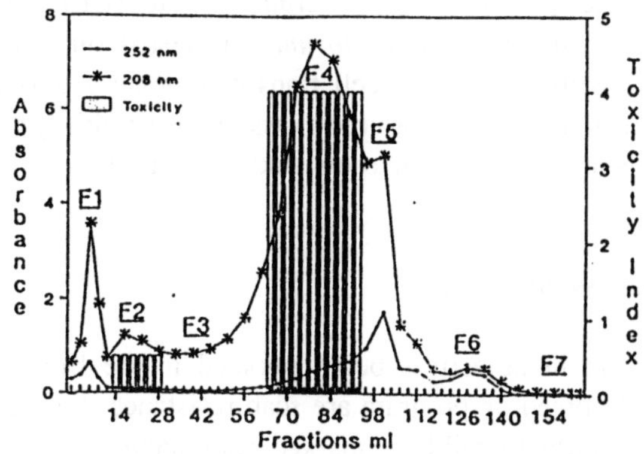

Fig. 1. Chromatographic fractions recorded by gel filtration of the non-permeate supernatant of the *P. cactorum* cultural filtrate. The histograms show the varying phytotoxicity level of the fractions detected via tomato seedling bioassay (Mezzetti et al. 1994, in press).

Somaclonal variation

This last part of the program, begun six years ago, is designed to induce somaclonal variation (regeneration from leaf callus) in rootstocks and to develop methods of detecting the susceptibility of apple tissue and cell cultures to *Phytophthora cactorum*. Leaf discs taken from resistant and susceptible rootstocks and incubated with culture filtrate (15% solution) of the fungus have shown a varying ion-release (measured by conductimeter) that is correlated to field response to the pathogen. Cells of the rootstocks MM106 and MM26 and their selected *in vitro* regenerants, when stained with merocyanine 540, checked by optical probe to detect changes in cell-membrane potential and incubated in cultural filtrate of virulent fungus strains, have shown variations in cell fluorescence (variations of membrane potential). By contrast, no changes in fluorescence emission have been found in MM111 cells (highly resistant to *P. cactorum*) (Mezzetti et al. 1992, 1993).

The more susceptible rootstocks like MM106 show a diminution of fluorescence and increases of ion-release from leaf discs induced by the toxin within the fungi's culture filtrate. A comparison of the ion-release and optical probe approaches suggests that the former appears to be a viable method for screening a large population and the latter useful in studying the host-parasite relationship. Current work is focused on finding the metabolites (part of the toxin) produced by the pathogen and released during its growth (Mezzetti et al. 1994, in press).

The phytotoxicity of crude filtrate (CF) (Fig. 1) of *Phytophthora cactorum* was tested on tomato seedlings, and the cell trans-membrane electrical potential is being measured to assess susceptibility and resistance of the apple rootstocks to the fungi.

Discussion

The main objectives being pursued in the Bologna apple breeding program are scab resistance, low susceptibility to mildew, marked spur and/or compact habit and high fruit quality. The accomplishments of these last fifteen years include about a hundred very promising selections for the second stage of testing and the findings registered for the inheritance of resistance.

Although most of this time has been spent working exclusively with heterozygote carriers of the Vf gene, genotypes carrying the Va gene ('Freedom', 'Antonovka') have been gradually introduced. While the former has registered a heritability rate in progeny of close to 50%, the latter, which is polygenic (Freedom), has recorded a more than 90% rate in resistant seedlings (higher than expected), a result presumably in relation to the crossing of parents with different resistance genes.

The progeny carrying the Va gene also appear to offer greater certainty both against new scab strains that overcome the Vf gene (Parisi et al. 1993) and against the eventual breakdown of resistance when it is vertical (Lespinasse 1989). Recent isozyme investigations indicate that PGM 1 is associated with the Vf gene (Manganaris 1989, Pancaldi et al. unpublished data) and seems also to be a marker of the Va gene (Pancaldi et al. unpublished) a fact that may lead to new interpretations of the phylogenetic dynamics of the resistance trait via investigation by molecular markers (RAPD, PCR).

Early selection of rootstocks proved to be viable using culture filtrate and, even better, purified fungal toxins to evaluate *in vitro* apple resistance to *Phytophthora cactorum*. This new approach should be of particular benefit in accelerating breeding work and, because of its pathogen specificity, in the study of the host-parasite relationship.

Paper supported by the Italian Ministry of Agriculture, Food and Forestry Resources, Project 'Frutticoltura', Rome.

References

Alston, F.H., 1983. Progress in transferring mildew (*Podosphaera leucotricha*) resistance from *Malus* species to the cultivated apple. IOBC Bull. Integrated Control in Orchards, INRA, Angers, pp. 87–95.

Buscaroli, C. & S. Sansavini, 1991. Le forme spur nelle antiche varietà di melo. Frutticoltura 1: 69–74.

Crosby, J.A., J. Janick, P.C. Pecknold, S.S. Korban, P.A. O'Connor, S.M. Ries, J. Goffreda & A. Voordeckers, 1992. Breeding apples for scab resistance: 1945–1990. Fruit Varieties Journal 46(3): 145–166.

Faust, M. & S.W. Zagaya, 1984. Prospect for developing low vigor fruit tree cultivars. Int. Workshop Controlling Vigor in Fruit Trees. Acta Hort. 146: 21–27.

Gallott, J.C., R.C. Lamb & H.S. Aldwinckle, 1985. Resistance to powdery mildew from some small-fruited *Malus* cultivars. HortScience 20(6): 1085–1087.

Lamb, R.C., H.S. Aldwinckle & D.E. Terry, 1985. 'Freedom', a disease-resistant apple. HortScience 20(4): 774–775.

Lamb, R.C. & J.M. Hamilton, 1969. Environmental and genetic factors influencing the expression of resistance to scab (*Venturia inaequalis* Cke. Wint.) in apple progenies. J. Amer. Soc. Hort. Sci. 94: 554–557.

Lapins, K.O., 1974. Spur type growth habit in 60 apple progenies. J. Amer. Soc. Hort. Sci. 99(6): 568–572.

Lespinasse, Y., 1989. Breeding pome fruits with stable resistance to diseases. 3. Genes, resistance mechanisms, present work and prospects. In: C. Gessler, D.J. Butt & B. Koller (Eds) Proc. Workshop Integrated Control of Pome Fruits Diseases, Vol. II, Brissago, Switzerland, 30/10–4/11/1988, IOBC Bull, pp. 100–115.

Manganaris, A.G., 1989. Isozymes as genetic markers in apple breeding. Ph.D. Thesis, University of London.

Mezzetti, B., R. Capasso, A. Evidente, F.A. Hammerschlag, R.H. Zimmerman, G. Cristinzio & P. Rosati, 1994. Interaction of partially purified phytotoxins from *Phytophthora cactorum* on tomato seedlings and apple cell membranes. J. Phytopathology (in press).

Mezzetti, B., P. Rosati, R.H. Zimmerman & F.A. Hammerschlag, 1993. Determination of resistance to *Phytophthora cactorum* culture filtrate in apple clonal rootstocks, cultivar and leaf regenerants, using in vitro proliferation and optical probe methods. II. Int. Symp. In vitro Culture and Horticultural Breeding, Baltimore 1992, Acta Hort. 336, pp. 93–100.

Mezzetti, B., R.H. Zimmerman, C. Mischke, P. Rosati, F.A. Hammerschlag, 1992. Merocyanine 540 as an optical probe to monitor the effects of culture filtrates of *Phytophthora cactorum* on apple cell membranes. Plant Sci. 83: 163–167.

Parisi, L., Y. Lespinasse, J. Guillaumes, & J. Krüger, 1993. A new race of *Venturia inaequalis* virulent to apples with resistance due to the Vf gene. Phytopathology 83: 533–537.

Sansavini, S., 1993. Il miglioramento genetico del melo per la resistenza alle avversità biotiche. Frutticoltura 5: 61–73.

H. Schmidt and M. Kellerhals (Eds.), Progress in Temperate Fruit Breeding, 117–121.

Aims of the apple breeding programme at Wädenswil

Markus Kellerhals & Madeleine Meyer
Swiss Federal Research Station for Fruit-Growing, Viticulture and Horticulture, CH-8820 Wädenswil, Switzerland

Key words: Apple, breeding, disease resistance, fruit quality, yield

Summary

The apple varieties 'Schweizer Orangenapfel', 'Maigold', 'Goro', 'Arlet', 'Iduna' and 'Marina' were released from the Swiss Federal Research Station Wädenswil. The requirements for future apple varieties are: outstanding quality, good yielding performance and durable disease resistance. Disease resistant apple varieties constitute an important component of a fruit production with a low input of chemicals. However, disease resistant varieties will only be commercially successful if fruit quality and yield capacity are similar or superior to well-known commercial varieties. The programme for breeding disease resistant apple varieties at Wädenswil started in 1986. Each year about 8,000 apple seeds are produced. The seedlings are screened for scab resistance in the glasshouse and for mildew susceptibility and growth habit in the nursery. From about 500 selected seedlings annually one tree on M27 is budded to evaluate fruit and tree characters. The most promising selections are included in our variety testing scheme. The first advanced selections will soon be available for grower trials.

In the breeding programme it is important to enlarge the spectrum of resistance sources. Our breeding strategy follows two lines. First we are breeding new varieties with commercial quality. Second we are searching for and trying to incorporate new sources of resistance.

Besides the resistance our main attention is focused on fruit quality components. Quality criteria which are of importance to the consumer comprise properties related to sensory, nutritional and technological quality as well as wholesomeness and image.

Introduction

The apple breeding programmes at the Swiss Federal Research Station Wädenswil have resulted in varieties such as 'Schweizer Orangenapfel', 'Maigold', 'Goro', 'Arlet', 'Iduna' and 'Marina' (Aeppli, 1984; Kellerhals, 1991; Kellerhals & Rusterholz, 1993). Maigold has become the third most important apple variety in Swiss fruit growing after 'Golden Delicious' and 'Idared'. 'Iduna' and 'Marina' are the most recent releases.

The programme for breeding disease resistant apple varieties at Wädenswil started in 1986 in collaboration with HRI East Malling, England, Hort + Research, Havelock North, New Zealand, Institute for Fruit Breeding, Dresden-Pillnitz, Germany and Federal Institute of Technology (ETH) Zürich, Switzerland.

Our actual breeding strategy follows two lines. Firstly we are breeding new varieties with commercial quality. Secondly we are searching for and trying to incorporate new sources of resistance.

In cooperation with ETH Zürich and foreign institutes we are trying to establish 'marker assisted breeding' using modern methods of molecular biology to select more precisely the desired genotypes at an early stage (Gianfranceschi et al., 1994). The host-parasite interactions are observed in orchards and plots not treated with fungicides at different locations in Switzerland (Kellerhals & Meyer, 1993; Kellerhals & Furrer, 1994).

Besides disease resistance our main attention is focused on fruit quality components. Quality criteria which are of importance to the consumer comprise properties related to sensory, nutritional and technological quality as well as wholesomeness and image

Table 1. Apple varieties released from the Swiss Federal Research Station Wädenswil

Variety	Parentage	Cross made in	Released
Schweizer Orangenapfel	Ontario × Cox Orange	1935	1955
Maigold	Fraurotacher × Golden Delicious	1944	1964
Goro	Golden Delicious × Schweizer Orangenapfel	1951	1973
Arlet	Golden Delicious × Idared	1958	1984
Marina	Kidds Orange × Idared	1970	1993
Iduna	Golden Delicious × Glockenapfel	1971	1993

Table 2. Results of the crossing programme 1982 (Kellerhals, 1991)

Combination No.	Parentage	Seeds sown	Seedlings planted to nursery	Selections planted in 1. step	Selections planted in 2. step
8201	Maigold × Arlet	800	726	75	6
8202	Golden Del. × (Idared × (Jonathan × Berlepsch))	800	736	108	2
8203	(Cox Orange × Maigold) × Elstar	800	726	122	2
8204	(Idared × (Jonathan × Berlepsch)) × Gala	800	768	131	10
8205	(Glockenapfel × Spartan) × Arlet	800	781	153	4
8206	Arlet × Gloster	800	763	143	2
8207	Arlet × (Cox Orange × Maigold)	800	713	129	2
8208	Arlet × (Idared × (Jonathan × Berlepsch))	590	509	69	2
8209	(Idared × Maigold × Elstar	800	761	119	17
8210	(Spartan × Maigold) × Elstar	300	244	53	1
Total		7290	6727	1102	48

(Höhn, 1990). A consumer panel of 80 households at Wädenswil is regularly testing new varieties and selections. This gives valuable information about consumer preferences and new trends.

Materials and methods

To ensure the appropriate choice of parents, due attention is paid to the main breeding aims which are fruit quality, productivity and disease resistance. Crosses are made in the field. Branches of the female parent are bagged with nylon tissue to prevent pollination by insects. Pollen from the male parent is collected in the lab from cut branches. Pollination is carried out with a brush. The seeds are extracted from the fruits after harvest and stratified at 4° C for three month in humid sand. After stratification the seeds are sown in plastic seed trays with 104 pressblocks of substrate each containing one apple seed. The young seedlings are screened for scab resistance in the glasshouse according to the method described by Kellerhals & Furrer (1994). Seedlings considered scab resistant are planted to the field. In the summer of the following year they are selected considering mildew susceptibility, juvenility, growth habit and possible escapes of the glasshouse screening for scab resistance. From the selected seedlings one tree is budded on the dwarf

Table 3. Yield performance and fruit quality in cross combinations with the cv. Rubinette compared to the parental varieties

Combination or variety	No of trees per combination or variety (on M 27)	% flowering 1990 (second leaf)	% flowering 1991 (3rd leaf)	Yield 1992 (g/tree 4th leaf)	Fruit weight 1992	Fruit colour 1992 (1 = no over colour 9 = 100% over colour)	Texture 1992 (1 = very soft 9 = very firm)	Juiciness 1992 (1 = very dry 9 = very juicy)	Flavour 1992 (1 = very poor 9 = very good
Gala × Rubinette	17	29	82	6208	101	4.4	6.5	6.1	5.8
Gala	6	100	100	5903	116	6.5	7.0	5.9	5.8
Rubinette	10	100	100	3668	100	5.2	6.2	6.2	6.9
Rubinette × Maigold	9	22	89	6465	118	5.0	6.2	6.3	5.9
Maigold	6	100	100	4637	161	7.0	7.7	7.5	6.5
Maigold × Rubinette	13	31	100	4340	134	5.5	6.7	6.3	6.6
Rubinette × Golden Del.	7	14	100	5180	95	4.4	6.7	6.3	6.6
Golden Del.	7	86	100	8328	171	1.9	6	6	5.9
Golden Del. × Rubinette	9	0	56	3407	147	4.1	5.6	5.6	5.4
Rubinette × Idared	10	40	90	6482	142	5.9	4.4	5.7	4.9
Rubinette × Gloster	8	38	88	7835	137	4.4	5.4	6.1	6.0
Rubinette × Glokkenapfel	13	38	85	7605	140	3.5	5.8	6.2	6.2

Table 4. Progeny analysis of the 1986 crosses for yielding capacity, mildew susceptibility and fruit quality (adapted from Kellerhals & Meyer, 1993)

Cross combination	% fruiting in the 2nd leaf	% fruiting in the 3rd leaf	Yield (g per tree in the 3rd leaf)	Mildew score (1 = no attack 9 = heavy attack)	Firmness (1 = very soft, 9 = very firm)	Juiciness (1 = very dry, 9 = very juicy)	Flavour (1 = poor, 9 = very good
Arlet × Freedom	21	83	1246	3.6	6.2	5.6	5.8
Greensleeves × Liberty	50	86	2245	2.7	6.0	5.2	5.1
Gala × Florina	19	80	1713	3.9	6.7	5.7	5.4
Elstar × Jonafree	11	79	2528	3.8	8.0	6.3	6.4
Rubinette × Florina	44	92	1947	3.4	6.4	5.4	5.3
Florina × Rubinette	14	72	1632	4.2	6.2	5.4	5.2
Florina × Elstar	20	85	1900	3.8	6.5	5.4	5.1
Golden Del. × A 849/5	38	100	2872	1.2	6.7	5.4	5.2
Gloster × A 849/8	20	87	2018	1.2	6.8	5.3	5.5
Ingrid Marie × A 849/6	13	91	1639	1.9	5.3	5.2	5.5
Jonathan × A 849/7	18	78	1900	1.6	7.6	5.6	5.4

rootstock M27. The one year old trees are planted at a distance of 4 × 0.6 m as spindle trees with a minimum of pruning and fruit thinning and no fungicide application. The tree and fruit characters are regularly observed on these trees. Most scores for the fruit and tree characters follow a 1 to 9 scale with 1 as the lowest and 9 as the highest score. The whole crop is harvested and weighed and a sample put in cold store at 4° C and 92% relative humidity. Eating quality is assessed according to the picking time once from October to December. Promising selections are kept till February or March and assessed a second time.

Results

The apple varieties released from the Federal Research Station Wädenswil are shown in Table 1. 'Maigold' and 'Arlet' are modern commercial varieties with excellent storage capacity. The most recent releases, 'Iduna' and 'Marina' also have a very good storage life. The apple crosses made by Aeppli (1984) in 1982 have resulted in 48 advanced selections which are currently under final evaluation. Table 2 shows that the combinations 8209, 8204 and 8201 gave a high number of promising selections. We aim to release 1 or 2 commercial varieties from this programme.

In 1985 we made crosses with the Swiss variety 'Rubinette' (Kellerhals & Höhn, 1987). The progenies were vegetatively selected in the nursery and the promising ones budded on M27. In the first step of testing on the rootstock M27 some parents were planted among the hybrids for comparison. Table 3 shows that the parental varieties were more precocious than the mean of the progenies. When comparing the yield per tree in the 4th leaf, 'Golden Delicious' was most productive, followed by the combinations 'Rubinette' × 'Gala' and 'Rubinette' × 'Gloster'. The variety 'Rubinette' itself had the lowest yield. The average fruit weight was highest in 'Golden Delicious', 'Rubinette' × 'Empire' and 'Maigold' and lowest in 'Rubinette' × 'Golden Delicious' and 'Rubinette' × 'Arlet'. The parental variety 'Maigold' was the most coloured and 'Golden Delicious' had the least amount of cover colour. 'Maigold' reached the highest scores for texture and juiciness. 'Rubinette' was rated best for flavour. 'Rubinette' in combination with 'Idared' resulted in poor texture and poor flavour in the progeny. We have selected 15 hybrids out of these combinations for further evaluation and for use in the breeding programme.

In 1986 we started the programme aimed at breeding disease resistant varieties with excellent fruit quality and yield. The donors for resistance to scab and mildew mainly originate from HRI East Malling, the Coop programme in the USA and from INRA Angers. From the 8,000 apple seeds produced annually only about 500 hybrids remain after the stringent selection in the glasshouse and the nursery. The early selection consists of two steps: the glasshouse screening for scab resistance and the preselection for mildew resistance, juvenility and growth characters in the nursery. The whole selection procedure is carried out without application of fungicides. Under the growing conditions of Wädenswil with more than 1300 mm annual rainfall a high natural pressure of scab and mildew is present. Table 4 shows an analysis of the 1986 progeny for yield capacity, mildew susceptibility and fruit quality. Unfortunately many donors of scab resistance are highly susceptible to mildew under our conditions. The A 849-selections which carry Pl1 mildew resistance segregate for mildew resistance in the progeny.

One aspect which becomes increasingly important in modern high density orchards is precocity. In the second leaf the percentage of fruiting trees has increased by up to 50% in precocious combinations such as 'Greensleeves' × 'Liberty' and 'Rubinette' × 'Florina'. It is surprising that the reciprocal cross 'Florina' × 'Rubinette' was the least precocious. However, all the progenies shown in Table 4 are satisfactorily precocious with 70–100% of the trees fruiting in the third leaf. More striking differences show up when considering the yield capacity. 'Golden Delicious', 'A 849-5', 'Elstar' and 'Jonafree' seem to be better performing parents whereas 'Florina' and 'Freedom' seem to be less effective.

An impression of fruit quality is given by considering texture, juiciness and flavour. Good parents for texture are 'Gala' and 'Jonafree'. The combinations with 'Greensleeves' and 'Ingrid Marie' gave the lowest scores. Table 4 also shows that texture and juiciness are often interrelated. Good parents for texture are also good parents for juiciness. Parents with a good flavour such as 'Elstar', the A 849 series, 'Ingrid Marie' and 'Arlet' had a positive effect on flavour. The influence of 'Florina' and 'Liberty' on flavour was less favourable, even in combination with highly flavoured varieties such as 'Rubinette' and 'Elstar'.

Promising selections from the disease resistance programme will be further evaluated and some of them released if they perform well.

Discussion

Apple breeding at Wädenswil is aimed at offering high quality productive varieties to the fruit industry and the consumer (Kellerhals, 1991). Since 1986 the scope has widened to breeding varieties which can be grown with a minimum of plant protection sprays by incorporating disease resistances. Careful evaluation of the value of parental varieties is a prerequisite of a successful breeding programme. Efficient pre-selection techniques for disease resistance as well as yield and quality determining factors are also essential (Hough, 1944; Alston & Bates, 1979; Alston & Kellerhals, 1990). Molecular-aided selection (Gianfranceschi et al., 1994) and new approaches in respect of durable disease resistance (Kellerhals & Furrer, 1994) are challenges which should lead to more efficient breeding procedures and durable resistance. Breeding is one component in developing integrated fruit production towards a low input fruit orchard ecosystem. We are developing our breeding strategy and our breeding aims in view of a productive but sustainable orchard ecosystem which allows an environmentally safe production of high quality apples appreciated by the consumer.

References

Aeppli, A., 1984. Rückblick und Ausblick auf die schweizerische Apfelsortenzüchtung. Erwerbsobstbau 26: 56–62.

Alston, F.H. & J.W. Bates, 1979. Selection for yield in apple progenies. Proceedings Eucarpia Fruit Breeding Section Meeting Angers, 1979: 15–27.

Alston, F.H. & M. Kellerhals, 1990. Breeding apples to develop integrated production. Acta Horticulturae 285: 135–141.

Gianfranceschi, L., J.M. McDermott, N. Seglias, B. Koller, M. Kellerhals & C. Gessler, 1994. First results in developing a marker assisted breeding for resistance against scab. In: H. Schmidt and M. Kellerhals. Progress in Temperate Fruit Breeding. pp. 293–296 (this volume). Kluwer Academic Publishers, Dordrecht.

Höhn, E., 1990. Quality criteria of apples. Acta Horticulturae 285: 111–118.

Hough, L.F., 1944. A survey of the scab resistance of the foliage on seedlings in selected apple progenies. Proc. Am. Soc. Hort. Sci. 44: 260–272.

Kellerhals, M. & E. Höhn, 1987. The apple variety 'Rubinette'. Erwerbsobstbau 29: 241–242.

Kellerhals, M., 1991. Apple breeding at Wädenswil. Erwerbsobstbau 33: 219–224.

Kellerhals, M. & P. Rusterholz, 1993. Iduna und Marina – zwei neue Apfelsorten der FAW. Schweiz. Zeitschrift für Obst- und Weinbau 129: 321–323.

Kellerhals, M. & M. Meyer, 1993. Combining stable disease resistance with high fruit quality and good yielding capacity in apple. Norwegian Journal of Agricultural Sciences, in press.

Kellerhals, M. & B. Furrer, 1994. Approaches for breeding apples with durable resistance. In: H. Schmidt and M. Kellerhals. Progress in Temperate Fruit Breeding. pp. 95–99 (this volume). Kluwer Academic Publishers, Dordrecht.

H. Schmidt and M. Kellerhals (Eds.), Progress in Temperate Fruit Breeding, 123–125.

Four decades of apple breeding at the Fruit Research Institute at Čačak

Ž.V. Tešović, M.M. Stanisavljević & M.J. Srećković
Agricultural Research Institute 'SRBIJA', Fruit and Grape Research Centre Čačak, Yugoslavia

Key words: Apple, breeding, new cultivars

Summary

From 1952–1992 431 parental combinations with 111 cultivars and hybrids were used in apple breeding and selection. 15 cultivars were used only as female parents, 40 only as male parents and 56 as both parents. Over 100,000 hybrid seedlings were produced and planted in the field. 135 promising hybrids were selected among the population – 126 of them with known parents and 9 promising hybrids created by open pollination of six cultivars. 66 promising hybrids were chosen and selected for further trials, two of them were named and released as cultivars: Hybrid J/10/43/59 (Starking × Jonathan) – Čačanska pozna; and Hybrid J/5/8/63 (G. Delicious × Jonathan) – Čadel. The promising hybrid J/5/2/63 (G. Delicious × Jonathan) is currently under test and is expected to be named as new cultivar. 69 promising hybrids are left to be studied.

Introduction

Existing apple cultivars are not satisfactory when production and consumption are taken into consideration. There is a lack of good quality apple cultivars of different ripening time which can be planted with minimal pesticide use. These problems indicate the necessity for continuing work on breeding and selection of new apple cultivars in Yugoslavia.

The principal reason for apple breeding at the Fruit Research Institute at Čačak from 1952–1992, and in future is the development of fertile and relatively resistant, mostly winter cultivars of good quality suitable for successful cold storage.

Material and methods

While working on the development of new apple cultivars at the Fruit Research Institute in Čačak over the above-mentioned period (1952–1992), the following cultivars and hybrids were used:

1. Ananasrenette, 2. Auvil Spur, 3. Bancroft, 4. Batulenka, 5. Baumanns Renette, 6. Benoni, 7. Black Jon, 8. Budimka, 9. Champagner Renette, 10. Charlamowsky, 11. Cox's Orange Pippin, 12. Cortland, 13. Croncels, 14. Čadel, 15. Čačanska pozna, 16. Delicious, 17. Djulica, 18. Earlyred Bird, 19. Elisson's Orange, 20. Ernst Bosch, 21. Giant Wealthy, 22. Gloster, 23. Golden Delicious, 24. Goldparmäne, 25. Granny Smith, 26. Grosse Kasseler Renette, 27. HAR 5T2, 28. HAR 5T8, 29. HAR 5T55, 30. Idared, 31. James Grieve, 32. Jefferis, 33. Jonadel, 34. Jonathan, 35. J/12/52 (Ananasrenette × Jonathan), 36. J/22/47 (Budimka × Jonathan), 37. J/38/65 (G. Delicious O.P.), 38. J/93/65 (G. Delicious O.P.), 39. J/35/2/56 (Delicious × Jonathan), 40. J/23/3/59 (G. Delicious × Ch. Renette), 41. J/35/39/62 (G. Delicious × Jonathan), 42. J/60/7/63 (G. Smith × G. Delicious), 43. J/27/127/62 (Idared × M. Worcester), 44. J/27/130/62 (Idared × M. Worcester), 45. J/35/12/56 (Jonathan × Delicious), 46. J/3/4/57 (Jonathan × G. Delicious), 47. J/8/52 (Jonathan × Kiseljaja), 48. J/19/51 (Jonathan × šumatovka), 49. J/19/54 (Jonathan × šumatovka), 50. J/4/7/56 (Jonathan × Cox's Orange Pippin), 51. J/54/53/59 (Red Cox's O.P.), 52. J/18/51 (šumatovka × Delicious), 53. J/5/4/59 (Weisser Klarapfel × Vidovača), 54. Kiseljaja, 55. Landsberger Renette, 56. Laxton's Epicur, 57. Lodi, 58. Londoner Pepping, 59. Lord Lambourne, 60. Luxembourg, 61. Malus silvestris, 62. Malus zumi, 63. Mantet, 64. McIntosh, 65. Melba, 66. Melrose, 67. Merton Worcester, 68.

Minjon, 69. Newton Yellow Pippin, 70. Nugget, 71. Ontario, 72. Orleans, 73. Prima, 74. Queen Cox's Orange, 75. Red Boscop, 76. Red Grieve Neuman, 77. Red Jonathan Alexandra, 78. Red Jonathan Geisenheim, 79. Richared, 80. Ringstand, 81. Roter Bellefleur, 82. Roter Pogatscher, 83. Rosamund, 84. Ruby, 85. Sanspareil de Peasgood, 86. Seržan, 87. Shay 56, 88, Shay 63, 89. Shay 64, 90. Shay 73, 91. Shay 74, 92. Shay TNR10T23, 93. Shay TNR8T151, 94. Shay Hough OR-53-T-53, 95. Shay Hough OR-54-T-170, 96. Starkrimson Delicious, 97. Stark Earliest, 98. Starking, 99. Šerbetka, 100. Šumatovka, 101. Tetovka, 102. TSR23T181, 103. Turley, 104. Vidovača, 105. Vizaljka, 106. Wagener, 107. Wealthy, 108. Weisser Klarapfel, 109. Weisser Winter Kalwill, 110. Winesap, 111. Worcester Pearmain.

Over the forty-years period of developing new apple cultivars, planned intraspecies hybridization was the main method.

Results and discussion

431 combinations were realised by planned intraspecies hybridization. Cross-pollination was performed among the above-mentioned cultivars and hybrids.

Within 431 parental combinations Jonathan was most often used as parent (41 times), then Idared (24 times), Golden Delicious (19), Melrose (17), Newton Yellow Pippin and Starking (16), Goldparmäne, Granny Smith, Richared (10) and the rest were used 2–9 times.

More than 100,000 seedlings were bred in the field. Out of 431 hybrid combinations, 126 promising hybrids originated from only 43 combinations, while 388 combinations did not give promising hybrids. The most promising hybrids were selected out of the Jonathan × G. Delicious combination – 14, then Prima × Melrose combination – 13, Idared × Melrose – 12, G. Delicious × Jonathan – 11, Starking × Jonathan – 7, G. Delicious × Ch. Renette – 6, Melrose × Prima – 5, Čačanska pozna × Idared – 4.

Combinations between the following cultivars gave three, two or one promising hybrid: Ananasrenette, Budimka, Champagner Renette, Cox's Orange Pippin, Čadel, Delicious, G. Delicious, Goldparmäne, Granny Smith, Idared, James Grieve, Jonathan, J/35/39/62 (G. Delicious × Jonathan), Landsberger Renette, Laxton's Epicur, Lord Lambourne, McIntosh, Melrose, Merton Worcester, Ontario, Prima, Richared, Ruby, Seržan, TNR10T23, Starking, Šumatovka, Vidovača and Weisser Klarapfel.

Besides cross pollination, open pollination gave a large number of promising hybrids. Three promising hybrids were developed by open pollination of Golden Delicious, two by open pollination of Lord Lambourne, while open pollination of Earlyred Bird, Jonathan, Red Cox's O.P. and Weisser Klarapfel gave one promising hybrid, respectively.

Among the selected promising hybrids 69 are under test, while the research is finished with 66 hybrids. Two promising hybrids were named as new cultivars: hybrid J/10/43/59 (Starking × Jonathan) – Čačanska pozna (1972) and hybrid J/5/8/63 (G. Delicious × Jonathan) – Čadel (1984). Hybrid J/5/2/63 (G. Delicious × Jonathan) is currently under test and is expected to be named as new cultivar.

Disregarding the fact that a certain number of promising hybrids were developed, each parental combination has a series of drawbacks. For example the combination Jonathan × G. Delicious and vice versa gave hybrids with satisfactory fruit shape and flavour, but the skin colour (pale patched-red), ripening time of fruits (most hybrids ripen before both parents) and susceptibility to *Venturia inaequalis* and *Podosphaera leucotricha* are unsatisfactory. The combination Prima × Melrose gave hybrids practically resistant to *V. inaequalis* and *P. leucotricha*; fruits have a satisfactory flavour, colour and shape, but their ripening time is much earlier than Melrose which is a serious disadvantage. The combination Idared × Melrose gave hybrids mostly with flat-shaped fruits, while other pomological characteristics are satisfactory. Other combinations gave less promising hybrids and they are mostly unsatisfactory for their pomological and technological characteristics.

Economic-biological properties of new apple cultivars and hybrids

Čačanska pozna It is derived from a cross between Starking × Jonathan made in 1959 by Dr P. Mišić and Dr M. Gavrilović, and named and released in 1972 (Mišić & Gavrilović 1973). A winter vigorous cultivar recommended for growing grafted on M9 and M26 rootstocks. Fruits are large, averaging 200 g, roundish-flat. The ground colour is green-yellow, blushed dark red on 80% of fruit surface. Fruit skin is medium thick, tough, glossy and waxy. The flesh is pale yellow, medium acid, with pronounced aroma. Fruits are suitable for transporting and successful cold storage until April.

Table 1. Fruit weight and yield per hectare of Čadel, Čačanska pozna and Golden Delicious, grafted on M9, M26 and MM106 rootstocks.

Cultivar	Rootstock	Fruit Weight (g)			Yield per hectare (t)		
		1987	1988	average	1987	1988	Average
Čadel	M9	177	192	184	21.3	29.3	25.3
	M26	189	196	193	27.5	31.2	29.3
	MM106	178	190	184	26.5	28.4	27.4
Average		181	193	187	25.0	29.6	27.3
Čačanska pozna	M9	224	218	221	26.8	31.3	29.1
	M26	240	231	235	29.9	26.5	28.2
	MM106	241	235	238	24.7	29.2	27.4
Average		235	228	232	23.8	29.0	26.4
Golden Delicious	M9	135	142	139	20.0	30.3	25.1
	M26	147	151	149	29.4	38.5	33.9
	MM106	135	139	137	16.8	34.3	25.6
Average		139	144	142	22.1	34.3	25.6

It is a diploid cultivar, with mid-late flowering time, together with Melrose. Ripens in late September-early October. It is not susceptible either to *V. inaequalis* or to *P. leucotricha*.

Čadel It is derived from a cross between Golden Delicious × Jonathan and was named and released in 1984 (Tešović 1988). It is a winter cultivar recommended for growing, grafted on weak and medium vigorous vegetative rootstocks. Fruits are medium large to large (200 g), round-conical, slightly ribbed. Skin is thin, smooth and very tough. The ground colour is yellow-green, with 50 – 80% red blush, which is more intense on the sunny side. Flesh is firm, crispy, pale creamy white, fine-textured, juicy, acid, with a pronounced characteristic aroma. Successful cold storage is possible until May. It is a diploid cultivar, with early flowering time, together with Idared. Ripens in late September-early October. Susceptible to *V. inaequalis* like G. Delicious, but not susceptible to *P. leucotricha*.

Elite hybrid J/5/2/63 Derived from the same population as Čadel. It is currently under test and is expected to be named as new cultivar. Fruits are medium large to large (200–250 g), roundish-cylindrical (Tešović 1986). Fruit skin is thin, smooth, very tough. The yellow-green colour is 50–70 % blushed red, which is more intense on the sun-exposed side. Fruits are suited to transporting and can be successfully kept in cold store until January. Flesh is pale yellow, subacid and fine-textured. If grown on vegetative rootstocks of weak and moderate vigour it produces high yields. This hybrid is very late flowering. It ripens at the end of August-early September. Fruits do not ripen uniformly. It is less susceptible to *V. inaeqalis* than G. Delicious, and is not susceptible to *P. leucotricha*.

In Table 1 are presented data on fruit weight and yield per hectare for Čadel and Čačanska pozna compared to Golden Delicious. The fruit weight is significantly higher with both cultivars compared to Golden Delicious, grafted on all three rootstocks. The yield per hectare is higher with Čadel and Čačanska pozna in 1987 (a drought year), and lower in 1988 (a year with enough rainfall).

References

Mišić, P.D. & M.D. Gavrilović, 1973. Perspektivni hibrid jabuke 10/43/59 (Starking × Jonathan). Journal of Yugoslav Pomology 17–18: 325–331.

Tešović, Ž.V., 1986. Stvaranje i osobine novih sorti i perspektivnih hibrida jabuke. Proceedings of Yugoslav Symposium on Fruit Breeding and Selection: 23–29.

Tešović, Ž. V., 1988. Čadel – nova zimska sorta jabuke. Journal of Yugoslav Pomology 84–85: 163–166.

H. Schmidt and M. Kellerhals (Eds.), Progress in Temperate Fruit Breeding, 127–129.

Apple and pear breeding in Bulgaria

V. Djouvinov
Fruit Growing Research Institute, 4004 Plovdiv, Bulgaria

Key words: Apple, diseases, *Malus*, pear, *Pyrus*, resistance

Summary

A new apple and pear breeding programme was started in 1986. The main goal of this programme is to produce very early and late maturing apple and pear cultivars resistant to the main diseases and with range of tree habits – standard, compact or columnar and weeping. Hybrid 142 (Golden Delicious × *M. zumi*) which is resistant to scab mildew and San José scale is very effective as donor for powdery mildew resistance. To incorporate resistance to scab parents as Prima, Redfree, Liberty, Florina carrying the Vf resistance were included. Hybrid 2350 (Williams × *P. ussuriensis*) × Clapp's Favourite is used as donor for scab resistance and will be studied for resistance to *Psylla pyricola*. Hybrids with columnar habit have been selected from the crosses McIntosh Wijcik × Redfree and McIntosh Wijcik × Liberty, and with weeping habit from progenies Elise Rathke × Prima and Elise Rathke × Florina. Embryoculture is applied to very early ripening apple (early July – Geneva Early and Vista Bella) and pear (second part of June – Ranna bolyarka and Trapezitsa) cultivars.

Introduction

The apple is the main fruit crop in Bulgaria. From 1976 to 1989 the proportion of trees of the major commercial cultivars Golden and Red Delicious decreased from 82 to 72%. During this period the cultivars Melrose, Granny Smith, Prima, Mollie's Delicious came to occupy 4,2% of the orchards.

During the last 20 years four new apple cultivars have been introduced in Bulgaria, Roumyana (Delicious × Jonathan), Trakijska slava (Golden Delicious × Northern Spy), Chervena jubileyna (Golden Delicious × Jonathan) and Trakijska kasna (Jonathan × McIntosh). These new and two main cultivars are susceptible to scab and usually need 12–16 fungicide treatments from April to the and of July to control the disease. Moreover Melrose, Granny Smith, Mollie's Delicious and most of the new Bulgarian cultivars are susceptible to mildew, another economically important disease of apples in our country. Many of the scab resistant cultivars originating from USA, Canada, United Kingdom, etc., have been found to be unsuitable under our climatic conditions.

The pears Williams, Curé, Passe Crassane, Beurré Giffard, Beurré Bosc are most commonly grown in orchards. New cultivars are: Trapezitsa (B. Giffard × Tserovka), Hebar (Dr Jules Guyot × B. d'Hardenpont), Kasna Vilyamova (Max Red Bartlett bud sport), Ranna bolyarka (B. Giffard × Tserovka), Kyustendilska maslovka (B. d'Hardenpont × Doyenne d'hiver). During the last several years Pear Psylla is the most important pest of pears on main commercial and new cultivars. These problems prompted a new breeding programme in 1986, apple and pear breeding being discontinued in 1972 (Djouvinov 1989).

Materials and methods

Plovdiv is situated in the middle part of Southern Bulgaria, 160 m above sea level. The average annual temperature is 12 °C, the humidity RH 73%, precipitations 516 mm and frost-free period lasts 184 days, i.e., from 13 April to 15 October.

The apple collection includes 733 accessions, the pear collection 245 accessions. Sources for apple scab (*Venturia inaequalis* (Cke.) Wint.) resistance are Pri-

Table 1. Percentage of seedlings with and without mildew infection in the nursery (1990 and 1991).

Cross	Number of plants	Scale (in %) 0	1	2	3	4	5
Starkrimson × 142	56	44,6	51,8	3,6	–	–	–
Russet free Golden × 142	122	40,9	36,4	–	13,6	5,0	4,1
Gloster × 142	32	31,9	59,0	9,1	–	–	–

ma, Redfree, Liberty and Florina all carrying the Vf resistance gene and Freedom with Vf and polygenic resistance from Antonovka (Lespinasse 1989).

Two other sources of resistance are also included: hybrid 142 (Golden Delicious × *M. zumi*) and a hybrid from G. Delicious × *M. niedzwetzkyana* with resistance to scab and mildew (*Podosphaera leucotricha* (Ell. & Ev.) Salm.) (Baev 1974).

In 1986 and 1987 some of these selection (No. 36) also showed resistance to San José scale (*Quadraspidiotus perniciosus*, Comstock) in laboratory and field conditions (Mitkov and Baev 1988).

McIntosh Wijcik is used as a source of compact habit (Lapins & Watkins 1973) and Elise Rathke for weeping habit (Malichenko & Roudenko 1972). Apple seedlings are inoculated in the greenhouse with *Venturia inaequalis* (Brown 1975, R.P. Penev unpublished data).

Two years mildew infection is assessed in the nursery on a scale (0 = no visible symptoms, 1 = very slight infection, 2 = slight infection on leaves and shoots, 3 = moderate, 4 = heavy, and 5 = very heavy infection on leaves and shoots) (Djouvinov & Slavov 1979).

Hybrids from *P. ussuriensis* are used for resistance to pear scab *Venturia pirina* Ederh. and some of which (No. 546 (Williams × *P. ussuriensis*) × Clapp's Favourite) have also shown resistance to San José scale (Mitkov & Baev 1988).

We used embryoculture, following Kolova (unpublished) for seed from the very early ripening (early July) apple cultivars Geneva Early and Vista Bella and the pears, Ranna Bolyarka, Trapezitsa and hybrid 5932 (Beuré Giffard × Echmenka) which ripen between June 15–30 (Iliev et al. 1984, Komitov et al. 1989).

Table 2. Inheritance of compact habit from McIntosh Wijcik to progenies.

Cross	Number of plants	Compact plants (in %)
McIntosh Wijcik × Redfree	119	42,1
McIntosh Wijcik × Liberty	114	37.5
McIntosh Wijcik o.p.	149	46,9

Results and discussion

Apple During the first two seasons in the orchard, eight seedlings of progeny 8642 (Democrat × Melrose) were not attacked by *Aphis pomi* (De Greer). At the same time four seedlings from 8640 (Democrat × Smoothee) showed some slight attack. Resistance to Green Apple Aphid was found on all 9 plants of progeny 8620 (Krimskoe selektsionnoe × Florina).

Investigations in the nursery on mildew susceptibility in progenies 8930 (Starkrimson × 142), 8912 (Gloster × 142) and 8936 (Russet free Golden × 142) showed a high percentage of resistant seedlings, 44,6, 40,9 and 31,9%, respectively. This result is very close to that of Kopan et al. (1989) using hybrid 142 in Kiev. The ratio of resistant:susceptible seedlings is close to 1:1 (Table 1).

In progenies 9127 (McIntosh Wijcik × Redfree), 9129 (McIntosh Wijcik × Liberty) and McIntosh Wijcik-open polinated, the ratio of seedlings with a normal tree habit to those with a compact habit was about 1:1 (Table 2).

Plants with a weeping habit have been selected from progenies 8805 (Elise Rathke × Prima) and 8806 (E. Rathke × Florina). The inheritance of these characteristic will be studied.

Pear A promising seedling, 3155 (Buttira Precoce Morettini × Willams) has attractive fruits, a size similar to B.P. Morettini and very good fresh quality. The ripening time is several days before Willams. Fruit of 2350 F2 (Williams × *P. ussuriensis*) × Clapp's Favourite an F2 derivative of the Pear *Psylla* resistant species *P. ussuriensis*, has satisfactory size (about 100 g). New seedlings at the stage F3 (2350 × Williams, Max Red Bartlett, Packam's Triumph and Passe Crassane) will be evaluated for resistance to Pear *Psylla* following the methods of Bell and Stuart (1990).

Conclusion

The first results of the new breeding programme for disease resistance are very promising. For the stable and durable resistance to the main diseases of apple and pear, we shall use a combination of different genes from various germplasm sources at Plovdiv and from abroad as suggested by Lespinasse (1989) and Parisi et al. (1993). Evaluation for resistance to Green Apple Aphid on plants from hybrids 8642 and 8620 will be continued.

McIntosh Wijcik and Elise Rathke can be successfully used in breeding of new cultivars resistant to scab and mildew with columnar and weeping habit, respectively.

Embryoculture is a successful aid to breeding very early apple and pear cultivars with disease resistance.

Acknowledgements

The author thanks his collaborators, L. Kolova and R. Penev, for the development of the embryoculture method and inoculation in a greenhouse. Acknowledgments are due to the Organizing Committe of Eucarpia Fruit Breeding Section Meeting 1993 and especially to the local organizer Dr M. Kellerhals for financial support for the participation at the Meeting in Switzerland

References

Baev, H., 1974. Investigations on some biological characters of interspecific hybrids in *Malus*. Ph.D. Thesis, Plovdiv, pp. 108–109 (in Bulgarian).

Bell, R.L. & L.C. Stuart, 1990. Resistance in eastern European Pyrus germplasm to Pear Psylla nymphal feeding. Hort. Science 25(7): 789–791.

Brown, A.G., 1975. Apples. In: J. Janick & J.M. Moore (Eds) Advances in Fruit Breeding, pp. 3–37. Purdue University Press, W. Lafayette, Indiana.

Djouvinov, V., 1989. Fruit breeding – Some results and trends of development. Agric. Science 27(1): 54–60 (in Bulgarian).

Djouvinov, V. & N. Slavov, 1979. Apple. In: N. Nedev & J. Grigorov (Eds) The Methods of Investigation on Fruit Genetic Resources, pp. 24–32. Fruit Growing Res. Institute, Plovdiv (in Bulgarian).

Iliev, I., S. Popov, T. Angelov & V. Djouvinov, 1984, Apple cultivars and pear cultivars. In: I. Iliev (Ed) Small Pomology, pp. 36–148. H.G. Danov publ. house, Plovdiv (in Bulgarian).

Komitov, R., V. Djouvinov & T. Tanev, 1989. Ranna bolyarka. Plant Science 26(10): 57–59 (in Bulgarian).

Kopan, V.P., K.N. Kopan & L.D. Boldizeva, 1989. Interspecific hybridization in plant breeding at Ukrainian Fruit Res. Institute. In: G.A. Koursakov (Ed) Interspecific Hybridization and Intensification of Fruit Growing, pp. 28–34. Central Genetical Lab., Michourinsk, (in Russian).

Lapins, K.O. & R. Watkins, 1973. Genetics of compact growth habit. Rep. E. Malling Res. Stn. for 1972: 136.

Lespinasse, Y., 1989. Breeding pome fruits with stable resistance to diseases: 3 Genes resistance mechanisms, present work and prospects. In: C. Gessler and D.J. Butt (Eds) Integrated Control of Pome Fruit Diseases, Vol. II, pp. 100–115. IOBC, Brissago, Switzerland.

Malichenko, V.V. & V.F. Roudenko, 1972. Apple trees for ornamental plantings. Proc. of Appl. Bot. Genetics and Plant Breed. 49: 137–144 (in Russian).

Mitkov, A. & H. Baev, 1988. Resistant hybrids of apple and pear. Nature 37(4): 58–60 (in Bulgarian).

Parisi, L., Y. Lespinasse, J. Guillaumes & J. Krüger, 1993. A new race of *Venturia inaequalis* virulent to apples with resistance due to the Vf gene. Phytopathology 83(5): 533–537.

H. Schmidt and M. Kellerhals (Eds.), Progress in Temperate Fruit Breeding, 131–133.

Results of mutation breeding of apples at RBIP Holovousy

F. Paprštein, J. Blažek & J. Vondráček
Research and Breeding Institute of Pomology, 507 51 Holovousy, Czech Republic

Key words: Apple, irradiation, mutation breeding, cultivars

Summary

Two promising clones provisionally selected in a programme of mutation breeding at Holovousy were evaluated in a trial since 1976–1992 in comparison with standard cultivars. 'Lord Lambourne Red', registered in 1990, proves to be superior to standard 'Lord Lambourne' by better colour of fruit skin and bigger fruit. A dark red washed blush covers nearly the whole surface of the fruit. Fruit can be consumed just after harvest but it stores well till December. This new cultivar has been successfully tested both on the dwarf rootstock M 9 and the vigorous one A 2. Trees are precocious, very productive and their yields are regular. They are, however, as sensitive to Apple rubbery wood virus and to late spring frosts as the original cultivar. 'James Grieve Double Red', registered in 1993, has very attractive fruit covered with bright red nearly all over the surface. In comparison to standard cultivar 'James Grieve' this new mutant has larger and more elongated fruit with a shorter stalk. It flowers about two days later and it is more productive with smaller trees.

Introduction

First experiments on a study of effects of gamma (Co^{60}) irradiation using fruit tree species in Czech Republic were established in 1962 (Vondráček & Ullmann 1963). Later on a programme of mutation breeding concentrated on methodological aspects e.g., identification and isolation of mutations in fruit characters (Blažek & Vondráček 1974), selection of compact growth habits (Paprštein 1988) and improvement of selection techniques (Paprštein 1990).

In the spring of 1965, a mutation breeding programme with apple trees was started by placing potted trees on M 9 rootstock into the gamma field at Prague-Strnady. A chronic irradiation from Cobalt 60 source was applied during the growing season (from June 7th till September 18th). Total doses of irradiation 101 Gy for 'Lord Lambourne' and 62 Gy for 'James Grieve' were applied at the rate of $2.48 \cdot 10^{-5}\,\text{Gy} \cdot \text{s}^{-1}$. The irradiated trees were planted into an experimental orchard at Holovousy in the spring of 1966. There, a selection of mutations was done during the subsequent 10 years. The best mutants were multiplied to establish clones.

From the material that was obtained in the programme two promising mutants of apple (*Malus pumila*) cultivars 'James Grieve' and 'Lord Lambourne' were selected. The evaluation of these mutants in comparison with the original cultivars is the aim of this contribution.

Material and methods

A trial orchard was established in 1976 in a spacing 4 × 2 m on the M 9 rootstock. From every clone 6 trees (not randomised) were planted. Clean herbicide stripes were kept in the rows of trees and grass mulch was maintained between the rows. Trees were trained like freely growing dwarfs. Yields were recorded.

Samples of 10 fruits were cold stored, for evaluation of fruit quality. The characters evaluated, using rating scales from 1 (worst) to 9 (best), were taste, appearence, red colour and juiciness.

Table 1. Mean values of selected yield characters for 1980–1992.

Cultivar	Flower set	Fruit set	Annual yield (kg · tree^{-1})
Lord Lambourne Red	5.5	5.2	17.1
Lord Lambourne	5.7	5.3	16.3
James Grieve Double Red	7.3	6.9	24.7
James Grieve	7.5	6.1	21.4

Results and discussion

The evaluation of yield characteristics for both mutants and their standards is shown in Table 1. The average annual yield per tree was about 0.8 kg higher with 'Lord Lambourne Red' than with the standard. The difference was greater in the case of 'James Grieve Red' which gave about 3.3 kg of fruits per tree more than the standard. 'Lord Lambourne Red' has also very good performance and yields on the vigorous rootstock A 2 (unpublished data).

Average values of selected fruit characters are given in Table 2. Both mutants had bigger fruit than the standards. In 'Lord Lambourne Red' the fruit weight was about 16.6 g (12.7%) and in 'James Grieve Double Red' about 14.4 g (9.7%) greater than in the standard. 84% of fruits of 'Lord Lambourne Red' exceeded 60 mm in the diameter whereas only 57% did so in standard cultivar. However, the greatest advantage of the mutants is the amount of red cover colour on their fruits. Both mutants have the colour distributed nearly all over the surface. Therefore their appearance is more attractive than that of the standards. There were no clear differences between mutants and standards in the other fruit characters: taste and juiciness.

Both mutants are promising for commercial growing in Czech Republic as fruit growers already have good experience of the original cultivars. The acreage of 'Lord Lambourne Red' has been growing significantly during recent years and this cultivar, the product of mutation breeding, will soon be an important early winter season cultivar in this country.

Both selected mutants from 'James Grieve' and 'Lord Lambourne' distinguished themselves by regular high yield and very attractive fruits from the very beginning. Both clones were sent to the State Variety Trials in 1983 under a designation HL 111 A ('Lord Lambourne') and HL 166 B ('James Grieve'). The first mutant was named 'Lord Lambourne Red' and it was registered for a commercial use and legal protection in 1990. The latter was named 'James Grieve Double Red'. It was registered for a commercial use and legal protection in 1993.

Description of the new cultivars

'Lord Lambourne Red'

Fruit Size is intermediate or, if well grown, intermediate to large. Shape is globose. Green-yellow ground colour of skin is covered by dark-red over nearly the entire surface.

Tree Medium vigorous, spreading. Productivity is better than that of the standard and yields are regular. Trees are very precocious, first fruits appearing very often in the second year in the nursery. The productivity is very good on all common apple rootstocks.

The average time of ripening at Holovousy is about mid September, but the harvest period is quite short as ripe fruits soon drop. In cold store (2 °C) fruits keep well till the end of December. Ability to transport is very good.

Trees are medium susceptible to scab but they are tolerant to mildew. This cultivar like its standard is sensitive to the Apple rubbery wood mycoplasm disease. Trees are also somewhat sensitive to late spring frosts and therefore the cultivar is not suitable for the more severe climatic conditions of high altitudes.

The cultivar has not any special demands for training. With ageing, stronger pruning is recommended to encourage new growth of shoots.

The flowering period is early and its duration is intermediate. 'Akane', 'Diadém', 'James Grieve Double Red' and 'Oldenburg' are recommended as suitable pollinators. Cross-incompatibility was found with 'Šampion'.

'James Grieve Double Red'

Fruit Size is intermediate to large. Shape is globose conical, somewhat more elongated than with standard 'James Grieve'. Appearance is improved to a great extent by solid washed red over the whole surface. The stalk is shorter and flesh taste a little sweeter in comparison to the standard.

Tree Medium vigorous to weak growth and very precocious, very productive with regular yields.

Table 2. Mean values of selected fruit characters for 1980–1992.

Cultivar	Fruit weight (g)	Taste	Appearance	Cover colour distribution	Juiciness
Lord Lambourne Red	146.8	7.3	7.9	8.1	6.3
Lord Lambourne	130.2	7.2	6.3	5.4	6.2
James Grieve Double Red	162.5	6.9	7.4	7.6	7.4
James Grieve	148.1	7	5.9	4.7	7.5

The time of ripening at Holovousy is at the beginning of September. Fruits are suitable for short-time storage till October.

Trees are medium susceptible both to scab and mildew like the standard cultivar. The period of flowering is about 2 days later than the standard. 'Alkmene', 'Idared' and 'Lord Lambourne Red' are recommended pollinators for the cultivar.

References

Blažek, J. & J. Vondráček, 1974. Studium účinkůgama záření na vyvolání mutací. ZZ VÚO Holovousy: 1–123.

Paprštein, F., 1988. Improvement of selection methods for the compact growth character in the mutation breeding of apple cultivars. Acta Horticulturae 224: 395–412.

Paprštein, F., 1990. Improved selection technique in mutational breeding. Tag.-Ber., Landwirtsch. Wiss. DDR, Berlin 292: 187–191.

Vondráček, J. & J. Ullmann, 1963. Primární účinky gama paprsků na jabloně. Věd. práce ovoc. VÚO Holovousy 2: 197–211.

H. Schmidt and M. Kellerhals (Eds.), Progress in Temperate Fruit Breeding, 135–139.

Fruit varieties from Belarus

Zoiya A. Kozlovskaya
Belorussian Research Institute for Fruit Growing, Samokhvalovitchy, Minsk Region 223013, Belarus

Key words: Apple, Belarus, cherry, cherry-plum, pear, plum, selection, varieties

Summary

The main aim in fruit breeding (apple, pear, plum, cherry-plum, sour cherry and sweet cherry) in Belarus is winterhardiness and high fruit yielding. Great success was achieved in apple and pear breeding for resistance to fungi and long term storage, bacterial canker and scab. Winterhardy varieties of stone fruit crops were selected. The most important varieties of apple, pear, plum, cherry-plum, sour cherry and sweet cherry are described.

Introduction

Belarus occupies 207'000 km^2 between longitude 23° 11′–32° 54′ E and latitude 54° 14′–56° 10′ N. Soils are rather poor mainly sands and loamy sands. The climate is moderately warm and moist and is influenced by atlantic and arctic air masses. Thus in spring and autumn the weather is unstable. Winters are mild with long-lasting thaws (30–60 days for 3 winter months).

The growing season between May and October is 180–190 days. As a rule there is a frostless period of 4 months and a half. Summer is not hot (mean temperature in July is 17–19 °C), and it often rains. Total of active temperatures above 10 °C ranges between 2000–2200° (in cold years – 1650–1850°, in hot ones – 2400–2600°). But in general climatic conditions of the republic differ greatly from year to year and within a year.

These conditions determine the aims of fruit breeding. A.E & E.P. Syubarovs started this work in our Institute at the beginning of the 20th Century. For many years they and other researchers bred a number of varieties of apples, pears, plums, cherry plums, sour cherries and sweet cherries (Syubarova 1964, Syubarov 1968, Matveev et al. 1982, Syubarova & Semashko 1983, Kovalenko 1989). Apple varieties such as Alesya, Antei, Bananovoye, Belorusskii Sinap, Belorusskoye Malinovoye, Pamyat Sikory, Pamyat Syubarovoi and Charaunitsa have long term storage, high resistance to low temperatures and polygenic resistance to scab (Kovalenko & Evdokimenko 1986, Semashko personal communication). G.K. Kovalenko developed promising apple selections carrying genes Vf, Vm, Vr and having ecological adaptation. The second direction of apple breeding is the development of spur type habit and fruit of definite colour, size and taste.

Pear growing is very important for farmstead orchards in our republic. Cvs. Belorusskaya Pozdnyaya, Belorusska, Bere Loshitskaya, Dukhmyanaya have valuable agricultural characters such as high yield, high winterhardiness and good quality (Myalik 1984).

At present the main attention is paid to the development of pear varieties resistant to bacterial canker (*Pseudomonas syringae* van Hall) and scab. M.G. Myalik & N.A. Konovalova have already got promising hybrids (Myalik & Konovalova 1993).

Plum is widely used in belorussian orchards. As a result of breeding programmes a number of plum varieties were developed by E.P. Syubarova, V.A. Matveev & Z.A. Kozlovskaya among them Narach, Vitebskaya Pozdnyaya, Kroman. Cherry-plum is a new fruit crop for our republic. But V.A. Matveev & M.P. Malyukevitch have bred the first winterhardy varieties such as Mara, Naidyona and Vetraz (Matveev & Malyukevitch 1987).

E.P. Syubarova obtained the first winterhardy forms of sweet cherry in the climatic conditions of the republic. The varieties Narodnaya, Severnaya, Zhurba, Pobeda are today of importance. R.M. Sulimova & M.I. Vyshinskaya obtained new varieties of sour cherry and sweet cherry which have good taste, resistance to low temperatures and to cherry leaf spot (*Coccomyces hiemalis* Higg.).

Material and methods

The original parent material consisted of local and widely used varieties and later a number of valuable hybrids developed by the researchers were used for crossing.

Results

The important fruit varieties growing in Belarus are describeded below. Some fruit characters are given in Table 1.

Apple

Alesya (Belorusskoye malinovoe × Bananovoe)
Vigour of the tree is medium, the shape of the crown is rounded. Fruit shape is oblate with smooth mid-thick skin of yellow ground colour and red intensive over-colour, white flesh. Sour-sweet taste. Late harvest maturity, medium vegetative period, regular fruit bearing. Winterhardy, resistant to apple scab. Dry matter content 12–13%, titratable acidity 0.59–0.97%, total sugar 10.1–10.98%, ascorbic acid 11.8 mg%.

Antei (Belorusskoye malinovoe × 59–13/9)
Vigour of the tree is medium, shape of the crown is rounded. Fruit shape is inversely conical with smooth mid-thick skin of green ground colour, dark red intensive over colour, light green flesh. Sour-sweet taste. Late ripening, long vegetative period, regular fruit bearing, highly resistant to low temperatures, moderately resistant to apple scab, highly productive. Dry matter content 13.2%, titratable acidity 0.65%, total sugar 9.4%, ascorbic acid 20.3 mg%.

Belorusskii sinap (Antonovka × Pippin Litovski)
Vigour of the tree is strong, shape of the crown is wide-pyramidal. Fruit shape is conical with dry thick skin with green ground colour and weak blush brick over colour, light green flesh. Sour-sweet taste. Ultra-late harvest maturity, long vegetative period, regular fruit-bearing, absence of fruit drop, extremely resistant to low temperatures, highly resistant to apple scab, very productive. Dry matter content 12.5%, total sugar 9.3%, titratable acidity 0.82%, ascorbic acid 8.3 mg%.

Belorusskoye malinovoye (Antonovka × Lavfam)
Vigour of the tree is strong, the shape of the crown is rounded. Fruit shape is oblate with smooth thin skin of light green ground colour, bright red over colour of striped pattern, light green flesh. Sour-sweet taste. Late ripening, medium vegetative period, regular fruit- bearing, absence of fruit drop. Extremely resistant to low temperatures, susceptible to scab. Highly productive. Dry matter content 12.8%, total sugar 9.6%, titratable acidity 0.78%, ascorbic acid 14,5 mg%.

Charaunitsa (Belorusskii sinap × Cox's Reinette)
Vigour of the tree is medium, the shape of the crown is rounded. Fruit shape is flat with smooth thick oily skin of yellow ground colour and light red striped over colour, yellow flesh. Sweet taste. Late harvest maturity, long vegetative period, regular fruit bearing, absence of fruit drop. Extremely resistant to low temperature and moderately resistant to scab. Highly productive. Dry matter content 13.99%, total sugar 11.35%, titratable acidity 0.47%, ascorbic acid 3.08 mg%.

Minskoye (Wealthy o.p.)
Vigour of the tree is medium, the shape of the crown is rounded. Fruit shape conical with smooth mid-thick skin of yellow ground colour and light red striped over colour, light yellow flesh. Sour-sweet taste. Medium harvest maturity and vegetative period, regular fruit bearing. Extremely resistant to low temperature and to apple scab. Highly productive. Dry matter content 12%, total sugar 9.8%, titratable acidity 0.84%, ascorbic acid 12.3 mg%.

Pamyat Sikory (Folvel × Simirenko's Renette)
Vigour of the tree is weak, the shape of the crown is rounded. Fruit shape is oval-conical with smooth mid-thick skin of green ground colour and weak blush brick over colour, light green flesh. Sour-sweet taste. Late ripening, long vegetative period, regular fruit bearing, absence of fruit drop. Highly resistant to low temperatures and to apple scab. Very productive. Dry matter content 13.13%, total sugar 10.31%, titratable acidity 0.47%, ascorbic acid 8.3 mg%.

Table 1. Mean of fruit characters.

Cultivar	Fruit weight g	Yield (1–9) a	Storage life, b month	Dry matter %	Titratable acidity %	Total sugar %	Ascorbic acid mg %
Apple							
Alesya	130	7	4	12,5	0.97	11,0	11,8
Antei	228	9	3	13,2	0,65	9,4	20,3
Belorusskii sinap	123	7	5	12,5	0,85	9,3	8,3
Belorusskoye malinovoye	137	6	3	12,8	0,78	9,6	14,5
Charaunitsa	142	6	3	14,0	0,47	11,4	3,1
Minskoye	235	9	12	12,0	0,84	9,8	12,3
Pamyat Sikory	115	8	5	13,1	0,47	10.3	8,3
Pamyat Syubarovoi	135	8	4	13,0	0,78	9,5	10,4
Zaslavskoye	215	7	4	12,7	0,62	9,8	8,5
Pear							
Belorusskaya pozdnyaya	110	9	2	12,3	0,13	9,0	13,4
Duchmyanaya	190	9	9	12,3	0,13	9,2	13,0
Plum cherry							
Mara	23	9		14,8	1,53	10,1	
Naidyona	27	9		10,5	1,18	8,1	
Plum							
Kroman	35	7		18,6	1,10	12,1	
Narach	28	8		14,8	1,0	9,0	
Sour cherry							
Venok	4,7	9		11,4	1,10	7,8	5,2
Zhyvitsa	4,3	7		13,9	0,81	8,8	4,2
Sweet cherry							
Gronkavaya	5,2	9		15,8	0,55	12,3	4,6
Pobeda	7,8	7		15,0	0,53	11,2	1,7

a) scale: 1 = very low to 9 = very high.
b) months: 1 = January to 12 = December.

Pamyat syubarovoi (Seruel × Belorusskii Sinap)
Vigour of the tree is medium, the shape of the crown is rounded. Fruit shape is oval-conical with smooth oily mid-thick skin of light yellow ground colour and red over colour of striped pattern, white flesh colour. Sour-sweet taste. Late harvest maturity long vegetative period, regular fruit bearing, absence of fruit drop. Highly resistant to low temperatures and to apple scab. Highly productive. Dry matter content 13.0%, total sugar 9.53%, titratable acidity 0.78%, ascorbic acid 10.4 mg%.

Zaslavskoye (Wealthy × 11/9) × (Jois × Babushkino)
Vigour of the tree is medium, the shape of the crown is rounded. Fruit shape is globose-conical with smooth oily thick skin of green ground colour and dark red over colour, light green flesh. Sweet-sour taste. Ultra-late harvest maturity, long vegetative period, regular fruit bearing, absence of fruit drop. Extremely resistant to low temperature and moderately resistant to scab. Medium productive. Dry matter content 12.7%, total sugar 9.8%, titratable acidity 0.62%, ascorbic acid 8.5 mg%.

Pear

Belorusskaya pozdnyaya (Bonne Louise d'Avranches)
Vigour of the tree is weak, the shape of the crown rounded. Fruit is pear-shaped with dry-scabrous thick skin of yellow-orange ground colour, absence of over

colour, white flesh colour. Sweet taste. Late ripening, early and regular fruit bearing, absence of fruit drop. Extremely resistant to low temperatures and susceptible to pear scab. Highly productive. Dry matter content 12.3%, total sugar 8.96%, titratable acidity 0.13%, ascorbic acid 13.4 mg%.

Duchmyanaya (Alexandrovka × Favorite de Clapp)
Vigour of the tree is medium, the shape of the crown is rounded. Fruit is pear-shaped with smooth oily mid-thick skin of green ground colour and weak blush red over colour, white flesh colour. Sour-sweet taste. Late-summer maturing, extremely short vegetative period, early and regular fruit bearing. Highly resistant to low temperatures and highly resistant to pear scab. Highly productive. Dry matter content 12.26%, total sugar 9.18%, titratable acidity 0.13%, ascorbic acid 13.0 mg%.

Cherry plum

Mara (18/1 (Pr. cerasifera × Pr. ussuriensis) O.p.)
Vigour of the tree is medium, the shape of the crown is rounded. Fruit shape is rounded with yellow skin, soft yellow flesh, small stone. Sour-sweet taste. Late uniform ripening, early fruit bearing. Winterhardy and resistant to *Monilia* and hole leaf spot. Highly productive. Dry matter content 14.83%, total sugar 10.13%, titratable acidity 1.53%.

Naidyona (Scoroplodnaya × Desertnaya)
Vigour of the tree is weak, shape of the crown is rounded. Fruit is large and eggshaped with red skin and yellow flesh, mid-sized stone. Sour-sweet taste. Early ripening and fruit bearing, highly productive. Winterhardy, resistant to *Monilia* and hole leaf spot. Dry matter content 10.50%, total sugar 8.14%, titratable acidity 1.18%.

Plum

Kroman (Perdrigon × d'Agen)
Vigour of the tree is medium, the shape of the crown is oblate. Fruit is rounded with dark red ground colour and blue over colour, firm yellow flesh, mid-sized stone. Sour-sweet taste. Mid-early uniform ripening, winterhardy, resistant to *Monilia* and hole leaf spot, highly productive. Dry matter content 18.6%, total sugar 12.1%, titratable acidity 1.1%.

Narach (Altesse Ordinaire × Althana)
Vigour of the tree is strong, the shape of a crown is oval. Fruit is oval-rounded with light green ground colour of skin and brick-red intensive over colour, yellow flesh colour, intermediate-firm flesh, mid-sized stone. Sour-sweet taste. Late ripening, highly winterhardy, very productive. Resistant to *Monilia* and hole leaf spot. Dry matter content 14.8%, total sugar 8.97%, titratable acidity 1.0%.

Sour cherry

Venok (Novodorskaya O.p.)
Vigour of the tree is strong, the shape of the crown is pyramidal. Fruit shape is rounded with dark red skin, flesh and juice, small stone. Sour-sweet taste. Winterhardy, highly productive, mid-early uniform ripening. Resistant to cherry leaf spot and *Monilia*. Dry matter content 11.4%, total sugar 7.82%, titratable acidity 1.1%, ascorbic acid 5.2 mg%.

Zhyvitsa (Cerise d'Ostheim × Bigarreau Jaune de Doenissen)
Vigour of the tree is medium, the shape of the crown is rounded. Fruit is round with dark red skin, flesh and juice, small stone. Sour-sweet taste. Early and uniform fruit ripening, winterhardy. Resistant to *Monilia* and cherry leaf spot. Highly productive. Dry matter content 13.9%, total sugar 8.73%, titratable acidity 0.81%, ascorbic acid 4.2 mg%.

Sweet cherry

Gronkavaya (Severnaya o.p.)
Vigour of the tree is medium, the shape of the crown is wide-pyramidal. Fruit is heart-shaped with dark red skin, flesh and juice, mid-firm flesh, small stone. Sweet taste. Early and uniform fruit ripening. Winterhardy and resistant to cherry leaf spot. Highly productive. Dry matter content 15.84%, total sugar 12.3%, titratable acidity 0.55%, ascorbic acid 4.6 mg%.

Pobeda (Bigarreau Gaucher o.p.)
Vigour of the tree is strong, the shape of the crown is wide-rounded. Fruit shape is wide-heart with dark red skin, flesh and juice, firm flesh, small stone. Early and uniform fruit ripening, highly productive, resistant to cherry leaf spot and *monilia*. Dry matter content 15.0%, total sugar 11.24%, titratable acidity 0.53%, ascorbic acid 1.69 mg%.

References

Kovalenko, G.K. & V.M. Evdokimenko, 1986. Stepen ustoichivosti sortov yabloni k parshe v Belorussii. In: A.V. Kruglyakov (Ed) Plodovodstvo. Minsk, Vyp. 6, pp. 19–22 (in Russian).

Kovalenko, G.K., 1989. Novye sorta yabloni. Selskoye khozyaistvo Belorussii 1: 39–40 (in Russian).

Matveev, V.A., G.K. Kovalenko & R.M. Sulimova, 1982. Zimostoikost plodovyh cultur v usloviyah Belorussii. In: V.A. Gryazev (Ed) Sbornik nauchnyh trudov VNIIS, vyp. 35, pp. 54–58 (in Russian).

Matveev, V.A. & M.P. Malyukevitch, 1987. Selektia diploidnyh vidov slivy na zimostoikost. In: A.F. Bogdanovskii (Ed) Plodovodstvo, vyp. 7. Minsk, pp. 3–5 (in Russian).

Myalik, M.G. 1984. Pear. In: A.G. Voluznev (Ed) Sorta plodovyh i yagodnyh cultur v Belorusskoi SSR, pp. 51–77. Urozhai, Minsk (in Russian).

Myalik, M.G. & N.A. Konovolova, 1993. Selektia grushi na ustoichivost k boleznyam v Belarusi. In: E.N. Sedov (Ed) Otdalyonnaya gibridizaciya i polyploidiya v selektii plodovyh i yagodnyh cultur. Orel, p. 38 (in Russian).

Syubarova, E.P., 1964. Chereshnya. Minsk, Urozhai: 108 (in Russian).

Syubarova, E.P. & E.V. Semashko, 1983. Novyi sort yabloni. Selskoye khozyaistvo Belorussii 3: 44 (in Russian).

Syubarov, A.E., 1968. Yablonya. Minsk, Urozhai: 280 (in Russian).

H. Schmidt and M. Kellerhals (Eds.), Progress in Temperate Fruit Breeding, 141–145.

The Pillnitz apple rootstock breeding programme results

Manfred Fischer

Genbank Obst Dresden-Pillnitz am Institut für Pflanzengenetik und Kulturpflanzenforschung Gatersleben, D-01326 Dresden, Germany

Key words: Apple rootstock breeding, new varieties, testing programme

Summary

The Pillnitz apple rootstock breeding programme was concluded with the following results:

1. Three dwarfing rootstocks were submitted for breeder's rights
Pillnitzer Supporter 1 (Pi-AU 7–33): M 9 × *M. baccata* (L.) BORKH. (*M. pumila* × *M. baccata ?*);
Pillnitzer Supporter 2 (Pi-AU 9–16): M 9 × *M. micromalus* MAKINO;
Pillnitzer Supporter 3 (Pi-AU 9–82): M 9 × *M. micromalus* MAKINO;
2. Four dwarfing to semi dwarfing rootstocks were prepared for application for breeder's rights:
Pi-AU 36–2: M 9 × (M 9 × *M. baccata* (L.) BORK.);
Pi-AU 51–4: M 4 open pollinated;
Pi-AU 51–11: M 4 open pollinated;
Pi-AU 56–83: M 11 open pollinated.
3. The rootstock Pi 80 (Pillnitzer Supporter 4) formerly bred at Pillnitz is at present being propagated by the French nursery Demol/Mondragon.
4. Numerous methodical findings were integrated into the breeding process.

The aims of the rootstock breeding programme were, the improvement of propagation ability, dwarfness, resistance to biotic and abiotic damage, better stability and a positive influence on the yield and fruit quality of the cultivars. Not all aims could be realized in one genotype. Propagability is better in the M 4-progenies but not so good in the wild species populations, for which we recommend propagation by softwood cuttings. Stability could not be obtained together with dwarfing. Different rootstocks give indications of resistance against scab, winterfrost, fireblight or mildew. – A completly resistant new rootstock must be the aim for the future. The most important results of our breeding programme are the much improved fruit quality and earlier and higher yield of cultivars on the dwarfing Pillnitzer Supporter rootstocks.

Introduction

There has been a long tradition of rootstock breeding at Pillnitz. In 1911, Schindler (Schindler 1921) started testing apple rootstocks and later those of pears. He corresponded with Hatton (Hatton 1935) of Great Britain on the systematic revision of *Malus* rootstocks. This included the breeding work of Maurer of Germany (Maurer 1939). Schindler worked with European rootstocks and sought to improve them by breeding and clonal selection. By 1920, he had begun his selection involving M 9. One result was the rootstock 'Pillnitzer Reihe no. 80', named 'Pi 80'. It was selected from a M 4 × M 9 cross and was tested in the GDR by Müller. He described Pi 80 as a semi-dwarfing rootstock like M 26 (Müller 1956). It has good compatibility with cultivars and the mechanical strength needed by vigorous scion varieties. The French nursery DEMOL at Mondragon has had good results with Pi 80 and owns the propagation rights.

Börner & Gollmick of Naumburg (Börner & Seeliger 1941) worked using *Malus* species as donors of genes for scab, mildew, woolly aphid and frost resistance. Their rootstock breeding was continued by

Table 1. Populations of the Pillnitz apple rootstock breeding programme.

Cross Combination	Selected Clones		
	No.	On field testing	Promising
M 9 × *M. micromalus* MAKINO	125	46	6
M. pumila (MILL.) × M 9	7	3	0
M 9 × *M. baccata* (L.) BORKH. (M. pumila MILL.-bastard ?)	38	13	1
M 9 × *M. baccata* (L.) BORKH.	2	1	0
M 9 × (M 9 × *M. baccata* (L.) BORKH:)	17	8	1
(M 9 × *M. baccata* (L.) BORKH.) × M 9	33	13	0
M 9 × *M. baccata* (L.) BORKH. var. mandshurica (MAXIM.) SCHNEID.	1	0	0
M 9 × *M. prunifolia* (WILLD.) BORKH.	3	0	0
M. prunifolia × M 9	8	0	0
M 9 × *M. baccata* (L.) BORKH. var. himalaica (MAXIM.) SCHNEID.	2	1	0
M. baccata (L.) BORKH. var. himalaica (MAXIM.) SCHNEID. × M 9	1	1	0
M 9 × *M. prunifolia* (WILLD.) BORKH. 'xanthocarpa'	2	1	0
M. baccata (L.) BORKH. var. himalaica (MAXIM.) SCHNEID. × *M. prunifolia* (WILLD.) BORKH.	1	0	0
M. prunifolia (WILLD.)BORKH. × *M. floribunda* VAN HOUTTE	7	2	0
M. prunifolia (WILLD.)BORKH. × Charlamovsky	2	1	0
M. prunifolia (WILLD.) BORKH. × Weißer Klarapfel	3	1	0
M. prunifolia (WILLD.) BORKH. × Cox Orangen	2	0	0
M. sieboldii (REGEL) REHDER × Oldenburg	2	0	0
M. sieboldii (REGEL) REHDRE × Ontario	5	5	0
M. sieboldii (REGEL) REHDER × Dülmener Rosenapfel	2	1	0
M. sieboldii (REGEL) REHDER × Cox Pomona	2	1	0
Northern Spy × (*M. pumila* MILL. × *M. sieboldii* (REGEL) REHDER)	2	1	0
M 4 × Antonovka	65	2	1
M 4 × Weißer Klarapfel	60	1	1
Weißer Klarapfel × M 4	84	5	2
M 4 × Roter Trierer Weinapfel	40	0	0
Roter Trierer Weinapfel × M 4	74	0	0
Früher Victoria × M 4	16	0	0
M 4 open poll.	51	5	2
M 11 open poll.	190	11	2
Northern Spy open poll.	27	1	1
M 9 × Früher Victoria	8	2	0
M 9 × Antonovka	1	0	0
Antonovlka × M 9	1	1	0
M 9 × Weißer Klarapfel	9	0	0
Weißer Klarapfel × M 9	2	0	0
M 9 × Anurca	2	0	0
Anurca × M 9	2	2	0
M 9 × Roter Trierer Weinapfel	6	1	0
Simirenko × M 9	4	0	0
MM 106 × Antonovka	59	0	0
Antonovka × MM 106	24	1	0
MM 106 × Roter Trierer Weinapfel	14	0	0
MM 106 × Annurca	2	0	0
MM 106 open poll.	5	0	0
	1013	129	17

Schmadlak (1970) at Pillnitz. In 1965 Fischer assumed control of the material and had concluded the selection of existing and new crosses by 1983 (Fischer 1979, 1986, 1989, 1990, Fischer & Böttcher 1978). More than six test trials with ten different scion varieties were carried out. The last trial was planted in 1991/92. The outcome of earlier work resulted in the selection of seven new rootstocks. Applications for breeder's

Table 2. Stoolbed performance of selected rootstock clones (average of 4 sites, 4 years' results).

Rootstock	Rooted Layers for field no/year	% Rooted Layers for field of all rooted layers	Evaluation of rooting (5 – 1)*	Evaluation of feathering of layers (9 – 1)**
Supporter 1	3.75	68	2.7	6.6
Supporter 2	3.80	59	2.1	4.8
Supporter 3	3.95	63	2.2	5.9
Pi-AU 36–2	2.68	44	2.9	6.7
Pi-AU 51–4	8.73	61	3.7	7.4
Pi-AU 51–11	4.80	43	2.2	7.4
Pi-AU 56–83	5.26	61	3.3	8.0
M 9	2.13	7	2.4	7.6
M 4	4.35	62	3.0	7.9

* 5 = very good rooting; 1 = without roots.

** 9 = without feathering; 1 = total feathering.

rights have been made for the following three dwarfing rootstock varieties:

- Pillnitzer Supporter 1 (R): M 9 × *M. baccata* (L.) BORKH. (otherwise *M. pumila* × *M. baccata* – bastard);
- Pillnitzer Supporter 2 (R): M 9 × *M. micromalus* MAKINO;
- Pillnitzer Supporter 3 (R): M 9 × *M. micromalus* MAKINO.

Also the following dwarfing to semi-dwarfing types have been prepared for breeder's rights:

- Pi-AU 36–2: M 9 × (M 9 × *M. baccata* (L.) BORKH.);
- Pi-AU 51–4: M 4 open pollinated;
- Pi-AU 51–11: M 4 open pollinated;
- Pi-AU 56–83: M 11 open pollinated.

The objectives of this last Pillnitz apple rootstock breeding programme were to improve propagation and size-controlling potential. Other aims were to enhance resistance to biotic and abiotic stresses, to improve mechanical strength and promote cultivar yields and fruit quality.

Material and methods

Initially, crosses were made between *Malus* species and M 9. These were followed by crosses involving M- and MM-rootstocks and robust varieties such as: Antonovka, Annurca, Early Victoria, Northern Spy, Roter Trierer Weinapfel, Yellow Tranparent and others. Many seeds were collected from open pollinated rootstock plants. The cross populations are shown in Table 1.

Early selection methods are essential for efficient rootstock breeding. These involve seedling, stoolbed, nursery and orchard testing. All stages have to be carefully reviewed.

All seedlings were first tested in stoolbeds. Using artificial inoculation techniques, seedling plants were tested for resistance to scab (*Venturia inaequalis*), fireblight (*Erwinia amylovora*) and collar rot (*Phytophthora cactorum*). Resistance to mildew (*Podosphaera leucotricha*) and frost was recorded in stoolbeds. (Fischer & Fischer 1989, Fischer 1992) The softwood cutting propagation potential was assessed in progenies of wild species. In the nursery the budded rootstocks were tested for graft compatibility and vigour by measuring the one-year old maiden trees. At that stage, the rootstocks were tested for natural infestation by woolly apple aphids. The rate of elimination was more than 70%.

Pomological testing was done under a minimal pruning regime. After four harvests the cropping potential of rootstock/cultivar combinations was evaluated. Yield was determined in relation to tree size (Fischer & Pätzold 1991). Promising clones were quickly multiplied using in vitro propagation.

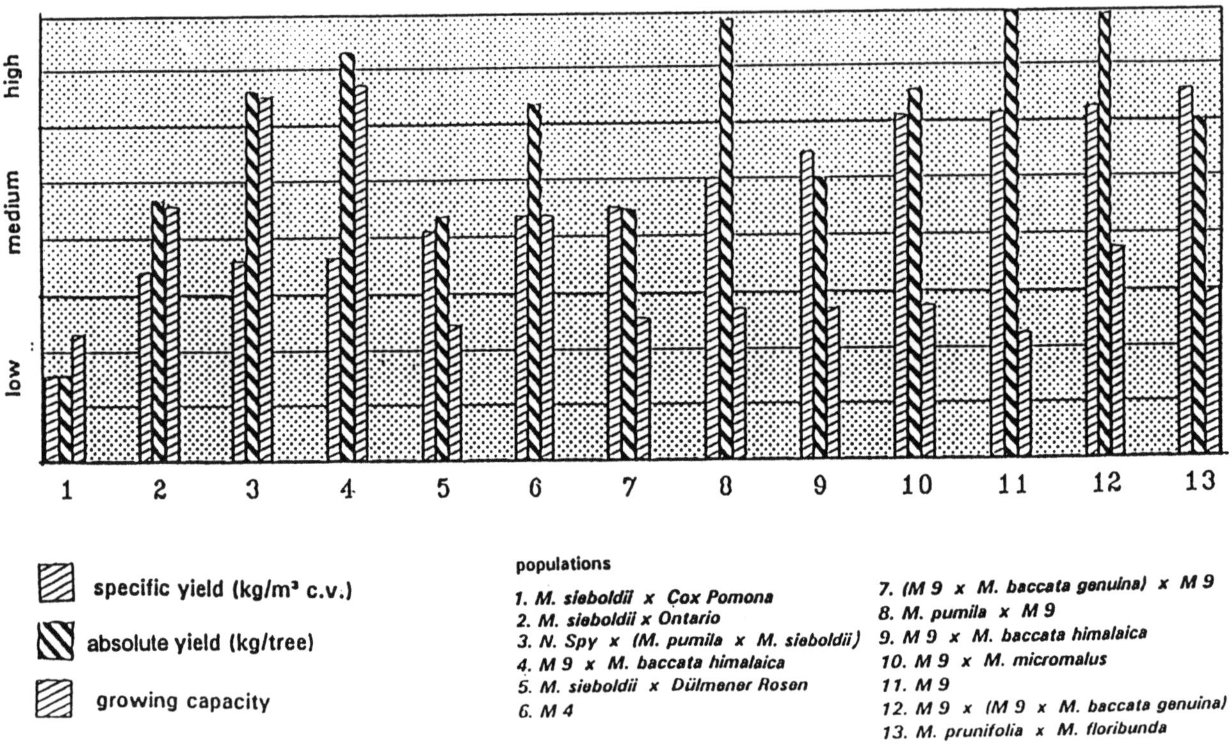

Fig. 1. Specific and absolute yield and growth capacity of progenies of interspecific *Malus* crosses, budded with 3 cultivars after 10 years.

Table 3. Propagation by softwood cuttings (average of 10 years, standard rootstocks for 5 years).

Rootstock	% Rooted Cuttings		% Plants growing in the field
	Average	Best Result in any year	
Supporter 1	58.7	82	93.2
Supporter 2	63.8	89	83.8
Supporter 3	60.2	90	86.5
Pi-AU 36–2	66.1	88	85.1
M 4	44.9	75	90.3
M 9	19.0	52	86.7
M 26	31.7	48	74.3
MM 106	30.0	38	90.1

Table 4. Yield and growing capacity of selected rootstock clones (3 cultivars, 6 sites, 9 years) (crown volume, average of 4 and 5 years before pruning).

Rootstock	Specific Yield		Crown volume % of M 9
	(kg/m^3) crown volume	% of M 9	
Supporter 1	7.49	110	78
Supporter 2	8.13	119	84
Supporter 3	7.81	115	92
Pi-AU 36–2	6.39	94	129
Pi-AU 51–4	6.18	91	126
Pi-AU 51–11	7.58	111	94
Pi-AU 56–83	8.27	122	68
M 4	3.99	59	187
M 26	7.12	105	118
MM 106	5.65	83	177
M 9	6.80	100	100

Results and discussion

The results of the Pillnitz rootstock breeding programme are summarized in Tables 2–4. Table 2 contains the results of the stoolbed evaluations. Table 3 shows the results of propagation by softwood cuttings whilst Table 4 gives details of growth capacity and yield capacity (specific yield) in kg/m^3 crown volume.

It is impossible to predict the growth potential of grafted trees from the growth capacity of unworked rootstock plants when wild species are used. The left hand columns of Fig. 1 shows specific yield, those in the middle are yield per tree whilst the right hand columns display the growing capacity of the thirteen progenies. The three characters are not correlated.

The dwarfing Pillnitzer Supporter rootstocks developed in this breeding programme are outstanding because of the excellent fruit quality they induce in cultivars and their ability to produce earlier and higher specific yields.

Not all aims could be realized in a single genotype. The propagation potential is better in M 4 progenies than in the wild species populations. Propagation by softwood cuttings is recommended for the latter. Mechanical strength and dwarfing potential was not ideally combined in any of the selections. Rootstocks varied in their resistance to scab, winter injury, fireblight and mildew. Completely resistant new rootstocks must be the objective in future breeding programmes.

The three recommended dwarf rootstocks have the following characteristics:

1. *Pillnitzer Supporter*$^{(R)}$ *1*

Breeding no.: Pi-AU 7–33.

Breeder: Institut für Obstforschung Dresden-Pillnitz/M. Fischer.

Origin: M 9 × *Malus baccata* (L.) BORKH. (*M. pumila* MILL. × M. *baccata ?*).

Field testing: since 1965 with 8 cultivars at 8 sites.

Main characteristics:

- dwarfing rootstock;
- vegetative propagation under mist (softwood cuttings) better than layering;
- layering efficiency of mother plants similar to M 9;
- early and high yield of the cultivars, 110% compared to M 9;
- vigour 80–90% compared to M 9;
- increased colour intensity of fruits.

2. *Pillnitzer Supporter*$^{(R)}$ *2*

Breeding no.: Pi-AU 9–16.

Breeder: Institut für Obstforschung Dresden-Pillnitz/M. Fischer.

Origin: M 9 × *Malus micromalus* MAKINO.

Field testing: since 1965 with 8 cultivars at 10 sites.

Main characteristics:

- dwarfing rootstock;
- vegetative propagation under mist better than layering;
- layering efficiency of mother plants 90% compared with M 9;
- early and high yield of the varieties, 110–120% compared to M 9;
- vigour 90–100 % compared to M 9;
- increased colour intensity of fruits.

3. *Pillnitzer Supporter*$^{(R)}$ *3*

Breeding no.: Pi-AU 9–82.

Breeder: Institut für Obstforschung Dresden-Pillnitz/M. Fischer & G. Pätzold.

Origin: M 9 × *Malus micromalus* MAKINO.

Field testing: since 1972 with 6 varieties at 5 sites.

Main characteristics:

- dwarfing rootstock;
- vegetative propagation under mist better than layering;
- layering efficiency of mother plants 90% compared to M 9;
- yield of the cultivars 110% compared to M 9;
- vigour 90% compared to M 9;
- well coloured fruits

References

Börner, C. & R. Seeliger, 1941. Die Naumburger Sämlingsanzuchten von Reben und Äpfeln. Mitt. Biol. Reichsanstalt 63: 70–71.

Fischer, M. 1979. Wuchsleistung und Ertrag von Apfelsorten auf Malus-Artbastard-Unterlagen. Tag. Ber. Akad. Landwirtsch.-Wiss. DDR, Berlin, Nr. 174: 247–256.

Fischer, M. 1986. Untersuchungen zur rationellen Bewertung von Apfelunterlagen für Produktion und Züchtung. Promotion B, Akad. Landwirtsch.-Wiss. Berlin, 2 Bd.

Fischer, M. 1989. Frühzeitige Ermittlung der obstbaulichen Leistungsfähigkeit von Apfelunterlagen-Neuzüchtungen. Archiv Gartenbau 37: 57–72.

Fischer, M. 1990. Vorschlag für ein System zur rationellen Prüfung neuer Apfelunterlagen. Tag. Ber. Akad. Landwirtsch.-Wiss. DDR, Berlin, Nr. 292: 101–108.

Fischer, M. 1992. Frostresistenzuntersuchungen an *Malus*-Artbastard-Unterlagen. Erwerbsobstbau 34: 194–197.

Fischer, M. & S. Böttcher, 1978. Analyse der Mutterbeetleistung von Malus-Unterlagen unterschiedlicher Abstammung. Archiv Züchtungsforsch. 8: 119–126.

Fischer, M. & C. Fischer, 1989. Testung biotischer Resistenzfaktoren in der Apfelunterlagenselektion. Archiv Gartenbau 37: 31–43.

Fischer, M. & G. Pätzold, 1991. Ergebnisse aus Unterlagenprüfungen III: Apfelunterlagenneuzüchtungen des Instituts für Obstforschung Dresden-Pillnitz. Erwerbsobstbau 33: 140–142.

Hatton, R.G., 1935. Apple rootstock studies. Effect of layered stocks upon the vigour and cropping of certain scions. J. Pomol. Hort. Sci. 13: 293–350.

Maurer, E. 1939. Die Unterlagen der Obstgehölze. Parey Verlag, Berlin.

Müller, H. 1956. Über neue Pillnitzer Apfelunterlagen. Archiv Gartenbau 4: 370–378.

Schindler, O., 1921. Ein Beitrag zur Unterlagenfrage im Obstbau. Obstbauliche Zeitfragen, Sonderheft der Deutschen Gartenges. 1921. Eigenverl.

Schmadlak, J. 1970. Zur Selektion vegetativer Apfelunterlagen: I. Archiv Gartenbau 18: 253–265; II. Archiv Gartenbau 18: 327–359.

H. Schmidt and M. Kellerhals (Eds.), Progress in Temperate Fruit Breeding, 147.

Stone-fruit selection in Transcarpathia

Short communication

V. Zayats
Uzhgorod State University, 294000 Uzhgorod, Ukraine

On the southern slopes of the Carpathians around the cities of Uzhgorod and Mukachevo, there are especially favourable conditions for fruit growing. Different species and forms, undoubtedly worth protecting as a valuable genepool can often be found in the wild state. These plants have been studied in the field and used in selection. Nobody has yet practised stone-fruit selection in Transcarpathia, though in Ukraine and neighbouring Moldova Ryabov & Kostina (1969), Shaitan (1967), Sokolova (1987), and Smykov (1989) have been successful in this work. Over the last 30 years we have done a great deal of work on the introduction of new germplasm, variety testing and selection of stone-fruits (Zayats 1988).

Thousands of seedlings have been grown from open pollination. From these seedlings 8 peach, 2 plum, 1 apricot and 2 cherry varieties have been selected.

A current programme is considering two new characters in peach selection: peach varieties having dark red, almost black leaf colour and those having red fruit pulp are being bred.

During photosynthesis the red-leaved varieties assimilate solar energy more intensively. They are more productive than those with green leaves. The mechanism of photosynthesis in the red-leaved peach varieties, and the role of anthocyanins in this process have not yet been studied completely.

As a result of the work, hundreds of unique red-leaved hybrids have been bred which deserve to be thoroughly studied. Among them, 'Negus' is the first red-leaved fruit variety in the world. The trees of this variety are vigorous and high yielding. Their leaves are dark red and medium-sized. Their fruits are round with shallow, but well-marked abdominal sutures. The average fruit weight is 125 g with maximal weight of 138 g. The pulp is greenish white, fibrous, with a pleasant flavour and moderately expressed aroma. The fruit ripen by August, 20.

The red-leaved varieties are also of great interest as root stocks.

It has been stated that red colouring of the leaves in peach is a dominant trait being inherited according to the laws of Mendel (Zayats 1992).

The newly bred peach varieties have fruits with dark red pulp, containing large amounts of biologically active substances, which positively influence human physiological processes and favour discharging of radioactive isotopes. From numerous hybrids of this group 16 varieties have been selected. Biochemical analysis of their fruits showed that varieties MP 10–8, MP 4–8, MP 11–8 and P 6–5 are notable in containing larger than usual amounts of anthocyans, polyphenols, flavonols and pectins. Alle the selected varieties have downy fruits. The fruit weight varies from 100 g to 180 g. The stone is easily separated from the pulp. The fruits can be used fresh or cooked as jam, juice, etc.

Specific protein analysis showed that original peach varieties inherited the red colour of the pulp from *Prunus salicina*.

This new direction in peach selection allows diversification of peach varieties and development of varieties with new features.

References

Ryabov, I.N. & K.F. Kostina, 1969. Selection of stone-fruits & subtropical fruits: Works: Nikitsky Botanical Gardens. Simferopol. XL: 7–65.
Shaitan, I.M., 1967. Culture of Peach-Tree. Urozhai, Kiev.
Sokolova, S.A. & B.V. Sokolov, 1987. A Peach-Tree. Cartea Moldovenească, Chisinău.
Smykov, V.K., 1989. An Apricot. Agropromizdat, Moscow.
Zayats, V.A., 1988. A Carpathian Peach-Tree. Karpaty, Uzhgorod.
Zayats, V.A., 1992. Genetic nature of red colouring of the leaves of some trees and shrubs. Theses of the reports of the 6th Congress of the Ukrainian Society of Geneticists and Selectionists, Vol. 3, Kiev, pp. 81–82.

H. Schmidt and M. Kellerhals (Eds.), Progress in Temperate Fruit Breeding, 149–156.

Temperate fruit crop improvement in India

R.L. Sharma & K. Kumar
Department of Fruit Breeding, University of Horticulture and Forestry, Nauni-Solan (HP) 173 230, India

Key words: Almonds, apples, fruit breeding, fruit production, scab resistant, cultivars, germplasm, *Malus*, *Prunus*

Summary

In India, temperate fruit cultivation is spread over 350,000 hectares with an annual production of nearly 1.6 million MT. Apple occupies 50% of this area. It is followed by pears, Japanese plums, peaches, apricots, seedling walnuts and almonds. The varietal improvement of these crops in the past has been solely through introduction and selection of better performing cultivars in a particular climatic region. These have a good adaptation and are grown on a commercial scale. As far as improvement through hybridization is concerned, sporadic attempts have been made. Some improved cultivars of apple, apricot, peach and almond have been evolved through hybridization but these have not gained much popularity. The possible reasons of limited success in fruit breeding are considered to be the lack of continuity and specific gene sources and mis-directed strategy. In this paper, an attempt has been made to delineate crop specific objectives, gene sources and breeding strategy to evolve improved fruit and nut cultivars.

Introduction

Diverse climate and soils provide environments conducive for growing of almost all kinds of fruits in India. Currently over 28 million MT of fruits (Singh & Pandey 1993) are produced from 3.8 million ha. Of these, 1.6 million MT are temperate fruits from 0.35 million ha (Table 1). The apple is the predominant crop covering more than half of the area and three-quarters of the production of temperate fruits (Table 2). Temperate fruit cultivation extends from Jammu and Kashmir and sub-tropical plains in the north to Arunachal Pradesh in the east within the latitudes of 27° to 36° N and longitudes of 75°–99° E and, in a small hilly area in the south, around 10° N and 75° E. The main region is however, the hilly areas of the three northern provinces Jammu and Kashmir, Himachal Pradesh and Uttar Pradesh.

Production level

Fruit production figures may sound impressive because India is ranked second in world after Brazil in this regard (Chadha 1993). But very low productivity relegates us to near bottom on a per hectare basis . Grape is the only fruit whose productivity is more than international level 15–20 MT as against 10–12 MT/ha. Although temperate fruit production has increased from a meagre 0.08 million MT in 1950–51 to 1.6 million MT in 1989–90, the productivity increases have been balanced out by a large number of less productive orchards.

Factors of low productivity

Several reasons can be advanced for low productivity of temperate fruits. Some important ones may be as follows:

1. Use of low productive cultivars, seedling/rootstocks and insufficient and non- synchronizing pollinizers.
2. Generally poor shallow soils with low water holding capacity on steep slopes.
3. Mostly non-existent or primitive irrigation facilities.
4. Many climatically marginal plantation sites.
5. Poor, insufficient roads and transportation.

Table 1. Area, production and productivity of fruit and nut crops in India (adapted from Singh & Pandey (1993)).

Sr. No.	Fruit/ Nut	Area (million ha)	Production (million MT)	Productivity (MT/ha)	Percent Share	
					Area	Production
I. Tropical and sub-tropical						
1.	Banana	0.33	6.17	18.21	8.54	21.29
2.	Cashew[1]	0.53	0.30	0.56	13.73	1.03
3.	Citrus	0.34	2.75	7.99	8.80	9.48
4.	Grapes	0.02	0.40	16.26	0.51	1.38
5.	Guava	0.10	1.01	9.90	2.59	3.48
6.	Mango	1.01	8.50	8.42	26.16	29.33
7.	Papaya	0.03	0.90	23.00	0.77	3.10
8.	Pineapple	0.05	0.78	14.14	1.29	2.69
9.	Sapota	0.02	0.33	14.19	0.51	1.13
10.	Others	1.08	6.23	5.73	27.97	21.49
	Total	3.51	27.37	6.66	90.87	94.40
II. Temperate						
11.	Apple[2]	0.19	1.27	6.56	4.92	4.38
12.	Nuts[3]	0.07	0.02	0.33	1.81	0.06
13.	Pear	0.02	0.14	7.00	0.51	0.48
14.	Others	0.07	0.18	2.57	1.81	0.62
	Total	0.35	1.61	4.60	9.05	99.94

[1] Chadha & Bhaskara Rao (1993).
[2] Dwivedi (1992).
[3] Kaul (1990).

Table 2. Acreage, production and productivity of temperate fruits and nuts in major growing provinces (Dwivedi 1992).

Province	Apple			Total		
	Area (000ha)	Produ. (000MT)	Productivity (MT/ha)	Area (000ha)	Produ. (000MT)	Productivity (MT/ha)
Jammu & Kashmir	69.9	726.0	10.38	132.6	774.6	5.84
Himachal Pradesh	62.8	394.9	6.28	104.4	437.9	4.19
Uttar Pradesh	61.1	151.7	2.48	101.8	228.8	2.24
Punjab	–	–	–	9.9	114.5	14.71
North-east	NA	NA	NA	7.5	29.2	3.91

6. Delayed replacement of less productive varieties due to long-lived seedling rootstocks and small land holdings.
7. Shortage of foodgrains necessitating low priority to fruit production at various levels.
8. Disease and pest epidemics.

There may be many more reasons for this plight. Yet, the role of improved varieties which has been to the extent of 60–65 per cent in any enhanced crop productivity during the past 30–40 years has been least exploited in temperate fruits in India.

Table 3. Temperate fruit varieties evolved through hybridization in India.

Fruit Variety	Parentage	Features/Remarks	Reference
APPLE			
Lal Ambri	Red Delicious × Ambri	Trees upright with moderate vigour, fruits large, conical to oblong conic, skin striped red, flesh white, sweet, free from scab incidence	Farooqi & Dalal (1976)
Sunehri	Ambri × Golden Delicious	Trees spreading with moderate vigour, fruits medium, roundish to conical, skin sulphur yellow flesh greenish white, sweet, free from scab	Farooqi & Dalal (1976)
Ambred	Red Delicious × Ambri	Trees tall and spreading, fruits and conical, skin with bright red stripes on yellow ground, flesh white, crisp, good keeping quality, ripens mid-September, less susceptible to powdery mildew, sooty blotch and scab	Sharma et al. (1984)
Ambrich	Rich-a-Red × Ambri	Trees semi-dwarf, semi-spur and spreading, fruit round, conical, skin crimson, flesh whitish, subacid and juicy, ripens mid-September, free from scab	ditto
Ambroyal	Starking Delicious × Ambri	Trees semi-dwarf and spreading, fruits conical, skin with red streaks on yellow ground, flesh white, sweet, good dessert quality, ripens 3rd week of September	ditto
Ambstarking	Starking Delicious × Ambri	Trees vigorous, tall and spreading, fruits round to conical, skin with streaks over chrome yellow, flesh white, crisp, juicy, ripens mid-September	ditto
Chaubattia Anupam	Early Shanburry × Red Delicious	Trees vigorous, upright, fruits medium to large, conical, skin with shining red streaks, flesh yellow to creamy white, very sweet, ripens mid-July	Seth & Ghildyal (1983)
Chaubattia Princess	Early Shanburry × Red Delicious	Trees with moderate vigour, upright, fruits medium, conical, skin with deep red streaks, flesh creamy white, very sweet, ripens end-June	Seth & Ghildyal (1983)
Hybrid 7–4/1	King David × Red Delicious	Trees with moderate vigour, spreading, fruits medium, conical, skin red, flesh whitish cream, juicy, sweet, ripens after mid-July	ditto
Hybrid 11–1/12	Red Delicious × Winter Banana	Trees vigorous, spreading, fruit medium, oblong to conical, skin with red streaks, flesh creamy white, sweet, keeping quality good, ripens mid-August, low chilling (700 h)	ditto
APRICOT			
Chaubattia Alankar	Kaisa × Charmagz	Fruits with moderate quality, suitable only for fresh market	Seth et al. (1984)
Chaubattia Madhu	Turkey × Charmagz	ditto	ditto
Chaubattia Kesri	St. Ambroise × Charmagz	ditto	ditto
ALMOND			
Hybrid-15	Thin shelled × Sloh (peach × almond)	High yielding, thin shelled, large-sized nuts, suitable for warmer areas	Uppal & Gill (1977)
AH-258	Pethick's Wonder × Sloh (peach × almond)	Self-fruitful, late maturing	Uppal et al. (1984)
PEACH			
Saharanpur Prabhat	Sharbati × Flordasun	Early maturing, better fruit quality	Solanki et al. (1988)

History of improvement

Temperate fruits have been known to exist for over 2000 years in some of the temperate areas of India yet their cultivation on commercial scale was started by European settlers and missionaries sometime after 1860 (Sharma 1993a). The names of pioneers like Capt. A.A. Lee, A.T. Banon, Mr Allen, Mr Smith, Col. Rennick and subsequently of Mr S.N. Stokes are held in high esteem in remote hill villages. One Mr Alexander Coutts had in 1889 more than 100 apple varieties in his private estate near Shimla (Singh 1969). Some varieties and species are indigenous but the commercial varieties are exotic. The first systematic testing was done for apples by a Frenchman, M. Pychard during his tenure as Director of Agriculture, Kashmir State from 1910–1920 after introducing a number of varieties from his country. Later on, as a result of recommendations of Prof R.W. Hodgson of University of California, about 200 varieties of apple, plums, apricots, peaches and Chinese jujube from California, France and Italy were introduced by the Maharaja of Patiala State. In 1931, ten experimental orchards were established at different elevations in the hills of that state to study the adaptability of various temperate fruit varieties (Verma 1949). The period from 1930–1970 can be regarded as the 'golden era' in the history of temperate fruit improvement during which the selection and large scale multiplication of present day commercial cultivars 'Starking Delicious', 'Red Delicious', 'Richared' and 'Golden Delicious' apples; 'Santa Rosa' and 'Mariposa' plums; 'Bartlett' pears; 'New Castle' and 'Royal' apricots and 'Elberta' and 'J.H. Hale' peaches took place. Scientific evaluation started at newly established government research stations and agricultural universities after 1950. All this led to a set of recommended varieties for cultivation in specific zones. Temperate fruit breeding work which started in 1956 proceeded slowly and intermittently. The few varieties bred in India have not been tested widely and adopted on commercial scale (Table 3).

Present status of commercial cultivars

By and large, only a few selected varieties of each of the fruits have found a place in commercial plantations leading to more or less of a situation of 'monoculture' (Table 4). These varieties do not fulfil the present day need of high yield, diversity in ripening and resistance to pests and diseases.

In apple, old Delicious group comprising of Red Delicious, Starking Delicious and Richared continues to be grown over 80–85 per cent area. These are highly susceptible to apple scab. Spur-types and two of scab-resistant apple varieties are in demand for the last couple of years on the basis of selection (Kumar & Sharma 1992, Sharma et al. 1988).

In pears, 'Bartlett' is late in maturity and 'red' soft pears which command premium are only with a few growers and in experimental stage after introduction from Oregon (Anonymous 1992). Likewise different strains of local Asian pear 'Patharnakh' are preferred due to their capacity to withstand rough handling in long distance non-refrigerated transport.

'Santa Rosa' which dominates (90–95%) plum culture suffers from soft pulp and sourness near the pit. There is no earlier maturing plum to spread the harvest. However, extension in harvest by another fortnight has now been found possible with the recommendation of a later maturing cultivar Frontier (Anonymous 1992). Still later maturing plums such as Nubiana, Friar, Queen Ann which do not tolerate rains in normal plum growing belt are now being tested for dry-temperate region.

In peaches, low chilling cultivars are not of desired quality while 'Elberta' is susceptible to leaf curl and being late, to fruit fly also. An apricot cultivar with wide climatic adaptation is not available.

Almonds plantations are mostly of early flowering and late maturing, thus suffering from poor fruit set due to spring frosts and from non-drying due to the monsoon rains. So in the monsoon belt, almonds are harvested at pre-hull split stage for green market.

In walnut and pecan, lateral bearing is known to have short juvenility with regular fruit bearing. However, there is not even a single such selection/cultivar in the country. This requires extensive surveys of the walnut and pecan growing areas.

Fruit breeding

Exotic introductions constituted the only component of fruit and nut crops worth the name in the crop improvement strategy until recently. In H.P., despite the constraints, some useful information has emerged after evaluation and selection work. The cultivars being identified from a germplasm collection of around a thousand varieties at the University of Horticulture and Forestry would improve productivity and quality (Table 5). In some indigenously sexually propagat-

Table 4. Major commercial cultivars of temperate fruits in India.

Fruit Crop	Cultivar(s)	
	Cooler Areas	Warmer Areas
Apple	Starking (Syn. Royal) Delicious, Red Delicious, Rich-a-Red, Golden Delicious	–
Pear	Bartlett, China Pear	Patharnakh (*Pyrus pyrifolia*)
Peach	Elberta	Shan-i-Punjab, Flordasun
Plum	Santa Rosa, Mariposa	Alubokhara
Apricot	Newcastle, Royal	–
Sweet cherry	Black Heart	–
Almond	Seedling Selections, Non-Pareil, Drake	–
Walnut	Seedling Selections	–

ed crops, comprehensive surveys to explore the vast germplasm could have shown quick results.

Hybridization programmes suffered in the past because of:

a) lack of correct appraisal of the objectives and genetic potential of selected parents;

b) lack of staff and their continuity on specific jobs;

c) insufficient and inappropriate gene sources;

d) non-availability of facilities e.g., glass house, incubation chambers.

Still some hybridization programmes have been initiated to incorporate certain desirable traits in the commercial cultivars. The objectives and gene sources being used in hybridization are given in Table 6. The resulting hybrid progenies are being raised for further evaluation.

The facts which have emerged from apple hybridization so far are:

1. Hybrids of better storage character are undesirably late in maturity.
2. Ambri parent tended to impart to its progeny its upright growth habit and long chill requirement.
3. The seed germination was better when scab resistant varieties were used as seed parent.

In plumcot breeding programme, it was the plum x apricot and not reciprocals which were successful including the cross involving Methley.

Future strategy

Conventional selection and breeding efforts will continue as the major emphasis. A necessary component in such efforts is the availability of appropriate gene sources. A beginning has been made by incorporating scab resistant varieties from the PRI series in a cross breeding programme and introducing after 'on the spot' evaluation by the senior author about 100 hybrids from France under the ICAR/INRA workplan and from the Czech Republic (Sharma 1993b), during the past two years. These hybrids are sweeter, less acidic, mostly red fruited and which might have got eliminated because of different taste requirements in Europe. These will be either released for commercial cultivation after multilocation testing or used in breeding. A similar collaboration with other breeding programmes of Europe may be helpful in introducing advance generation apple hybrids containing polygenic sources of scab-resistance.

Some of the problems faced by temperate fruit industry may be at least partially solved by undertaking systematic surveys of indigenous germplasm of walnut, pecan, almonds, apricots, pomegranate and apple plantations for elite trees. With the identification of self-fertile almonds in indigenous germplasm the possibility of large scale occurrence of such seedling trees in the dry temperate region appears bright. A new scab resistant gene source also seems to exist amongst the

Table 5. Promising temperate fruit and nut varieties.

Fruit/Character	Cultivars	Reference
APPLE		
Better yield & improved fruit quality	Spur types: Starkrimson, Redspur, Oregon Spur, Red Chief.	Kumar & Sharma (1992)
	Standard: Topred, Vance Delicious, Skyline Supreme	
Early maturing	EC38683, Co-op 12, Early McIntosh, Anna	Anonymous (1992) Sharma et al. (1988)
Scab resistance	Co-op 12, Liberty, Redfree, Prima	Sharma et al. (1988)
Low chilling	Michal, Anna, Maayan	Anonymous (1990)
PEAR		
Regular bearing and better yield	Laxton's Superb, Nouveau Pointeau	Anonymous (1992)
Early maturation	Junska Zlato, Smajkanska Pozdniaza, Sentya Brpnskya	Sharma (unpublished), Sharma et al. (1992)
PEACH		
High yield & better quality	Kanto-5, Shimizu Hakuto	Anonymous (1989)
Cold hardiness	Prairie Dawn, Prairie Rose	Anonymous (1987)
PLUM		
Firm flesh & uniform sweetness	Frontier, Nubiana, Friar, Burmosa, Queen Ann	Sharma (1993a)
Early maturation	Burmosa, Red Beaut	ditto
APRICOT		
Late flowering	Farmingdale, Alfred	Anonymous (1990)
Early maturation	Baiti, Beladi	ditto
ALMOND		
Regular bearing & better yield	Starbasin, Hybrid-15, Nikitskyi, Tree No. 125, White Brandis	Kumar et al. (1992)
Late flowering & early maturation	Nikitskyi, Prianyi, Tree No. 125, Tree No. 2, White Brandis	Kumar et al. (1993)
Well-sealed thin shelled nuts	Merced	Sharma (1993)
WALNUT		
Better yield with moderate shelling	Natra Attiro, Gobind, Blackmore	Anonymous (1989, 1990)
Kernel flavour	Blackmore	Anonymous (1992)

Himalayan crab apple strains. This will increase the germplasm base for facilitating transfer of resistance to disease, pests and environmental stresses. Sources of other horticulturally important traits might enhance the progress of the ongoing breeding programmes.

Since it may not be possible for a single centre to handle large sized F1 and back-cross populations, sub-centres for evaluation of hybrid progenies may have to be earmarked.

Table 6. Objectives and gene sources being used in hybridization.

Fruit	Objectives	Gene Sources/Cross Combination		Reference
APPLE i)	Better storage	Delicious strains (Red, Starking and Rich-a-red)	× Ambri,	Farooqi & Dalal (1976)
ii)	Scab resistance	Delicious strains (Hardeman, Topred, Starkrimson, Red Spur, Golden Spur, Tydeman's Early	× Co-op 12, Redfree, Liberty, Priscilla	Anonymous (1987)
iii)	Low chilling	Hardeman, Spur type Red Delicious, Skyline Supreme (and their reciprocals)	× Michal, Maayan, Shlomit, Vered	Anonymous (1990)
APRICOT	Low chilling & wider adaptability	Baiti, Alfred, Farmingdale	× Benazir, Baiti Beladi, Parmarosa	Anonymous (1992)
ALMOND	Self fertility	Non-Pareil, Merced, Ne-Plus-Ultra, White Brandis	× AH 258,	ditto
PEACH	Late ripening	JH Hale, Quetta, Alton, Saharanpuri, Elberta, Kanto-5	× Kateroo,	ditto
PLUM i)	Better fruit quality	Monarch, Kanto-5, Yellow Plum, Burbank Elephant Heart	× Frontier, Methley, Early Transparent Gage	Anonymous (1990)
ii)	Plumcots	Methley, Alubokhara	× Parmarosa, Baiti, Beladi, Parmarosa	Anonymous (1993)

References

Anonymous, 1987. Annual Report 1986–87. Regional Hort. Res. Stn. Bajaura, Kullu, H.P.

Anonymous, 1989. Annual Report 1988–89. Dept. Fruit Breeding & Genetic Resources, Univ. Hort. & Forestry, Solan, India.

Anonymous, 1990. Annual Report 1989–90. ibid.

Anonymous, 1992. Annual Report 1991–92. ibid.

Anonymous, 1993. Annual Report 1992–93. ibid.

Chadha, K.L. 1993. Horticultural crops – Present status future thrusts. Souvenir HSI Golden Jubilee Symposium, Horticultural Research – Changing Scenario, Bangalore, India, pp. 1–3.

Chadha, K.L. and E.V.V. Bhaskar Rao, 1993. Cashew industry in India. ibid. pp. 65–68.

Dwivedi, A.K. 1992. Development of temperate horticulture in India – present status and future strategies. Paper presented in workshop on Development of Temp. Hort. at Univ. Hort. & Forestry, Solan, India, Aug. 1992 (in press).

Farooqi, K.D. & M.A. Dalal, 1976. Improvement of apple by hybridization. Proc. Workshop on pome, stone and nut-fruits, Ranikhet, U.P., November, 1976, pp. 136–139.

Kaul, G.L. 1990. Our dry fruit basket. Indian Hort. 35(1): 46–48.

Kumar, J. & R.L. Sharma, 1992. Evaluation of some genotypes of apples under low altitude conditions of Kullu Valley of Himachal Pradesh. Proc. Natl. Seminar on Emerging Trends in Temperate Fruit Production in India, Sept. 4–5, 1992 at UHF, Solan, India. Abstr. No. 9: 6.

Kumar, K., G. Sharma & R.L. Sharma, 1992. Evaluation of almond (*Prunus amygdalus* Batsch.) germplasm under mid-hills of north-west Himalayas. ibid. Abstr. No. 13: 9.

Kumar, K., R.L. Sharma & G. Sharma, 1993. Some late flowering and early maturing almond (*Prunus amygdalus* Batsch.) germplasm. Indian J. Pl. Gen. Resources (in press).

Seth, J.N. & P.C. Ghildyal, 1983. Some promising apple hybrids developed at Chaubattia. Progr. Hort. 15(4): 239–243.

Seth, J.N., P.C. Ghildyal & R.P. Joshi, 1984. Fr. Var. J. 38: 51–54.

Sharma, G., R.L. Sharma & K. Kumar, 1992. Genetic resources of pear under mid-hills. Proc. Natl. Seminar on Emerging Trends in Temperate Fruit Production in India, Sept. 4–5, 1992, at UHF, Solan, India. Abstr. No. 10: 7.

Sharma, R.L. 1993a. Genetic resources of temperate fruits. In: K.L. Chadha (Ed) Advances in Horticulture, Vol. 1, pp. 243–266. Malhotra Publishing House, New Delhi, India.

Sharma, R.L. 1993b. ICAR/INRA workplan visit report. Dr YS Parmar Univ. of Hort. & Forestry, Solan, HP, India.

Sharma, R.L., J. Kumar & Ved Ram, 1988. Performance of some scab resistant apple cultivars in Kullu valley of Himachal Pradesh. J. Tree Sci. 7(1): 45–49.

Sharma, Y.D., T.R. Chadha & G.K. Gupta, 1984. Breeding of apple varieties with better keeping quality and disease resistance. Proc. Natl. Symp. on Temperate Fruits at University of Hort. & Forestry, Solan in March, 1984, pp. 65–68.

Singh, H.P. & R.M. Pandey, 1993. Quest for higher production and productivity of fruits – a coordinated approack. In: Souvenir at HSI Golden Jub. Symp. Horticultural Research – Changing scenario, Bangalore, India, pp. 4–20.

Singh, Ranjit, 1969. Fruits. NBT, New Delhi.

Solanki, S.S., S.N. Singh, P.K. Saxena & I.C. Pandey, 1988. Saharanpur Prabhat, an early sweet peach. Fr. Var. J. 42(1): 4.

Uppal, D.K. & R.P.S. Gill, 1977. A new cultivar of almonds (*Prunus dulcis* Mill.). In: G.S. Nijjar (Ed) Fruit Breeding in India, pp. 99–101. Oxford & IBH Publishing Co., New Delhi.

Uppal, D.K., D.S. Dhillon, G.S. Dhaliwal & Y.R. Chanana, 1984. Selection of self fruitful hybrids in intervarietal crosses of almond. Indian J. Hort. 41(1,2): 80–82.

Verma, R.S., 1949. The Synopsis of the Biggest Selective Trials of the Age in the Patiala State Hills. B.S. Press, Patiala.

H. Schmidt and M. Kellerhals (Eds.), Progress in Temperate Fruit Breeding, 157–160.

Fruit quality in selfed and open pollinated progenies of two apple cultivars

Annemarie Karnatz
Rudeloffweg 28a, 14195 Berlin, Germany

Key words: Malus, apple breeding, cultivar, fruit quality, genetics, open pollination, selfing

Summary

External and internal fruit characters of selfed and open pollinated progenies from two apple cultivars, James Grieve and Landsberger Reinette were compared. The variation of most features was similar. Exceptions were, the degree of fruit russet, which was significantly worse after selfing in both cultivars, maximum storage life in James Grieve progenies and fruit flavour and firmness of fruit skin in Landsberger Reinette progenies. Sharp segregation was observed in the pattern of fruit colour. Both progenies of James Grieve produced blushed: striped fruits = 1:3. However, opposite segregations were found in Landsberger Reinette progenies, with blushed: striped fruits = 3:1. It is suggested that open pollinated progenies can be used to provide preliminary information on the genetic potential of apple cultivars.

Introduction

The knowledge of a cultivar's genetic background is an important pre-requisite before including it in a breeding programme. The analysis of selfed progenies gives the most reliable information. For example, the existence of genes for red fruit colour in Golden Delicious were first determined from the presence of red fruits in an inbred progeny (Karnatz, 1979, 1981). The production of inbred progenies in apple is difficult, as most cultivars are self sterile (Krylova et al., 1985). However, it is easy to produce progenies through open pollination, although it is doubtful whether such progenies can be used to provide information on the genetics of potential parents. Information on selfing and open pollinating apples is confined on fruit set, there are no reports on the resulting progenies (Kocheshkova et al., 1979; Kondrashova, 1981). This paper reports observations leading to the conclusions that there are very few differences in the genetic variability of the two types of progeny.

Materials and methods

The selfed progenies were derived from experiments on parthenocarpy at the Institute of Crop Science, Section Fruit Science, of the Technical University Berlin. As apples are mainly self sterile, flowers were not emasculated but only bagged in their balloon stage just before opening.

Seedlings from open pollinated flowers of James Grieve and Landsberger Reinette, derived from experiments on juvenility, were also raised. Those derived from James Grieve were grown in Hannover (Jonkers, 1971), while those from Landsberger Reinette were grown in Berlin along with the inbred seedlings.

The trees were raised at different times, valid comparisons were restricted to fruit characters (Table 1), recorded in the third or fourth year of crop. After harvest all fruit were stored at 3° C.

Results

Only a few significant differences were found. They concerned russet, fruit flavour, firmness of fruit skin and storage life (Table 2).

Table 1. Number of tested seedlings

Cultivar	Inbreeding	Open pollination
James Grieve	53	37
Landsberger Reinette	57	87

Table 2. Tested characters and result of X^2-test concerning differences between selfed and open pollinated progenies

Characters	James Grieve	Landsberger Reinette
External quality		
Attractiveness	n.s.	n.s.
Russet amount	0,002**	0,000***
Fruit diameter	n.s.	n.s.
Fruit weight	n.s.	n.s.
Ground colour	n.s.	n.s.
Overcolour-intensity	n.s.	n.s.
Overcolour-shade	n.s.	n.s.
Overcolour-pattern	n.s.	n.s.
Internal quality		
Fruit flavour	n.s.	0,05*
Firmness of fruit flesh	n.s.	n.s.
Firmness of skin	n.s.	0,002**
Season of ripening	n.s.	n.s.
Maximum storage life	0,05*	n.s.

n.s. = P > 0,05; * P ≤ 0,05; ** P ≤ 0,01; *** P ≤ 0,001.

Table 3. Russet amount

Characters	% seedlings of			
	James Grieve		Landsberger Reinette	
	Selfed	Open polli-nation	Selfed	Open polli-nation
Without russet	15.1	46.0	8.8	35.6
Russet				
in stalk cavity, light	28.3	27.0	36.8	48.3
in stalk cavity, strong	37.7	8.1	33.3	8.1
on fruit curve, light	7.6	18.9	7.0	2.3
on fruit curve, strong	11.3	0	14.0	5.8

Table 4. Fruit flavour

Characters	% seedlings of			
	James Grieve		Landsberger Reinette	
	Selfed	Open polli-nation	Selfed	Open polli-nation
Better than parents	5.7	13.5	10.5	12.9
Similar to parents	58.5	29.7	33.3	49.4
Little less than parents	18.9	18.9	28.1	25.9
Less than parents	11.3	18.9	21.1	8.0
Extremely less than parents	5.7	18.9	7.0	3.5

Table 5. Firmness of fruit skin

Characters	% seedlings of			
	James Grieve		Landsberger Reinette	
	Selfed	Open polli-nation	Selfed	Open polli-nation
Softer than parents	3.8	0	0	14.1
Similar to parents	73.6	64.9	59.6	67.1
Harder than parents	22.6	35.1	40.4	18.8

Strong russet in the stalk cavity and on the fruit cheek was found more often amongst the inbred seedlings of both cultivars. A higher proportion of seedlings without any russet appeared in the open pollinated progenies (Table 3).

Amongst the selfed seedlings from Landsberger Reinette 43.8% tasted similar to the parent or better, compared with 62.3% of the open pollinated seedlings (Table 4). There was a striking difference between the percentage of those with fruits of a poor taste in selfed seedlings (28.1%) compared with open pollinated seedlings (11.5%). Among the progenies from James Grieve, a high percentage (62.4%) had a good taste, a low percentage (17.0%) had a poor taste in the inbred progenies compared with 43.2% and 37.8% respectively in the open pollinated progenies.

Fruit skin was significantly firmer after selfing Landsberger Reinette (Table 5).

There was little variation in the maximum storage life of the inbred seedlings of James Grieve, while in the open pollinated progeny it varied between 2 and

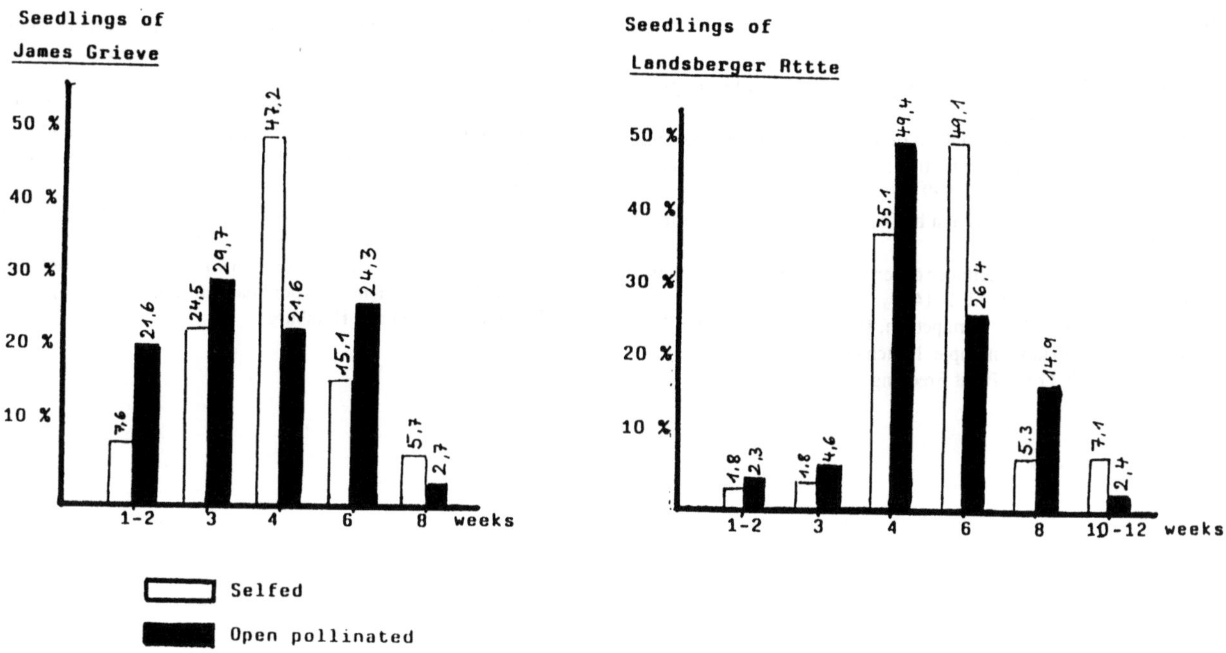

Fig. 1. Maximum storage life.

Table 6. Pattern of fruit colour

Characters	% seedlings of			
	James Grieve		Landsberger Reinette	
	Selfed	Open pollination	Selfed	Open pollination
Blushed	9	6	29	44
Striped	34	21	9	12

6 weeks (Fig. 1). The data for Landsberger Reinette showed a strong concentration around the mean in both progenies, inbred seedlings showing a peak at 6 weeks and open pollinated seedlings at 4 weeks, but this difference was not significant. Season of ripening followed the same trend in all progenies, but differences were also not significant.

Sharp segregation was only observed in the pattern of fruit colour (Table 6). Both progenies from James Grieve produced 1:3 segregations of blushed to striped fruits. However, opposite segregations were found in Landsberger Reinette progenies with blushed to striped = 3:1.

Discussion

Results showed that variability in selfed and open pollinated progenies is very similar. In connection with James Grieve, it is difficult to compare inbred progenies grown in Berlin with open pollinated ones grown in Hannover. But the differences found were less than those in the Landsberger Reinette progenies grown in Berlin. That means that the site factor appears to have little effect on the observed fruit characters.

Both, inbred and open pollinated, progenies from the striped cultivar James Grieve, segregated into 1 blushed: 3 striped seedlings. This corresponds with genetic control on the basis of a single dominant gene (Wilcox et al., 1936; Klein, 1958). But both progenies from the blushed cultivar Landsberger Reinette segregated into 3 blushed: 1 striped seedlings indicating that more than one gene is involved in apple skin pigment placement (Zwintscher, 1962).

The differences between inbred and open pollinated progenies were considered to be small. It is therefore supposed that an analysis of open pollinated progenies can provide some useful information on the genetic background of the seed parent.

References

Jonkers, H., 1971. An international experiment on juvenility in apple. Euphytica 20: 57–59.

Karnatz, A., 1979. Progeny analysis of self-fertilization in apple cultivars. Proc. Eucarpia fruit section symposium. Tree fruit breeding. Angers Sept. 3th–7th 1979.

Karnatz, A., 1981. Inzuchtlinien bei Apfelsorten. Erwerbsobstbau 23: 260–263.

Klein, L.G., 1958. The inheritance of certain fruit characters in apple. Proc. Amer. Soc. Hort. 72: 1–14.

Kocheshkova, T.V. & V.V. Vartapetyan, 1979. Study on self fertility and cross pollination in apple. Nauch dokl vyssh shkoly Biol n. No. 12: 83–87. From: Plant Breeding Abstr. 50: 9538.

Kondrashova, K.V., 1981. Inbreeding of some apple varieties. Nauch tr Voronezh s-kh in-t. 112: 71–76. From: Plant Breeding Abstr. 54: 7662.

Krylova, V.V. et al., 1985. Untersuchungen zum Grad der Selbstfertilität von Apfelsorten unter verschiedenen ökologischen Bedingungen der UdSSR und DDR. Archiv für Gartenbau. 33 (6): 359–366.

Wilcox, A.N. & E. Angelo, 1936. Apple breeding studies. 1. Fruit color. Proc. Amer. Soc. Hort. Sci. 33: 108–113.

Zwintscher, M., 1962. Die genetischen Grundlagen und Methoden der Züchtung. In: Handbuch der Pflanzenzüchtung, Bd. 6, S. 659, Paul Parey, Berlin und Hamburg.

H. Schmidt and M. Kellerhals (Eds.), Progress in Temperate Fruit Breeding, 161–164.

Shortening of the juvenile period in apple breeding

Christa Fischer
Bundesanstalt für Züchtungsforschung an Kulturpflanzen, Institut für Obstzüchtung Dresden-Pillnitz, Pillnitzer Platz 2, D-01326 Dresden, Germany

Key words: Apple, breeding, juvenile period, shortening

Summary

A very important aim for the acceleration of the breeding progress in apple is the development of methods for shortening the juvenile period. The period from sowing the seeds to the first bloom of the seedlings can take from 5 to 12 years. Only then we can begin with the complete evaluation of the progenies. Various cultivation methods were considered at the Pillnitz Institute. The basic idea is to grow the seedlings rapidly from the germination stage to the change from the juvenile to the adult stage. The best method appeared to be cultivation in the greenhouse under 16 hours light per day. At a minimum height of 180 cm graftsticks of each seedling were cut and grafted in April in the field on M9 rootstocks with Hibernal interstem. This way the growth was controlled and blossoming began earlier. Fruit size was normal and not smaller as on M 27. A few grafted seedlings began to blossom in the year of grafting. In the second year after grafting 20–25% blossomed, in the third year 65–85%, and in the fourth year nearly 95%. This high rate of blossoming seedlings enabled the completion of selection for all the important breeding objectives including the resistance to scab, mildew and fire blight within 6 years after sowing the seeds.

Introduction

An important prerequisite for the acceleration of the apple breeding progress is the development of methods for shortening of juvenile period. The period from sowing the seeds to the first bloom of the seedlings can take 5 to 12 years. Only then we can begin with the complete evaluation of the progenies. Various cultivation methods were considered at the Pillnitz Institute. The basic idea is to grow the seedlings rapidly from the germination stage to the change from the juvenile to the adult stage.

Material and methods

We used different crossing progenies. Various cultivation methods were tested in different periods. Three of the important methods were compared (Fig. 1):

– The classical method of cultivation of seedlings from the sowing until the beginning of blossoming:
  · sowing of the seeds in open ground in April (1st year);
  · growth of the seedlings in the nursery (2nd–4th year);
  · planting in the field and selection after beginning of blossoming (5th–12th year).

Of all tested methods in the following 2 acceleration methods will be described.

– First acceleration method:
  · sowing of the seeds in the greenhouse in April (1st year);
  · cultivation of the seedlings in the greenhouse (1st–2nd year);
  · grafting on rootstocks M9 in the nursery (3rd year);
  · cultivation in the nursery (3rd–4th year);
  · planting in the field and beginning of the selection with the first blossoming (5th–10th year).

– Improvement of the acceleration method mentioned before:
  · sowing of the seeds in the greenhouse in January instead of April (1st year);

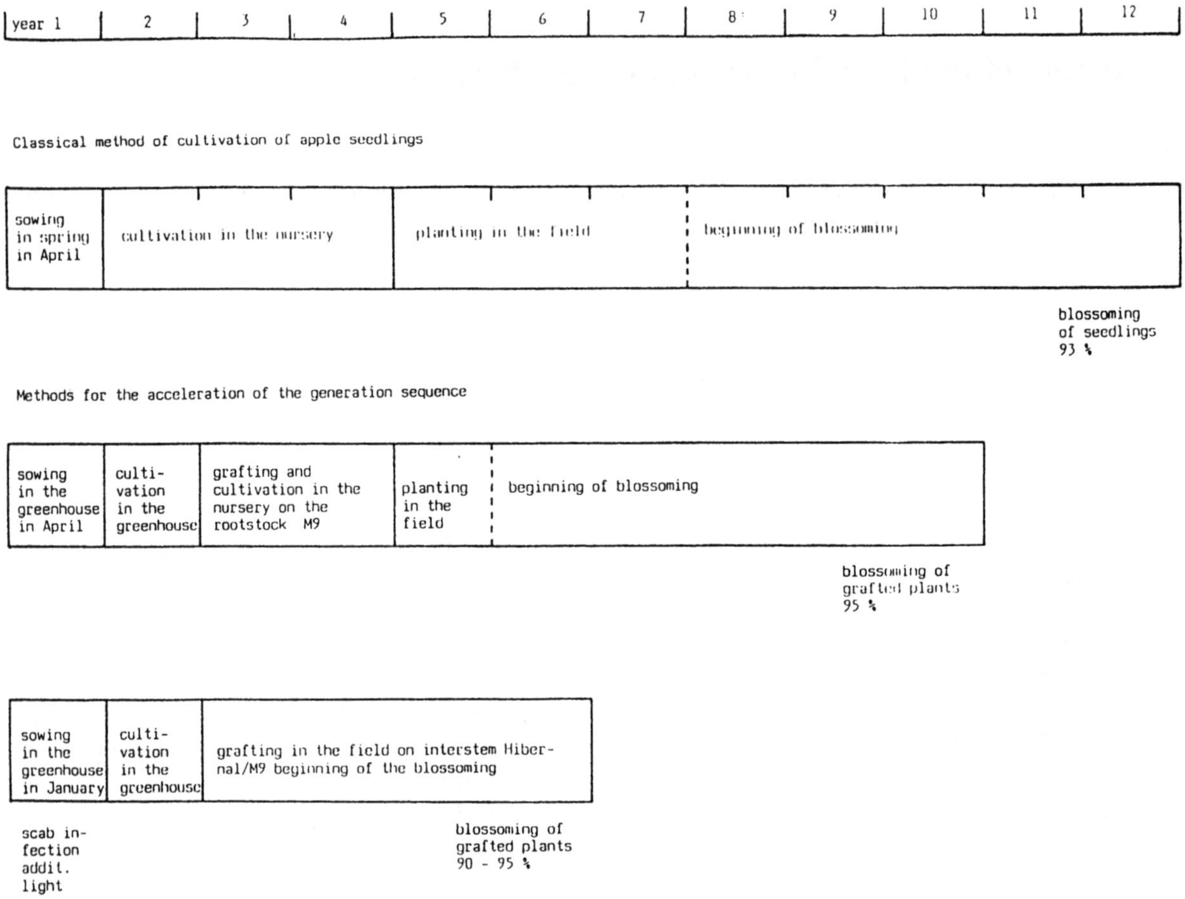

Fig. 1. Comparison and duration of the various cultivation methods in apple breeding until the beginning of blossoming

- scab infection and selection in the greenhouse in February to April (1st year);
- cultivation of the seedlings with additional light 12–16 h per day in April and May (1st year);
- cultivation of the selected seedlings in the greenhouse for growth acceleration (1st and/or 2nd year);
- grafting in April directly in the field on maiden trees (rootstock M9 and interstem Hibernal) – (2nd or 3rd year);
- beginning of the blossoming in the year of grafting and beginning of the selection (2nd or 3rd – 6th year).

The graftsticks are taken from the top of the seedlings higher than 180 cm at the adult stage.

Results

The length of the juvenile period is very different according to the methods used for shortening of juvenility. 2 methods are compared with cultivation of seedlings on own roots (Fig. 2). The standard method (method 1) included the following steps: sowing the seeds in April, scab infection and selection in the first year, in the 2nd to 4th year cultivation of seedlings in the nursery, after that planting into the field. The seedlings began to blossom in the 6th year after sowing. In the average of 268 progenies (8510 seedlings) the rate of blossoming amounted to 63% in the 8th year and 93% in the 12th year after sowing (Figs. 1, 2).

A shortening of juvenility is done by sowing the seeds in the greenhouse in April, the scab infection and selection in May. In the 2nd year the seedlings were cultivated in the greenhouse and grafted on M9 at the end of the 2nd year. After 2 years cultivation in the nursery, which is the 4th year after sowing, the trees

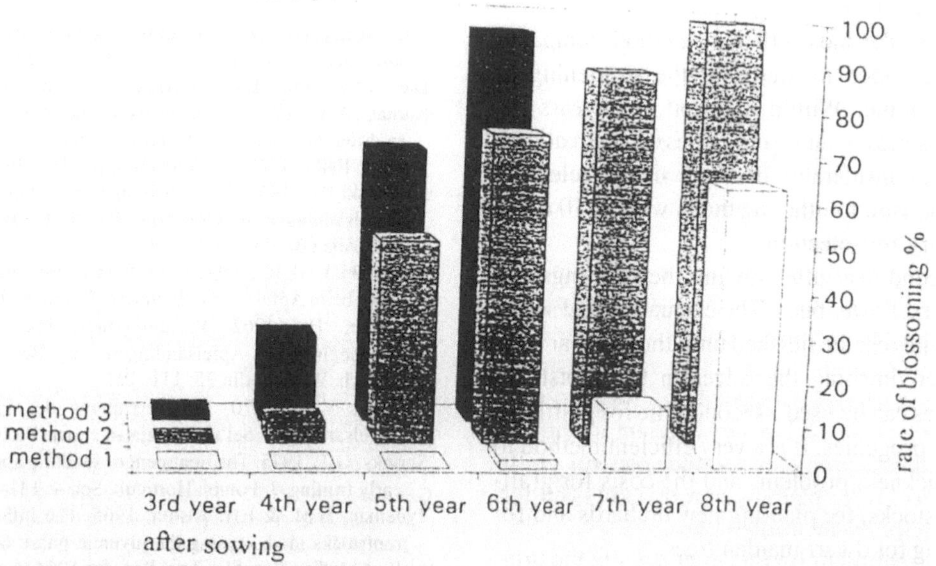

Fig. 2. Beginning and rate of blossoming in apple progenies with different cultivation methods for shortening the juvenile period

were planted in the field. The trees began to blossom in the 5th or 6th year after sowing and 95% of the grafted trees blossomed in the 10th year (Fig. 1). A variation of this method is to graft the seedlings on interstem Hibernal and rootstock M9 in the nursery (method 2). The flowering began in the 4th year and 99% of the trees blossomed in the 8th year after sowing (Fig. 2).

The best of the methods was method 3 with the following steps: sowing the seeds in the greenhouse in January, scab infection in February and selection in March/April, cultivation of the seedlings in the greenhouse for 2 years. In the spring of the 3rd year top of scions of the seedlings were grafted on interstem Hibernal/rootstock M9 directly in the field. The trees began to blossom in the same year. In the 6th year 95% of the trees blossomed in the average of 290 populations and 9355 trees (Fig. 2).

By grafting the seedlings on rootstock M9 and interstem Hibernal the growth is decreased and an earlier beginning of blossoming is obtained. During the period of 6 years the progenies were selected with regard to all important breeding objectives including resistance to scab, mildew and fire blight. The fruit size and growth on rootstock M9 were normal, similar to the apple orchards on M9 under production conditions. According to the comparison with rootstock M27 the plants on M27 grew slower and fruit size was smaller. The fruit size from seedlings on own root is between trees on M9 and M27 in our trials.

Discussion

Various methods have been developed for the acceleration of the breeding progress in different crops in order to reduce the juvenile period. In fruit trees, especially apple and pear, the period from sowing to first blossoming can last from 5 to 12 years with seedlings on own roots.

In the history of apple breeding many breeders tried to accelerate the beginning of blossoming (Diel 1804, Spinks 1925, Murawski 1957, Schander 1962, Tydeman & Alston 1965, Karnatz 1970, Smykov 1970, Visser 1970, Watkins & Alston 1973, Aldwinckle 1975, Murawski & Focke 1978). Methods were used like grafting in old adult trees, on dwarf rootstocks, intensive cultivation in the field and in the greenhouse and use of additional lighting in the greenhouse.

We got the best results in our combined method:

- early sowing in the greenhouse;
- 16 h light/day and an optimum of cultivation operations;
- cutting the graftsticks from the seedlings higher than 180 cm;

– grafting in spring directly on interstem Hibernal/M9 in the field.

This method is the most efficient method compared with other described methods for the shortening of juvenility on apple. Within a period of 6 years the selection of a series of apple progenies can be realized. This means a considerable decrease of the selection time in comparison to other methods with 8, 10 or 12 years necessary for selection.

In the second year after sowing the seedlings can develop the first flower buds. The evaluation and selection of progenies can be finished until the 6th year after sowing. After finishing the selection the rootstock-interstem-trees can be used a second time for grafting a new series of progenies. It is a very efficient method to reduce soil sickness problems and the costs for grafting new rootstocks, for planting new orchards and for frame building for dwarf maiden trees.

During this period of 6 years after sowing the progenies are selected with regard to all the important breeding objectives including the resistance and different horticultural properties. The breeding progress can be accelerated by this method in order to decrease the generation sequence.

References

Aldwinckle, H.S., 1975. Flowering of apple seedlings 16–20 months after Germination. Hort. Science 10: 124–126.

Diel, A.F.A., 1804. Die Obstorangerie. Cnobloch Verlag Leipzig.

Karnatz, A.M., 1970. Experiments for shortening the juvenile stage in different *Malus* species. Proc. Angers Fruit Breeding Symposium, INRA, C.N.R.A. Versailles, pp. 287–296.

Murawski, H., 1955. Untersuchungen zur Stadienentwicklung an Apfelsämlingen als Grundlage für die Obstzüchtung. Arch. f. Gartenbau Berlin 3: 346–352.

Murawski, H. & R. Focke, 1978. Beschleunigung der Generationsfolge beim Apfel. Arch. Züchtungsforsch. 8: 91–98.

Schander, H., 1962. Variationsstudie über den Eintritt des Blühbeginns bei Apfelsämlingen. Tag.-Ber. Dt. Akad. Landwirtsch. Wiss. Berlin 35: 111–121.

Smykov, V.K., 1970. Die Verfrühung des Fruchtansatzes bei Apfelsämlingen. Sel'skokhozjajstvennaja Biologija 5: 924–926.

Sprinks, G.T., 1925. The treatment of seedling apple trees to induce early fruiting. J. Pomol. Horticult. Sci. 4: 141–145.

Tydeman, H.M. & F.H. Alston, 1965. The influence of dwarfing rootstocks in shortening the juvenile phase of apple seedlings. East Malling Res. Sta. Ann. Rep. for 1964, A 48: 57–98.

Visser, T., 1970. Environmental and genetic factors influencing the juvenile period in apple. Proc. Angers Fruit Breeding Symposium, INRA, C.N.R.A. Versailles, pp. 101–106.

Watkins, R. & F.H. Alston, 1973. Breeding apple cultivars for the future Fruit. Present and Future. Royal Hort. Soc. London 2: 65–74.

H. Schmidt and M. Kellerhals (Eds.), Progress in Temperate Fruit Breeding, 165–172.

Inheritance of tree and fruit characters in progenies from crosses of sweet cherry (*Prunus avium* L.) cultivars

Robert Theiler-Hedtrich
Swiss Federal Research Station, CH-8820 Wädenswil, Switzerland

Key words: Sweet cherry (*Prunus avium* L.) breeding, inheritance of fruit set, fruit size, fruit colour, abscission layer formation, self fertility, bacterial canker (*Pseudomonas* spp.) resistance

Summary

Approximately 1000 seedlings from 20 combinations crossed in 1979, 1980 and 1981 (Theiler-Hedtrich 1985a) were tested for several characters: fruit set (yield), fruit size, fruit colour, formation of abscission layer and bleeding after fruit removal from fruit stalks, bacterial canker resistance, flowering and harvesting time. From progeny of crosses with Stella as pollinator, 56% (Vittoria × Stella) and 46% (Schüttler × Stella) of the seedlings were self-compatible, of which 14 were high yielding with good fruit size and quality. From the data recorded it can be concluded: *fruit set* is a recessive character; only 5 to 20% of very good yielding seedlings were obtained in different progeny, even if the parental plants were both very good croppers. *Fruit juice and skin colour* was in most progenies 'black' even if they were from combinations with 'white' varieties, e.g., Merton Glory or Schüttler. Only from the combination Schüttler ('white') × Stella ('black'), 50% of the seedlings were 'white'; Stella therefore is heterozygous for the character of fruit juice and skin colour. *Fruit size* is evenly distributed in progeny with respect to the fruit size of their parent plants. *Abscission layer formation and non-bleeding* is a genetically complex character. In combinations where both parent plants formed fruits with complete abscission layers and which were not bleeding after fruit removal from the stalk, this character was inherited only to 50% (Vittoria × Schüttler) or 85% (Vittoria × Frühe von der Weid) in the progeny. For the genetical control of this character further studies are necessary. *Bacterial canker susceptibility* was evenly distributed in seedlings from all combinations even if the highly resistant cv. Vittoria was used as one parent plant, thereby not confirming the expected results of a higher proportion of resistant seedlings from combinations with Vittoria. *Flowering and harvest time* of the seedlings from different combinations was within the range of the parent plants. Only in the combination of Vittoria × Stella (mid- to late-ripening season) one seedling out of 99 was found to form ripe fruits two weeks earlier than the parental plants. From the seedlings tested 40 have been chosen for further evaluation or genetical studies.

Introduction

Over the last 10 years the production of sweet cherries (*Prunus avium* L.) in Switzerland remained constant (approx. 18,000 t/year), whereas the number of cultivated plants declined from 1.2 to 0.8 million. A tendency towards intensive plantation systems has been observed. This was accompanied by a change in cultivated varieties. The climatic conditions in most of the cherry growing areas (high precipitation during ripening period), makes it difficult to cultivate high yielding cvs. like Hedelfinger, Büttners, Stella, Merpet, Merlat, Star, Kordia because of cracking. An improvement in high quality fruit production was achieved by covering the trees with plastic (Riesen et al. 1991). Old local varieties are much more adapted to the climatic conditions but are usually less productive (e.g., small fruit size).

In addition to table fruits there is a high demand for sweet, black coloured cherries for the canning industry. Fruits from these varieties should easily be harvested by hand-stripping or mechanically. One of the aims of this work was to breed and select new cultivars for the canning industry. Local varieties with a complete absicission layer and a low fruit removal force (FRF) are either too small in fruit size, Frühe von der Weid, or 'white' coloured (uncoloured fruit juice with a yellow-red skin), Schüttler, and therefore not acceptable for the canning industry. The new variety Vittoria (Bargoni 1970) has a complete abscission layer, is 'black', but has a high FRF value. As a consequence, these varieties were included into the breeding programme started in 1979–1981 (Theiler-Hedtrich 1985b).

Sweet cherry breeding has a long tradition mainly in the UK, Canada, Italy and the eastern countries. In Switzerland sweet cherry breeding was initiated by Kobel in 1956 and has continued until today (Theiler-Hedtrich 1985a). Except for some self-fertile cvs., e.g., Stella (Lapins 1971), Sunburst and Lapins (Sansavini & Lane 1983), sweet cherries are self-sterile and form incompatibility groups (Crane & Brown 1955, Krapf 1976, Way 1968). Crosses of varieties from the same incompatiblity group are excluded, and even more important, backcrosses are restricted to only one parent plant, if progeny from two varieties with e.g., S1S2 × S1S3 genes have been derived. Therefore detailed genetical studies by backcrosses are difficult. Matthews (1973) listed a number of tree and fruit characters which either are qualitative or quantitative, respectively.

Qualitative characters (one or two genes)
- self fertility
- fruit juice and skin colour
- albinism
- genetic dwarfs– vigour

Quantitative characters (polygenic)
- vigour, precocity
- fruit size
- flowering period
- ripening period
- cracking
- bacterial canker resistance

In general little information is available of progeny from backcrosses regarding tree- and fruit characters, which would permit more detailed information on sweet cherry genetics.

Material and methods

In Tables 1 and 2 the working scheme and aims for our breeding programme are listed.

To select the pollinators for the three parental plants with a complete abscission layer, complementary characters were sought (Table 2).

Crosses were always made by hand-pollination and seedlings were grown by embryo culture (Theiler 1971). All data records were obtained from seedlings. The results summarized here are based on two to three years of observation during the first test period. In Table 3 varieties which were crossed are listed. The number of progenies of each combination tested are given in the results (Figs. 2–4).

Abscission layer formation and non-bleeding was determined by hand-stripping of fruits and visual observation of the abscission layer, which was recorded as complete, incomplete or absent. Samples of which only a part of the fruits had a complete abscission layer were grouped into the lower grade (incomplete). Non-bleeding of sap at the transition of the fruit to the pedicel was detected by gently pressing the fruits after removal from the fruit stalk. A scale of 5 was used to distinguish between very good (vg), good, moderate (m) low and no (no) fruit set. Fruit size was measured by using a templet with holes of different diameters ranging from 17 to 23 mm. Usually a randomised sample of 10 to 20 fruits was measured and grouped as <17, 17–19, 19–21, 21–23, >23 mm. Fruit colour was defined as black, brown, red and yellow-red and the intermediate stages. For fruit juice only black (slightly red to dark red coloured) and white (uncoloured juice) in relation to skin colour was recorded. Self-fertility was detected by covering branches with a nylon fabric bag before flowers opened and the recording of fruit set. For bacterial canker resistance bark inoculations in late November were made and canker length recorded before bloom following the method and using the strains of *Pseudomonas* spp as described by Theiler-Hedtrich (1985b).

Results

Of special interest was the inheritance of the abscisson layer formation (fruit:pedicel) and the 'non-bleeding' of fruits after stripping off the fruit stalk, as well as fruit set, fruit size and fruit juice and skin colour.

Abscission layer formation and non-bleeding (Fig. 1) is a genetically complex character. In combinations where both parent plants formed fruits with complete abscission layers and which were not bleeding after fruit removal from the stalk, this character was inherited only to 50% (Schüttler × Vittoria (115)) or 85% (Frühe von der Weid × Vittoria (125)) of the progeny. If the cv. Späte Holinger (non abscission layer forming and bleeding) was crossed with Vittoria (133) or

Table 1. Concept of the breeding and evaluation cycle applied to progeny from crosses of sweet cherry cultivars.

Year	Procedure
1979–1981	crosses, embryo culture, nursery, double row plantation system (1 × 1 m with 2.5 m distance between double rows)
1983 + 1984	test of bacterial canker resistance by bark inoculation on one and two year old branches
1985–1987	recording of fruit and tree characters on all seedlings, test of self-fertility in progenies of crosses with Stella as pollinator, preselection and discarding of useless seedlings
1988–1990	Inbreeding of self-fertile seedlings and backcrosses with parental plants on preselected seedlings, chip budding on F-12/1 rootstocks in the nursery for further field trials
1991	removal of remaining trees

Table 2. Characterisation of the parent plants with a complete abscission layer formation and its complementary pollinator plants.

Parent plants with complete abscission layer	Characters	Crosses with cultivars of complementary characters
– Schüttler	skin colour: *yellow-red*	*black* skin colour
	flowering: *mid-late*	
	harvest time: *mid-late*	
	bacterial canker: *5/3**	*high* resistance
– Frühe von der Weid	skin colour: *black*	
	flowering: *mid-early*	*late* flowering
	harvest time: *early*	*late* harvest
	bacterial canker: *4/5*	*high* resistance
– Vittoria	skin colour: *brown-black*	*black* skin colour
	flowering: *mid-early*	*late* flowering
	harvest time: *late*	
	bacterial canker: *1/2*	

* Numbers indicate bacterial canker index (see Table 4).

Table 3. Cultivars used for crosses with the three varieties Schüttler, Frühe von der Weid and Vittoria in the years 1979–1981.

Schüttler ×	Frühe von der Weid ×	Vittoria ×
Magda (110)*	Delta (120)	Delta (130)
Delta (111)	Heidegger (121)	Basler Langstieler (131)
Basler Adler (112)	Basler Langstieler (122)	Stella (132)
Basler Langstieler (113)	Stella (123)	Späte Holinger (133)
Stella (114)	Vittoria (125)	Heidegger (134)
Vittoria (115)	Späte Holinger (127)	Merton Glory (135)
Frühe von der Weid (116)		
Späte Holinger (117)		
Total seedlings: 350	300	420

* In brackets: combination number, which is used in Figs. 1–4.

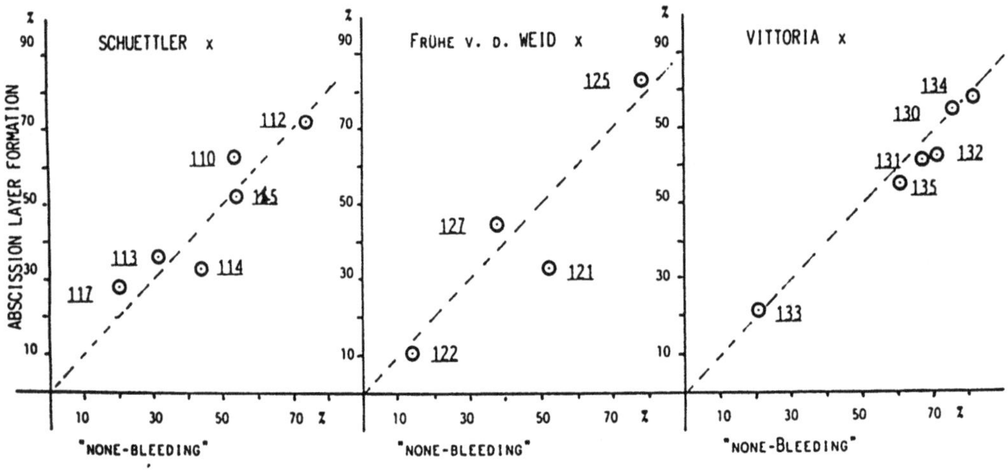

Fig. 1. Ratio of complete abscission layer formation and non-bleeding of fruits in progeny of the different combinations with Schüttler, Frühe von der Weid and Vittoria. Numbers refer to the combinations listed in Table 3. Note that for some of the progeny, e.g., combination No. 121 (Frühe von der Weid × Heidegger) fruits from only 30% of the seedlings formed a complete abscission layer but 50% did not bleed.

Schüttler (117), only 25% and 15% of the seedlings bore fruits with a complete abscission layer and were non-bleeding. But if Frühe von der Weid was crossed with Späte Holinger (127), 50% of the seedlings of this combination bore fruits with a complete abscission layer and were non-bleeding. For the genetical control of this character further studies are necessary.

Fruit set (Fig. 2) is a recessive character; only 5 to 20 % of very good yielding seedlings were obtained, even if the parental plants were both very good croppers. In some of the progenies 20% to 50% of the seedlings did not crop after 3 fruit-bearing seasons.

Fruit juice and skin colour (Fig. 3) was in most progenies 'black' even if they were from combinations with 'white' varieties, e.g., Merton Glory or Schüttler. Only from the combination Schüttler ('white') × Stella ('black') (combination No. 114), 50% of the seedlings were 'white'; Stella therefore is heterozygous for the character of juice and skin colour.

Fruit size (Fig. 4) is generally evenly distributed in progeny with respect to the fruit size of their parent plants. The highest proportion of seedlings (40%) with large fruits was obtained from combinations of parents with equally large fruits e.g., Vittoria × Heidegger (134) or Vittoria × Stella (132), with fruits in the range of 6–8 grams, which correlates to a fruit diameter of 23–25 mm (Theiler-Hedtrich 1990). Combinations of varieties of small-fruit sized × large-fruit sized parent plants, e.g., Vittoria × Späte Holinger (133), led to progeny of which more than 80% bore medium sized fruits. Fruit size also depends on seedling age; older seedlings yield larger fruits than younger ones (data not shown).

Self-fertility: from progenies of crosses with Stella as pollinator, 56% (Vittoria × Stella, 132) and 46% (Schüttler × Stella, 114) of the seedlings were self-fertile. From 60 self-compatible seedlings 14 were high yielding with good fruit size and quality and were further propagated for varietal tests.

Bacterial canker susceptibility was evenly distributed in seedlings from all combinations even if Vittoria (highly resistant cv., Theiler-Hedtrich, 1985b) was used as one parent plant (Table 4), thereby not confirming the expected results of a higher proportion of resistant seedlings from combinations with Vittoria.

Flowering and harvest time of the seedlings from different combinations was within the range of the parent plants. Only in the combination of Vittoria × Stella (mid- to late- ripening season) one seedling out of 99 was found to form ripe fruits two weeks earlier than the parental plants.

In progeny where Stella or Merton Glory was used as one parent plant a higher proportion of seedlings susceptible to fruit cracking was observed. In combinations with Frühe von der Weid a few seedlings were found with a high tolerance to fruit cracking but with a medium fruit size and in two of them combined with a strong bitter taste.

The occurrence of cherry leaf spot (*Coccomyces hiemalis*) and fruit monilia (*Sclerotinia* spp) depended on the climatic conditions in successive years. From the data collected no clear dependence of susceptibility

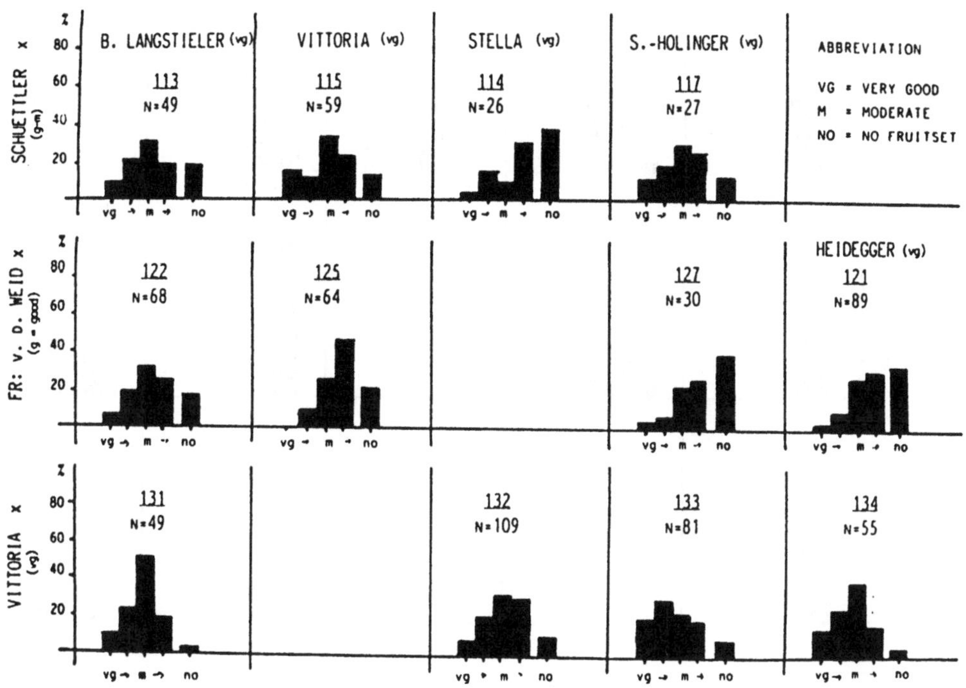

Fig. 2. Distribution of the ratio of fruit set on seedlings of different combinations recorded as very good (vg) to moderate (m) and low, as well as the seedlings which did not set fruits (no) after four cropping seasons.

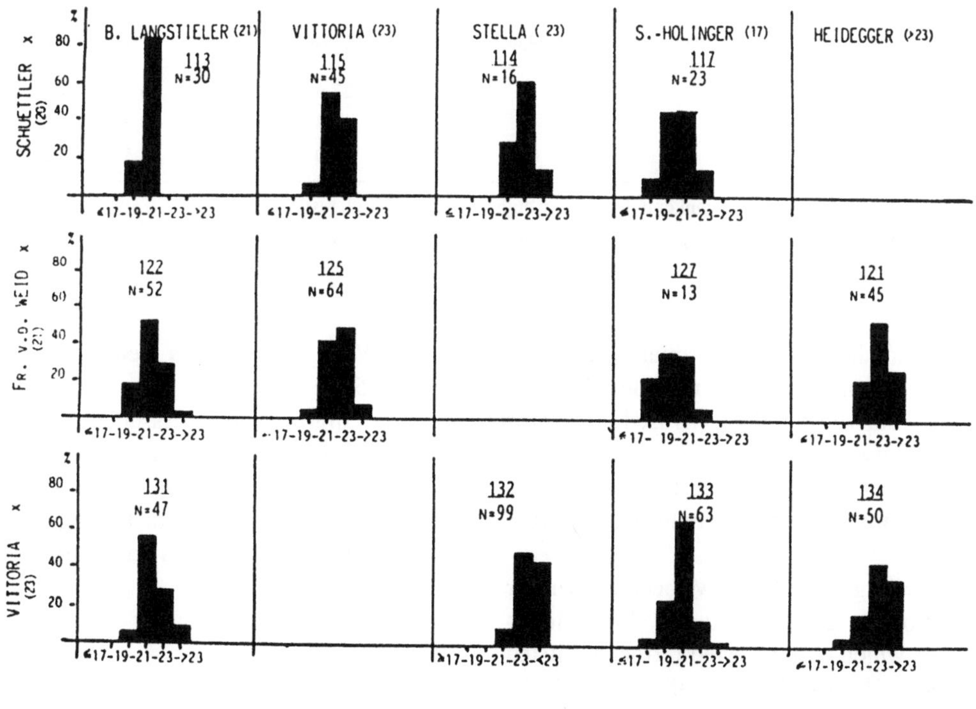

Fig. 3. Distribution of the ratio of fruit size of seedlings from different combinations defined as fruit diameter in mm. For the parent plants average fruit size is referred to in brackets.

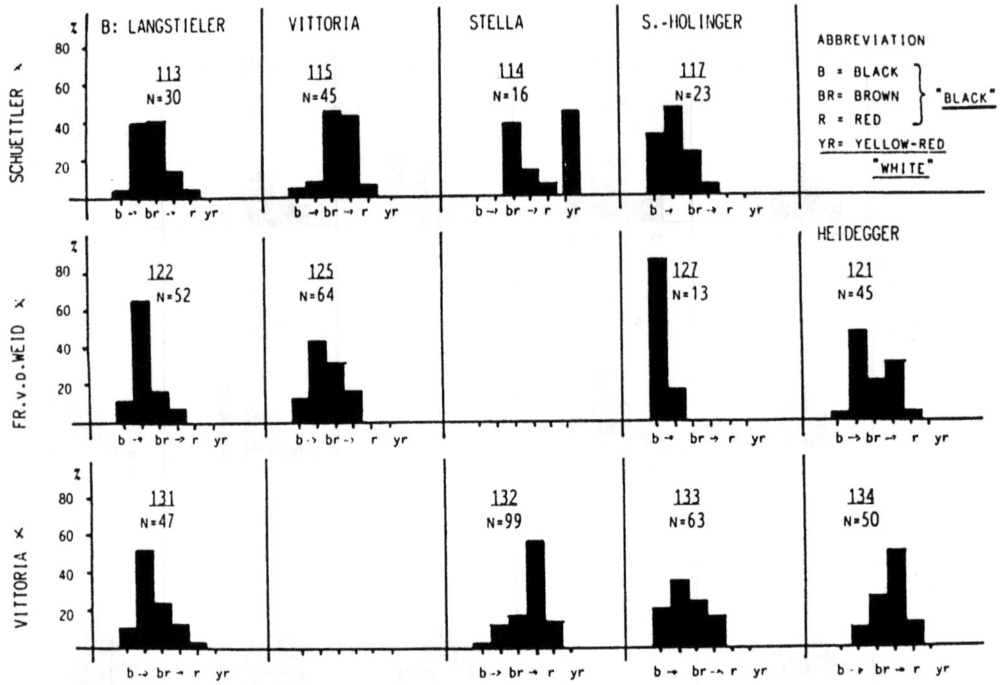

Fig. 4. Distribution of the ratio of fruit skin colour from seedlings of different combinations. Black and white refers to the fruit juice colour. Only in the combination Vittoria × Stella (114), 50% of the seedlings had white (yellow-red) fruits.

Table 4. Bacterial canker resistance of seedlings (in percent from number of plants) from different progenies of which one parent was either Schüttler, Frühe von der Weid or Vittoria.

Progenies from open pollinated cvs.	No. of plants	Class 1*	Class 2	Class 3	Class 4	Class 5
Schüttler	297	14.1	27.6	37.1	13.1	8.1
Frühe von der Weid	215	8.4	16.7	38.6	23.2	13.0
Vittoria	323	2.5	17.0	42.7	23.5	14.2

* Class indicates bacterial canker index: 1 very high resistance, 2 high resistance, 3 moderate resistance, 4 low resistance, 5 very low resistance (Matthews (1979)).

could be observed; in general a correlation between fruit cracking and monilia could be recorded. In some progeny where Frühe von der Weid was used as a parent plant this was not always the case. Frühe von der Weid seems to inherit a higher tolerance to fruit monilia but these observations have to be confirmed in future studies.

From the seedlings tested 40 have been chosen for further evaluations or genetical studies. Some interesting seedlings which will be tested further are listed in Tables 5 and 6. High-yielding seedlings with fruits with a complete abscission layer, dark skin and which were easy to harvest by shacking were selected from combinations in which Schüttler, Frühe von der Weid (Fvd Weid) and Vittoria were used as parent plants (Table 5).

For high quality dessert fruits, selections of seedlings with good fruit size and high yielding capacity were mainly from combinations in which Vittoria was one parent and Stella, Heidegger (Schaer 1977) and Merton Glory used as pollinators (Table 6).

Conclusion

The use of Stella as a pollinator has led to high yielding, self-fertile seedlings with medium to large fruit size. With respect to the genetics of *fruit juice and*

Table 5. Selected seedlings for further evaluation for the canning industry with a complete abscission layer, black skin and high yield.

Combination	Tree number	Yield	Fruit size in mm	Harvest time (week)	self-fertility
Schüttler × Magda	110.27	very good	21–22	2	–
Schüttler × Vittoria	115.3	very good	21–23	4	–
Schüttler × Fvd Weid	116.5	very good	19–21	2	–
Fvd Weid × Heidegger	121.54	very good	22–23	4	–
Fvd Weid × Vittoria	125.31	mod-good	23–24	4	–
Fvd Weid × Sp Holinger	127.46	good	22–24	5	–
Vittoria × Delta	130.14	very good	22–23	4	–
Vittoria × B. Langstieler	131.32	very good	22–23	4	–
Vittoria × Stella	132.97	very good	22–23	3	+
Vittoria × Heidegger	134.52	mod-good	22–23	5	–

Table 6. Selected seedlings for further evaluation as high quality dessert fruits.

Combination	Tree number	Yield	Fruit size in mm	Harvest time (week)	self-fertility
Vittoria × Stella	132.1	good	21–22	1	–
	132.74	very good	24–25	4	+
	132.75	very good	22–24	5	+
Vittoria × Heidegger	134.1	very good	22–24	4	–
	134.21	very good	23–24	4	–
	134.39	very good	24–25	4	–
	134.44	good	24–25	4	–
Vittoria × Merton Glory	135.32	very good	24–25	3	–

skin colour, Stella is heterozygous for 'black' (dark coloured skin and juice) and 'white' (yellow-red skin and colourless juice), whereas Schüttler is homozygous for 'white'. This confirms previous results from progeny of crosses of Heidegger × Sauerhäner (heterozygous) and Sauerhäner × Weisse Herz (heterozygous × homozygous for 'white'), from which 3:1 and 1:1 segregations of 'black' and 'white' fruits were obtained (Theiler-Hedtrich 1985a). From these results it is postulated that 'white' varieties, e.g. Merton Glory, Sue, Büttners späte Knorpelkirsche, are homozygous recessive for fruit juice and skin colour. On the other hand the genetics of colouration of the fruits might be much more complex, because in all varieties some pigmentation of the skin occurs (yellow-red). Further information on the inheritance of fruit juice and skin colour should be obtained from progenies derived from self-pollinated seedlings from Schüttler × Stella (homozygous 'white' × heterozygous) and Vittoria × Stella (homozygous 'black' × heterozygous).

The abscission layer formation seems to be a genetically complex character and needs further investigations by backcrossings.

Fruit size is a quantitative character, but if varieties with large fruits are combined, such as Stella, Heidegger, Merton Glory or Vittoria, a higher proportion of large-fruit bearing seedlings are obtained. This character can also be influenced in the reverse when small-fruit sized varieties are used for crosses. This was demonstrated in the combination of Vittoria × Späte Hollinger (Fig. 3) where only approx. 5% of the seedlings bore fruits which were as big as those of Vittoria.

For the grower only cherry cultivars which are precocious and high yielding are of interest. The character of fruit set seems to be polygenic and highly suppressed (Fig. 4). From all combinations tested, only a small proportion of the progeny were high yielding. Most of the seedlings had a moderate to very small or no crop, even after 4 fruit-bearing seasons.

Overall there are some promising perspectives by testing progeny from backcrosses and self-fertile seedlings to see, if there are more qualitative characters in this genus, which would allow to improve sweet cherry breeding work.

References

Bargioni, G., 1970. 'Vittoria' nuova cultivar di ciliegio dolce. Rivista della Ortoflorofrutticoltura Italiano 6: 3–12.

Crane, M.B. & A.G. Brown, 1955. Incompatibility and varietal confusion in cherries. Scient. Hort. 11: 53.

Krapf, B., 1976. Die Befruchtungsverhältnisse der Obstbäume. Flugschrift Nr. 30 der Eidg. Forschungsanstalt Wädenswil.

Lapins, K.O., 1971. Stella a self fruitful sweet cherry. Can. J. Plant Sci. 51: 252–253.

Matthews, P., 1973. Some recent advances in sweet cherry genetics and breeding. Proc. EUCARPIA Fruit Section Symp. 5. Topic Fruit Breeding. Canterbury: 84–107.

Matthews, P., 1979. Progress in breeding cherries for resistance to bacterial canker. EUCARPIA Fruit Section, Tree Fruit Breeding. Angers: 157–174.

Riesen, W., T. Meli, W. Zbinden & A. Widmer, 1991. Regendach für Tafelkirschen. Schweiz. Zeitschrift für Obst- und Weinbau 127: 360–372.

Sansavini, S. & W.D. Lane, 1983. 'Sunburst' e 'Lapins' ciliege autofertili durone-simili. Frutticoltura 45: 55–57.

Schaer, E., 1977. Die Kirschensorte 'Heidegger' (eine Erstbeschreibung). Schweiz. Zeitschrift für Obst- und Weinbau 113: 389–392.

Theiler, R., 1971. Embryonenkultur für die Anzucht neuer Kirschenhybriden. Schweiz. landw. Forschung. X: 65–93.

Theiler-Hedtrich, R., 1985a. Sweet cherry breeding programme at the Swiss Federal Research Station. I. Results of fruit characters and flowering period inheritance. Acta Horticulturae 169: 51–62.

Theiler-Hedtrich, R., 1985b. Sweet cherry breeding programme at the Swiss Federal Research Station. II. Results of bacterial canker resistance and seedling vigour. Acta Horticulturae 169: 63–71.

Theiler-Hedtrich, R., 1990. Beziehung zwischen Fruchtgewicht und Fruchtbreite bei Kirschen. Schweiz. Zeitschrift für Obst- und Weinbau 126: 590–598.

Way, R.D., 1968. Pollen incompatibility groups of sweet cherry clones. Proc. Amer. Soc. Hort. Sci. 92: 119–123.

H. Schmidt and M. Kellerhals (Eds.), Progress in Temperate Fruit Breeding, 173–176.

New plum varieties with adaptation to the Eastern Subcarpathian Region

V.I. Rati, F. Acatrinei, Dorina Brinza
Fruit Growing Research and Production Station, Bacău – Romania

Key words: Plum varieties, adaptability

Summary

Within the Eastern Subcarpathian Region of Romania plum cultivation is next in importance to apple, which is the most important fruit crop. It was necessary to improve the range of available plum varieties for this region. Thirteen varieties and 7 new hybrids were studied by the Fruit Growing Research Station from Bacău. With high yield (over 15 t/ha in the first fruit year), high quality of fruits (over 45 g/fruit, over 15% dry matter, low shedding), low vigour, high productivity index, resistance to specific diseases and with tolerance to plum pox, the varieties 'Centenar', 'Carpatin' and 'Piteştean' were introduced to the region. Many of the studied varieties and hybrids had valuable caracteristics ('Vâlcean' – over 80 g/fruit, 'Vânăt românesc' – 19.4% dry matter and so on). This study illustrates the agroproductive characteristics of the varieties and hybrids and their potential for improvement of the range of plum varieties adapted to the Eastern Subcarpathian Region.

Introduction

The soil climatic and relief conditions of the Eastern Subcarpathian Region are very favorable for plum tree cultivation. This species has a long history of cultivation.

The main varieties of the region were the old native varieties 'Vânat românesc' (very good for jam 'majun'), 'Gras românesc' (for 'tuica' liqueur) and 'Tuleu gras' (very good flavour), that was less wide spread because of its more difficult multiplication.

Between 1960 and 1970 we set up many new plum orchards, where besides the above mentioned varieties, we introduced also the varieties 'Anna Spath', 'D'Agen' and 'Reineclode'.

At present, demand for fresh fruits, of a variety of tastes, and available over a long period, is high, and has stimulated a study of new plum varieties for this region (Cociu 1981, Isac et al. 1989).

The main objective of the study was to create a range of varieties, capable of covering the market demand and suitable for intensive growing systems.

Material and methods

The experiment was done at the Fruit Growing Research Station Bacău, N-E of Bačau city, on the Bistriţa river meadow, 5 km from the confluence with the river Siret.

The material consisted of 20 plum varieties and hybrids, grafted on myrobolan rootstock. The varieties 'Piteştean', 'Carpatin'. 'Centenar', 'Tuleu timpuriu', 'Albatros', 'Dâmboviţa', were bred in Romania and have as a genitor the romanian variety 'Tuleu gras' (Cociu 1977, Cociu et al. 1986).

Planting distance was 4 × 3 m (833 trees/ha). The experiment was set up in 1982. The 20 varieties and hybrids were arranged in a randomised block, with 9 trees/variety. The crown was flattenedcup-shaped in the direction of the row. The soil had been fallow and was a typical alluvial. The texture was sandy, pH = 8,1–8,6, humus = 1,0–1,3%. The annual average temperature is 9,1 °C and the average rainfall is 553 l/m^2.

The following traits were observed: developmental phases, yield (t/ha), growth vigour (trunk cross section area (cm^2), measured at 30 cm from soil), index of productivity as the ratio of yield (kg/tree) and trunk cross section area (cm^2), fruit quality (dry matter, appear-

Table 1. Phenology of blossom and ripening season of varieties and hybrids. Study 1986–1992.

Variety	Blossom (Days) Start A* B*	Blossom (Days) Finish A* B*	The ripening season A* B*
1. Early Rivers	1 – 0	1 – 0	0 *(10 – 15 VII)*
2. Piteştean	3 – 1	1 – 1	13 – 10
3. Vâlcean	3 – 1	0 – 1	15 – 12
4. Silvia	2 – 0	2 – 0	15 – 12
5. Carpatin	5 – 3	5 – 4	15 – 12
6. Centenar	2 – 0	2 – 2	18 – 13
7. Tuleu timpuriu	7 – 3	9 – 5	21 – 17
8. Albatros	7 – 3	7 – 4	31 – 25
9. Dâmboviţa	6 – 3	9 – 4	37 – 32
10. Vânăt românesc	7 – 3	7 – 4	51 – 45
11. Verity	2 – 1	6 – 3	55 – 47
12. Vision	6 – 2	8 – 4	58 – 50
13. Bluefree	4 – 2	6 – 4	60 – 52
14. H 7.27	2 – 1	5 – 3	12 – 8
15. H 55.119	0 *(1 IV – 12 V*)*	0 *(8 IV – 16 V*)*	28 – 23
16. H 3.34	8 – 3	8 – 4	30 – 25
17. H 13.6	8 – 3	8 – 4	30 – 25
18. H 10.18	4 – 2	3 – 2	30 – 25
19. H 42.87	6 – 3	4 – 3	30 – 25
20. H 30.90	8 – 3	10 – 6	33 – 28

A* = days after the first variety manifested the phenophases in the year with the earliest vegetation (reference date marked).

B* = days after the first variety manifested the phenophases in the year with the latest vegetation (reference date marked).

ance), degree of attack (D.A. %) by diseases on the leaves (*Polystigma rubrum* and *Coryneum beijerinkii*).

$$D.A.\% = E(F \times I)/100 \qquad (1)$$

where F = Frequency of attack = percentage of leaves attacked; I = Intensity of attack = E (i × f)/100, where E = amount; i = % percentage of attack on the leaf (25, 50, 75, 100%) – corresponding to grades 1, 2, 3, 4; f = number of cases of attack at each percentage (grade). (Cociu & Oprea 1989).

Results

The varieties studied flowered at approximately the same time, but with important variations between years over a period of about 40 days. The start of blossomming varied from 1 April to 12 May (Table 1), depending on the climatic conditions of the region from year to year. There was a risk of flower damage during late spring hoarfrosts. Therefore, varieties with late blossom, such as 'Carpatin', Dâmbovita', 'Albatros' and 'Vânăt românesc' are recommended.

In the earlier years the difference in time between the phenophases of the varieties was larger than in the later years when the phenophases of the varieties were closer in time (Table 1).

Regarding the fruit maturity, all the varieties ripened about 10–15 days after the variety 'Early Rivers', ('Piteştean', 'Vâlcean', 'Silvia', 'Carpatin', followed at short intervals by the varieties and finished between 10–20 September with the varieties 'Verity' and 'Bluefree').

To establish the suitability of the varieties for intensive growing conditions the index of productivity, as a ratio of the yield and growth, was used. The highest val-

Table 2. Trunk cross section area, average annual yield, productivity index and fruit quality. Study 1986–1992.

Variety	Productivity Index kg/cm²	Trunk cross cm²	Weight of fruits %	Dry matter %	Yield Average (t/ha)	Difference Control	Significance
1. Early Rivers (C)	0,04	176,4	37,4	13,0	5,23	000	–
2. Piteştean	0,11	159,5	47,2	16,3	15,72	10,48	xx
3. Vâlcean	0,03	261,9	81,8	15,6	6,52	1,28	–
4. Silvia	0,09	152,1	45,6	15,1	11,10	5,87	–
5. Carpatin	0,16	148,4	56,7	15,0	19,83	14,60	xxx
6. Centenar	0,16	125,6	45,1	15,8	17,30	11,90	xxx
7. Tuleu timpuriu	0,04	224,2	47,3	14,8	7,85	2,62	–
8. Albatros	0,09	134,0	47,6	16,3	10,00	4,77	–
9. Dâmboviţa	0,06	226,0	74,8	15,9	11,55	6,32	–
10. Vânăt românesc	0,09	163,6	24,5	19,4	12,63	7,40	x
11. Verity	0,08	214,5	55,4	15,5	13,63	8,40	x
12. Vision	0,05	239,5	59,3	15,6	10,30	5,58	–
13. Bluefree	0,12	132,2	59,0	16,2	13,38	8,15	x
14. H 7.27	0,05	193,0	78,9	14,8	8,38	3,15	–
15. H 55.119	0,12	117,3	50,5	14,1	11,50	6,27	–
16. H 3.34	0,05	155,0	46,8	15,2	6,62	1,38	–
17. H 13.6	0,05	151,0	44,2	15,4	6,63	1,40	–
18. H 10.18	0,05	239,0	55,7	14,7	10,13	4,90	–
19. H 42.87	0,07	185,7	49,5	15,5	10,37	5,13	–
20. H 30.90	0,08	99,1	38,4	16,6	6,25	1,02	–

DL 5% = 6,36; DL 1% = 8,43; DL 0,1% = 10,88.
xxx – very significantly positive.
xx – distinct significantly positive.
x – significantly positive.
– – not significantly.
Significance given control test 't'.

ues were recorded from the varieties 'Centenar' (0.16), 'Carpatin' (0.16) and the hybrid H 55/119 (0.12).

The hybrid H 30–90 (99,1 cm²) and varieties 'Centenar', 'Albatros', 'Bluefree', 'Carpatin' had the smallest trunk cross section area.

The highest average yields in the first 6 years of production were obtained from the varieties 'Carpatin' (19.8 t/ha), 'Centenar' (14.1 t/ha), 'Piteştean' (15.7 t/ha), 'Verity', 'Bluefree', 'Vânăt romanesc' and from the hybrid H 55/119 (see Table 2).

Large fruit was obtained from the varieties 'Vâlcean' (average weight/fruit 81.8 g, but maximum weight recorded was over 100 g), 'Dâmboviţa' (74.8 g) and from hybrids H 7/27 (78.9 g/fruit). The highest dry matter content by refractometric determination, was shown by the variety 'Vânăt românesc' (19.4%) and by hybrid H 30/90.

The behaviour of the varieties in respect of diseases was also examined.

The most sensitive to *Coryneum beijerinkii* were the varieties 'Early Rivers', 'Tuleu timpuriu' and 'Silvia' (over 4%). A low attack (under 0.7%) was shown by varieties 'Dâmboviţa', 'Verity', and a majority of the hybrids (see Table 3).

The varieties 'Vânăt românesc', 'Vision' and 'Bluefree' were susceptible to *Polystigma rubrum* (degree of attack over 1.6%). The varieties 'Piteştean', Vâlcean', 'Carpatin', 'Tulelu timpuriu' and the hybrids H 10/18, H 42/87 were resistant.

Visual observations made on the leaves and fruits did not reveal the presence of the viral disease, plum pox on the varieties 'Carpatin', 'Centenar' and 'Dâmbovita'. These were considered to be tolerant to this disease (Cociu 1990).

Table 3. Coryneum beijerinckii and *Polystigma rubrum* on the leaves of 20 plum varieties and hybrids.

Varieties and hybrids	*Coryneum beijerinckii*	*Polystigma rubrum*
1. Early Rivers	6.4	0.5
2. Pitestean	1.1	–
3. Valcean	1.8	0.0
4. Silvia	4.2	0.95
5. Carpatin	0.8	–
6. Centenar	1.8	0.7
7. Tuleu Timpuriu	5.5	–
8. Albatros	0.85	0.95
9. Dambovita	0.6	1.1
10. Vinat Romanesc	0.7	5.7
11. Verity	0.5	–
12. Vision	2.1	1.8
13. Bluefree	0.8	1.85
14. H 7.27	3.0	0.7
15. H 55.119	0.6	1.5
16. H 3.34	0.7	0.7
17. H 13.6	0.75	0.75
18. H 10.18	0.75	–
19. H 42.87	0.6	–
20. H 30.90	0.8	0.9

Conclusions

The following new varieties were adapted to the regional conditions and were introduced as additions to the existing varieties.

'Pitestean': maturity 20–28 July, high yield, high productivity index, good quality fruit (47.2 g/fruit) and 16% dry matter, disease resistant.

'Centenar': maturity 25 July – 2 August, low vigour, high yield, high productivity index, good quality fruit (45.1 g/fruit), tolerant to plum-pox.

'Carpatin': maturity 20–30 July, low vigour, high yield, high productivity index, good quality fruit (56,7 g/fruit, 15% dry matter) disease resistant, tolerant to plum-pox.

For breeding, the following varieties and hybrids are valuable:

'Vâlcean' (81.8 g/fruit), 'Tuleu timpuriu' (good flavour fruit), 'Vânăt româesc' (fruit with high dry matter), 'Bluefree' (large fruit and late maturity), H 7/27 (large fruit and early maturity), H 55/119 (low vigour, productive), H 30/90 (very low vigour, fruit with high dry matter).

References

Cociu, V., 1977. The Results of Research Work Concerning the Obtaining of New Varieties in Romania. Ceres, Bucharest.

Cociu, V., 1981. Some aspects on fruit and small fruit varieties location in Romania. Scientific papers I.C.P.P. Pitesti IX: 9–23.

Cociu, V. et al. 1986. Present plum assortment and objectives for its improvement – Scientific papers I.C.P.P. Pitesti XVI: 63–83.

Cociu, V., 1990. New Varieties, Progress Factor in Fruit Growing. Ceres, Bucharest.

Cociu, V. & St. Oprea, 1989. Methods of Research in Improvement of Fruit Breeding. Dacia, Cluj/Napoca.

Isac I., R. Roman & I. Nicolaescu, 1989. The promising plum assortment related to the cultural areas and evaluation directions. Scientific papers I.C.P.P. Pitesti XIII: 125–133.

H. Schmidt and M. Kellerhals (Eds.), Progress in Temperate Fruit Breeding, 177–178.

Selecting apricots with good fruit quality and resistance to late spring frosts in Gürün

Yasar Akça & Seyit Mehmet Sen
Yüzüncü Yil University, Faculty of Agriculture, Department of Horticulture 65080 Van, Turkey

Key words: Apricot, breeding selection, late spring frosts, late flowering

Summary

Growth vigour, productivity, fruit characteristics, resistance to late spring frosts and phenological characteristics were studied in native apricot populations at Gürün in 1990–1992. There was much variation in the response of flowerbuds to extreme frost in winter and in flowers to late spring frost. The greatest resistance to late spring frost was shown by selection 226 which flowered at 1270 m in mid-May. Selection 226 flowered 8–10 days later than Hacihaliloglu. The lowest temperatures at flowering time were - 4.5° C and - 4.1° C in 1990 and 1992 respectively. Selection 117 ripened in mid-October, 63–73 days later than Hacihaliloglu.

Introduction

The most serious problem in apricot growing in Turkey is late spring frost. The timing of late spring frosts affecting apricot flowers vary from year to year. The most important aims in Turkish apricot breeding programmes are selection for resistance to late spring frosts and for late flowering. Güleryüz (1988) reported that selections were found with late flowering and resistance to frost in the Erzincan plain.

There is a rich apricot seedling population in Gürün, (1250 m), where late spring frosts occur almost every year.

This research was aimed at the development of late flowering plants with resistance to spring frosts.

Materials and methods

This study was conducted in Gürün between 1990 and 1992. The cultivar Hacihaliloglu was used as control variety. Selections avoiding spring frost damage were found by selecting late flowering forms. Pomological characteristics were measured in 10 samples that were chosen at random. Fruit characteristics investigated were, fruit weight, seed weight, soluble solid content, seed separation, flesh/seed ratio, fruit length, fruit structure, fruit diameter and resistance to speckle.

Results and discussion

The effect of frost on the cropping of trees was recorded after temperatures fell to about - 5° C. According to Passecker (1947) apricots are resistant to temperatures down to - 3° C. However, Duric (1981) recorded a large reduction in fruit set at - 2° C during flowering. In a study carried out in the Erzincan plain, Güleryüz (1988) found that - 4.8° C damaged apricots on a vast scale. In this study - 4.8° C and - 4.1° C were sufficient to select for resistance in 1990–1992. In comparison to Hacihaliloglu 76 seedlings were found to be resistant in 1990, 35 being promising from the point of view of yield. At that time, Hacihaliloglu cropped poorly. Selection 226 was very late flowering and gave the heaviest crop in 1990 and 1992. Late spring frost damage (- 4.1° C) was seen at flowering time in 1992. Selection 226 flowered 8–10 days later than Hacihaliloglu.

The latest time of harvesting was mid-October observed in selection 117 which ripened 63–73 days later than Hacihaliloglu.

Table 1. Important fruit characteristics of selected types

Type Nr	Fruit Weight (gr)	Seed Weight (gr)	Fruit Length (mm)	Fruit Suture (mm)	Fruit Diameter (mm)	Soluble Solids (%)
9	32.14	2.28	38.10	39.27	36.28	11.44
108	27.92	2.29	35.87	40.34	36.84	14.51
203	31.33	2.63	40.19	38.89	35.50	18.62
133	35.54	3.19	40.61	41.11	39.61	11.32
120	46.93	3.10	44.48	40.79	41.78	14.38
128	50.63	3.43	46.64	47.98	44.85	9.96
17	47.94	3.74	44.91	44.10	44.06	10.22
31	49.99	4.21	48.25	46.70	37.65	10.17
204	38.15	3.28	44.85	41.18	37.33	11.65
13	28.91	2.60	38.08	40.12	37.33	9.92
140	38.31	2.85	43.14	43.59	39.71	9.90
107	30.16	3.13	37.67	39.47	35.39	10.30
226	25.32	3.13	34.26	39.42	35.82	12.06
117	25.63	2.94	33.24	38.67	34.84	10.95
Haliloglu cv	48.57	2.13	43.15	45.27	41.59	23.27

Fruit weight in the best selections was between 25.32 g (selection 226) and 50.63 g (selection 128). These selections were more promising than those of Güleryüz (1988) from the point of view of fruit weight.

Seed weight of the best selections was between 2.28 g (selection 9) and 4.21 g (selection 31). The most promising selections had smaller seeds than those of Güleryüz (1988) but larger than those of Hacihaliloglu. Soluble solids ranged between 9.96 (selection 128) and 18.61 (selection 203) both lower than Hacihaliloglu.

Fruit length ranged between 35.87 mm (selection 108) and 48.25 mm (selection 31), fruit diameter between 35.39 mm (selection 107) and 44.85 mm (selection 128), and the fruit suture between 38.89 mm (selection 203) and 47.98 mm (selection 128).

All the selections had free seeds, were resistant to speckle and had a high to medium yield.

References

Passecker, F., 1947. Vermehrungs und Züchtungsfragen bei Aprikose. Der Züchter, 17–18. Jahrg. Heft 9: 227–284.

Duric, B., 1981. Contribution to the Investigation of the Susceptibility of Young Fruits of Apricot Varieties to Late Spring Frosts. Plant Breeding Abst. 51 (8): 7561.

Güleryüz, M., 1988. Erzincan Ovasinda Gec Ilkbahar Donlarina Karsi Mukavbim ve Kaliteli Zerdali Tiplerinin Seleksiyon Yoluyla Islahi Üzerinde Arastirmalar, Profösölük Taktim Tezi.

H. Schmidt and M. Kellerhals (Eds.), Progress in Temperate Fruit Breeding, 179–181.

Studies on selection of walnut (*Juglans regia* L.) in Gürün

Yasar Akça & Seyit Mehmet Sen
Yüzüncü Yil University, Faculty of Agriculture, Department of Horticulture, Van, Turkey

Key words: Juglans regia, lateral fruitfulness, nut quality, selection, walnut

Summary

This paper reports selection studies on walnut (*Juglans regia* L.) in Gürün, Turkey. 365 selected walnut trees were scored for tree characteristics during the three years 1990 to 1992 and samples were investigated for important fruit characteristics. At the end of this study 41 wild types were selected. In these types average nut weight varied between 10.4–19.6 g, average kernel weight varied from 5.8–9.4 g, kernel to nut ratio varied from 46.1 to 64.2%, light yellow kernel colour varied between 50–100%, nut length varied from, 38.1 to 50.8 mm protein content varied from 16.1 to 26.1%, fat content varied from 61.6 to 76.85% and lateral fruitfulness varied from 40 to 100%.

Introduction

Turkey is one of the most important centers of origin of *Juglans regia* L. which is used for walnut production all over the world. This species has been grown economically for thousands of years in Anatolia and it has mostly been propagated from seed in Anatolia, with more than four million walnut trees is thus a botanical garden, a gene bank and a walnut museum (Sen, 1988).

Turkey supplies 12.8% of the world's walnut demand and is third among walnut producers of the world. In order to select superior types of walnut, selection work started in western Turkey and 20 promising types were selected for production in that region (Ölezx, 1971). Another selection work started in 1977 in the North Eastern Anatolia and the Eastern Black See and selected 26 superior types for production (Sen, 1980).

Our work can be considered as an extension of selection work to the eastern regions of Turkey, with the aim of finding high quality seedlings. We aimed to find new walnut types reaching the following standards:

70% fat content, 20% protein content, 15–20 g nut weight, 8–10 g kernel weight, 55% kernel to nut ratio, thin shell (1 mm) and 65% lateral fruitfulness. These standards are higher than those in other countries: 65% fat content, 16% protein, heavy nut (14.7 g), 45–54% kernel ratio, thin shell (1 mm), 40–50% lateral fruitfulness (Serr, 1962; Pandele, 1968; Nenjuhin, 1971).

Gürün is situated on the south of Sivas. Its altitude is 1150–1250 m and the average annual temperature is 9.5° C. There are more than 10.000 walnut trees all of them produced by seed.

Material and method

365 wild walnut trees from around Gürün, a town of Sivas, were evaluated for tree and nut characters from 1990–1992. Shape of nuts, size, density, volume, weight, kernel ratio, kernel colour, shell thickness, smoothness of shell, adhesiveness over shell surface and clinginess of kernel to shell were measured on ten nuts chosen at random (McGranahan & Ramos, 1990; McGranahan & Ramos, 1992). In the second year samples were analysed chemically. The nuts were dried until the moisture content of the whole walnuts was reduced to less 5%. The principle criteria for selecting the most promising walnuts are as follows: 1. Late leafing; 2. High yield and uniform distribution of fruit over the tree with 50% lateral fruitfulness and minimum 2–3 fruits in each cluster; 3. Non alternate bearing; 4. Resistance to cold, insects and diseases; 5. Healthy, well developed trees; 6. Uniform ripening; 7. Average fruit weight should not be less than 15 g and with a

Table 1. Observations made for 8 nut characteristics of selected Gürün walnuts (mean of 10 nuts)

Type Nr	Nut Weight (gr)	Kernel Weight (gr)	Kernel Ratio (%)	Nut Length (mm)	Nut Suture (mm)	Nut Cheek (mm)	Shell Thickness (mm)	Protein Content %
02	19.61	9.41	47.99	49.63	40.14	41.09	1.43	19.32
01	18.69	9.32	49.87	50.84	37.58	38.78	1.45	26.10
08	16.95	8.63	50.92	40.99	39.32	38.73	1.19	21.70
32	15.17	8.46	55.77	42.07	34.88	36.08	1.12	16.90
19	14.92	8.37	56.10	46.16	37.54	36.87	1.19	21.90
15	16.31	8.28	50.77	42.49	37.73	38.21	1.18	24.90
22	15.99	8.28	51.78	38.91	36.07	38.07	1.12	17.15
03	14.68	8.27	56.34	46.55	35.96	37.51	0.97	18.00
11	15.71	8.24	52.45	43.81	35.94	38.96	1.44	18.96
07	15.83	8.23	51.99	45.12	38.30	37.77	1.37	22.29
06	15.02	8.18	54.47	47.90	35.07	36.26	1.29	19.60
61	14.81	7.99	53.95	45.11	34.72	34.79	1.26	18.67
33	14.26	7.98	55.96	44.94	36.32	35.29	1.16	16.85
04	13.45	7.95	59.11	45.31	35.08	35.70	0.95	18.50
10	14.76	7.82	52.98	51.40	36.60	36.22	1.26	19.15
37	14.64	7.92	53.42	40.84	35.24	34.63	1.45	19.67
25	14.72	7.73	52.51	44.99	35.71	35.07	1.35	18.16
17	14.53	7.71	53.07	41.52	35.51	37.14	1.19	18.73
12	14.17	7.65	53.98	47.40	38.74	36.64	1.28	19.28
09	13.88	7.35	52.96	43.78	41.23	35.45	1.24	25.10
13	14.17	7.27	51.31	44.13	36.12	36.24	1.30	20.00
38	13.17	7.23	54.89	39.55	32.82	33.85	1.27	21.10
05	13.94	7.18	51.51	41.42	37.38	36.63	1.07	17.52
27	13.35	7.15	53.55	40.29	34.30	34.78	1.30	22.15
46	13.33	7.07	53.04	40.06	36.22	36.72	1.29	17.50
48	12.78	7.06	55.24	45.84	34.42	34.67	1.10	21.41
20	13.22	6.98	52.79	42.57	38.74	37.16	1.00	19.87
41	10.99	6.72	61.14	38.15	38.24	33.95	0.95	21.61
44	11.85	6.67	56.29	40.20	35.66	36.00	1.16	17.95
40	10.36	6.65	64.19	41.07	34.03	33.62	0.59	22.30
54	12.18	6.48	53.20	41.28	34.15	34.12	1.32	22.16
56	14.03	6.47	46.12	45.70	33.18	34.85	1.31	19.77
21	12.14	6.38	52.55	44.84	34.54	34.65	1.15	17.65
51	13.15	6.31	47.98	41.40	34.98	34.43	1.47	21.70
23	12.22	6.31	50.64	42.12	34.70	36.51	1.24	18.30
45	12.50	6.29	50.32	40.34	33.61	33.64	1.29	24.30
14	12.43	6.28	50.53	42.03	37.85	36.51	1.08	16.08
43	12.12	6.24	51.48	39.45	32.70	33.57	0.87	20.04
47	11.59	6.23	53.76	41.06	33.52	34.33	0.97	21.30
31	12.47	6.21	49.80	39.78	31.57	34.08	1.14	20.18
16	11.35	5.77	50.84	41.10	34.42	33.51	1.14	19.67

kernel to nut ratio of at least 50%; 8. Light colour, full, easily separated, and 9. Smooth and thin shell.

Results and discussion

From the 365 seedling evaluated as described, 41 superior types were selected as promising and data for some of the nut characters is given in Table 1.

High yield capacity is an important character which determines the economy of walnut production. Walnut productivity is highly dependent on the number of fruit buds on annual shoots the number of fruit per cluster, the percentage lateral fruitfulness, nut weight, nut size, kernel ratio and the time blooming.

The percentage of lateral fruitfulness of selected types varied between 40–100%. Average percentage of lateral fruitfulness of selected types is 66%. In 83% of selected types percentage of lateral fruitfulness was more than 50%. The number of fruit per cluster of selected types varied between 2–5.

The mean nut weight of selected types varied from 10.4–19.6 g and kernel weight varied from 5.8–9.4 g (Table 1). Average nut weight of selected types is 13.9 ± 0.3 g and average kernel weight 7.4 ± 0.1 g. In our types, nut weight and kernel weight were higher than is selected types of Ölez (1971) and Sen (1980). In our types kernel to nut ratio varied from 46.1 to 64.2% and average kernel ratio of selected types is 52.9%. In the work of Ölez (1971) kernel ratio of 8 selected walnut types were less than 50%. But in the study of Sen (1980) there are only two types with kernel ratio less 50%. Our results and results of Sen (1980) are similar with respect to the kernel ratio. In the work of Malik (1964) the walnut type Pestisani 154, selected from 400 types, has a nut weight 14.7 g and a kernel ratio of 53.7%; in our work, 14 types were found to have a kernel ratio more than 53.7%. Average nut length of selected types is 43.2 ± 0.2 mm, nut suture of selected types is 35.8 ± 0.1 mm and average nut cheek is 35.9 ± 0.2 mm (Table 1).

The percentage of clean yellow kernel should be at least 50% in promising cultivars. In this work in 48% of selected types had kernels 100% yellow, in 29% had kernels 90% yellow, in 12% yellow colour was 80% and in 11% it was 70%.

Pandele (1968) said the protein content must be at least 16%, kernel ratio least 45% and fat content least 65%. In our work protein content varied from 16.1 to 26.1% and fat content varied from 61.6 ro 76.9%.

Zhadan & Strukov (1977) accept nuts which have shell thickness between 0.7 and 1.5 mm as the promising types; 0.92 mm is the value according to Nenjuhin (1971). In our work the shell thickness of selected types varied between 0.59 mm and 1.45 mm and average shell thickness was 1.2 ± 0.02 (Table 1).

According to other research, protogynous walnut trees are more fertile than protandrous trees. 21 of selected types were protogynous, 18 were protandrous and 2 were homogamous.

There was an opportunity to assess the cold resistance of selected types during the two years period in the years 1990 and 1992. Many trees were damaged from late spring frosts, but none of the selected types were damaged. All of the selected types were late leafing or very late.

Harvesting date of selected types varied from mid September to mid October. Selected types cropped regularly every year, and were resistant to *Xanthomonas* and *Gnomonia*. The mean leaf area of selected types varied between 96 and 248 cm^2.

Some of these selected types in Gürün have better nut characteristics than much walnut cultivars as Pestisani, Plovdiv, Sibisel, Mirskovskii, Franquette, Hartley, Payne, Serr, Ashley, Chico, Vina, Amigo, Howard, Pedro, Tulare, Chandler & Cisco.

References

Malik, T., 1964. Quality types of walnut in the southwest region of Slovakia, Hort. Abs. 34: 681.

McGranahan, G. & D.E. Ramos, 1990. 'Cisco' persian walnut, HortScience 35: 372.

McGranahan, G. & D.E. Ramos, 1992. 'Tulare' persian walnut, HortScience 27: 186–187.

Nenjuhin, V.N., 1971. Selection of plus trees of walnut in the Ukraine, Pl. Br. Abst. 41: 187.

Ölez, H., 1971. Marmara Bölgesi cevizlerinin seleksiyon yolulya islahi üzerinde arasitirmalar, doktora tezi, basilmamis, Yalova.

Pandele, I., 1968. Biochemical characterization of the principal varieties and types of walnut, almond and hazel in Rumania and determination of general metabolic correlations specific to nuts, Pl. Br. Abstr. 38: 871.

Serr, E.F., 1962. Selecting suitable walnut varieties, California Agricultural Experiment Station, Daves, California, Leaf 144.

Sen, S.M., 1980. Kuzeydoğku Anadolu ve Doğku Karadeniz cevizlerinin seleksiyon yoluyla islahi üzerinde arastirmalar, doçentlik tezi, basilmamis, Atatürk Üniversitesi Erzurum.

Sen, S.M., 1988. Anatolia as a walnut garden, International conference on walnuts, Atatürk central horticultural research institute, Yalova, Turkey.

Zhadan, V.M. & M.V. Strukov, 1977. Breeding walnut for fruit size, Pl. Br. Abstr. 47: 918.

H. Schmidt and M. Kellerhals (Eds.), Progress in Temperate Fruit Breeding, 183–187.

Breeding strawberries for Eastern Central Canada

Shahrokh Khanizadeh
Agriculture Canada, Research Station, 430 Boul. Gouin, St-Jean-sur-Richelieu, Quebec, Canada J3B 3E6 and Department of Plant Science, McGill University, Ste-Anne-de-Bellevue, Quebec, Canada H9X 3V9 *

Key words: Breeding, Canada, 'Chambly', 'Oka', pedigree, strawberry

Summary

Canada imports more than 21,000 tonnes of strawberries annually, of which approximately 5,000–6,000 tonnes are frozen berries used for processing at a cost of more than $ 10 million. These could be replaced by Canadian grown fruit if improved cultivars, and suitable for mechanical harvesting, processing and fresh market adapted to canadian climatic conditions would be available.

For many years producers in Quebec have grown the cultivar 'Redcoat' (Sparkle × Valentine) released in 1957 by Agriculture Canada, Ottawa. Redcoat is now rapidly being replaced by 'Kent', ('K68-58 × Raritan') released in 1981 by Agriculture Canada, Kentville. Despite Quebec's preeminant position in Canada as a strawberry producing region, no strawberry cultivars have ever been bred specific for Quebec conditions. In 1982 a breeding program was initiated at Agriculture Canada St-Jean. This program was terminated in 1986 without any material having been tested sufficiently for release. In 1987 a joint project between McGill University and Agriculture Canada initiated to continue the breeding program and develop cultivars particularly suitable for mechanical harvesting and processing, resistant to disease and adapted to local conditions.

History of the breeding program

The Agriculture Canada program. In 1983 and 1984 a pool of the best adapted parental material was established to be used in the breeding program. Twenty-five cultivars, chosen on the basis of their performance under local conditions, or as potential donors of valuable characteristics, comprised the parental pool. Forty new strawberry genotypes from crosses or self-pollinations carried out at St-Jean were selected from the F_1 plants evaluated in 1984 at L'Acadie and made up the first series of new lines.

Also in 1984, 71 strawberry cultivars (23 of which were in a replicated trial) were evaluated at L'Acadie with the aim of identifying those which performed best under Québec conditions and using them as standards and as parents in future crossing programs. Yield and quality were determined with the cooperation of the Food Research Institute, Ottawa. Measurements were made at each harvest, of yield, fruit size and fruit size range. Yield components were studied by measuring the number of stems per plant and the number of flowers per stem, in an effort to select genotypes with more concentrated fruit production. The interaction between yield characters and planting density, as well as most other data, have been analyzed.

In 1985 seeds resulting from crosses made in 1984 were planted out, 2,500 F_1 seedlings were established in the field for later evaluation. Crosses in 1985 produced 4,500 F_1 seedlings which were planted in the field in 1986. Forty three seedlings selected from the crosses made in 1983 were planted in replicated plots for further evaluation. In 1986 the breeding program at St-Jean was ended.

The McGill-agriculture Canada program. A joint project between McGill University and Agriculture Canada started in 1987 to continue the breeding program.

Data collected from previous experiments, including the 1986 field trials, were pooled and analyzed to

* Agriculture Canada, St-Jean-sur-Richelieu, Contribution No. 335/93.12.02R

Fruit Characteristic	Rank values				
Season rippening	1 Very early	2 early	3 Midseason	4 late	5 Very late
Neck size	1 	2	3	4	5
Sepal form	1 	2	3	4	5
Achene position	1 	2	3	4	5
Skin colour	1 Very pale	2 Pale	3	4 Dark	5 Very dark
Flesh colour	1 Very pale	2 Pale	3	4 Dark	5 Very dark
Flavour	1 Very bad	2 Bad	3 Average	4 Good	5 Very good
Firmness	1 Very soft	2 Soft	3 Average	4 Firm	5 Very firm
Leaf diseases Leaf spot [*Mycosphaerella fragariae* Schw.] & Leaf Scorch [*Diplocarpon earliana* Ell. and Ev.]	1 Very Susceptible	2 Susceptible	3 Average	4 Resistant	5 Very resistant

Fig. 1. Score sheet used to evaluate individual line.

obtain a general idea of the performance of St-Jean selections. Twenty two selections from 1983, 1984 and 150 from 1985, 1986 selections were chosen for further, based on their performance relative to standard cultivars. Many more crosses were made and evaluated since 1987.

Selection criteria used in breeding program

The overall selection criteria for the breeding program were: high yield, suitability for mechanical harvesting (raised neck, elevated sepals, shape, skin texture, interior and exterior colour, size, firmness), suitability for processing (flavour and texture, organic acids, sugars, etc.), root and foliar disease resistance, resistance to herbicides, and hardiness (Khanizadeh et al., 1992a). A score sheet for evaluating some important characteristics is used to rank individual genotypes (Fig. 1). Some other techniques were also developed for genotype evaluation: 1) a computer program to trace genotype characteristics and draw family trees (Khanizadeh et al., 1991), 2) three mathematical indices to estimate potential for early dollar return, harvest concentration, and trait similarity (Khanizadeh & Fanous, 1992).

Achievements

Two selections were named and released since 1988:

Table 1. Yield, fruit size and other characteristics of two recently released strawberry cultivars ('Chambly' & 'Oka') compared to some commonly grown strawberry cultivars in eastern Canada[z]

Cultivar	Yield (kg/2 m)	Fruit size (g)	Season[y] ripening	Neck[x] size	Sepal[x] form	Achene[x] position	Skin[x] colour	Flesh[x] colour	Flavour[x]	Firmness[x]	Leaf[x] diseases
Bounty	3.81	7.6	L	3.0	3.9	3.0	3.3	3.0	3.8	2.0	3.0
Chambly	5.51	9.2	EM	4.0	4.7	3.0	4.0	4.0	3.6	3.8	4.2
Glooscap	5.01	7.6	M	3.4	4.4	2.0	3.0	3.1	3.1	2.6	4.0
Honeoye	4.95	9.2	LM	3.0	2.0	2.0	3.0	3.0	3.0	3.0	3.3
Kent	5.02	8.8	M	3.5	3.2	2.3	2.5	2.5	3.0	3.3	3.6
Oka	4.95	9.8	M	2.7	2.5	2.3	3.2	2.2	3.3	3.0	3.2
Redcoat	3.45	6.7	EM	3.8	3.0	2.8	3.0	2.8	3.0	2.8	3.5
Sparkle	3.55	6.6	LM	2.4	3.4	2.3	3.0	2.0	3.4	2.0	3.5
LSD[w]	0.7	0.5		0.1	0.1	0.1	0.1	0.1	0.2	0.1	0.4

[z] Average of six years (1987–1992), minimum of four replicates per year, data were transformed to arcsin prior to analysis of variance (SAS 1988).
[y] Season ripening: L = late, LM = late-midseason, M = mid-season, EM = early-midseason, E = early.
[x] See Fig. 1 for details.
[w] Least significant difference at 95% confidence.

Table 2. Yield, fruit size and other characteristics of two recently released selections 'SJ8416-1' and 1SJ83OR-2' compared to some commonly grown strawberry cultivars in eastern Canada[z]

Cultivar	Yield (kg/2 m)	Fruit size (g)	Season[y] ripening	Neck[x] size	Sepal[x] form	Achene[x] position	Skin[x] colour	Flesh[x] colour	Flavour[x]	Firmness[x]	Leaf[x] diseases
Bounty	3.28cd[w]	8.3ab	L	3.1c	4.1b	3.1a	3.6a	3.1abc	3.6a	2.4bc	3.1ab
Glooscap	5.55ab	7.6ab	M	3.9ab	4.5a	2.7abc	3.6a	3.4ab	3.8a	3.1ab	3.9a
Kent	5.45ab	8.9ab	M	3.8ab	3.6b	2.6abc	2.9b	2.9bc	3.4ab	3.3a	3.6ab
SJ8416-1	3.95bcd	8.9ab	EM	3.1c	3.1c	2.4abc	3.6a	3.6ab	3.4ab	3.1ab	3.1ab
SJ83OR-2	5.68a	9.0ab	EM	3.5b	3.6b	2.0abc	3.6a	3.3abc	3.4ab	2.9abc	3.9a
Sparkle	3.52cd	6.5b	LM	2.7d	3.4b	2.6abc	3.1b	2.4c	3.9a	2.4bc	3.9a

[z] Average of six years (1988–1993), minimum of four replicates per year, data were transformed to arcsin prior to analysis of variance (SAS 1988).
[y] Season ripening: L = late, LM = late-midseason, M = mid-season, EM = early-midseason, E = early.
[x] See Fig. 1 for details.
[w] Mean separation within column by Duncan's new multiple range test, 5% level.

'Chambly' strawberry

'Chambly' is the first June bearing early-midseason strawberry cultivar (*Fragaria* × *ananassa* Duch.) bred specifically for Southern Québec conditions. 'Chambly' resulted from a cross between 'Sparkle' and 'Honeoye' (Khanizadeh et al., 1992b). 'Chambly' was released because of its high yields of firm, deep red fruit with raised neck, elevated calyx and uniform well coloured flesh. It is recommended for both fresh market and processing (Khanizadeh et al., 1990). Yield, fruit size and some fruit characteristics are presented in Table 1.

Table 3. Yield, fruit size, plant and fruit characteristics of five advanced Quebec strawberry selections developed for fresh market compared to five standard cultivars during 1990–1993 field trials in Lavaltrie, L'Acadie and St-Foy in Quebec and Bouchtouch in New-brunswick[z]

Cultivar	Yield (kg/2 m)	Fruit size (g)	Plant and/or fruit characteristics observed in the field
Bounty	3.27c[y]	9.7c[y]	Late ripening, vigourous, susceptible to mildew, sensitive to zinbar, good flavour, medium firm, good for fresh market and processing, easy to dehull.
Glooscap	3.41b	9.3c	Midseason, very easy to dehull, medium firm, good flavour, susceptible to Botrytis fruit rot, dard red skin colour
Honeoye	4.18a	11.5b	Late-midseason, reflex calyx, bright red skin colour, medium firm, tender skin
Kent	3.56ab	11.9b	Midseason, very susceptible to zinbar, mildew, leaf spot, scorch and Botrytis fruit rot, firm fruit, attractive, glossy red, very vigourous.
SJ8518-11	4.17a	13.6	Midseason, irregular large fruit, pale colour, goof for fresh market, resistant to leaf diseases.
SJ89264-6	3.45b	14.1a	Late-midseason, large fruit, easy to dehull, light shiny pale colour, long conic with good size neck, very good flavour, very firm, excellent shelf life quality, reflex calyx, very good for fresh market and P.O.Y., susceptible to leaf spot.
SJ89288-2	4.27a	14.2	Mid-season, easy to dehull, shiny pale skin colour, round-conic, good flavour, firm, resistant to leaf diseases.
SJ8976-1	2.37cd	14.5a	Late-midseason, very similar to SJ89264-6, large fruit, easy to dehull, light shiny pale colour, conic, no neck, very good flavour, very firm, excellent shelf life quality, very good for fresh market and P.O.Y., susceptible to leaf spot.
SJ89700-1	3.41bc	14.0a	Very-late, shiny red skin colour, very attractive, large fruit.
Veestar	2.31d	7.7d	Early-midseason, easy to hull, vigourous, good for fresh market, susceptible to leaf diseases.

[z] Average of 16 replicates for three locations. Mean separation within column by Duncan's new multiple range test, at 5% level.

'Oka' strawberry

'Oka' is a mid-season, high yielding June-bearing strawberry (*Fragaria* × *ananassa* Duch.) cultivar. It has excellent productivity, flavour, relatively firm and is recommended for fresh or pick-your-own markets. Skin colour of 'Oka' is similar to 'Bounty' but tend to be darker in warm climate and sandy soil. 'Oka' resulted from a cross between K75-13 and 'Honeoye' (Khanizadeh et al., 1992b). Yield, fruit size and some fruit characteristics are presented in Table 1.

Advanced new selections

Two old lines (SJ8416-1 and SJ83OR-2) from 1983–84 which have potential for high yield and good fruit characteristics (Khanizadeh et al., 1992a), were purified and freed of diseases through tissue culture in 1991–1992 and are under multiplication for 1994 large scale field trials (Table 2).

In addition, one selection from 1985 and four from 1989 were selected for further evaluation in 1993/1994. These five lines (SJ8518-11, SJ89264-6, SJ89700-1, SJ8976-1, SJ89288-2) have excellent fruit characteristics with good yield and large fruit size and have potential for naming in near future (Table 3). Approximately 1000 plants of each were prepared and planted in large scale for commercial evaluation. Detail characteristic of each lines can be obtained from the author (SK) through Agriculture Canada (St-Jean-sur-Richelieu) or Plant Science Department of McGill University.

Present breeding program (1992–)

At present time the breeding program is supported by McGill University, Agriculture Canada Research Station and MAPAQ (Ministère de l'Agriculture des Pêcheries et de l'Alimentation du Québec). The main objective of the program is to develop hardy, pale skin colour, red-stele and leaf-disease-resistant cultivars suitable for fresh market. Approximately 300 new crosses have been made using parents selected based on their fruit and/or plant characteristics (Table 4). These seedlings will be transplanted to the field for visual evaluation in 1994.

Table 4. Selections and standard cultivars used currently in Quebec breeding program

Genotypes	Desirable plant and/or fruit characteristics
Aberdeen	Red stele resistant
Allstar	Red stele, verticellium wilt and botrytis fruit rot resistant
Annapolis	Red stele resistant
Arking	Red stele, leaf spot, leaf scorch and powdery mildew resistant
Bogota	Late cultivar, with good fruit characteristics
Blackmore	Mite, leaf blight resistant
Bounty	Good fruit flavour and colour
Cambridge Favourite	Red stele, Alternaria leaf disease and verticellium wilt resistant
Chambly	Leaf spot and leaf scorch resistant, very hardy and well adapted in Quebec condition
Delite	Red stele, verticellium, leaf spot, leaf scorch and leaf blight resistant
DelNorte	Red stele resistant
Earliglow	Red stele, verticellium wilt, leaf spot, leaf scorch and botrytis fruit rot resistant
Empire	Leaf spot and leaf scorch resistant
Guardian	Red stele, verticellium wilt and leaf scorch resistant
Honeoye	Very good fruit quality, fruit size and well adapted to northeast
Hood	Resistant to red-stele, verticellium, leaf spot and scorch
Jewel	Excellent fruit size, skin colour and other fruit quality
K75-13	Good fruit quality
Lateglow	Resistant to red stele, leaf spot and leaf scorch
MD683	Red stele resistant
Marshall	Red stele, verticellium wilt, leaf spot, leaf scorch, powdery mildew, alternaria leaf spot and mite resistant
Sparkle	Excellent good fruit flavour, well adapted to northeast
Stelemaster	Red stele and mite resistant
Sunrise	Red stele, leaf blight and powdery mildew resistant
Yaquina A	Red stele resistant

The present breeding objectives will be in effect for next three years, in accord with the priorities published by the MAPAQ.

References

Khanizadeh, S., M. Lareau & D. Buszard, 1992a. Evaluation of advanced strawberry selections in Quebec. Fruit Varieties Journal 46(1): 53–57.

Khanizadeh, S., M. Lareau, D. Buszard & D. Bagnara, 1992b. 'Oka' strawberry. HortScience 27(4): 374–375.

Khanizadeh, S. & M.A. Fanous, 1992. Mathematical indices for comparing small fruit crop genotypes for ripening season, harvest concentration and trait similarity. HortScience 27(4): 346–348.

Khanizadeh, S., M. Lareau & D. Buszard, 1991. Pedigree: a program to trace and draw family tree by microcomputer. HortScience 26(8): 1088.

Khanizadeh, S., D. Buszard, M. Lareau & D. Bagnara, 1990. 'Chambly' strawberry. HortScience 25(8): 984–985.

Statistical Analysis System, 1988. SAS/STAT, SAS/BASIC guide for personal computers. Version 6.04 (ed.). SAS Institute, Inc. Cary, N.C.

H. Schmidt and M. Kellerhals (Eds.), Progress in Temperate Fruit Breeding, 189–191.

Progress in breeding decaploid *fragaria* × *vescana* hybrids

Annelise Bauer
Hechlstraße 11, D-83254 Breitbrunn, Germany

Key words: Fragaria × *vescana*, breeding, interspecific crosses, decaploid, adaptation, varieties

Summary

The decaploid F2 hybrids of *Fragaria* × *vescana* represent interspecific crosses of *F.* × *ananassa* (8×) with *F. vesca* (4×). Breeding work started in 1949 and was improved step by step. Two *F.* × *vescana* cultivars have been released for cultivation, SPADEKA in 1977 and FLORIKA in 1989. Their cultivation is of increasing interest. The advantages they offer, in ecological and economic terms, are based on their hybrid character. Phenotypically they demonstrate heterosis. This is characterized by a strong production of runners during the first year, resulting in a dense stand of runnerplants which can be used as a 'meadow-culture'. Physiologically, they combine the short-day sensitivity of *F.* × *ananassa* with the day-neutral reaction of *F. vesca*. Compared with *F.* × *ananassa*, the vegetative growth of the F. × *vescana* hybrids – expressed in the strong runner production – ceases only with low temperatures. Thus, there is only a very short period for sidecrown formation, compared with *F.* × *ananassa*, and the senescence is strongly retarded. Contrary to the short-living *F.* × *ananassa* cvs the *F.* × *vescana* 'meadow-culture' can be cropped for many years with a minimum of care. Spraying against botrytis is not required. The current breeding programme concentrates on the use of selected 6x-parents which form unreduced gametes at a high frequency. Special aims are adaptation to mechanical harvesting, everbearing, and red stele root rot resistance. Selections under test for commercial use offer improvement in fruit quality by combining the fine flavour of the *F. vesca* parent with the firmness of the *F.* × *ananassa* parent.

Introduction

The genus *Fragaria* includes more than 20 species, with polyploid levels of 2×, 4×, 6×, and 8×. Species of the same polyploid level are interfertile. Thus, *F.* × *ananassa* originated from the spontanous combination of two octoploid wild species – the North American *F. chiloensis* and the South American *F. virginiania* – during the 17th century in France. Both parent species, which come from extremely different climatic regions, have opposite characteristics, providing great variability for breeding and high sensitivity to changing environmental conditions.

The ultimate goal of interspecific hybridization consists in the improvement of the cultivated strawberry either by incorporation of a single gene or by full genome complements. Attempts to incorporate features directly from 2×, 4× or 6× species into 8× varieties have failed. Methods involving manipulation of ploidy led to the development of fertile decaploid *Fragaria* hybrids. Among those who succeeded in breeding fertile 10× hybrids between *F.* × *ananassa* and *F. vesca* were D.H. Scott, USA, and R. Bauer in Germany, Max-Planck-Institut für Züchtungsforschung, (Scott 1951, Bauer 1959, 1960a). Three steps were necessary for the development of fertile decaploid F2 hybrids:

1. Chromosome doubling of 2× *vesca* by colchicine treatments.
2. Crosses of 8× *F.* × *ananassa* cvs with 4× *F. vesca* to obtain 6× seedlings (only partly fertile).
3. Backcrosses of 6× hybrids with *F.* × *ananassa* cvs to obtain fertile decaploids, originating from unreduced gametes of the 6× hybrids with 4 gametes of *F.* × *ananassa*. Bauer named these hybrids, which combine 2 *vesca* gametes with 8 of *F.* × *ananassa*, *F.* × *vescana* (Bauer & Weber 1989).

Chronology of breeding *F.* × *vescana* hybrids at the Max Planck Institute

1949–1956: In the first series of crosses in Voldagsen old *F.* × *ananassa* cvs were used. After 10 years of experiments Bauer (1960) demonstrated that the decaploids showed a similar high range of phenotypic variation as the 8× parents, thus confirming D.H. Scott's (1951) results.

Positive characters were: hybrid vigour, high winter hardiness, resistance to drought, a very high sensitivity to viruses (self elimination – one of Bauers first breeding aims (Bauer 1966)), intensive *F. vesca* flavour and easy capping of the calyx. However, very soft flesh and susceptibility to mildew prevented any commercial use (Bauer 1959, 1960b).

1957–1975: In the second breeding cycle in Köln with high yielding and firm-fleshed advanced *F.* × *ananassa* cvs, selection was concentrated on the following characters: heterosis for high runner formation leading to the '*F.* × *vescana* strawberry meadow', *F. vesca* flavor, high yield, good fruit quality, resistance to botrytis and mildew, and suitability for mechanical harvest. During that period breeding techniques to obtain decaploids were further developed and improved.

As a result of this breeding cycle the first two cvs of *F.* × *vescana* were released, SPADEKA at the Max Planck Institut, Koeln-Vogelsang in Germany and, in cooperation with the Balsgard Fruit Breeding Institute, ANNELIE in Sweden (Bauer & Bauer 1967, 1979).

During this period parallel experiments in greenhouses and under field conditions were conducted in order to determine the physiological adaptability of the *F.* × *vescana* hybrids.

The adaptation of *F.* × *ananassa* is strongly influenced by temperature and daylength. This high sensitivity to climatic factors limits the breeding of widely adaptable strawberry varieties. Adaptation of the decaploid *F.* × *vescanas* is not so strongly limited by climatic conditions. The combination of the short-day sensitivity of *F.* × *ananassa* with the day-neutral reaction of *F. vesca* leads to changes in the growing pattern of the *F.* × *vescana* hybrids compared to their parents. The vegetative growth, expressed as runner formation, ceases only with temperatures below 5 °C in autumn, reducing the time for side crown formation, typically for *F.* × *ananassa* cvs. Consequently, senescence is strongly retarded and a longer cultivation possible. At temperatures of 5 to 10 °C in the greenhouse without additional light in winter *F.* × *vescanas* behaved like the *F. vesca* parent: they had already formed new leaves, flowers and fruits by February, and continued this throughout the following season.

1975–1982: Since 1975 all breeding was done on a private basis in Breitbrunn/Chiemsee and it was necessary to reduce the breeding material. By selection of 6× hybrids which produce unreduced gametes quite readily, the frequency of fertile 10× hybrids could be increased. Progenies of (8× cv SPARKLE × 4× *F. vesca*) and (8× cv VALENTINE × 4× *F. vesca*) proved outstanding in this respect.

1982–1985: A special breeding project with 6× selections of (8× cv SPARKLE × 4× *F. vesca*) and (8× cv VALENTINE × 4× *F. vesca*) crossed with everbearing *F.* × *ananassa* cvs (HUMMI GENTO, KLETTERERDBEERE HUMMI and OSTARA) was carried out at the Department of Horticulture, TU Muenchen-Weihenstephan, by Birgid Schlindwein and Gertrud Spiegler (1983, 1985, 1986).

These investigations gave no clear results with respect to inheritance of everbearing. In all combinations the percentage of single crop types was higher than that of everbearing. No everbearing type for commercial use could be selected. The combination of (8× cv SPARKLE × 4× *F. vesca*) × KLETTERERDBEERE HUMMI, which had most plants with positive characters, resulted in the selection of the single crop variety FLORIKA. With its high yielding capacity, easy picking, excellent flavour and no need for sprayment, FLORIKA is a step forward in the *F.* × *vescana* breeding.

1984–1991: Recent breeding aims in Breitbrunn follow two lines: Mechanical harvest needs a concentrated ripening, hand picking a prolonged ripening period. Besides crossing 6× hybrids with *F.* × *ananassa* cvs, cross combinations were made on the 10x level for combining positive characters. Improved selections are under test at the Spitzl fruit farm, Grabenstaett, and at the Research Station Weinsberg (Spitzl 1986, Buchter-Weisbrodt 1992).

Red stele root rot has become a serious problem in German strawberry cultivation. Breeding for resistance is in progress. Strawberry cvs reported to be as resistant to red stele root rot as DARROW, SCOTT, and EARLIGLOW, have been used for backcrosses with hexaploid (8× cv SPARKLE × 4× *F. vesca*).

References

Bauer, R., 1959. In: Rudorf, W. (Ed) Dreissig Jahre Züchtungsforschung, pp. 210–211. G. Fischer Verlag, Stuttgart.

Bauer, R., 1960a. Aspekte der Polyploidiezüchtung in der Gattung *Fragaria*. Extr. Bull. Inst. Rech. Gembloux, Hors Série II: 994–1006.

Bauer, R., 1960b. Grundlagen und Methoden bei der Züchtung der Gartenerdbeere. Zeitschrift der Pflanzenzüchtung 44: 73–100.

Bauer, R., 1966. Aspekte der Züchtung zur Lösung der Virusprobleme bei Beerenobst. Acta Agric. Scandinavica Suppl. 16: 144–147.

Bauer, R., 1976. New pattern of heteroses in *Fragaria* hybrids. Hungarian Academiy of Sciences EUCARPIA VII: 63–68.

Bauer, R. & A. Bauer, 1967. Neue Wege in der Erdbeerzüchtung. Erwerbsobstbau 5: 83–85.

Bauer, R. & A. Bauer, 1969. Smulgubben – en ny linje inom jordgubbs förädlingen. Bärodlaren 3: 5–7.

Bauer, R. & A. Bauer, 1979. Hybridzüchtung in der Gattung *Fragaria*: 'Spadeka' eine neue Sorte mit dem Aroma der Walderdbeere. Erwerbsobstbau 7: 151–161.

Bauer, A. & H.E. Weber, 1989. *Ribes* × *nidigrolaria* R. & A. Bauer und *Fragaria* × *vescana* R. & A. Bauer – Beschreibung zweier Hybridarten. Osnabrücker Naturwiss. Mitt. 15: 49–58.

Buchter-Weisbrodt, H., 1992. 'Florika': *Vescana*-Erdbeerwiese. Obstbau 6:300–304.

Schlindwein, B., 1983. Untersuchung zur Erzeugung von dekaploiden *Fragaria*-Bastarden. Diplomarbeit, TU Muenchen-Weihenstephan, Institut für Zierpflanzenbau.

Scott, D.H., 1951. Cytological studies on polyploids derived from tetraploid *Fragaria vesca* and cultivated strawberries. Genetics 36: 311–331.

Spiegler, G., 1985. Untersuchungen zur Selektion von dekaploiden *Fragaria*. Diplomarbeit, TU Muenchen-Weihenstephan, Institut für Zierpflanzenbau.

Spiegler G., B. Schlindwein & H. Schimmelpfeng, 1986. Untersuchungen zur Selektion von dekaploiden *Fragaria* × *vescana*. Erwerbsobstbau 28: 220–221.

Spitzl, A., 1986. Erfahrungen mit feldmässigem Anbau von *Vescana* Erdbeeren im süddeutschen Raum. Erwerbsobstbau 8: 222–223.

H. Schmidt and M. Kellerhals (Eds.), Progress in Temperate Fruit Breeding, 193–197.

Selection in seedling populations and clonal progenies of strawberry (*Fragaria ananassa* Duch.)

Viola Hanke
Federal Centre of Breeding Research on Cultivated Plants, Institute for Fruit Breeding Dresden-Pillnitz, Pillnitzer Platz 2, D-01326 Dresden, Germany

Key words: Clone, early selection, *Fragaria* × *ananassa*, phenotypic correlation, seedling, strawberry

Summary

To determine the efficiency of selection in seedling populations and their vegetative progenies the relationship between characters of seedlings, of clones and between seedlings and clones were computed by the use of phenotypic correlations. 17 seedling populations with 1,625 seedlings and 231 clones from 8 parents were chosen for the selection of clones and the evaluation of plant and yield characters. The vigour of the seedling at the beginning of its development was not decisive for its further productivity. The yield of the strawberry seedling could be estimated by the expression of vegetative characters in autumn of the planting year and in spring of the year of harvest. For clonal plants the correlations between yield components of two cropping years were compared. On the base of correlation coefficients large differences between seedlings and clonal plants in their phenotypic characters could be found. Against this background, an efficient selection for yield and yield components has to be carried out in the clonal progeny.

Introduction

The main problem in strawberry breeding is that the number of seedlings and clones which can be evaluated during the growth and cropping period is limited by space and time. The estimation of plant characters on the basis of features assessed at an early stage of plant development may greatly increase the efficiency of the breeding work.

To eliminate plants that are unlikely to achieve a reasonable yield it is possible to use vegetative characters, which were shown to be correlated with yield (Lacey 1973). Several authors were studying phenotypic and genotypic correlations, associated with yield and its components, mainly on clones or selected genotypes. Only Hortynski et al. (1976) investigated the relationships between features of the seedlings and their vegetative progeny.

The experiments reported here were designed to examine a large population of seedlings and clones from a range of combinations typical of a strawberry breeding programme for correlations between several plant characters in seedlings and in clones as well as between seedlings and their vegetative progeny to make the breeding process as effective as possible.

Material and methods

17 seedling progenies with 1,625 seedlings and 231 clones from 8 parents (Korona, Tenira, Sivetta, Comet, Holiday, Francesco, Savio, Sequoia) were chosen for the selection of clones and the evaluation of plant and yield features. The methods of seedling and clonal plant cultivation are described earlier (Hanke 1990). Seeds were sown in February after one year cold storage. Seedlings were cultivated in the glasshouse and planted in the field in late July 83.5 cm × 40 cm apart. Clones were obtained from runners of a single seedling in early July the following year. Seven propagules per seedling were rooted under mist and planted in the field in late August 83.5 cm × 25 cm apart. The evaluation of characters studied was accomplished as follows: Leaf size – by multiplying the lengths of a leaf

Table 1. Correlation coefficients between vegetative characters and yield of seedlings.

Character	Date	Correlation coefficient between the vegetative character and yield	
		1984	1985[1]
Leaf size 1. leaf	outpricking	−0.04	0.08
Plant size	potting	0.04	0.05
Number of leaves	planting	0.16*	0.09
Diameter of crown	planting	0.13*	0.04
Number of leaves	August	0.32*	0.17*
Plant vigour	October	0.41*	0.16*
Number of leaves	November	–	0.38*
Plant vigour	November	–	0.24*
Plant size	April	0.27*	0.24*
Number of leaves	April	–	0.25*
Number of crowns	May	0.34*	0.31*

[1] Years of crossing; * Correlation coefficients are significant at $P < 0.05$
n for seedlings: 1984 = 814; 1985 = 525.

Table 2. Correlation coefficients between different vegetative characters of clones in the first cropping year.

		1	2	3	4	5	6	7	8
Plant size at planting	1								
Diameter of crown at planting	2	0.47*							
Number of leaves at planting	3	0.51*	0.50*						
Number of leaves in November	4	0.46*	0.22	0.25*					
Plant vigour in November	5	0.65*	0.42*	0.37	0.49*				
Plant size in April	6	0.60*	0.19	0.25*	0.59*	0.47*			
Number of leaves in May	7	0.36*	0.14	0.13	0.65*	0.44*	0.51*		
Number of crowns in May	8	0.28*	0.08	0.13	0.52*	0.38*	0.32*	0.76*	
Yield	9	0.30*	0.14	0.00	0.33*	0.32*	0.34*	0.51*	0.41*

* Correlation coefficients are significant at $P < 0.05$; n for clones (7 plants per clone): 1984 = 76.

at right angles; plant size – by multiplying the width of the plant at right angles; number of leaves, number of crowns, diameter of crown, number of fruit trusses by counting or measuring every plant; plant vigour by evaluating in a scale from 1–9 (9 – the most vigorous plant); yield and number of fruits by weighting and counting the total number of fruits from each plant, including those below marketable size and quality; single fruit weight by computing yield and number of fruits. The ripe fruits were picked every three days.

Results

Correlations in seedlings

The germination of the seeds began two to three weeks after sowing. Large differences between seeds of each progeny were expected concerning the percent and the date of germination. The seedlings of the progenies were ranked in groups in the order of the date of their pricking out which was realized at the stage of two opened true leaves. The seedlings of the first pricking date showed the best records for vegetative characters at the beginning of seedling development based on sizes of their first and second true leaves. However,

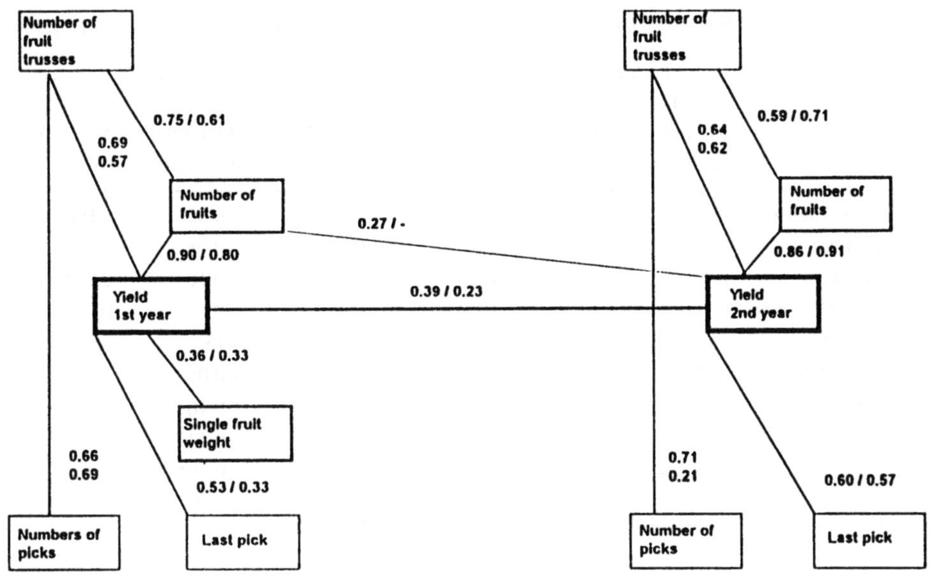

Fig. 1. Significant correlations between yield components of clones in strawberry. Numbers on lines: coefficients of correlations for progenies crossed in 1984 and 1985; number of paired values: 1984 = 76; 1985 = 123.

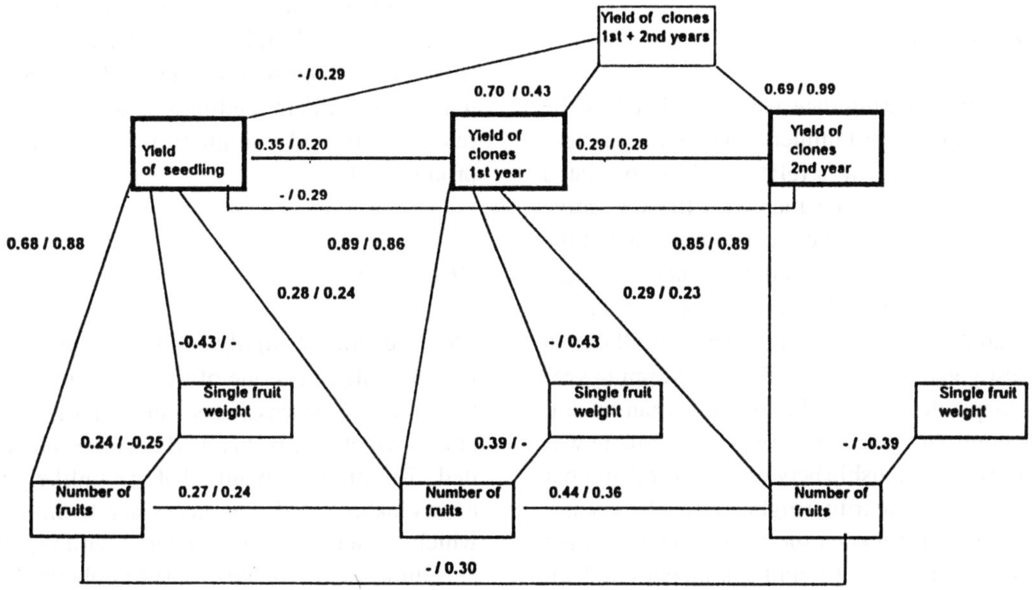

Fig. 2. Significant correlations between yield components of the seedlings and their vegetative progeny. Numbers on lines: coefficients of correlations for progenies crossed in 1984 and 1985; number of paired values: 1984 = 76 (all progenies investigated); 1985 = 123 (the whole progeny Comet × Korona).

seedlings which were pricked earlier and showed a more pronounced vegetative development did not reach a higher yield.

Between vegetative characters recorded from the beginning of the seedling growth up to its planting in the field low correlations were obtained (Table 1). The expected yield of the seedling can be estimated by vegetative characters in autumn of the planting year and in spring of the year of harvest. Within a population of seedlings the number of fruits showed a more pronounced variability compared with single fruit weight. The yield was more dependent on fruit number than on single fruit weight. Normally in seedlings the correla-

Table 3. Correlation coefficients between characters of seedlings and characters of clones in the first cropping year; progeny Comet × Korona.

Seedling plants	Clonal plants							
	No. of leaves at planting	Plant size December	No. of leaves April	No. of crowns May	Beginning of flowering	No. of fruit trusses	No. of fruits	Yield
Number of leaves August	0.21	−0.03	0.03	−0.08	−0.02	−0.12	−0.10	−0.17
Plant size September	0.34*	0.25*	0.06	0.13	−0.08	0.01	0.01	0.02
Plant vigour October	0.31*	0.26*	0.16	0.08	−0.16	0.27*	0.01	0.03
Number of leaves November	0.15	0.05	0.09	0.00	−0.18	0.08	−0.22*	−0.19
Number of leaves April	−0.03	−0.02	0.02	0.11	0.15	−0.09	−0.17	−0.19
Number of crowns May	0.10	0.04	−0.08	0.05	−0.05	−0.11	−0.20	−0.31*
Number of fruit trusses	0.26*	0.11	−0.01	−0.21	−0.16	0.11	0.06	0.03
Number of fruits	0.17	0.04	−0.05	−0.16	−0.23*	0.20	0.19	0.13
Yield	0.18	0.12	−0.04	−0.11	−0.30*	0.22	0.21	0.16

* Correlation coefficients are significant at $P < 0.05$; number of paired values = 79

tion between single fruit weight and fruit number was negative.

Correlations in clones

Plant size at the planting date was considered as the character with the highest influence on the further vegetative productivity but it is relatively low correlated with yield (Table 2). All characters of the vegetative productivity in spring and the yield components of the clonal plants were considerably dependent on the vegetative characters recorded in autumn/winter. Fig. 1 shows the analysis of yield components for all clonal plants investigated. The yield of the first cropping year was more dependent on number of fruits than on single fruit weight. In the second cropping year there was obtained only a relationship between yield and number of fruits not with single fruit weight. A statistically significant correlation between the fruit number and the single fruit weight was not found. An increase in fruit number per plant was not associated with a change of single fruit weight. The correlation between the yields of two cropping periods was not strong.

Correlations between seedling and clone

From our investigations it can be seen that the characters of seedlings were not well correlated with the characters of the clones (Table 3). Some significant correlations were found, but they are not strong enough to be used as criteria for practical selection.

The analysis of yield components between seedlings and propagated clones showed that the yields of the clones in the first and in the second cropping year were not strongly related to the yield of the seedling (Fig. 2). High correlations existed between yield and number of fruits in seedlings as well as in clones. The relationship yield – single fruit weight was not clearly pronounced.

Discussion

The selection of highly productive seedlings from a large population is one of the most difficult phases of the breeding process. The number of seedlings and clones which can be handled during evaluation is limited. To eliminate plants that are unlikely to achieve a reasonable yield it is necessary to find characters, which are shown to be correlated with yield.

Guttridge & Anderson (1981) obtained an important correlation between size of the first and the second true leaves and yield in a selfed population of 'Red Gauntlet'. In our investigations from 16 progenies only 4 progenies showed a significant relationship. In general, the yield of the seedlings was not influenced by the pricking date.

The calculated phenotypic correlations within the seedlings are in agreement with the values obtained by Lacey (1973) and Hortynski et al. (1976). The expression of vegetative growth characters in autumn is of great importance for the potential plant development in spring and for the potential yield. As shown from our results plant vigour and number of leaves in October/November are the main traits which influence the yield of the seedling.

In agreement with Hortynski et al. (1976) the plant vigour in autumn can be used as an early criterion for selection on yield and single fruit weight in clones. To reach a maximum yield capacity the period of flower bud differentiation in autumn, which is affected by the development of vegetative plant characters, is important. For this reason attention should be payed to the cultural conditions in autumn. The investigations emphasize the main importance of the number of fruits as yield component. Similar observations were reported in previous studies (Murawski 1968, Strik & Proctor 1988). In contrast, Hortynski et al. (1976) found in clones a high effect of the single fruit weight on yield. The negetative correlation between fruit number and single fruit weight, found in seedlings, was not confirmed in the clones, as stated by Hortynski et al. (1976). For practical breeding work it should be kept in mind, that an increase of number of fruits and single fruit weight at the same time is always possible until an optimum is reached. Hortynski et al. (1976) stated, that the most vigorous clones in the first year of cropping gave also the highest yield in the next year. In our investigations a low correlation between the yields of both cropping seasons was found, what indicates the two-year-assessment of yield in clones.

The prerequisite for a successful selection in seedling populations is a wide correspondence between the development of characters or productivity of the seedling plant and its vegetative progeny, the clonal plant. According to Bringhurst & Voth (1970) the clonal plant represents the seedling in vegetative and in generative features. In agreement with Hortynski et al. (1976) a low correlation between seedling and clone in yield, number of fruits per plant and single fruit weight was found.

On this basis of all investigated correlation coefficients it cannot be recommended to select for yield and yield components in seedlings.

Acknowledgements

The assistance received from I. Schau and G. Klotzsche is gratefully acknowledged.

References

Bringhurst, R.S. & V. Voth, 1970. Relationship between original seedling performance and subsequent performance of strawberry selection. Hort. Sci. 5: 310.

Guttridge, C.G. & H.M. Anderson, 1981. Assessing fruit yield characteristics and potentials in strawberries. Hort. Res. 21: 83–93.

Hanke, V., 1990. Untersuchungen zur effektiveren Bewertung von Zuchtmaterial in der Erdbeersortenzüchtung. Diss. Institut für Obstforschung Dresden-Pillnitz.

Hortynski, J., I. Flis, T. Hulewicz, 1976. Phenotypic correlations concerning generative and vegetative characters in strawberry (*Fragaria ananassa* Duch.). Z. Pflanzenzüchtung 77: 112–132.

Lacey, C.N.D., 1973. Phenotypic correlations between vegetative characters and yield components in strawberry. Euphytica 22: 546–554.

Murawski, H., 1968. Untersuchungen über den Erbwert von Erdbeersorten. Arch. Gartenbau 16: 291–318, 453–476.

Strik, B.C. & J.T.A. Proctor, 1988. Yield component analysis of strawberry genotypes differing in productivity. J. Amer. Hort. Sci. 113: 124–129.

H. Schmidt and M. Kellerhals (Eds.), Progress in Temperate Fruit Breeding, 199–202.

Breeding black currants, and characteristics of promising hybrids

Miloljub Stanisavljević & Žarko Tešović
Agricultural Research Institute 'Srbija' Fruit and Grape Centre Čačak, Yugoslavia

Key words: Ribes nigrum L., breeding, major biological and economic characteristics of the hybrids

Summary

A programme of breeding new black currant cultivars was started in 1982 at the Fruit and Grape Research Centre at Čačak to promote the culture of this crop. The objective was to develop cultivars with improved hardiness to late spring frosts and disease resistance, higher yields and fruit quality (anthocyanins and ascorbic acid) and long strigs. Special attention has been paid to developing cultivars with erect growth habit and ease of berry detachment from the stalk, suited to mechanized harvesting. The breeding programme involved 17 cultivars and 3 hybrids in 40 crosses. Some 100,000 hybrid seeds were obtained and 15,000 hybrid seedlings planted out. Twenty promising hybrids were singled out, and one of them, I/75, has been registered with the Commission for Cultivar Approval and Release. This hybrid is suited to hand picking because of its long strigs and large berries. The ease of berry separation from the stalk enables mechanized harvesting of simply shaking the berries onto a cover, which largely facilitates harvesting compared to hand picking. Besides hybrid I/75, the hybrids R/I/IX/87 and 4/II/IX/87 have also shown outstanding characteristics so far. They both have long strigs, and hybrid R/I/IX/87 has by 7–10 days later flowering time compared to standard cultivars.

Introduction

Black currant ranks fourth in importance and area among the small fruit species grown in SR Yugoslavia, after raspberry, strawberry and blackberry (Oblak et al. 1984).

Bearing in mind the nutritional value of black currants, favourable agrocological conditions for its successful growing in the hilly-mountainous regions on small holdings and the existence of new cultivars which may serve as donors of desirable characters, a breeding programme was started at the Fruit and Grape Research Centre at Čačak in 1982 aimed at developing new black currant cultivars which are adapted to our conditions and cost less to produce (Stančević et al. 1986). A number of researchers have been engaged in the work on breeding new black currant cultivars at several scientific research institutions worldwide (Pisani & Baldini 1961, Tamas 1961, Mather et al. 1980, Dale 1987, Bauer 1989; Gwozdecki 1989). Several new cultivars developed abroad have been included in our breeding programme depending on their characters.

Material and method

The work on breeding black currants using planned hybridization has been conducted at the Agricultural Research Institute 'Serbia', Fruit and Grape Research Centre at Čačak since 1982. Seventeen cultivars have been used in this work (Amos Black, Baldwin, Ben Lomond, Ben More, Ben Nevis, Black Reward, Cotswold Cross, Daniels' September, Fertodi 1, II/1, Malling Jet, Mendip Cross, Silvergieters, Tenah, Tsema, Wellington XXX, Viola) and three promising hybrids (I/75, R/I/IX/87 and 4/II/85/89) in 40 crosses. The parental cultivars most frequently used were Ben Nevis (47 times), Ben Lomond (39), Malling Jet (32), Tenah (22) and Tsema (22).

Our programme of black currant breeding was focused on reducing production costs through developing new cultivars better adapted to cultural conditions in SR Yugoslavia. The aim was to develop cultivars adapted to particular modes of harvesting, yielding large crops consistently from year to year and having desirable technological properties. Also, they should show resistance to the adverse effects of major biotic and abiotic factors.

Work has been conducted on the development of cultivars suited to hand picking or to mechanical harvesting. The cultivars for hand picking were given priority because of the conditions in which this crop is grown, i.e., mainly on small holdings of 5 to 20 acres, usually on terrain unsuitable for mechanical harvesting. The ease of berry detachment from the stalk is a desirable character so that they can be shaken off by hand or in some other way onto a cover. For this purpose, the cv Malling Jet, which has exceptionally long strigs, was used as a donor of characters desirable for hand picking.

Another breeding objective was to develop cultivars which would yield heavy crops consistently from year to year through a higher cropping potential. This is affected by the number of strigs/bush, number of berries/strig and berry size (Tamas 1961). The number of strigs per bush depends on the internode length, branch length, bush vigour and number of strigs per node. The cultivars used as parents having high cropping potential were Fertodi 1, Tenah, Tsema and Viola. Abundant and regular crops are also affected by hardiness to late spring frosts (Mather et al. 1980, Gwozdecki 1989). This hardiness may be improved by developing cultivars with later flowering time whereby the hazard of frost damage is reduced. The cultivars used as donors of later bud burst and flowering were Ben More, Ben Nevis, Ben Lomond and Amos Black (Dale 1987).

With regard to the resistance of new cultivars to major pests and diseases, special attention has been paid to a higher degree of resistance to powdery mildew (*Sphaerotheca morsuvae*) (Bauer 1989). The cultivars used as donors for this character were Ben Lomond, Ben Nevis and Ben More, which show significant resistance to this disease.

As for the fruit qualities of new black currants, they should be at least equal to, if not superior to the present day cultivars. The juice must have good flavour, an intense, stable colour and a high vitamin C content.

The selected hybrids were tested in comparative trials with three standard cultivars in an experimental planting established in spring 1988, except for the hybrid II/4/85/89 which was planted later, in autumn 1990 at the locality Ljubič belonging to the Fruit and Grape Research Centre, Čačak. The standard growing methods for this fruit species were applied (Lantin 1966, Pisani & Baldini 1961).

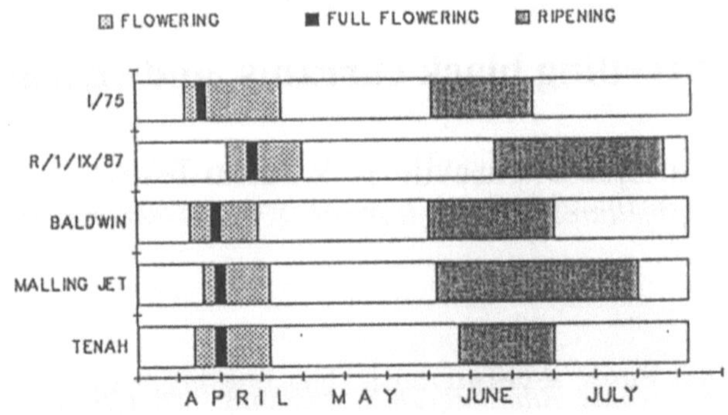

Fig. 1. Phenological data on flowering and ripening of promising hybrids and three standard black currant cultivars (1989–1992).

Results and discussion

The 10-year work on breeding black currants by crossing 17 cultivars and 3 hybrids in 40 different combinations. The total number of parental crosses was greater, since some of them were repeatedly used over several years. Over that period, some 100,000 hybrid seeds were obtained and 15,000 seedlings raised. Twenty promising seedlings were singled out, among which hybrids I/75, R/1/IX/87 and 4/II/85/89 stand out. Hybrid I/75 has been registered with the Commission for Cultivar Approval and Release.

I/75 Open pollinated seedling of Malling Jet obtained in 1980 and selected in 1984. The bush is vigorous, having at first erect growth and then spreading under the weight of the crop. Leaves are large, dark green, with pronounced venation. Buds are large and elongated. It flowers almost simultaneously with Baldwin, but its flowering period is six days longer (Fig. 1). It begins to ripen with Baldwin, but its ripening season is six days shorter. The long central axis of its racemes bears on average 8.7 large flowers with green-yellow petals. There are on average 2.05 strigs and 17.0 flowers per node (Table 1). The internode length averages 1.7 cm. These data indicate a high cropping potential of this hybrid. If the data are compared with those for the standard cultivar Tenah, it can be observed that hybrid I/75 has shorter internodes, a larger number of racemes and thus more flowers per node. Fruit setting through open pollination in hybrid I/75 amounted to 76.6% and the value obtained by self-pollination controlled via isolation was 48.5%, which approximates the value for the standard cultivar Tenah. This shows I/75 to be satisfactory self-fertile. Over the period stud-

Table 1. Cropping potential and yields in some promising hybrids and standard black currant cultivars (1989–92).

Cultivars/ Hybrids	No. Strigs/ node	No. Flowers/ raceme	No. Flowers/ node	Internode length (cm)	% Berries set via o.p.	% Self-fertility	Yield t/ha
I/75	2.0	8.7	17.0	1.7	76.6	48.5	10.5
R/1/IX/87	2.4	10.9	26.7	2.7	–	27.1	9.1
Baldwin	–	–	–	–	67.2	–	3.5
Malling Jet	–	–	–	–	80.2	–	8.3
Tenah	1.2	11.6	14.2	2.5	83.8	49.1	6.4

Table 2. Fruit quality characters of promising hybrids and some standard black currant cultivars (1989–92).

Cultivars/ Hybrids	Strig mass (g)	No. Berries/ strig	Berry mass (g)	No. Seeds/ berry	Soluble solids (%)	Total acids (%)	pH
I/75	8.1	8.5	1.2	42.3	14.6	2.9	2.7
R/1/IX/87	10.0	12.8	1.0	–	13.7	–	–
4/II/85/89	10.0	12.6	0.6	–	15.5	3.2	3.0
Baldwin	3.0	6.0	0.6	36.0	15.5	4.0	3.0
Malling Jet	3.0	8.1	0.6	19.6	13.5	3.2	2.8
Tenah	5.1	7.4	0.8	43.1	14.5	–	–

ied, this hybrid was the highest bearing of all (10, 480 kg/ha), which is by far superior to the standard cultivars Baldwin and Tenah (Table 1).

The strigs are long, with an average mass of 8.1 g and 8.5 berries. The berries are large, fairly uniform (1.2 g), averaging 42.3 seeds (Table 2). The berries easily separate from the stalks when fully ripe, but do not drop to the ground. This character is particularly desirable for harvesting by shaking, either by hand or by machine onto a cover. This greatly facilitates harvest and lowers the costs involved.

As for technological properties, the fruits of hybrid I/75 are superior to the parental cultivar Malling Jet. They contain on average 14.6% soluble solids and 2.88% total acids (Table 2). The fruits are suitable both for processing (especially for juices) and for fresh use.

So far this hybrid has not shown marked susceptibility to major diseases or damage caused by late spring frosts.

R/1/IX/87 Obtained from a cross of unknown parentage (during planting, parentage combination was lost) in 1984 and selected in 1987. The bush is moderately vigorous to vigorous and compact, with strong and stout basal branches. It does not require severe pruning since there is no elongation or bare portions on the branches.

This hybrid is distinguished by late bud burst, flowering and ripening (Fig. 1). It flowers ten days later than the standard cultivar Baldwin, and six days later than Malling Jet. Berry ripening also occurs considerably later compared to the standard cultivars.

Besides later flowering, this hybrid has an outstanding cropping potential. It averages more than 2.5 racemes per node, 10.9 flowers per raceme and 26.7 flowers per node (Table 1). The fruit set from controlled self-fertilization of 27.1% showed it to be self-fertile, but not to a high degree. It was superior in bearing to the standard cultivars.

Hybrid R/1/IX/87 has long strigs, averaging 10 g, with 12.8 berries (Table 2). The berries are large (0.98 g) and contain on average 13.7% soluble solids. They are suitable for processing. This hybrid is also suited to hand picking because of its long strigs.

II/4/85/89 Derived from a cross of Mendip Cross × Malling Jet made in 1985 and selected in 1989. It is distinguished by exceptionally large and long strigs, with an average mass of 10.03 g and with 12.6 berries per strig (Table 2). It contains on average 15.5% soluble solids and 3.2% total acids. The other important characters are under study, but preliminary results indicate that it might be highly suitable for hand picking due to its very long strigs.

References

Bauer, A. 1989. Developments in Ribes breeding with regard to mildew and gall mite resistance. Acta Horticulturae 262: 141–144.

Dale, A., 1987. Some studies in spring frost tolerance in black currants (*Ribes nigrum* L.). Euphytica 36: 775–781.

Gwozdecki, J. 1989. Frost injury assessment in currant collection at Skiernewice. Acta Horticulturae 262: 151–156.

Lantin, B. 1966. Étude de variétés de cassis. Annale de L' Amelioration des plantes, No. 4, Paris.

Mather, P.J.C., I. Modlibowska & E. Keep, 1980. Spring frost resistance in black currant (Ribes nigrum L.) Euphytica 29.

Oblak, M., Z. Tešović, D. Veleva, A. Soškić & M. Petronijević, 1984. New Yugoslav assortment of black and red currants. Jug. voć 18(67–68) 39–50.

Pisani, Z. & M. Baldini, 1961. Contributo allo studio di alcune cultivar di ribes nero. Rivista dell. Ortoflorofrutticultura Italiana, No. 6, Firenze.

Stančević, A., V. Bugarčić, Z. Tešović, M. Stanisavljević & M. Srećković, 1986. Breeding of small fruits and characteristics of cultivars and new selected hybrids. Proceedings of Yugoslav Symposium on Fruit Breeding and Selection, Čačak, pp. 121–131.

Tamas, P., 1961. Yield analysis in black currants. Sver. Pomol. For. Arsskr. 62: 117–126.

H. Schmidt and M. Kellerhals (Eds.), Progress in Temperate Fruit Breeding, 203–206.

Polyploid and mutation breeding in *Cornus mas* and *Ribes*

Štefan Soják, Július Ivanička & Jozef Cvopa
Fruit Research Breeding Institute, Bojnice, Slovakia

Key words: Cornus mas, polyploid breeding, *Ribes* sp., induced mutations

Summary

Colchicine treatment of cornelian cherry seedlings (*Cornus mas*) produced polyploids, mixoploids and diploid mutants which were identified by macroscopic and microscopic evidence. Tetraploids and numerous mixoploids showed very low flower bud intensities and fruit sets in comparison with diploids. Their fruits had no better pomological characteristics than diploid cultivars. However, we have found several diploid mutants which produced clones of different fruit size, colour, shape and ripening time.

The fruits of mutants derived from gama ray irradiated and dimethylsulphonate treated winter scions of red currants (*Ribes rubrum*) and gooseberries (*Grossularia uvacrispa*) showed different stability for individual fruit characters. On the basis of variation coefficients we determined the most stable characters to be the diameter of berries and the cluster length in red currants and the berry width and length in gooseberries. These fruit characters should be the most reliable for the evaluation of fruit changes induced by mutagens. Although no new variety was directly developed we obtained clones with increased genetic variability useful for further breeding work.

Introduction

Induced mutation of vegetatively propagated plants is an important way of increasing genetic variation and creating valuable breeding materials. The genetic and commercial value of induced mutants in fruit crops depends predominantly on compact fruit habit, self-fertility, disease resistance, fruit colours and productivity (Lapins, 1973; Lacey & Campbell, 1982; Paprštein & Blažek, 1985; Theiler-Hedtrich, 1990). In small-fruit species, the two black currant cultivars 'Westra' (Bauer, 1957) and 'Burga' (Lantin, 1979) originate from induced mutations. Colchicine-induced polyploidy has given many valuable tetraploids, aneuploids and diploid mutants (Dermen, 1954; Lapins, 1975; Perry & Lyrene (1984).

Induced mutations have been covered in the breeding programme at the Fruit Research Breeding Institute at Bojnice in the years 1980–1990. The present contribution reviews results in cornelian cherry (*Cornus mas*) using colchicine and in red currant (*Ribes rubrum*) and gooseberry (*R. grossularia*) using irradiation and a chemical mutagen.

Materials and methods

For evaluation of reproductive characteristics of cornelian cherries we used one hundred 8–10 years old colchiploids which had been chosen in V_0–V_4 generations after treatment of shoot tips by the aqueous colchicine solutions at the concentrations of 0.1, 0.25 and 0.5%. As control plants diploid cultivars 'Devin' and 'Titus' were used. Some macroscopic and microscopic assays were described in an earlier paper (Ivanička et al., 1988). In this work we enlarged the microscopic evidence with the estimation of cell nuclei size in the histogenic layers L I, L II and L III. However, we laid the main stress on the evaluation of flower and fruit characteristics (weight, size, shape and colour of fruits). Flower intensity and fruit set we determined using 1–9 and 1–5 point scales, respectively.

Winter cuttings (5–6 buds) of red currants (cv. Detvan) and gooseberries (cvs. Triumph Lowetts & Produkta) had been exposed to acute gamma radiation with a dose 2 kR (33.3 R min^{-1}). Other cuttings had been treated with an aqueous solution of dimethylsulphonate at the concentration of 0.025% for 12 hours. The treat-

Table 1. The reproductive characters of colchiploid and diploid cornelian cherries

Mutants cultivars	Flower set (1–9)	Fruit set (1–5)	Weight of fruit, g x	Weight of stone, g x	Shape of fruit	Colour of fruit
Tetraploids 4 x						
$V_0$8 and its derivates	1	1	1.80	0.40	round-oval	dark red
V_2 42/7	2	2	3.37	0.47	cylindric	bright red
Mixoploids						
V_0 42	5	4	2.85	0.47	oval-round	dark red
V_2 42/2	3	3	2.80	0.39	oval-round	deep dark red
V_4 42/5/3	8	4	3.04	0.29	oval-elongated	dark red
Controls 2n						
'Devin'	9	5	2.20	0.31	oval	bright red
'Titus'	9	5	2.07	0.30	oval-pyriform	dark red

ed materials were grafted into *Ribes aureum* rootstocks and then repropagated into V_1–V_5 generations. In the first V_0–V_4 generations growth changes were evaluated. On 353 selections of V_4 and V_5 we evaluated the following quantitative fruit characteristics: the total yield per bush, the weight of 10 cluster, the width and length of gooseberry fruits, the diameter of red currant berries, the length of cluster and the number of berries in one cluster. For comparison of values of analyzed quantitative characteristics we calculated their % coefficients of variation.

Results and discussion

Cornelian cherry

The observations of the histogenic layers confirmed our previous cytological analyses of stomata, chromosomes and pollen that clones derived from the colchiploid plant, $V_0$8, were wholly tetraploid of the cytological type 4-4-4. The size of their nuclei was 11.5 ± 0.38 μm in comparison with 7.0 ± 0.15 μm for the diploid type. Most colchiploid clones were determined as mixoploids (chimaeras). For example, the type 4-2-2 was identified representing epidermal polyploidy. Another selection in V_0 generation identified as mixoploid after repropagation into V_1–V_4 clones produced also tetraploids and diploid mutants. The differences between diploids, aneuploids and polyploids were also proved from pollen analyses (Lapins, 1975; Dudukal, 1984; Ivanička et al., 1988).

The cornelian cherry tetraploids and numerous mixoploids had delayed flowering and very low flower intensity. While the diploids formed their first flower buds on the 2nd year after budding, polyploids did not do so till the 5th year. The flower bud intensity was only 0.8–10.0% of that in the diploid controls with 100% of set (Table 1). Tetraploids formed bigger flowers but uneven in morphology because some anthers were transformed into petals. Thus their flowers produced much less pollen than diploids.

Tetraploids and aneuploids had extremely low fruit set. The fruits were small, round-oval in shape having the mean fruit shape index 1.05 in comparison with diploids with oval-cylindric shaped fruits which were 1.5. Thus tetraploids did not have neither better cropping nor more attractive fruits. On the other side, we have found some interesting colchiploids which after repropagation produced plants resembling diploids but earlier ripening or with bigger fruits but low yield or with good crops of oval to pyriform fruits with smaller stones in comparison to untreated diploid controls (Table 1).

These results showed that colchicine induced some interesting cytological and pomological mutants useful for breeding work at the polyploid level and for growing as well.

Red currant and gooseberry

The growth changes of the irradiated and DMS-treated red currants and gooseberries have been already described. In general, in relation to the experimen-

Table 2. The quantitative fruit characteristics of red currants after mutagen treatment expressed in % of variance

Variant	Yield of fruits per plant (g)		Weight of 10 clusters (g)		Length of cluster (cm)		Number of berries (in cluster)		Diameter of berry (mm)	
	x	v%	x	v%	x	v%	x	v%	x	v%
Control n = 10	842	36.7	122	20.4	9.8	10.7	18.2	6.6	10.5	4.1
250 ppm DMS n = 18	1205	32.6	125	20.3	10.7	10.2	17.7	14.6	10.5	5.8
250 ppm DMS n = 42	1086	49.7	118	24.8	10.5	9.4	16.3	17.1	10.6	6.2
250 ppm DMS n = 61	1286	29.4	125	19.4	10.8	10.0	19.4	10.7	9.9	6.7
2 kR n = 47	1309	38.1	135	17.8	10.7	8.4	19.2	12.1	10.8	5.2
2 kR n = 48	1817	32.4	132	19.2	11.1	9.8	18.2	11.8	10.4	6.3
2 kR n = 127	657	46.8	144	22.3	10.6	9.9	18.4	16.8	10.3	4.7
Total n = 353	1172	31.8	128	6.8	10.6	3.8	18.2	5.7	10.4	2.7

tal variants we have obtained mutants or selections of about 30% weaker growth (shortened shoots and internodes) in frequencies from 0.0 to 2.4% (Ivanička et al., 1988).

In V_4 and V_5 generations we chose several hundred selections to investigate their quantitative fruit characters. In the evaluated set of currants 45 selections had better and 46 worse values of the fruit features at the significance level P = 0.5. At the high significance P = 0.1 only two selections were better and nine worse. Within the whole evaluated set the total frequency of these characters represented 29.73%. Our results confirmed findings of other researchers that mutagens induce mainly negative changes of quantitative characters.

For investigation of variability of individual fruit characters we calculated the coefficients of variation (Table 2). These values for the total yield per plant was in the range from 29.4 to 49.7 in currants. Although the coefficients for gooseberries are not here tabulated they had the values from 26.1 to 50.5%. This character was the most variable. The width of gooseberry fruits and the diameter of red currant berries were the most stable characters with values from 5.1 to 9.3% and 2.7 to 6.7, respectively.

On the basis of calculated coefficients of variation we determined the following order of increasing stability of the fruit characters of red currants: the yield per plant, the weight of 10 clusters, the number of berries in a cluster, the length of cluster and the diameter of berries. The order of increasing stability of gooseberry fruit characters was as follows: the yield per plant, the weight of 10 berries, the berry length and the berry width.

According to the results we state that in the investigation of possible mutagenic effects in red currants it will be necessary to analyze mainly the diameter of berries and the length of cluster because of their values of coefficients of variation. In gooseberries we should take into account the width and length of berries as the most stable characteristics. The calculated data can be useful for determining of fruit character stability with similar species after an experimental mutagenesis.

Acknowledgement

We are grateful to Mrs. Bernadeta Ivaniçková for her valuable technical assistance.

References

Bauer, R., 1957. Induction of vegetative mutation in *Ribes nigrum*. Hereditas 43: 323–337.

Dermen, H., 1954. Colchiploidy in grapes. J. Hered. 45: 150–172.

Dudukal, G.D., 1984. Chromosome morphology of *Cornus mas*. In: A.R. Žučenko (Ed.) Genetical basis of breeding of agricultural plants and animals, p. 190.

Ivanička, J., I. Cagaňová & S. Soják, 1988. Inconventional breeding techniques for fruit crop improvement. Acta Horticulturae 224: 429–435.

Lacey, C.N.D. & A.I. Campbell, 1982. Progress in mutation breeding in apples (*Malus pumila* Mill.) at Long Ashton Research Station, Bristol, UK. In: Induced mutations in vegetatively propagated plants II. IAEA, Vienna, pp. 11–28.

Lantin, B., 1979. Deux obtentions nouvelles de INRA pour une production précossé de qualité. Bull. Pet. Fruits 15: 10.

Lapins, K.O., 1973. Induced mutations in fruit trees. In: Induced mutations in vegetatively propagated plants. IAEA, Vienna, pp. 1–19.

Lapins, K.O., 1975. Polyploidy and mutations induced in apricot by colchicine treatment. Can. J. Genet. Cytol. 17: 591–599.

Paprštein, F. & J. Blažek, 1985. The induction of mutations in apples. Mutation Breeding Newsletter 26: 11.

Perry, J.L. & P.M. Lyrene, 1984. *In vitro* induction of tetraploidy in *Vaccinium darrowi, V. elliottii,* and *V. darrowi* × *V. elliottii* with colchicine treatment. J. Amer. Soc. Hort. Sci. 109: 4–6.

Theiler-Hedtrich, R., 1990. Induction of dwarf F-12/1 cherry rootstocks by *in vitro* mutagenesis. Acta Hort. 280: 367–374.

H. Schmidt and M. Kellerhals (Eds.), Progress in Temperate Fruit Breeding, 207–210.

A research on the selection of cornel

Atilla Eriş, Arif Soylu, Erdoğan Barut & Zeynel Dalkiliç
Uludağ Universitiy, Faculty of Agriculture, Department of Horticulture, P.O. Box 44, 16036 Bursa, Turkey

Key words: Cornel, selection

Summary

Fruit samples were taken from 20 cornel seedlings and cultivars which were considered as having high economical value depending on the inquiries and observations conducted in Bursa vicinity where the cornel has been grown intensively. Harvest in the vicinity began from August onwards and continued till the end of September depending on the cultivar and environment. Harvest time, fruitfulness and fruit characteristics such as fruit weight, fruit form, flesh/seed ratio, seed form and size, flesh-seed clinging, the juice colour, pH, acidity degree, the level of water soluble solids and the sugar content exhibited significant differences among the seedlings and cultivars examined. Almost all of the sugar in the fruit juice consisted of invert sugar, sucrose was present in minute quantities. Considering the results of the selection study 'Değirmendere-I', 'Memeli-I', 'Yuvarlak Bardak', 'Buğur', 'Kantartopu', 'Aşi', 'Malatya' and the seedlings number 008, 015 and 016 were selected as superior with respect to the characteristics considered.

Introduction

Anatolia is an important centre of origin of cornel. Therefore, it has rich cornel populations (Ülkümen 1973). There were a total of 1,634,000 cornel trees in Turkey according to 1988 statistics and production of 18,000 tons was reported (Anonymous 1990a). The greatest production (10,764 tons) is obtained from Black Sea Region and this is followed by Middle North (2,639 tons) and Aegean (1,407 tons) Regions. Marmara Region rankes 7th with a tree number of 76,980 and a production of 633 tons (Anonymous 1990b). Production in Bursa vicinity was 192 tons obtained from 20,600 trees (Anonymous 1991).

Cornel forms a small or medium-sized tree. The fruit is a stone fruit with one seed. Fruits are similar to sour cherry except fruit shape, which is rather eliptical. Skin and juice colour are various shades of red. Various characteristics of fruit (size, shape, taste, etc.) and fruitfulness of the trees show differences among the seedlings and the cultivars. The fruit is processed to juice, jam or marmalade, rather than fresh consumption. Since the juice has a pleasant taste, this will be a new and promising species in the future for the juice industry.

Most of the cornel trees in Bursa consist of non-grafted seedling types. However, in some areas the growers have grafted cornel types with high yield and quality; thus, the local cultivars have been improved by vegetative propagation. Nevertheless, it is likely that types which are superior to these cultivars are available among the existing populations. The aim of this research is to select the cornel seedlings and cultivars with superior yield and quality among those which are grown in Bursa vicinity.

Material and methods

The material of the study consists of cornel seedlings and local cultivars (cvs. Değirmendere-I, Memeli-I, Yuvarlak Bardak, Buğur, Değirmendere-II, Kantartopu, Aşi, Malatya, Memeli-II) which are grown in various places of Bursa.

Cornel growing areas were determined at the beginning of 1991 and local observations and collections were made in the central villages of Bursa and some

Table 1. Some fruit characteristics of cornel types and cultivars studied.

seedling	local name	fruit length/ width ratio	fruit weight (g)	flesh/ seed ratio	flesh-seed clinging
001	Değirmendere-I	1.5 ef	6.5 a	6.0 bc	clinging
002	Memeli-I (U.-Bar)	1.7 b	4.2 e	6.2 bc	clinging
003	Yuvarlak Bardak	1.3 hi	4.3 d	5.7 cd	clinging
004	Buğur	1.6 bc	4.7 c	6.3 bc	clinging
005	–	1.7 b	1.2 m	5.5 cd	non-clinging
006	Değirmendere-II	1.4 gh	6.0 b	6.4 b	clinging
007	Kantartopu	1.4 gh	4.3 d	5.5 cd	non-clinging
008	–	1.4 gh	6.1 b	7.4 a	non-clinging
009	Aşi	1.5 ef	4.0 f	5.1 de	non-clinging
010	Malatya*	1.8 a	–	–	non-clinging
011	–	1.5 de	1.9 k	2.9 fg	non-clinging
012	–	1.3 i	2.9 i	3.1 f	clinging
013	–	1.5 ef	1.7 l	2.7 fg	clinging
014	–	1.5 ef	1.9 k	2.5 fg	clinging
015	–	1.5 ef	3.9 g	5.6 cd	non-clinging
016	–	1.4 gh	4.0 f	4.1 e	clinging
017	Memeli-II	1.6 bc	3.2 h	–	clinging
018	–	1.5 ef	2.0 k	2.1 g	clinging
019	–	1.3 hi	3.0 i	2.6 fg	non-clinging
020	–	1.6 bc	2.3 j	2.2 f	clinging

* Some characteristics are not given, due to insufficient samples.

villages of Keles, Kestel and Orhaneli counties. Moreover some fruit samples brought to the local markets were also evaluated. In this way, 20 seedlings and cultivars which have high economical value were identified. Observations and analyses were made in 1991.

Harvest time, to be grafted or non-grafted, local name and fruitfulness of the types were noted in the orchard.

Fruitfulness

The trees were classified as low, middle, high and very fruitful according to the crop load on the tree.

The following physical and chemical analyses were made on about 0.5 kg of fruit taken randomly from each tree.

Physical analyses

The length and width of fruits were measured on randomly selected 30 fruits (10 × 3) and the ratio of length/width was calculated for each sample. Fruit weight was calculated as the average of 50 fruits. 100 g of fruit was weighed, their seeds separated, and flesh/seed ratios were calculated. Juice colour was determined according to Mearz & Paul (1950) and classified as light red and red. Flesh-seed clinging (considered an important character for processing to jam and marmalade) and seed size were also determined.

Chemical analyses

Water soluble solids was read on refractometer. Acidity was found by titration and expressed as citric acid, and pH was measured by digital pH-meter. The dinitrophenol method was used for the analyses of invert sugar and sucrose (Ross 1959).

The results of physical and chemical analyses were analized according to a randomised block design with three replicates, and the mean values were compared by Duncan's Test at $P = 0.01$ level.

Table 2. Chemical composition of fruits of cornel seedlings and cultivars studied.

seedling	local name	water soluble solids (%)	acidity (%)	pH	invert sugar (g/100 ml)
001	Değirmendere-I	11.03 cd	4.30 bcde	2.85 g	7.08 def
002	Memeli-I (U. Bar)	12.27 abc	4.48 bcd	3.00 cd	10.33 ab
003	Yuvarlak Bardak	12.50 abc	4.99 abc	3.05 abc	9.96 ab
004	Buğur	12.33 abc	4.95 abc	2.98 cde	10.00 ab
005	–	13.13 abc	6.82 a	2.91 efg	9.29 abc
006	Değirmendere-II	13.93 abc	5.39 ab	2.96 def	8.85 abcd
007	Kantartopu	14.23 ab	5.73 ab	3.06 abc	10.72 a
008	–	13.87 abc	6.61 a	3.02 bcd	9.84 ab
009	Aşi	14.97 a	6.03 ab	3.12 a	9.88 ab
011	–	13.07 abc	2.30 ef	2.98 cde	8.47 bcdef
012	–	12.47 abc	3.15 cdef	3.00 cd	7.21 cded
013	–	14.17 ab	4.33 bcde	3.10 ab	10.57 ab
014	–	9.23 d	2.17 f	2.98 cde	6.25 f
015	–	12.00 bcd	2.67	2.89 fg	5.35 g
016	–	12.83 abc	2.75 def	2.84 g	7.28 cdef
017	Memeli-II	11.40 bcd	–	3.00 cd	8.49 bcde
018	–	11.53 bcd	2.60 def	3.01 cd	7.35 cdef
019	–	11.37 bcd	2.69 def	3.00 cd	7.56 cdef
020	–	9.17 d	2.24 ef	3.00 cd	6.47 ef

Results

The harvest began on 20 August 1991 and continued until 14 September 1991. The maturation periods of each seedling and cultivar were compared. Among the seedlings and cultivars 'Değirmendere-I', 'Memeli-I', 'Yuvarlak Bardak', 'Buğur' matured earlier. Harvest began late in localities at higher altitudes such as Keles and showed differences among the seedlings. 'Memeli-I' and 'Yuvarlak Bardak' were very highly fruitful, whereas the others were high or middle fruitful cultivars. Some fruit characteristics of the seedlings and cultivars are given in Table 1. Fruit weights ranged between 1.2 and 6.5 g and the differences were statistically significant. The fruits of grafted cultivars such as 'Değirmendere-I', 'Memeli-I', 'Yuvarlak Bardak', 'Buğur', 'Kantartopu' and 'Malatya' were large. Fruit size of the non grafted type No. 008 ranked second after cv. 'Değirmendere-I'. Fruit length varied from 1.6 to 2.8 cm and width from 1.1 to 1.9 cm. Fruit length/width ratio varied between 1.3 and 1.8. This ratio can be used as the determinant of the fruit shape.

Seed length varied from 1.3–1.8 cm; and width varied from 0.5–0.9 cm. Length/width ratios varied from 1.9 to 2.6 among the selections. The seeds of big fruited cultivars were generally large and the correlation coefficients between fruit length and seed length, and fruit width and seed width were: $r_1 = 0.918$ and $r_2 = 0.700$, respectively. Flesh/seed ratios showed more differences among the seedlings than the cultivars and differed from 2.1 to 7.4. Generally the cultivars had high value in this respect except seedling No. 008 (Table 1).

Flesh-seed clinging differed among the cultivars and the seedlings. Fruit flesh of non-grafted seedlings 008 and 015 and grafted cultivars such as 'Kantartopu', 'Aşi' and 'Malatya' could easily be separated from the seed (Table 1).

There were significant differences in the chemical composition of the seedlings and cultivars. Water soluble solids varied from 9.17 to 14.97%, acidity from 2.17–6.82%, pH from 2.84–3.12 and invert sugar from 5.35–10.72%. Nearly all of the sugars consisted of invert sugar, with no or only low levels of ($\leq$ 1%) sucrose. For this reason, only invert sugar contents are given in Table 2.

Chemical composition of the fruits are likely to be affected by environment. For example, water soluble solids and invert sugar content of the fruits cv. 'Değirmendere' taken from the 1st locality were 11.03% and 7.08%, respectively whereas they were

13.93% and 8.85% for the fruits taken from the 2nd locality. An increase in sugar content was parallel to the increase in water soluble solids. A similar situation was observed with cv. 'Memeli' (Table 2). The fruit and juice colours of the cultivars and seedlings showed various shades of red. But they were classified in to two main groups as light red (Değirmendere-I, Memeli-I, Yuvarlak Bardak, Buğur, 005, Değirmendere-II, Kantartopu, 008, Aşi, 015, 020) and red (011, 012, 013, 014, 016, Memeli-II, 018, 019). This character is important for the juice industry.

Discussion

Significant differences were determined with respect to fruitfulness, fruit size and other quality features among the cornel types and cultivars investigated. The growers have extended their range of cultivars by the selection of seedlings in their environments and the efforts led to the development of new local cultivars. There is always the possibility of finding high quality types among seedling populations. Thus type No. 008 was similar to the grafted cultivars in respect to size, flesh-seed clinging and other fruit charcateristics; it is even superior with respect to the flesh/seed ratio. Again, type no 015 and 016 approach the grafted cultivars in either fruit size or other quality features.

The results of this study show that cultivars such as 'Değirmendere-I', 'Memeli-I', 'Yuvarlak Bardak', 'Kantartopu', 'Aşi', 'Malatya', 'Buğur' and type No. 008, 015 and 016 were the best cultivars and types as regards their fruitfulness, size of the fruit, flesh-seed clinging, flesh/seed ratio and juice colour. Extension of the selection studies to other localities will raise the possibility of finding further high quality cornel types.

References

Anonymous, 1990a. The Summary of Agricultural Statistics. State Institute of Statistics, Prime Ministry Republic of Turkey. No. 1406. Ankara, Turkey.

Anonymous, 1990b. Agricultural Structure and Production. State Institute of Statistics, Prime Ministry Republic of Turkey. No. 1450. Ankara, Turkey.

Anonymous, 1991. Statistical Records. Agricultural Administration of Bursa Province, Bursa, Turkey.

Mearz, A. & M.R. Paul, 1950. A Dictionary of Color. McGraw Hill Book Company, Inc., New York.

Ross, F.A., 1959. Dinitrophenol method for reducing sugars. In: W.F. Talbart & O. Smith (Eds) Potato Processing, 469–470. AVI Publishing Co., Connecticut.

Ülkümen, L., 1973. Viticulture and Horticulture. Atatürk University, Faculty of Agriculture, Publ. No. 67. Erzurum, Turkey, p. 415.

H. Schmidt and M. Kellerhals (Eds.), Progress in Temperate Fruit Breeding, 211–213.

Heritability estimates in raspberry breeding

Rolf Nestby
Kvithamar Research Station, N-7500 Stjørdal, Norway

Key words: Cane traits, fruit traits, heritability, raspberry

Summary

From data collected in two three year old red raspberry (*Rubus idaeus* L.) populations, located at Kvithamar Research Station at Stjørdal in Norway, heritability estimates were calculated based on observations in one year for each population. Broad-sense heritability (H^2) was estimated for fruit firmness, freeze tolerance, dry matter of fruit juice (Brix), fruit colour and firmness. H^2 was low for freeze tolerance, at a medium level for Brix-level and fruit size and high for fruit colour and firmness. Some families had a higher average for one important trait than the cultivar 'Veten', and a few had a higher average than 'Veten' in several traits.

Introduction

A main object in the Norwegian raspberry breeding programme is to breed for freeze tolerance both for fresh and processing market cultivars. Also priority is given to easy loosening from the receptacle and fruit firmness.

In spite of the narrow genetic base of present day cultivars (Øydvin 1970, Dale et al. 1989), it was, since the Norwegian breeding programme is new, considered as the best strategy for germplasm enhancement to introduce cultivars and advanced selections from foreign breeding programmes. The first crosses were made in 1985, but the basic crosses were made in 1986, when freeze tolerant Norwegian cultivars were crossed with advanced selections from the programmes at Research Station Vancouver, BC, Canada, and at Puyallup Research Station, Washington, USA.

However, in the literature there is little information on heritability estimates of important characters in raspberry. To get a systematic improvement of the breeding stock a strategy to study the inheritance of important characters was initiated in 1988 in the Norwegian programme.

This report summarizes the information on heritability estimates yielded by the programme in the years 1986 to 1988.

Material and methods

Information of breeding values are based on registrations in progeny fields in 1989 and 1991 after crosses in 1986 and 1988. This fields will later be refered to as F-86 and F-88. The F-86 field was not replicated, while the F-88 field was replicated twice with the families randomly planted within each replication, All crosses were made on the basis of mass selection. In the F-86 field the data were sampled from all seedlings of each family, and in the F-88 field from 10 randomly selected plants in each plot.

The broad sense heritability was estimated as

$$H^2 = \sigma_g^2/(\sigma_g^2 + 1/2\sigma_e^2), \quad (1)$$

where $\sigma_g{}^2$ = variance of genotype; $\sigma_e{}^2$ = Environmental variance.

Table 1. Resistance (0–9) to cane botrytis (*Botrytis cinerea*) and cane disease (*Didymella applanata*) and freeze tolerance (0–9) of raspberry families, all registrated in 1989 as averages of N plants. Score 9 is no and 0 is 91–100% injury.

Family	Bot.	Did.	Freeze t.	N
WSU 608 × BC 72-6-104	9	9	8.3 ± 1.2	17
Norna × Nootka	9	8.4 ± 0.9	7.4 ± 1.5	236
Norna × Algonquin	8.6 ± 0.9	8.6 ± 0.6	7.0 ± 1.9	86
Norna × Chilliwack	8.4 ± 1.2	8.3 ± 1.0	6.9 ± 1.8	432
Norna × Comox	8.7 ± 0.9	8.2 ± 1.0	4.8 ± 2.7	88
Norna × BC 82-2-57	8.3 ± 1.3	8.0 ± 1.1	5.7 ± 2.5	135
Norna × BC 82-3-8	8.1 ± 1.5	8.7 ± 0.6	5.6 ± 2.3	59
Norna × WSU 608	8.6 ± 1.0	8.8 ± 0.5	7.8 ± 1.5	46
Veten × Skeena	8.8 ± 0.7	8.8 ± 0.5	7.2 ± 2.0	122
Veten × Chilliwack	8.4 ± 1.5	8.7 ± 0.6	7.8 ± 1.5	19
WSU 608 × Chilliwack	8.9 ± 0.4	8.8 ± 0.5	8.1 ± 1.1	207
Chief × Chilliwack	8.7 ± 0.8	9	8.4 ± 0.9	12
Chief × Comox	9	8.8 ± 0.4	8.3 ± 0.8	6
Veten	–	–	8.4	96
Mean of families	8.7	8.6	7.2	113

Statistical evaluation

The data were analysed with the SAS procedure GLM, and standard deviations were produced with the SAS procedure TABULATE (SAS Inst. Inc. 1987).

Results and discussion

In the F-86 field (Table 1) resistance to cane botrytis, cane disease and freeze tolerance was registered. The freeze injury of the control cultivar 'Veten' was low and several of the families produced averages of freeze tolerance equal to 'Veten'. Offspring of 'Norna' crossed either with 'Comox', BC 82–3–8 or BC 82–2–57 expressed the lowest freeze tolerances. However the standard deviations show that even the family with the lowest average freeze tolerance (Norna × Comox) had individuals with satisfactory freeze tolerance. The registrations of freeze tolerance indicate that WSU 608 and 'Chief' will be contributing parents to a high level of freeze tolerance, at a relatively low freeze stress in late winter and early spring. It is surprising that 'Norna', which is a relatively freeze tolerant cultivar (Thorsrud & Hjeltnes 1963), gave relatively freeze susceptible offspring.

The table also shows that WSU 608 and 'Chief' produced offspring almost without cane fungi, but the standard deviations showed that the other families also produced single offspring without cane fungi. This indicates that WSU 608 and 'Chief' could be sources of resistance to cane botrytis and cane disease, as well as for freeze tolerance, but that resistant offspring may be found also in more susceptible families.

The F-88 field was the first in the breeding programme with a design that made it possible to estimate broad sense heritability (H^2) (Table 2). The H^2 values for the fruit characters size and dry matter were medium high, while firmness and colour expressed high values. That indicates clonal selection for these characters to be successful in developing clonal cultivars, and that the progress in population improvement would be fast. However, the data are unbalanced and dominated by registrations on offspring from N71–2709 and 'Asker'. That makes it impossible to indicate to what degree the different parent clones have influenced the characters registered, for instance by calulation of general combining ability (g.c.a.). The most reliable would be to estimate specific combining ability (s.c.a.). The two best combinations for dry matter were N71–2709 crossed with N84–2–26 or 'Asker'. For berry size the cross N80–35 with 'Asker' was outstanding, in spite of

Table 2. Fruitcolour (0–9), fruit dry matter (Brix), fruitsize in g/fruit, fruitfirmness measured as resistance towards pressure in g/cm^2 and freeze tolerance (0–9) in offspring and clones, and estimates of broad sense heritability (H^2) registrated in 1991 for offspring of red raspberry. High score for colour indicates more anthocyanin. Score 9 for freeze tolerance is no and 0 is 91–100% injury.

Family	F. colour	Brix	F. size	F. firm	Freeze t.
N71–2709 × Vene	5.9	10.7	3.5	116	5.9
N71–2709 × Asker	5.4	10.9	3.5	93	4.6
N71–2709 × N84–2–26	5.4	11.3	3.5	143	6.1
N71–2709 × N84–2–70	5.9	10.2	3.5	123	5.0
N71–2709 × N84–2–14	6.3	9.3	2.7	81	5.4
N71–2709 × N84–1–45	5.3	10.6	2.9	131	5.8
N84–2–70 × Asker	4.7	9.9	3.2	113	5.4
N84–2–26 × Asker	6.5	10.4	3.4	89	5.4
N84–1–45 × Vene	5.1	10.4	3.3	99	7.3
N84–2–70 × N71-B3	6.3	9.6	3.4	102	4.0
N80–35 × Vene	4.3	10.7	2.6	86	7.1
N80–35 × N84–2–70	6.7	10.0	3.0	83	7.8
N80–35 × Asker	6.5	9.7	4.5	155	6.1
Glen Prosen × Asker	5.6	9.5	3.4	135	3.9
Glen Isla × Asker	5.4	9.3	3.8	123	5.4
Skeena × Asker	4.8	9.1	3.7	110	5.0
Veten	7.0	9.9	3.7	91	6.6
N71–2709	5.1	9.2	3.5	82	7.7
N71-B3	4.8	7.7	3.5	82	5.8
H^2	0.59	0.39	0.42	0.78	0.27
Mean	5.6	10.0	3.4	113	5.6
LSD1%	0.7	0.8	0.6	30	1.4

the small berries of 'Asker'. For firmness the crosses of N71–2709 with N84–2–26, N84–2–70 or N84–1–45 gave offspring with significant improvements over 'Veten', as did N80–35, 'Glen Prosen' and 'Glen Isla' crossed with 'Asker'. The H^2 for freeze tolerance was low, but high enough to expect a significant effect of clonal selection. The freeze tolerance were especially high in offspring from the cross between N80–35 and N84–2–70; however, the fruits were relatively soft and would not be suited for the fresh market. The cross between N80–35 and 'Asker' is interesting since it produced offspring with large, firm and relatively freeze tolerant offspring.

The data from these fields have given some information of heritability estimates, and the F-88 design was the most useful since it made it possible to calculate broad sense heritability. However, the design was too unbalanced to give a reliable indication of g.c.a. To improve the prediction of the value of phenotypes for breeding, the program will still use mass selection as a basis, but it will be combined with balanced family tests. This concept was used for the crosses in 1991 and 1992, where especially resistance and fruit quality characters will be studied.

References

Dale, A., R. McNicol, P. Moore & T.S. Sjulin, 1989. Pedigree analysis of red raspberry. Acta Horticulturae 262: 35–39.

SAS Institute Inc., 1987. SAS User's Guide: Statistics, 1987 Edn., SAS Inc., Cary, NC.

Thorsrud, J. & A. Hjeltnes, 1963. Undersøkelser over frostherdighet hos bringebær. Forskning og forsøk i landbruket 14: 99–117 (English summary).

Øydvin, J., 1970. Viktige foreldrelinjer og cultivarer i foredling av bringebær. Statens forsøksgard Njøs: 42 pp. (English summary).

H. Schmidt and M. Kellerhals (Eds.), Progress in Temperate Fruit Breeding, 215–217.

Sharp shifts exist in phase change of woody plants

Zai-long Li
Department of Horticulture, Zhejiang Agricultural University, Hangzhou, Zhejiang, 310029, The People's Republic of China

Key words: Phase change, transition point, woody seedlings

Summary

Woody seedling trees pass through several stages from seed germination up to the appearence of first flowering and change in morphological characters and growth habits. However, evidence found from pear seedlings has revealed a sharp shift of some internal physiological conditions determining phase changes. This is a brief report of how two transition points were found, by which three development zones were distinguished. This finding illustrates a spatial pattern of seedling development.

Two transition points important in phase change of pear seedlings (*Pyrus spp.*) seem to exist widely in woody plants. They have recently been named the M and T° points, respectively.

The M point in the lower part of the seedling marks the physiological maturation of apical meristems by which the apical meristem gains its potential to switch from the juvenile to the transition state. However, the M point on the seedling cannot be located until some years after first flowering when the downward movement of the T point, the site of the lowest flower in the canopy of seedlings, has ceased.

The T° point in the upper part of the seedling is the intrinsic site of first balance of endogenous hormones. In rare cases a few flowers are formed on the terminal of the central leader at the time of first flowering of seedlings.

Before understanding the significance of M and T° points, some other phenomena need to be presented.

The T point is the position at which a shift occurs from the vegetative to the reproductive state. In monitoring T points on pear seedlings over 13 years, it was found that they moved downward as the trees grew; the speed of movement was rather fast at the beginning, then slowed down; finally, the T points stopped their downward movement somewhere in the lower parts of the seedlings overlapping the sites of the M point.

The T point is considered as an indicator of the site of hormone balance of the seedling. Based on circumstantial evidence obtained from various investigations (Jaffe et al. 1986, Lang 1987, McDaniel et al. 1992, Vince-Prue 1985, Wareing & Frydman 1976), it is assumed that a decreasing gradient of hormone levels exists from the tree base upward. Hormone(s) produced by juvenile organs, such as juvenile leaves, buds, and/or root system, act as inhibitors to flower initiation. On the other hand, promotor(s) to flower initiation may be released from adult organs, forming a decreasing gradient of concentration from the tree top to the base. The downward movement of the T point would be most likely caused by ratio changes of concentration in two sets of hormones resulting from shoot growth occuring mostly in the upper part of the tree as a result of apical dominance.

Thus, increase in tree size by means of annual growth pushes the T point down. However, the downward movement cannot pass beyond the M point, because apical meristems in the juvenile zone are physiologically immature and have no response to flowering.

As the T point is measured as the position from the top of the plant (H), it was found out that extension of HT area varied greatly. From various comparisons conducted between H and T, it is now understood that the variation of HT areas was mainly due to the different time durations between the two events, occurence of hormone balance and cessation of annual shoot growth. If the two events happen synchronously, a few flowers may appear on the terminal of the central leader only. The later the occurence of hormone balance in annual growth, the shorter the time duration between the two events, and the smaller the extension of HT area. In contrast, the earlier the hormone balance occured in the season, the longer the time duration, and the greater the HT area.

Different time durations between the two events result from the differences in genotype and in growth rate of seedlings. Additional growth above the T° point, the site of original hormone balance, moves the site of hormone balance down. The distance of HT is then greatly extended from both directions. As hormone balance occurs shortly after the beginning of shoot flush in spring time, the distance of HT even in the first season of flowering can be considerably extended beyond the length of annual shoot growth. It is certain that this happens in some seedlings, since extreme distances (HT) as long as 1.5–2.0 m have been observed in pear seedlings.

Various comparisons of positional heights between H and T points confirmed that there exists a latent T° point in each pear seedling, and the T points mostly observed were lower than the T° point; this was also due to from downward movement of hormone balance.

As reported earlier from our pear breeding programme (Li 1990), the positional height of T points i.e. the length of the juvenile span, varied depending on genotype and plant height of the seedlings. Juvenile span is an important biological character of seedlings. It is also highly correlated to juvenile period. Therefore, it is reasonable to suppose that both M and T° points should have specific positional heights for each seedling genotype.

By observation of the T point, it is easy to partition a seedling tree into two parts, the lower juvenile part and the upper adult part. However, with the discovery of downward movement of the T point, a seedling can be practically partitioned into three development zones (Fig. 1), the juvenile zone, the transition zone, and the adult zone. Without this understanding, we were unable to understand such phenomena as flowering accompanied by some juvenile characters, and rejuvenation occuring in apparently adult zones.

No matter how time elapes, flowers cannot occur below the M point, the true juvenile zone. Apical meristems developing from G (seed germination) to the M point progress out of juvenility. Therefore, research

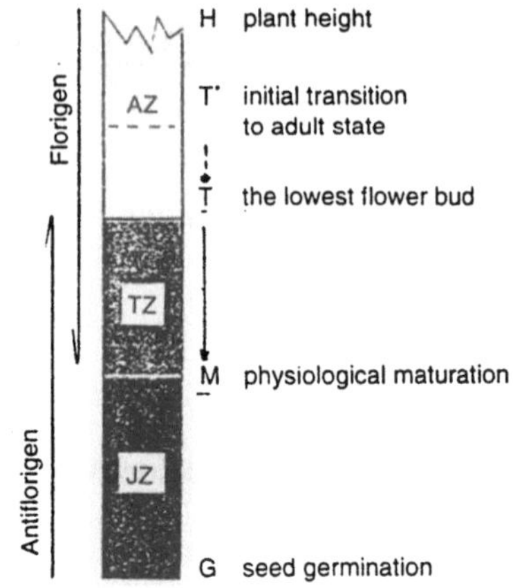

Fig. 1. A spatial pattern of ontogenesis in woody plants.

seeking the mechanism of juvenility must be focused with high priority on the true juvenile zone.

Potential response to flower stimuli seems to commence at the M point, and increases gradually until the full flowering response occurs at the T° point. As the seedling continues to increase in size, the T point moves down, and finally stops at the M point. This phenomenon suggests that the apical meristems in the transition zone are of competence to flowering, but, whether or not this occurs is most likely dependent on the balance of endogenous hormones. Application of various horticultural practices, such as spray of growth regulators, stem girdling, fertilization with too much nitrogen, and heavy pruning, can greatly affect the balance of endogenous hormones and hence, either enhance or delay the occurence of flowering.

Discovery of M and T° points leads to the establishment of a spatial pattern of seedling development, and allows to distinguish the physiological difference between juvenile and transition zones, and improving our understanding of many puzzling phenomena observed on seedlings.

However, many questions remain. For example, by what means does the apical meristem gain its competence to flower? Does a major transition really occur at the M point? If so, what changes happen at the M point in cellular and molecular level? Does hormone balance involve only the known hormones or other unknown substances? At what ratio of concentration levels is the balace reached?

References

Jaffe, M.J., K. Bridle & J. Kopcewicz, 1986. Flowering Newsletter 1: 6.

Lang, A. 1987. Perspectives in flowering research. In: J.L. Key & L. McIntosh (Eds) Plant Gene Systems and their Biology, pp. 3–24. A.R. Liss, New York.

Li, Z.L., Shen, T.X. & S.Q. Zheng, 1981. On the 'Juvenile span', fruiting and inheritance of pear seedlings. Journal of Zhejiang Agricultural University 7(4): 74–83.

Li, Z.L., 1990. Advances of pear breeding program at Zhejiang Agricultural University. Acta Horticulturae 279: 167–184.

McDaniel, C.N., S.R. Singer & S.M.E. Smith, 1992. Developmental state associated with the floral transition. Developmental Biology 153: 59–69.

Vince-Prue, D. 1985. Photoperiod and hormones. In: R.P. Pharis & D.M. Reid (Eds) Encyclopedia of Plant Physiology, New Series, Vol. 11, pp. 308–351. Springer-Verlag, Heidelberg/New York/Tokyo.

Wareing, P.F. & V.M. Frydman, 1976. General aspects of phase change, with special reference to *Hedera hedera helix* L. Acta Hort. 56: 57–69.

CHAPTER 4

Breeding for specific aims

H. Schmidt and M. Kellerhals (Eds.), Progress in Temperate Fruit Breeding, 221–224.

Combining apetalous parthenocarpy with columnar growth habit in apple

K.R. Tobutt
Horticulture Research International, East Malling, West Malling, Kent, ME19 6BJ, U.K.

Key words: Apetaly, apple, genetics, growth habit, *Malus pumila*, parthenocarpy

Summary

'Wijcik', a sport of 'McIntosh' with columnar growth habit, was crossed with 'Wellington Bloomless', which has apetalous flowers and bears parthenocarpic seedless fruit if not pollinated. The seedlings segregated for columnar versus normal habit but all had normal flowers and fruit. Four columnar seedlings were crossed with 'Spencer Seedless', another apetalous cultivar, and the resulting seedlings segregated not only for plant habit but also for apetalous versus normal flowers, approximately 1 : 1. Thus apetaly is controlled by a recessive gene, for which the symbol *ape* is proposed. Apetalous columnar apples may be useful for planting in very high density orchards, cropping without pollination and thus not dependent on bees, pollinator varieties and warm weather at flowering time; moreover, being seedless they may avoid biennial bearing tendencies that are attributable to developing seeds inhibiting fruit bud formation.

Introduction

The cultivars of apple (*Malus pumila*) 'Spencer Seedless' and 'Wellington Bloomless' have peculiar flowers and fruit. The flowers lack petals and stamens, and have two whorls of sepals and up to fifteen styles. The flowers do not attract bees but can set seedless fruit without pollination, or seeded fruit if hand pollinated. The fruits – of poor or mediocre quality – contain two cores, one above the other. Stout (1929) illustrated fruit of these two cultivars, and clear descriptions of more or less similar types were provided by Crandall (1926) and Evreinoff (1935). Chan and Cain (1967) found that seeded fruits in 'Spencer Seedless' following hand pollination greatly reduced flowering the following year, whereas seedless fruits did not inhibit flower bud formation; they concluded that hormonal processes concerned with seed development, rather than nutritional competition by developing fruits, lead to biennial bearing.

Tobutt (e.g. 1985, 1989) has described the breeding of apples inheriting the remarkable columnar growth habit of 'McIntosh Wijcik', a natural sport of 'McIntosh'. The six cultivars released from East Malling so far, 'Maypole', 'Telamon', 'Tuscan' and 'Trajan' (Tobutt, 1989) and 'Charlotte' and 'Obelisk' (formerly SA54–81) (Tobutt, 1991), are intended for amateur gardeners. However, the main aim of the breeding programme is to breed columnar apples suitable for planting in very intensive orchards of maybe 10,000 trees per hectare.

For two reasons it seemed worth trying to combine apetalous parthenocarpy with columnar habit. It might allow columnar cultivars to be planted in solid blocks rather than as mixtures of cultivars, perhaps of different vigours, and dependent on bees and warm weather for fruit set. Also it might introduce a component of regular cropping, as parthenocarpic types, being normally seedless, should be less prone to biennial bearing, a condition to which 'McIntosh Wijcik' and many columnar selections are susceptible. Apetaly is not mentioned in the recent reviews of apple genetics (Brown, 1992) and apple genetic resources (Way et al., undated) and is mentioned only obliquely, under parthenocarpy, in the review of apple breeding by Brown (1975). This paper draws attention to the potential use of this interesting character, gives data relating to its genetic control and reports its combination with columnar growth habit.

Table 1. Segregation for columnar habit and normal v. apetalous flowers in progenies derived from 'Wellington Bloomless' and 'Spencer Seedless'

Year of cross	Progeny	Parentage	Fraction of seedlings with columnar habit	No. of selections observed		
				Normal flowers	Apetalous flowers	Total
1978	SA339	'Wellington Bloomless' × 'Wijcik'	46/140	22	0	22
1983	SA577	'Spencer Seedless' × SA339–4	7/18	0	5	5
1986	SA684	'Spencer Seedless' × SA339–12	122/234	29	20	49
1986	SA685	'Spencer Seedless' × SA339–13	30/66	5	6	11
1986	SA689	'Spencer Seedless' × SA339–30	138/286	31	20	51
		Pooled data from SA577, SA684, SA685, SA689	297/604	65*	51*	116
1985	SA633	'Spencer Seedless' × 'George Carpenter'	0	35	0	35
1987	SA736	'Spencer Seedless' × 'Falstaff'	0	21	0	21
1987	SA737	'Spencer Seedless' × 'Fiesta'	0	48	0	48
1987	SA712	'Wellington Bloomless' × SA679–2	0	26	0	26

* This pooled segregation approximates to a 1 : 1 segregation with a χ^2 value of 1.69, $p > 0.1$.

Materials and methods

Initially 'Wellington Bloomless' was crossed with 'McIntosh Wijcik'. Later 'Spencer Seedless' was crossed with four seedlings from the original cross having columnar habit and normal flowers. In addition, the two apetalous cultivars were crossed with several cultivars or selections of normal habit: thus 'Spencer Seedless' was crossed with 'George Carpenter', a cultivar reputed to root easily, and with 'Falstaff' and 'Fiesta', two good quality cultivars bred at East Malling, and 'Wellington Bloomless' was crossed with A679–2, a selection resistant to scab and mildew. Table 1 lists the crosses.

Crossing technique differed slightly from usual. As the apetalous types lack stamens and petals, pollen could be applied to the styles after removing at most a few sepals to provide access. With two cores per fruit, up to 18 seeds per fruit developed.

The seedlings were scored for growth habit in their second year, columnar seedlings being recognised as relatively sturdy with few side shoots. Healthy seedlings were selected for further evaluation, and in the case of progenies segregating for columnar habit all selected seedlings were columnar. Selections were budded on to M27 and the resulting maiden trees were planted at 0.6 m × 2.5 m, unstaked in the case of columnar seedling selections and as cordons in the case of normal ones. Seedling selections generally came into flower just over four years from germination and were scored for flower and fruit characters for about three years.

Results

The segregations for columnar habit in the original seedlings and for apetaly within the selections are given in Table 1. The progenies having a columnar parent segregated more or less 1:1 for columnar habit, as expected for a dominant trait. The selections from the original cross, 'Wellington Bloomless' × 'McIntosh Wijcik', all had normal flowers. The selections resulting from crossing four of these to 'Spencer Seedless' segregated for apetalous versus normal flowers, the pooled segregations fitting a 1:1 segregation. Selections from crossing 'Spencer Seedless' with 'George Carpenter', 'Falstaff' and 'Fiesta', and 'Wellington Bloomless' with A679–2 all had normal flowers. Fruits from apetalous selections were generally seedless, occasionally with one or two seeds, and of poor quality.

Discussion

The absence of apetaly from the first generation progenies and the segregations for normal versus apetalous flowers approximating 1:1 in the backcrosses indicate control by a recessive gene. The symbol *ape*, for apetaly, is proposed. 'Spencer Seedless', 'Wellington

Bloomless' and the apetalous seedling selections are then homozygous *ape ape*.

Although apetaly is not mentioned in the recent reviews of apple genetics, genetic resources and breeding there have been occasional reports of it in the literature relating to apple breeding which are worth mentioning here.

Stout (1929) wrote that apetalous apples were being used in breeding at Geneva, New York, and Dennis (1970) mentioned two seedless apetalous New York selections, 19697 and 19726, the parentage of which was given by Way et al. (1976) as 'Wellington Bloomless' × NY 11466, (which itself came from the cross 'Wellington Bloomless' × 'Cortland' (R.D. Way, personal communication)). Crandall (1926) referred to seedlings of an apetalous apple that were growing at Illinois.

There are also several accounts in German or Russian, such as those of Ewert (1929), Kobel (1931), Chernenko (1953), Cuprinjuk (1969), Eliseev (1979) and Ponomarenko (1980). These reported that apetalous seedlings could result from crosses between cultivars with normal flowers and/or that seedlings of apetalous cultivars had normal flowers, indicating apetaly was recessive but with little or no segregation data.

Apetaly seems to be the only potentially useful 'non-wild-type' character in apple to have been ascribed so far to a recessive gene. It would aid further breeding if a linked molecular marker could be identified, to tag the gene in the heterozygous condition. However, apetaly is not one of the agronomic characters included in the European Apple Genome Mapping Project (King et al., 1991).

It is interesting that such a complicated character should be under simple genetic control. Apetaly in apple appears to be an example of a homeotic floral mutation (Coen, 1991, Acquaah, 1992). It resembles those mutants in *Antirrhinum* and *Arabidopsis* that Coen classes together as affecting the identities of organs in both the second and third floral whorls, with sepals instead of petals in the second and carpels instead of stamens in the third. In *Antirrhinum* and *Arabidopsis* homoetic mutants are proving useful for studying the molecular genetics of flower development.

Apetalous types could also be useful subjects for developing techniques of genetic transformation in apple while there are fears about transgenes escaping via pollen or seed into wild populations: the apetalous types do not produce pollen and are unlikely to produce seeds without hand pollination.

Although apetaly has been successfully combined with columnar habit, we have got to check if the apetalous segregants crop more reliably than the normal ones. It is salutary to note that of five apetalous cultivars or selections growing as single trees Dennis (1970) said only 'Spencer Seedless' set a heavy crop consistently, and that Ponomarenko (1980) described some apetalous types which never set fruit and others that set fruit only after artificial pollination.

Leaving aside any doubts about cropping, it is unlikely that any of the apetalous columnar selections raised so far will be useful commercially. Fruit quality is poor. However, some selections, of normal habit and heterozygous *Ape ape*, from progenies SA633, SA736 and SA737 have better quality and will be useful as parents. And it is hoped that some selections from SA633 will prove to be easy rooting and some from SA712 to be resistant to scab and mildew; these too would be useful parents. Because of the recessive nature of apetaly, breeding progress will be relatively slow.

Acknowledgements

The breeding of columnar apples at East Malling is financed by the Ministry of Agriculture, Fisheries & Food.

References

Acquaah, G., J.W. Saunders & L.C. Ewart, 1992. Homeotic floral mutations. Plant Breeding Reviews 9: 63–99.

Brown, A.G., 1975. Apples. In: J. Janick & J.N. Moore (Eds). Advances in Fruit Breeding, pp. 3–37. Purdue, Indiana.

Brown, S.K., 1992. Genetics of apple. Plant Breeding Reviews 9: 333–366.

Chan, B.G. & J.C. Cain, 1967. The effect of seed formation on subsequent flowering in apple. Proceedings of the American Society for Horticultural Science 91: 63–68.

Chernenko, E.S., 1953. (Apple trees bearing flowers without petals.) Priroda 12: 105–108.

Coen, E.S., 1991. The role of homeotic genes in flower development and evolution. Annual Review of Plant Physiology and Plant Molecular Biology 42: 241–279.

Crandall, C.S., 1926. Apple breeding at the University of Illinois. Bulletin of Illinois Agricultural Experiment Station 275: 337–600.

Cuprinjuk, V., 1969. (Apple trees with apetalous flowers). Sadovodstvo 7: 19.

Dennis, F.G., 1970. Effects of gibberellins and naphthaleneacetic acid on fruit development in seedless apple clones. Journal of the American Society for Horticultural Science 95: 125–128.

Eliseev, I.P., 1979. (The origin, polymorphism and evolutionary importance of apetalous apple forms.) Trudy Gorkov Selskokhozyaistvennyi Institut 136: 101–120.

Evreinoff, V.A., 1935. Malus pumila Mill. f. apetala Munchh. Review Horticole Paris 107: 369–371.

Ewert, R., 1929. Blühen und Fruchten der insektenblütigen Garten- und Feldfrüchte unter dem Einfluss der Bienenzucht. Neumann, Neudamm.

King, G.J., F.H. Alston, I. Batlle, E. Chevreau, C. Gessler, J. Janse, P. Lindout, A.G. Manganaris, S. Sansavini, H. Schmidt & K.R. Tobutt, 1991. The 'European Apple Genome Mapping Project' – Developing a strategy for mapping genes coding for agronomic characters in tree species. Euphytica 56: 89–94.

Kobel, F., 1931. Lehrbuch des Obstbaus. Springer, Berlin.

Ponomarenko, V.V., 1980. (Clarification of the origin, systematics and morphology of apetalous apples.) Byulleten Vsesoyuznogo Ordena Lenina i Ordena Druzhby Narodov Instituta Rastenievodstva Imeni N.I. Vavilova 103: 14–18.

Stout, A.B., 1929. The development of seedless fruits by breeding. Journal of the New York Botanic Garden 30: 270–277.

Tobutt, K.R., 1985. Breeding columnar apples at East Malling. Acta Horticulturae 159: 63–68.

Tobutt, K.R., 1989. La creation d'arbes colonnaires pour le nouveaux systemes de vergers. Le Fruit Belge 57: 101–106.

Tobutt, K.R., 1991. Columnar apple breeding. Annual Report of Horticulture Research International 1990–91: 19–20.

Way, R.D., R.C. Lamb, C. Pratt & J.N. Cummins, 1976. Pale green lethal gene in apple clones. Journal of the American Society for Horticultural Science 101: 679–684.

Way, R.D., H.S. Aldwinckle, R.C. Lamb, A. Rejman, S. Sansavini, T. Shen, R. Watkins, M.N. Westwood & Y. Yoshida, undated. Apples (*Malus*). In: J.N. Moore & J.R. Ballington (Eds). Genetic Resources of Temperate Fruit and Nut Crops 1, pp. 3–62. ISHS, Wageningen.

H. Schmidt and M. Kellerhals (Eds.), Progress in Temperate Fruit Breeding, 225–228.

Improvement of seed set and seed germination in apple

Johan Keulemans, Rik Eyssen & Geert Colda
Fruitteeltcentrum K.U. Leuven, Willem de Croylaan 42, B-3001 Leuven-Heverlee, Belgium

Key words: Apple, fruit set, June drop, pollination, seed germination, seed set

Summary

To improve seed set and seed germination in the apple cultivar Elstar, the influences of the number of pollinated stigmas per flower, the number of flowers per cluster and the pollinator were studied.

From 1 or 3 flowers per cluster 1, 3 or 5 stigmas were pollinated with pollen of the cultivars James Grieve, Cox's Orange Pippin, Idared or Gloster. Fruits were harvested one week before or after the commercial picking date.

The seed set, expressed as the number of seeds per pollinated flower, increased with the number of pollinated stigmas on the flower. This increase, from 0.40 to 1.38, was caused by a higher fruit set after June drop and a higher number of seeds per fruit. The positive correlation between seed number per fruit and fruit set suggest that the sink strength of an individual fruit was influenced by seed number. Compared to 1 flower per cluster, seed set was slightly reduced when 3 flowers per cluster were left, due to a lower fruit set, although the seed number per fruit was somewhat higher. Since the pollinator has no clear effect on final fruit set, the influence of the pollinator on seed set was determined by the number of seeds per fruit, ranging on average from 2.0 to 3.2. Seed germination was lowest when Gloster was used as pollinator (58%) and highest in the case of Cox's Orange Pippin (89%). For all pollinators seed germination was negatively correlated with seed number, probably due to competition between seeds in the fruit. A delayed picking time also reduced seed germination.

Introduction

In large-scale breeding programs a lot of flowers have to be pollinated within a short period. The efficiency of this pollination depends strongly on the final seed set and the subsequent seed germination. Surprisingly, in breeding research little attention has been paid to the pollination technique to maximize seed set and seed germination. Seed set can be seen as the result of fruit set and the seed number per fruit. Both can be influenced by manipulating the number of pollinated styles, as mentioned by Visser & Verhaegh (1987) in apple and pear. But as far as we know no quantitative data are available neither on the influence of number of flowers per cluster on fruit or seed set nor on the relation between seed number per fruit and seed germination.

Materials and methods

The pollination experiments were carried out on branches of adult trees of the cultivar Elstar, grafted on M9 rootstock. The branches, each containing 5 clusters, were bagged and the clusters were thinned to 1 or 3 flowers per cluster. The flowers were emasculated at the balloon stage and styles were thinned to 1, 3 or 5 styles per flower. The stigmas were pollinated with pollen of cultivars Gloster, Idared, James Grieve or Cox's Orange Pippin. For all treatments (combination of pollinator & style number per flower & flower number per cluster) 12 comparable branches were chosen at random.

Fruit set was recorded before and after the June drop (initial and final set respectively). Fruits were harvested on 6th September and 22th September, the commercial picking time being 14th September.

For each treatment 20 fruits, originating from all 12 branches, were collected and the number of normal

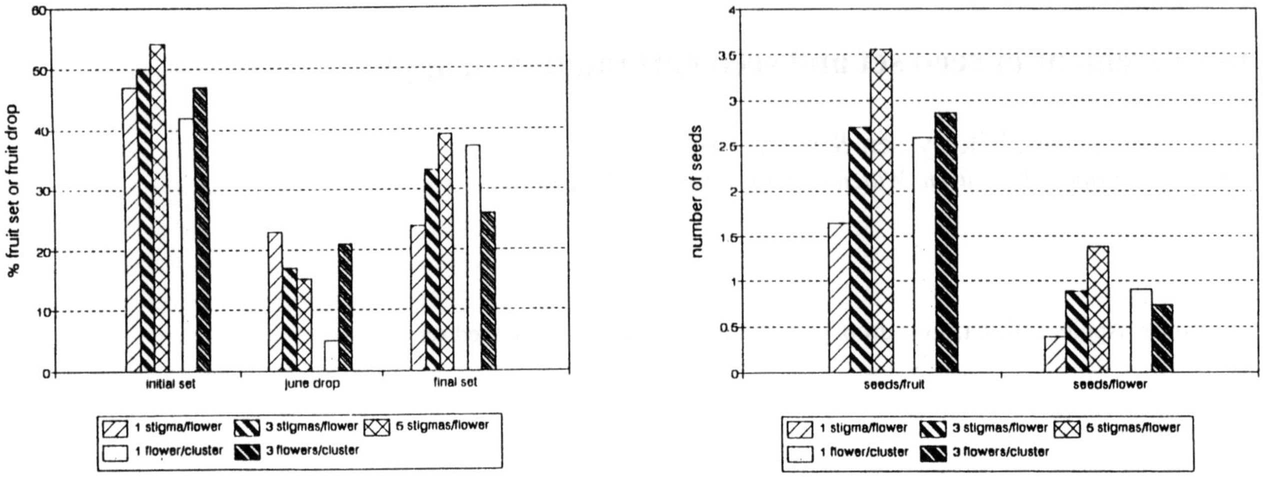

Fig. 1. Influence of the number of pollinated stigmas per flower and of the number of flowers per cluster on initial fruit set (%), June drop (%), final set (%), number of seeds per fruit and seed set (seeds per pollinated flower). Variety Elstar pollinated with James Grieve, Cox's Orange Pippin, Idared or Gloster.

developed seeds in each fruit was counted 10 days after harvest. The seeds were sterilised in Na hypochlorite for 10 minutes and stratified aseptically on wet filter paper in petri-dishes at 3° C. Seed germination was evaluated at regular intervals until maximum germination was reached.

Results

Fruit set

As illustrated in Fig. 1 the final set increased significantly from 24% to 39% with increasing number of pollinated stigmas. These increases with number of stigmas pollinated resulted from both an increased initial set and a reduced June drop. When 1 or 3 flowers per cluster are pollinated little effect on the initial set was observed, although June drop was dramatically increased from 5% (1 flower) to 21% (3 flowers).

Significant differences for the initial set were found among pollinators, ranging from 45% (Cox's Orange Pippin) to 56% (Idared), but since high initial set was correlated with high June drop no reliable differences occurred for the final set (Table 1).

Seed set

As expected, the number of seeds per fruit was correlated with the number of pollinated styles (Fig. 1) but when expressed in relation to the theoretical maximum per fruit the correlation is negative: 82% of the maximum for 1 style, 45% for 3 styles and 36% for 5 styles. The number of seeds per fruit was also influenced by the pollinator and varied from 2.0 (Cox's) to 3.2 (Gloster) but no relation with fruit set was observed (Table 1).

The seed set (number of seeds per pollinated flower) slightly decreased from 0.94 to 0.74 respectively when 1 or 3 flowers per cluster were pollinated. The number of pollinated styles had a more pronounced influence on seed set: 0.4 (1 style), 0.83 (3 styles) and 1.38 (5 styles) (Fig. 1). The seed set varied also among

Table 1. Influence of pollinators on fruit set (before and after June drop, %), June drop (%), seeds per fruit or per pollinated flower, seed weight (g) and seed germination (percentage and per pollinated flower) in Elstar

	James Grieve	Cox's O.P.	Idared	Gloster
initial set	51	45	56	48
June drop	22	14	21	16
final set	29	31	35	32
seeds/fruit	2.9	2.0	2.8	3.2
seeds/flower	0.84	0.62	0.98	1.02
seed weight	0.067	0.070	0.073	0.068
seed germination %	82	89	73	58
germinated seeds/flower	0.69	0.55	0.72	0.59

Table 2. Influence of the number of pollinated stigmas per flower and of the number of flowers per cluster on seed germination in Elstar. Fruits harvested at 6 September (1) and 22 September (2). Flowers pollinated with James Grieve, Cox's Orange Pippin, Idared or Gloster

	Seed germination percentage		Seed germination per pollinated flower	
picking time	1	2	1	2
stigmas per flower				
1	82	77	0.33	0.31
3	73	67	0.65	0.60
5	72	54	1.00	0.75
flowers per cluster				
1	87	77	0.82	0.72
3	71	60	0.52	0.44

pollinators from 0.62 to 1.02 as a result of differences in seed number per fruit (Table 1).

Seed germination

In all cases delayed picking (16 days) had a negative effect on seed germination (Table 2). Also an influence of the number of pollinated stigmas per flower or flowers per cluster was observed. In general the seed germination seems to be correlated negatively with the number of seeds per fruit, but this correlation was pollinator-dependent. This means that pollinators giving a relative high seed number per fruit can also have a high seed germination: compared to Cox's Orange Pippin seed number was high in James Grieve (2.0 and 2.9 respectively) and also seed germination was comparable (89% and 82% respectively), while for Gloster, seed number was high (3.2) and seed germination was low (58%) (Table 1).

Discussion and conclusions

A positive correlation between fruit and seed set and the number of pollinated stigmas was also found by Visser & Verhaegh (1987) in some apple and pear cultivars but in other cultivars no correlation could be proved. They suggested that this independence versus dependence was based on the property of a cultivar to have or not to have more than one ovule per style fertilized. Since the mean seed number per fruit in Elstar was 1.65 when 1 style was pollinated, this cultivar appears to have the above mentioned property and thus an independence of fruit and seed set and number of pollinated styles. Our data prove the opposite. Dennis (1986) mentioned a reduced fruit drop when more seeds per fruit are present. A reduced fruit drop corresponds with a higher final fruit set. The fact that fruit set seems to be regulated by seeds in the fruit is not surprising. Goldwin (1992) demonstrated in Cox's apple that fruit set is regulated by plant growth regulators and that parthenocarpic fruit set intensity is correlated with the number of styles, treated with PGR's. PGR's are also produced in seeds like diffusible IAA and this explains the relation between sink strength and seed number as was mentioned by Ho (1992). So a high seed number per fruit implies, probably through PGR's, a strong sink strength and a reduced fruit drop.

Compared to 3 flowers per cluster, we found that final set was improved when 1 flower per cluster was pollinated. This is in accordance with results obtained in other fruit species like olive, where thinning of flowers induced higher final set (Paz Suarez et al., 1984). This higher final set in our experiments was caused by a reduced June drop, indicating a more intensive competition in the case of 3 flowers/cluster for the same pool of assimilates compared to 1 flower/cluster. Also Dennis (1986) and Ho (1992) reported an increased competition between fruits for assimilates in the case of high fruit numbers.

Differences in fruit set among pollinators are generally accepted in apple (Owens, 1992) and also in other fruit species like plum (Keulemans & Van Laer, 1989). Whether these differences are caused by gametophytic incompatibility or by genotypic sensitivity of pollen germination or pollen tube growth to temperature is not clear. The latter was proven in several fruit species, including apple (Milutinovic, 1983), apple and pear (Deckers & Porreye, 1984) and plums (Keulemans, 1984). We found differences among pollinators only for the initial fruit set, but these differences largely disappeared during the June drop, independent from the number of seeds per fruit, which was also influenced by the pollinator. This suggests probably a qualitative effect (not related with seed number) of the pollinator on sink strength and fruit drop.

From our data we conclude that seed germination was different among pollinators, seed number per fruit and picking time of the fruits. Whether reduced germination was related to poorly developed (non-viable) embryos or to dormancy is not clear. An increase of non-viable embryos with increasing seed number seems reasonable since we may expect higher competition between seeds, even leading to embryo abortion. This latter is suggested by the fact that the seed number per ovule was lower when 3 or 5 stigmas were pollinated.

Why seed germination was lower when fruits were harvested later in the season is not clear, but the same influence of picking time on seed germination was found in plums (Grzyb & Czynczyk, 1990).

To improve the number of germinated seeds per pollinated flower, it seems that all stigmas of a flower have to be pollinated and that in our experiments a reduction of the number of flowers per cluster to less than 3 results also in a positive effect. The influence of picking date was unexpected and should be investigated further. Although the choice of pollinators is sometimes limited in breeding programmes, this choice influences seed set and subsequent germination.

Acknowledgement

Part of this work was supported by the 'Instituut voor aanmoediging van Wetenschappelijk Onderzoek in Nijverheid en Landbouw – I.W.O.N.L.' – Brussels.

References

Deckers, T. & W. Porreye, 1984. Influence of temperature on pollen germination of different cultivars of apple and pear: trials *in vitro*. Acta Horticulturae 149: 123–129.

Dennis, F.G., 1986. Apple. In: S.P. Monselise (Ed.) Handbook of Fruit Set and Development, pp. 1–44. CRC Press Inc., Boca Ratou, Florida.

Goldwin, G.K., 1992. Environmental and internal regulation of fruiting, with particular reference to Cox's Orange Pippin apple. In: C. Marshall & J. Grace (Eds) Fruit and seed production, pp. 75–100. Cambridge University Press, Cambridge.

Grzyb, Z.S. & A. Czynczyck, 1990. Attempts to improve seed germination and seedling growth of Wangenheim Prune generative rootstock. Fruit Science Reports 17: 101–110.

Ho, L.C., 1992. Fruit growth and sink strength. In: C. Marshall & J. Grace (Eds) Fruit and seed production, pp. 101–124. Cambridge University Press, Cambridge.

Keulemans, J., 1984. The effect of temperature on pollen tube growth and fruit set on plum trees. Acta Horticulturae 149: 95–101.

Keulemans, J. & H. Van Laer, 1989. Effective pollination period of plums: the influence of temperature on pollen germination and pollen tube growth. In: C.J. Wright (Ed.) Manipulation of fruiting, pp. 159–171. Butterworths, London.

Milutinovic, M., 1983. Correlational dependence of the degree of functionality of embryonic sack and pollen and fruit set in apples. Acta Horticulturae 139: 97–101.

Owens, S.J., 1992. Pollination and fertilization in higher plants. In: C. Marshall & J. Grace (Eds) Fruit and seed production, pp. 33–55. Cambridge University Press, Cambridge.

Paz Suarez, M., R. Fernandez-Escobar & L. Rallo, 1984. Competition among fruits in olives II. Influences of inflorescence or fruit thinning and cross-pollination on fruit set components and crop efficiency. Acta Horticulturae 149: 131–143.

Visser, T. & J.J. Verhaegh, 1987. The dependence of fruit and seed set of pear and apple on the number of styles pollinated. Gartenbauwissenschaft 52: 13–16.

H. Schmidt and M. Kellerhals (Eds.), Progress in Temperate Fruit Breeding, 229–233.

Genotypic variation in apple × quince progenies

I.S. Rudenko & I.I. Rudenko
Botanical Institute, Academy of Science, Lesnaya 18, Kishinev, 277018, Moldova

Key words: Allopolyploid, apple, distant hybridisation, quince

Introduction

Intergeneric hybridisation can be used to reveal the basis for the development of novel types of plants. It is important to identify the evolutionary stages of crop plants thus enabling an extention of the genetic basis leading to a wider range of varieties. The hybridization between *Cydonia oblonga* Mill. and *Malus domestica* Borkh. is very important for establishing genetic affinity between these taxon, creating new commercial types which combine the best parental characters and providing a genetic source of various new features important for pome fruit breeding. Burbank (1955) was the first creating apple × quince hybrids. They turned out to be sterile. Ryabov (1970) has also raised *Malus* × *Cydonia* hybrids. They were shrubbery with quince characters, weak and slowly growing.

Material and methods

The breeding work was carried out at the Botanical Garden of the Academy of Science in Kishinev, Moldova. Diploid forms of quince × apple hybrids were obtained from V. Panov in Bulgaria as basis for the programme. At Kishinev about 600 seeds from open and controlled pollination were produced. The seedling trees started to bear fruits after 1980. The collection is planted at a distance of 4 × 6 m. It comprises 47 quince × apple F2 hybrids obtained by I.S. Rudenko (Rudenko, 1978). The hybrids were grafted on quince rootstock. Morphological and pomological descriptions were made according to Hedrik (1937).

Results

Only allodiploid viable hybrids were obtained after quince × apple hybridization in F1. The characteristics of vegetative and generative parts including vigour and fertility differ according to the cross combination. Their pollen shape is very variable and shows differences in viability also. Fruits of F1 hybrids are variable in form and rarely set seeds. Single normally developed seeds are set in some fruits originating from open and artificial pollination. Their germination is weak. Most F2 seedlings die during the early stages of growth. About 50% of the surviving seedlings developed normally after grafting on quince and started to flower and to set fruits after 10 years. All F1 seedlings grew much weaker when compared to the parents, and they are also distinct in fruit form, structure and quality, and morphological characters of vegetative organs.

Slightly more than 50% of the F2 seedlings were found to be allotriploids ($3x = 2n = 51$), the rest were tetraploids ($4x = 2n = 68$). Most of the allotriploids resembled quince morphologically. All these plants formed only single-flowered inflorescences and set mostly seedless quince-type fruits. About 15% of the allotriploids displayed apple-like characters and had seedless fruits with intermediate taste. Inflorescences of this group had up to five flowers, but the majority had two to three flowers. All allotriploids produced slightly viable pollen. Allotetraploid ($4x = 2n = 68$) hybrids position is intermediate between the diploids and the triploids. Their inflorescences are single-flowered and their pollen is more viable than that of the diploids and the triploids. It mainly consists of large pollen grains of probably diploid ($2x = 34$) genotype. The other part of pollen is smaller in size, maybe haploid ($x = 17$), but very small sterile grains also occurred. Germination percentage, on 15% saccharose solution was close to 50%. Fruits were flat-rounded, yellow with dense pulp, containing normally developed seeds (up to 25), most of which germinated well.

Various F1 and F2 hybrids reached a growth height under non-irrigation between 2.6 m and 4.65 m. Dwarf bushy F3 seedlings have been grown on their own roots. They originated from various cross combinations, self-pollination and isolation with and without pollination. At the age of 10 years they reached a maximum height of 1.5 m and do not yet bear fruits.

After growth starts, up to 50% of the seedlings can be eliminated. Some are stunted, forming minute branches and leaves and die at the age of 3–4 years after reaching 10–20 cm height.

F2 and F3 seedlings selected among allodiploid F1 and allotriploid F2 progenies have the most significant genotypic and phenotypic variability. F3 progenies of allotetraploids are similar and this makes it possible to presume that they are of apomictic origin.

According to leaf anatomical structure all quince × apple hybrids can be divided into a) resembling quince and b) resembling apple.

The content of biological active substances in hybrid fruits is 4–5 times higher than in parental species, polyphenolic substances content is 2–3 times higher than in quince varieties.

Description of four triploid apple-type hybrids:

1) 2-69

Obtained in 1969. Trees are 2–3 m high, with a lot of fruiting wood. Branches grow with almost right angles from the trunk. The bark is smooth, dark grey, with light brown lenticels. There are rudiments of aerial roots on the trunk. The annual shoots are brown-greenish, covered with grey, easily removable pubescence. They end with mixed terminal buds. The leaves are lengthened-oval, with rounded base, acuminate tip, weakly serrated edges. Petioles are 1.5–2 cm long, pubescent. Stipules long, saw-toothed.

Blossom starts when apples finish and quince begin to blossom. 58.4% of inflorescences have single flowers, 24.8% two, 10.8% three, 4.4% four and 1.6% five flowers, Fig. 1a. Fruits ripen at the end of June to July. The fruit weight varies from 120 to 320 g. Skin colour is yellow, sometimes a little pink on the sunny side. Flesh is yellowish, juicy, sour-sweet, almost no stone cells. The 5–6 seed chambers are rigid, rounded, usually seedless, Fig. 1b.

Fig. 1 a. Quince × apple F2 triploid 2-69. Flowers.

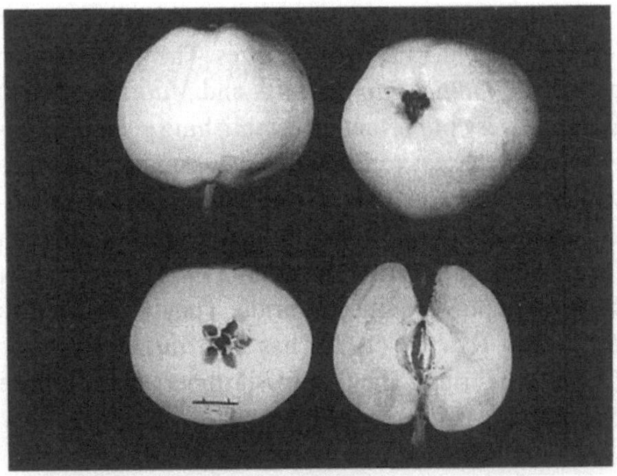

Fig. 1 b. Quince × apple F2 triploid 2-69. Fruits.

2) 4-72

Obtained in 1972. Trees reach 3 m height, Fig. 2a. Branches grow with almost right angles. The bark is grey, cracked, with brown lenticels. Burrknots are formed on the trunk and the leader branches. Shoots are covered with dense pubescence and have clear lenticels. The dark green leaves are oval-lengthened, with rounded base, acuminate tip, serrate edges. The lower side is densely pubescent, the upper side has sparse hairs along the ribs. Petioles are 1–2 cm, reddish, pubescent.

Fig. 2 a. Quince × apple F2 triploid 4-72. Tree.

Fig. 2 b. Quince × apple F2 triploid 4-72. Fruits.

89% of the inflorescences have one or two flowers. Petals easily drop off. Fruits are apple type, 180–200 g, Fig. 2b. Mature fruits have a wavy, bright yellow skin with white and brown hypodermic dots. Flesh is cream-coloured, quickly oxydizes when

Fig. 3 a. Quince × apple F2 triploid 7-72. Tree.

Fig. 3 b. Quince × apple F2 triploid 7-72. Fruits.

cut, quince flavour. The 4–5 seed chambers are narrow with rigid walls, single seeds are empty.

3) 7-72

Obtained in 1972. Tree height is 1.5–2.5 m, Fig. 3a. Branches have nearly right angles with the trunk, then grow downwards. The bark is grey, smooth, with large grey lenticels. Annual shoots are dark brown, covered with dense, easily removable pubescence and light-brown lenticels. The shape of leaves is oblong-ovate with acuminate tip and base, serrulated edge and wavy blade, the upper side is dark green.

The inflorescences consist of 1–5 flowers, mostly 1 and 2 flowers. Fruits ripen early in autumn (August–September). They are apple-type, 112–427 g, Fig. 3b. The skin is smooth yellow, with numerous hypodermic greenish or brown dots, oily.

Fig. 4 a. Quince × apple F2 triploid 15-74. Tree.

Flesh is cream-coloured, aromatic, sour-sweet, astringent or slightly bitter, juicy. Texture and flavour are similar to apple, pungency simar to quince. There are 5 seed chambers with rigid walls. Single seeds are large, brown, filled.

4) 15-74

Tree height 4.5–5 m, Fig. 4a. The angle of the branches varies from 60–90°. The bark is dark grey, with round lenticels. Annual shoots have green, closely downed bark, internodes 2–3 cm. The shape of leaves is broad-ovate, the blade is concave, wavy, tip acuminate or rounded. They are pubescent on both sides. Blossoming lasts 5–7 days. The flowers are similar to apple and the majority of the inflorescences have single flowers. The apple-type fruits are flat-rounded and mature in autumn (September–October), Fig. 4b. They look attractive with yellow colour, waxy. Fruit stalk is

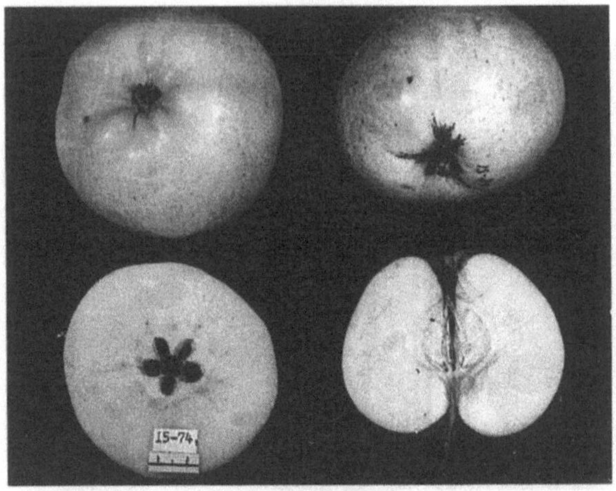

Fig. 4 b. Quince × apple F2 triploid 15-74. Fruits.

1–2.5 cm long. Flesh is white, tinted with greenish, no stone cells. Tastes sour-sweet with apple flavour. The seed chambers are rounded, with rigid walls. Seeds are sparse, 1–2 per fruit, dark brown, oblong. Yields are relatively low.

Discussion

Selected triploid apple-type hybrids are different from each other in respect to leave and fruit characteristics. They inherited characters of both parental species. All hybrid forms have to some degree single flower inflorescences. Within the Pomoideae subfamily this character is only present in *Cydonia oblonga*. The apple-type hybrids resemble apple in respect to shape, texture and flavour. Stone cells are lacking (Rudenko & Negru, 1987). They can be used for fresh consumption. It would be interesting to study the development of apple-type triploids. It is probable that after pollination of the egg cell of F1 hybrids with the restitution nucleus, triploid embryos arose.

Quince × apple hybrids can be intercrossed with apple, quince and pear. They are allopolyploids and can be used in breeding programmes. They should be considered as a new pome fruit. Triploid and tetraploid forms could be used for breeding apples with single flower inflorescences.

References

Burbank, L., 1955. Selected work. Foreign Literature Publishers, Moscow.

Hedrik, Y.P., 1937. Pomology. Agriculture Literature Publishers, Moscow.

Rudenko, I.S., 1978. Distant hybridization and polyploidy of fruit plants. Shtiinza Publishers, Kishinev.

Rudenko, I.S. & E.I. Negru, 1987. Pericarp anatomy of F2 quince × apple hybrids of various ploidy levels. In: Proceedings Vth Congr. Genet. Sel. Moldova, July 1987, p. 139. Shtiinza Publishers, Kishinev.

Ryabov, I.N., 1970. Hybrids of cultivated apple with quince. In: Proceedings Congr. Distant Hybr. Plants & Anim., 5–8 Febr., 1968 Moscow, v. 2, p. 49–54. Kolos Publishers, Moscow.

H. Schmidt and M. Kellerhals (Eds.), Progress in Temperate Fruit Breeding, 235–238.

Segregation for fruit characters in some crosses of European and Asian pears

A.G. White & H.I. Selby
HortResearch, Havelock North, New Zealand

Key words: Pyrus communis, Pyrus pyrifolia, pears, interspecific crosses

Summary

Segregations were observed for major fruit characteristics in populations of intra- and interspecific crosses of European and Japanese pear cultivars. A number of individuals in the populations were identified as having potential as commercial cultivars.

Introduction

European type pears (*Pyrus communis*) have soft flesh and short shelf-life when ripe which restricts both the availability and the quality in the market (La Rosa & Monnot 1984). In contrast Japanese pears (*Pyrus pyrifolia*) are crisp and have a shelf-life of 10–15 days (Lallu 1990). The Japanese cultivars are delicately flavoured and could have wider appeal in traditional European pear markets if they were more flavoursome (White 1986).

In 1983, a pear breeding programme was commenced which included intra and interspecific hybridization of European and Japanese pears (White 1986). Crosses within the European and Japanese pear types have been carried out to produce high quality cultivars of these types. The aim of the interspecific programme was to develop a range of pear cultivars which combined the rich flavour and aroma of European pears with the crisp textural characteristics of the Japanese types.

Hybridization has continued annually and a total of about 10,000 seedlings produced to date. Preliminary results from the evaluation of 767 seedlings from these progenies suggest that there is considerable potential for the use of wide crosses between quality pear cultivars of different pear species as well as crosses within both the European and Japanese pear types to develop a range of cultivars adapted to current and future market requirements.

Materials and methods

The observations in this paper were recorded on progenies from crosses made at Havelock North in 1984 and 1986, within and between cultivars of European and Japanese pear cultivars (Table 1). Seedling trees were planted in non-replicated blocks in 1986 and 1988 and fruited by 1990 and 1992 respectively. Progenies from the 8 populations were fruiting by the 4th year in the orchard in Havelock North. The results presented are from observations made on fruit harvested on three dates from the seedling trees after 6 weeks storage at 0–0.5 °C + 7 days at 20 °C in a conditioning room.

The fruit characteristics recorded were observed in the populations along a continuum from one extreme to the other. However arbitrary groupings were chosen for each character based on the lowest quality desired for selection using the parents as standards. Skin thickness, flesh characteristics and fruit shape were assessed subjectively. Whilst segregations can only be loosely compared between the different populations some general trends were apparent.

Skin thickness:
The fruit skin was considered thin if it was fragile and easily damaged as with 'Max Red Bartlett' and 'Kosui'. Skin was considered thick if it was tough and chewy as with 'Nijisseiki'. Medium thickness covered the range considered most preferred which included 'Packhams Triumph'. Skin thickness is important for pears grown

Table 1. Populations of pear seedlings grown at Havelock North.

Pear Crosses	No. of Seedlings
Packhams Triumph × President Heron	43
Packhams Triumph × 32–238	78
Packhams Triumph × 20–317	74
Packhams Triumph × Shinseiki	139
Packhams Triumph × Nijisseiki	99
Nijisseiki × Max Red Bartlett	165
Shinseiki × Hosui	67
Shinseiki × Kosui	102

in New Zealand as it is a factor in susceptibility to bird damage and puncture wounds during handling.

Flesh texture:
Fine texture was equated with 'President Heron' and 'Packhams Triumph' or finer, medium texture with 'Red Bartlett' and 'Kosui' and coarse texture with 'Nijisseiki' or coarser. Texture is related to relative cell size with small cells giving fine texture and large cells coarse texture. In our selection criteria we placed no preference on any of the categories although we considered coarse to medium texture to be better in combination with firm flesh and fine texture with soft flesh.

Flesh firmness:
This was graded with firm flesh equating to the Japanese cultivars and soft flesh to the European cultivars after ripening.

Flesh grittiness:
If grit cells occurred outside the coreline the fruit was classified as gritty. Those classed as not gritty were free of grit cells in the flesh but may have had them in the fruit core. None of the parents was considered gritty.

Fruit juiciness:
Fruit was classified as having dry flesh if there was no free juice when eaten. All the parents were considered juicy when ripe. The fruit was scored juicy if the level of juice was assessed as acceptable and dry if it was less than acceptable.

Fruit acidity:
Fruit was graded acid or low acid based on the perception of how sour the flesh tasted by the evaluation: Acid flesh was where the dominant sensation was sour, whereas as low acid referred to flesh which had little or no perceivable sourness.

Fruit flavour:
Fruit with flavour levels similar to 'President Heron' was classified as rich; 'Max Red Bartlett', 'Hosui', 'Kosui' and 'Packhams Triumph' medium; and 'Shinseiki' and 'Nijisseiki' delicate.

Fruit shape:
Fruit shape of the Japanese cultivars was considered round whereas oval to pyriform fruit was classed as pyriform.

Parental characters:

- 'Packhams Triumph' is a major European pear cultivar in New Zealand. It has a precocious and productive bearing habit. Fruit is of good quality, with sweet melting flesh when eating ripe.
- 32–238 and 20–317 were supplied as pollen by Dr Fred Hough from the breeding programme at Rutgers University, New Jersey, USA. They are selections from a cross between the European pear cultivars 'Carrick' and 'Seckel'. Both are reputed to have soft, highly flavoured flesh and a red blush on the skin.
- 'Max Red Bartlett' is a red European pear which produces fruit with a good characteristic aroma and moderate eating quality. Its inclusion in our programme was primarily as a donor of red colour.
- 'Nijisseiki' and 'Shinseiki' are Japanese pear cultivars with crisp,coarse mild flavoured flesh.
- 'Hosui' and 'Kosui' are two of the best quality Japanese cultivars with sweeter, richer flavour and finer texture than 'Nijisseiki' or 'Shinseiki'.

Results

Several individuals from the first populations showed desirable combinations of European and Asian pear characteristics.

Examination of the populations in total gave the following information (Table 2):

Table 2. The segregation of eight pear populations for important fruit characteristics.

Cross	Skin thickness			Texture			Firmness		Grittiness		Juiciness		Acidity		Flavour			Shape	
	Thin (%)	Med (%)	Thick (%)	Fine (%)	Med (%)	Coarse (%)	Soft (%)	Firm (%)	Not Gritty (%)	Gritty (%)	Juicy (%)	Dry (%)	Low Acid (%)	Acid (%)	Delicate (%)	Med (%)	Rich (%)	Round (%)	Pyriform (%)
Packhams Triumph × President Heron	64	19	17	60	30	10	37	63	89	11	79	21	89	11	47	53	0	0	100
Packhams Triumph × 32–238	77	15	8	50	49	1	59	41	90	10	59	41	95	5	41	50	9	0	100
Packhams Triumph × 20–317	54	35	11	35	56	9	60	40	79	21	66	34	100	0	44	47	9	27	73
Packhams Triumph × Shinseiki	63	24	13	30	48	22	36	64	80	20	75	25	83	17	90	10	0	25	75
Packhams Triumph × Nijisseiki	71	27	2	70	30	0	13	87	90	10	83	17	77	23	79	21	0	19	81
Nijisseiki × Max Red Bartlett	14	59	27	55	33	12	55	45	66	44	70	30	65	35	51	46	3	39	61
Shinseiki × Hosui	25	61	14	7	34	59	7	93	84	16	95	5	91	9	84	16	0	95	5
Shinseiki × Kosui	50	48	2	16	61	23	7	93	94	6	97	3	99	1	96	4	0	97	3

Fruit skin thickness: There was a high proportion of thin skinned progeny in the crosses where 'Packhams Triumph' was used as a parent.The 'Nijisseiki' × 'Max Red Bartlett' and 'Shinseiki' × 'Hosui' populations both had the majority of fruit in the preferred range. About half the 'Shinseiki' × 'Kosui' fruit were also classified in the preferred range.

Flesh texture: The European crosses had a high proportion of fine textured progeny, the Japanese crosses a very low proportion was fine textured. The interspecific crosses partitioned for texture similarly to the European populations. In the two crosses with 'Nijisseiki' as a parent the majority of the fruits was fine textured.

Flesh firmness: Flesh firmness varied continuously from hard to very soft. Seven percent of the seedlings in the Japanese × Japanese pear crosses had soft flesh. The three European × European crosses and the interspecific cross between 'Nijisseki' and 'Max Red Bartlett' partitioned about equally for firm or soft flesh. The interspecific crosses with 'Packhams Triumph' as a parent had the majority of fruit classified soft.

Flesh grittiness: The percentage of gritty progeny ranged from 44% for the 'Nijisseiki' × 'Max Red Bartlett' cross down to 6% for the 'Shinseiki' × 'Kosui' cross. Grittiness appeared to be more influenced by the parents than by the species.

Fruit juiciness: The Japanese crosses had more than 95% of the progeny classed as juicy. The European and interspecific crosses had similar levels and ranged between 60% and 80% in the juicy classification.

Acidity: Perceived acidity was very low in the European and Asian progenies with a higher proportion of fruit in the three interspecific crosses having an acid flavour.

Fruit shape: No round fruit were found in two of the European crosses although a number were oval. The majority of the progeny in the Japanese crosses was round. The remaining were oval rather than pyriform. The interspecific crosses ranged from strongly pyriform to round with about 30% of the fruit being oval.

Flavour: The European crosses had a higher proportion of good flavoured progeny than Japanese crosses.'Packhams Triumph' when used in combination with Japanese cultivars produced mainly delicate flavoured seedlings, 'Max Red Bartlett' appeared to confer much more flavour in a similar combination.

Discussion

The important pear fruit characters included in this study are considered to be inherited additively (Layne & Quamme 1975). The observations on the eight populations reported here would support this. Where both parents were complimentary for a character, such as firm flesh or round shape, little variation from that was seen in the progeny. Where parents were markedly different for a character, such as firm versus soft flesh or round versus pyriform shape, the progeny showed variation ranging between the two parents.

There was a higher proportion of acid types in the progenies of interspecific hybrids than the intraspecific hybrids. The perception of acidity was possibly influenced by the relative level of sugar in the flesh. In our experience European pears tended to have higher levels of both acid and sugar than do the Japanese pears. This higher proportion of acidic interspecific seedlings could have arisen from the combination of lower sugar traits of the Japanese cultivars with the higher acid of the European cultivars.

Our objective in presenting this paper has been to demonstrate that wide crosses between high quality cultivars of European and Japanese pears can be made with the expectation that the progenies will segregate for the desired characteristics in similar proportions to progenies of parents from the same species. Pear breeders should feel free to choose parents with the complimentary characters they want to reinforce; and diverse characters to exploit variation in the progeny.

References

Lallu, N., 1990. Storage and storage disorders. In: A.G. White (Ed) Nashi – Asian Pear in New Zealand, pp. 66–74. DSIR Publishing, Wellington.

La Rosa, A. & P. Monnot, 1984. Commercialisation de poires, marchè Francais et Europeen. Acta Horticulturae 161: 81–95.

Layne, R.E.C. & H.A. Quamme, 1975. Pears. In: J. Janick & J.N. Moore (Eds) Advances in Fruit Breeding, pp. 38–70. Purdue University Press, West Lafayette, Indiana.

White, A.G., 1986. European pear breeding strategy in New Zealand. In: T.A. Williams & G.S. Wratt (Eds) Plant Breeding Symposium DSIR, pp. 150–154. Agronomy Society of New Zealand Special Publication No. 5, Christchurch.

H. Schmidt and M. Kellerhals (Eds.), Progress in Temperate Fruit Breeding, 239–242.

Methods for evaluating fruit characters and incorporation in a final index in the Norwegian pear breeding programme

S.H. Hjeltnes
Ullensvang Research Station, Substation Njøs, N-5840 Hermansverk, Norway

Key words: Appearance, fruit quality, fruit size, index, pear breeding, *Pyrus communis*, shelf life

Summary

In the Norwegian pear breeding programme selection is based on evaluation of 20 fruit samples together with field observations of precocity, productivity and resistance to biotic and abiotic factors. Fruit weight is recorded in grams, and then transformed to a scale of 20 gram classes. Appearance, shelf life and flavour are recorded on a 0 to 9 scale, and observations of shelf life and flavour are made at 4 defined intervals following ripening at 20 °C. The final index is made by summarizing the scores of fruit size and appearance, mean score of internal breakdown, mean score of flavour, maximum score of flavour and scores of precocity, productivity and resistance. All parameters, except mean flavour, are given a weighting of one, mean fruit flavour is given a weighting of 2. Data for 160 seedlings evaluated in 1992 is presented.

Introduction

One of the most essential parts of the breeding programme is to define the main characters of the breeding goals. Then it is necessary to find reliable and rapid methods to record these characters and finally identify the genotypes which have the potential to become a new cultivar, or parent for a new generation, based on these records.

The breeding goals in the Norwegian pear breeding programme are attractive and large fruits with high quality and long shelf life. Furthermore the trees should be precocious, productive and have good resistance to biotic and abiotic factors. There is a need for new cultivars for the whole season, however mid-season cultivars with a good storage ability and early-maturing cultivars have the highest priority (Øydvin, 1980).

Most pear breeding programmes report of selection based on horticultural characters as healthy trees, vigour and productivity and quality characters as taste and shelf life (e.g. Bouma, 1988; Thibault et al., 1988). However very few reports exist on how genotypes are selected by means of the recorded parameters. In the USDA pear breeding programme Blake & Van der Zwet (1979) applied a quality index which included five parameters with different weights. Flavour was given weight 4, grit weight 3, texture weight 2 and appearance and russet weight 1 each. The scale applied was 1–9 for each of the characters.

Material and methods

In the Norwegian pear breeding programme, 20 mature fruits are sampled at random from each genotype. These fruits are weighed individually immediately after harvest, and then placed in a cold storage at −1 °C. In order to get a scale comparable to the other characters, the integer value of fruit weight divided by 20 is kept for further calculations.

After a certain period of storage, fruits are taken out of the cold storage, and the sample is evaluated for appearance. The evaluation is made by 2 persons, and a scale of 1 to 9 is applied, where 9 is the best.

Following a short period at 10 °C, the sample is divided into 4 sub-samples which are ripened at 20 °C. One sub-sample is taken out from the ripening room at each of 4 different dates. In 1992 the intervals chosen were 1, 8, 15 and 22 days.

Table 1. Percentage distribution of five fruit characters in 160 pear progenies evaluated in 1992. Scale 0–9 where 9 is best. For fruit weight: Score = int (weight/20). N = 160

Score	Internal breakdown	Appearance	Fruit size	Flavour	
				Maximum	Mean
0	3.7	0.0	0.0	3.7	13.8
1	3.1	0.8	0.0	1.3	31.3
2	22.5	5.0	7.7	5.6	25.6
3	11.9	10.8	17.8	20.6	18.8
4	19.4	19.2	26.0	32.5	10.0
5	12.5	23.3	16.0	26.9	0.6
6	10.6	22.5	14.8	8.8	0.0
7	4.4	15.0	8.9	0.6	0.0
8	7.5	3.3	3.0	0.0	0.0
9	4.4	0.0	3.6	0.0	0.0
10			1.2		
> 10			1.2		
Mean	4.6	5.0	4.9	4.2	2.3
S. dev.	2.22	1.53	2.03	1.35	1.18

At each date, the samples are evaluated for internal breakdown and flavour by two trained people. The pears are split longitudinally, and a scale of 0 to 9 is applied, where 0 is no sign of internal breakdown, and 9 is completely deteriorated. The mean value of all observations on a genotype is calculated, and then this value is subtracted from 9, in order to get the highest score for the most favourable genotypes. All sub-samples that have fruits which are rated better than 5 for internal breakdown are then evaluated for flavour at a 1 to 9 scale. In sub-samples where fruits are too deteriorated to be eaten, flavour score is set to 0. In order to avoid tasting of poor-flavour genotypes, genotypes rated below 3 in flavour are not tasted the next dates. The flavour score is then set to missing value or zero, depending on degree of internal breakdown. The maximum score for flavour in the ripening period and the mean score at all four dates are kept for further calculations.

In the present report, a brief presentation is given of the method applied at a seedling population in 1992 where a total of 160 genotypes were evaluated. The seedlings were planted in the field in 1986, and evaluations were carried out on progenies of 19 combinations represented by 1 to 20 genotypes.

Results and discussion

Fruit size

Table 1 shows that the score for fruit size had a mean of 4.9, and fitted a normal distribution even though there was a long tail towards large fruits. A total of 9 per cent of the genotypes had fruits that were larger than 160 grams. Rosati et al. (1979) found 6.9 per cent of the progenies in a three-year study of 5498 pear seedlings to exceed 150 g.

From the difference in latitude between Italy and Norway one would expect fruits to be generally larger in Italy. The present differences are probably caused by random effects due to the low number of observations in the present material.

Fruit size should reach a level of 5, corresponding to a mean fruit weight of 100 to 119 grams. However, slightly smaller fruits (level 4) could be accepted if the yield is large.

Appearance

Scores for appearance were normally distributed round a grand mean score of 5.0, with extremes of 1 and 8 (Table 1). Thibault et al. (1988) evaluated attractiveness at a scale of 1–5 in the pear breeding programme in Angers, and found a similar distribution of the different classes as in the present study. They observed a grand mean of 2.45, which is also in accordance with the present observations when the different scales are taken into account.

A score of 5 is defined as acceptable. However, a score of 4 for appearance should also be accepted, as we would assume large differences in what people like or dislike.

Flavour

Scores for maximum flavour and mean flavour are presented in Table 1. The mean score for maximum flavour was 4.2 and for mean flavour, 2.3. The mean of maximum flavour was nearly twice the score of mean flavour, and the score for maximum flavour was at the same level as the scores for appearance and fruit size. The values for both components fitted a normal distribution even if the distribution of scores of mean flavour had a bias towards zero. This is what could be expected, as the score of zero was frequently applied as fruits deteriorated.

Table 2. Percentage distribution of a quality index based on five fruit characters of 160 pear progenies evaluated in 1992. Index = Internal breakdown + appearance + fruit size + maximum flavour + 2*mean flavour. N = 160

Index	Percent of the progenies
0–4	0.0
5–9	4.3
10–14	11.1
15–19	25.6
20–24	23.1
25–29	25.6
30–34	9.4
35–39	0.9
Mean	21.4
S. dev.	6.35

Thibault et al. (1988) rated fruit quality of 9675 genotypes at a 1 to 7 scale, and found a grand mean for the progenies of 1.29. Nearly 96 per cent of the seedling genotypes were rated in class 1 and 2, which were the two poorest classes. The distribution and grand mean are comparable to the mean score of flavour in the present study even if the scales applied are not quite similar. Vondracek (1988) applied a 1–9 scale, and found average fruit taste ranging from 4.9 to 6.2 in 544 genotypes from hybrids between cultivars of *Pyrus communis*. The percentage of genotypes with score of 6 and above was high, and our level of acceptance seems to be different.

A score of 5 for maximum flavour should be attained for the genotype to be accepted. Mean flavour should also be in the upper part of the scale.

Shelf life

Scores for shelf life expressed by internal breakdown averaged 4.6 and were normally distributed (Table 1). The mean of the scores was comparable to the other quality parameters in this report.

Index for selection

The most important parameters for the marketability in pears are appearance, fruit size, shelf life and fruit quality. The most complex of these parameters are appearance and fruit quality, and in a previous proposal of a quality index in pears (Blake & Van der Zwet, 1979), both were split in sub-characters. In the present report, appearance is a joint record of russet, fruit form, colour, etc., while flavour is a joint record of texture, juiciness, firmness, grit content, sugar/acid balance, aroma content, etc.

In order to find a method to segregate seedlings based on the breeding aims, the breeder has to consider two major aspects. First he has to decide which records that should be taken to define a specific breeding aim, i.e. whether a character should be split in sub-characters or not, and how the characters should be measured. Secondly he has to give the individual characters importance relative to each other, i.e. weighing the characters.

In the present report it is suggested that flavour should be given a weight of 3, and the other characters a weight of one. Flavour is split in two sub-components, maximum and mean flavour which are both recorded in the same way, but differently calculated. Mean flavour is given a weight of 2, maximum flavour a weight of 1. By this way of calculation notes both for high, and stable quality is retained. This is an alternative approach to evaluate fruit quality compared to splitting flavour in e.g. juiciness, texture, firmness and sugar/acid balance.

The distribution of the index for quality was normal (Table 2), and genotypes with the most favourable genes will be in the upper part of the scale.

To obtain a final index, horticultural characters as precocity, productivity and resistance should be included. These parameters would be evaluated in the field at a 0–9 scale, and given a weight of 1.

The experience with this index is up till now limited, however it provides an instrument for selection of genotypes that have a good overall expression of the recorded characters. To assure that no outstanding genotypes will be discarded one should always screen the genotypes with a lower index.

References

Blake, R.C. & T. van der Zwet, 1979. The USDA pear breeding program. III. Fruit evaluation. Fruit Varieties Journal 33(4): 131–136.

Bouma, J., 1988. Breeding of new pear varieties at the fruit breeding station Techobuzice. Acta Horticulturae 224: 257–262.

Øydvin, J., 1980. Norsk vekstforedling i frukt. NLVF-utredning 110.

Rosati, P., P. Gibertoni, L. Rivalta & C. Fideghelli, 1979. Study of some characters in different pear progenies. Proc. Eucarpia Fruit Section Symposium Tree Fruit Breeding. Angers, 3–7 September 1979: 83–90.

Thibault, B., L. Hermann, A. Belouin & B. Mangin, 1988. Inheritance of some agronomical traits in pear. Acta Horticulturae 224: 199–209.
Vondracek, J., 1988. Breeding of pear cultivars with late consumption maturity. Acta Horticulturae 224: 247–255.

H. Schmidt and M. Kellerhals (Eds.), Progress in Temperate Fruit Breeding, 243.

Eradication of *Erwinia amylovora* (fire blight) from apple propagating wood by hot water treatment

Herb S. Aldwinckle & Herb L. Gustavson
Cornell University, Dept of Plant Pathology, Geneva, NY 14456, USA

Summary

Our research objective was to develop a method for eradicating viable cells of *Erwinia amylovora* from symptomless scion budsticks of apple by immersion in hot water. The effect of duration of immersion in water at different temperatures on viability of dormant buds of apple cvs. 'Idared' and 'Empire' and on survival of *E. amylovora* were determined. Approximately 50% of buds of both cvs. survived 60 min at 48° C, but all buds were killed by 30 min at 50° C. Shoot tips of greenhouse-grown potted 'Idared' and 'Empire' trees were inoculated with *E. amylovora*. After lesions had ceased expanding 1 cm segments of the stem, directly below the lesion margin and including a node, were immersed in water at 48° C for 60 min. No viable *E. amylovora* was detected by direct planting of phosphat-buffer-ground tissue on CCT medium, or by plating of Kado 523 broth that had been inoculated with grindate and incubated for 18 hours. Unimmersed tissues yielded viable *E. amylovora*. Experiments to determine the time course of killing of *E. amylovora* cells infected apple tissue at 48° C indicated that after 60 min there was $< 1 \times 10^{-6}$ chance of a single viable cell remaining. A phytosanitary protocol for *E. amylovora*-free apple scions will be discussed.

H. Schmidt and M. Kellerhals (Eds.), Progress in Temperate Fruit Breeding, 245–248.

Flower quality and fruit quality in almond: Conflicting objectives?

R. Socias i Company & A.J. Felipe
Unidad de Fruticultura, SIA-DGA, Apartado 727, 50080 Zaragoza, Spain

Key words: Almond, double kernels, fruit quality, *Prunus amygdalus*

Summary

The presence of double kernels in almond (*Prunus amygdalus*, Batsch) decreases fruit quality. This is a cultivar trait, but its percentage shows large variation depending on the sample and on the year. Significant differences in double kernel percentage have been observed in 'Guara' fruits from the first opened flowers (18.51%), the middle flowers (6.75%) or the late flowers (3.55%). This suggests that the first opened flowers, in general of higher quality, have the two ovules viable in a higher proportion giving a larger proportion of fruits with two kernels. Besides, a good nutritional status and favourable temperatures at blooming may also increase the percentage of double kernels.

Introduction

Since Williams (1965) established the concept of flower quality, much attention has been paid to the factors which may improve it, assuming that higher flower quality results in higher fruit quality and therefore in better production. The exact definition of flower quality is rather difficult and different aspects must be taken into account for different species. Almond, as the other species of the genus *Prunus*, has perfect flower with a pistil made up of a single carpel containing two ovules. It is generally assumed that one of the two ovules degenerates and stone fruits contain a single seed or kernel.

Almond is the only stone fruit whose kernels have commercial interest, although apricot kernels are also marketed to a very limited extent. Consequently, the only species where the production of double kernels resulting from fertilization and development of the two ovules has attracted attention is almond, although our experience shows that double kernels can be found in all stone fruits.

The presence of double kernels in almond is considered a negative trait, lowering fruit quality rating depending on their proportion (Kester et al. 1980). This is due to the fact that when two kernels are produced, they are deformed and make the commercial processes of cracking, size selection and peeling difficult. Our experience, as well as that of others (Kester & Asay 1975, Spiegel-Roy & Kochba 1974), shows that the percentage of double kernels is a cultivar trait, but presents large variations depending on the sample and on the year. Some physiological and climatic reasons have been pointed out as possible causes of these variations, but none has been clearly defined. Particularly, low temperatures before blooming (Egea & Burgos 1994) or at blooming time (Grasselly 1972, Rikhter 1969, Spiegel-Roy & Kochba 1974) have been mentioned as promoting higher percentages of double kernels. Optimized pollination has also been suggested to increase the percentage of double kernels (Palasciano et al. 1993).

Transmission of this trait to the offspring has received more attention because of its relevance when evaluating seedlings in a breeding program (Dicenta López-Higuera 1991, Grasselly 1972, Grasselly & Gall 1967, Kester et al. 1977, Spiegel-Roy & Kochba 1974, 1977, 1981, Vargas & Romero 1988). However, no conclusions could be drawn from different behaviours of different parent cultivars. It has only been concluded that there is a transmissible quantitative component.

These observations only allow us to conclude that the presence of double kernels is a tendency of certain cultivars. Thus our objective was to assess if the time of flower opening can affect the percentage of double kernels in almond.

Materials and methods

'Guara' is a recently released almond cultivar (Felipe & Socias i Company 1987) showing a percentage of double kernels of 5 to 20%, which can be considered as medium for almond. Three trees of the field trials, grafted on the clonal "pollizo" plum 'Montizo'-646, were partially deblossomed on March 10, 1992, when blooming had just started. As a result each tree was left with a subpopulation of flowers according to the time of flower opening, of approximately one third of the total number of flowers: one tree was left with the earlier flowers, another tree with the middle flowers and the last tree with the later opening flowers. All trees were open pollinated.

In September fruits were harvested separately from each tree, and the percentage of double kernels was determined for each sample. Percentages were compared for equality by the t test (Sokal & Rohlf 1969).

Table 1. Percentage of double kernels in each population of flowers of the almond 'Guara'.

Flower Type	N° of Fruits	% Double Kernels
First flowers	389	18,51 a
Middle flowers	311	6,75 b
Last flowers	535	3,55 c

z: Percentage separation at $P = 0.05$.

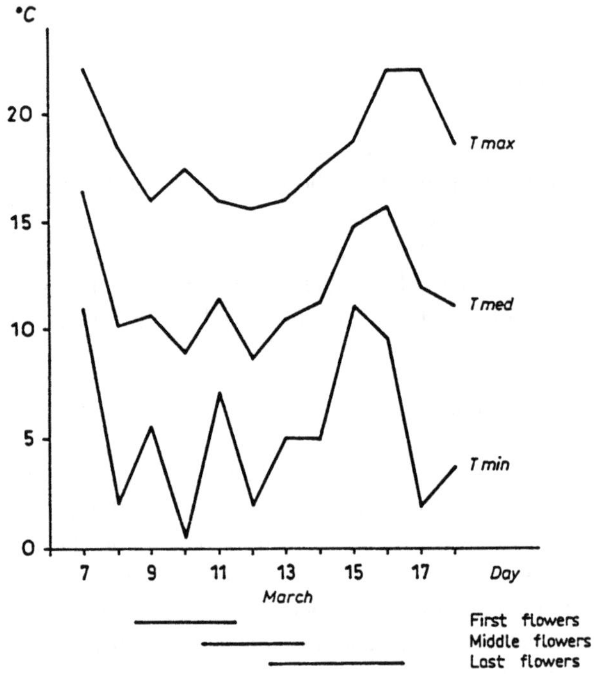

Fig. 1. Temperature evolution during blooming period of 'Guara'. Lines indicate the anthesis of the different flower populations.

Results

The percentage of double kernels was significantly different between trees (Table 1), being especially high (18.51%) for the first opening flowers, as compared to the middle flowers (6.75) and the later opening flowers (3.55). As low temperatures have been reported to induce higher percentages of double kernels, temperatures at blooming time were recorded (Fig. 1). However, their evolution shows that there were not large variations during the blooming period of 'Guara', which suggests that their effect on the three flower subpopulations was similar. This allowed us to discard the temperature effect and suggests that the differences in the percentage of double kernels are due to differences in the viability of the two ovules in the different flower subpopulations, where open pollination and fruit development happened similarly, as expected in clonal material, and therefore genetically identical.

Discussion

Blooming dynamics in almond follows a pattern whereby the first opening flowers are fertile in a higher proportion than later opening flowers (Bernad & Socias i Company 1994, Socias i Company 1983). 'Guara' shows the same behaviour (Socias i Company 1992), which can be related to flower quality. Williams (1965) observed that the most important difference between strong and weak flowers was ovule longevity and this could affect the presence or not of two kernels in almond. In apple, the two ovules of each carpel follow a parallel evolution and can produce two seeds provided proper pollination, but Williams (1965) did not refer to the number of seeds of strong or weak flowers.

The ovule is an organic structure with a limited life, which needs to be fertilized before degeneration in order to produce a seed, a requirement of primary importance in almond. Ovule degeneration has been studied in detail in 'Nonpareil' almond (Pimienta & Polito 1982), a cultivar showing only single kernels because one of the two ovules always degenerates. These authors distinguish between the two ovules. A primary one, whose embryo sac develops and is fertilized and a secondary one, abortive and degenerating in 'Nonpareil' during the progamic phase, after anthesis but before fertilization of the primary ovule. At anthesis no differences between the two ovules can be established because the embryo sac does not start development until the end of anthesis (Pimienta & Polito 1983). Only three to four days after pollination a megagametophyte is observed in one ovule. During the first stages of degeneration there was a normal translocation towards the two ovules, but this translocation decreased at the chalaza of the secondary ovule as degeneration advanced. This blockage could be due to the formation of callose deposits at the ovule integuments.

The embryo sac is already formed at anthesis in other stone fruits, such as apricot, plum, cherry and sour cherry, although the two ovules can follow a different evolution from the beginning (Cerovic & Ruzic 1992), as opposed to the observations in 'Nonpareil' (Pimienta & Polito 1983). These observations are limited to only one almond cultivar and possibly other cultivars may behave more similarly to other stone fruits and may have, as well, different kinetics of ovule degeneration. This together with the dynamics of flower opening (Socias i Company 1992) may suggest that the first opening flowers have a higher proportion of two ovules maintaining their viability, giving rise to two kernels, while the later opening flowers, sterile in a higher proportion, may maintain only a viable ovule, thus producing a single kernel.

Pimienta & Polito (1982) pointed at a blockage in nutrient translocation to the secondary ovule, whose degeneration could be due to defficient nutrition. Evidently, nutrition affects flower quality (Williams 1965) and its initiation (Carrera Morales 1975). Keeping this in mind, flower evolution in 'Guara' can be hypothesized as following the sequence of the first differentiated buds being initiated when the tree is in a better nutritional status, thus being of better quality, evolving ahead of other buds, blooming earlier and having more nutritional reserves available to maintain the viability of the two ovules. As differentiation proceeds, reserves are depleted, buds are of lower quality, evolving after the other buds, blooming later and having a single viable ovule in a higher proportion. This evolution would account for the higher percentages of double kernels in the first opening flowers.

This hypothesis may explain why higher percentages of double kernels have been reported in colder years. Temperatures affect pollen tube growth rate (Socias i Company et al. 1976), thus at low temperatures pollen tubes take more time to reach the ovule. On the other hand, the ovules also follow their evolution, and high temperatures shorten their viability, like in sweet and sour cherry (Cerovic & Ruzic 1992, Potsweiler et al. 1985). Low temperatures delay both the penetration of the pollen tubes into the ovary and ovule degeneration, but the extent of these delays is not known. Differential delays could explain the effect of low temperatures, which might prolong the effective pollination period by retarding pollen tube growth but even more so the rate of ovule degeneration, thus increasing the possibility of fertilization of the two ovules before the degeneration of the secondary one, and raising the proportion of double kernels. References to low temperatures (Grasselly 1972, Spiegel-Roy & Kochba 1974) actually imply only slightly lower temperatures than normal rather than cold weather, which would negatively affect fruit set. As in our work the three populations of flowers followed the same evolution during the winter, differences cannot either be attributed to temperatures before blooming, as suggested by Egea & Burgos (1993) when comparing results from different years.

The nutritional status of the tree may also affect the proportion of double kernels by increasing the proportion of high quality flowers, also distinguished by ovules with longer viability (Williams 1965), and consequently of flowers with two ovules, both of them with a higher probability of being fertilized.

As a conclusion we can state that both the presence of double kernels in almond and their heritability are complex traits needing further approaches for their elucidation, but our observations and the reviewed results suggest that each almond cultivar possesses a particular ability to produce two viable ovules. This ability being a quantitative trait, it can be partially inherited by the offspring, but its expression can be modified by the nutritional status of the tree, which may or may not result in the differentiation of two viable ovules, by the temperatures before or at blooming time, which shorten or lengthen the period of ovule viability, and by the flowers which would produce the crop, as frosts and other weather conditions can determine fruit setting as a function of the time of flower opening.

Acknowledgements

Research conducted under projects INIA 9567 and CICYT AGF 92–0294.

References

Bernad, D. & R. Socias i Company, 1994. Characterization of some self-compatible almonds. II. Bloom evolution and flower quality (submitted to the HortScience).

Carrera Morales, M., 1975. Estudio sobre el proceso de inducción floral en variedades de peral. Ph.D. Thesis, Univ. Madrid.

Cerovic, R. & D. Ruzic, 1992. Senescence of ovules at different temperatures and their effect on the behaviour of pollen tubes in sour cherry. Scientia Hort. 51: 321–327.

Dicenta López-Higuera, F., 1991. Mejora genética del almendro (*Prunus dulcis* Miller) por cruzamientos intervarietales: herencia de caracteres y selección. Ph.D. Thesis, Univ. Murcia.

Egea, J. & L. Burgos, 1993. Clima and double kernel in almond. I Int. Congr. Almond ISHS, Agrigento, May 17–19, 1993.

Felipe, A.J. & R. Socias i Company, 1987. 'Aylés', 'Guara', and 'Moncayo' almonds. HortScience 22: 961–962.

Grasselly, C., 1972. L'amandier; caractères morphologiques et physiologiques des variétés, modalités de leurs transmissions chez les hybrides de première génération. Ph.D. Thesis, Univ. Bordeaux.

Grasselly, C. & H. Gall, 1967. Étude sur la possibilité de combinaison de quelques caractères agronomiques chez l'amandier. Ann. Amélior. Plant. 17: 83–91.

Kester, D.E. & R. Asay, 1975. Almonds. In: J. Janick & J.N. Moore (Eds) Advances in Fruit Breeding, pp. 387–419. Purdue Univ. Press, West Lafayette, Indiana.

Kester, D.E., P.E. Hansche, V. Beres & R.N. Asay, 1977. Variance components and heritability of nut and kernel traits in almond. J. Amer. Soc. Hort. Sci. 102: 264–266.

Kester, D.E., W.C. Micke, D. Rough, D. Morrison & R. Curtis, 1980. Almond variety evaluation. Calif. Agric. 34(10): 4–7.

Palasciano, M., A. Godini & L. de Palma, 1993. Optimized self-pollination and production of double kernel in almond. I Int. Congr. Almond ISHS, Agrigento, 17–19 May, 1993.

Pimienta, E. & V.S. Polito, 1982. Ovule abortion in 'Nonpareil' almond (*Prunus dulcis* (Mill.) D.A. Webb). Amer. J. Bot. 69: 913–920.

Pimienta, E. & V.S. Polito, 1983. Embryo sac development in almond (*Prunus dulcis* (Mill.) D.A. Webb) as affected by cross-, self- and non- pollination. Ann. Bot. 51: 469–479.

Potsweiler, K., R. Stösser & S.F. Anvari, 1985. The effect of different temperatures on the viability of ovules in cherries. Scientia Hort. 25: 235–239.

Rikhter, A.A., 1969. Ways and methods in almond breeding. Tr. Gos. Nikit. Bot. Sada 43: 81–94 (in Russian).

Socias i Company, R., 1983. Flower sterility in almond. Acta Hort. 139: 69–74.

Socias i Company, R., 1992. Dinámica de la floración y cuajado en el almendro 'Guara'. Inf. Técn. Econ. Agrar. 88V: 193–199.

Socias i Company, R., D.E. Kester & M.V. Bradley, 1976. Effects of temperature and genotype on pollen tube growth in some self-compatible and self-incompatible almond cultivars. J. Amer. Soc. Hort. Sci. 101: 490–493.

Sokal, R.R. & F.J. Rohlf, 1969. Biometry. W.H. Freeman, San Francisco.

Spiegel-Roy, P. & J. Kochba, 1974. The inheritance of bitter and double kernel characters in the almond (*Prunus amygdalus* Batsch). Z. Pflanzenzucht. 71: 319–329.

Spiegel-Roy, P. & J. Kochba, 1977. Inheritance of some kernel characters in the almond. III Coll. GREMPA, 3–7 Oct. 1977, Bari, pp. 117–129.

Spiegel-Roy, P. & J. Kochba, 1981. Inheritance of nut and kernel traits in almond (*Prunus amygdalus* Batsch). Euphytica 30: 167–174.

Vargas, F.J. & M.A. Romero, 1988. Comparación entre descendencias de cruzamientos intervarietales de almendro en relación con la época de floración y la calidad del fruto. VII Coll. GREMPA, Rap. EUR 11557: 59–72.

Williams, R.R., 1965. The effect of summer nitrogen application on the quality of apple blossom. J. Hort. Sci. 40: 31–41.

H. Schmidt and M. Kellerhals (Eds.), Progress in Temperate Fruit Breeding, 249–254.

Sensory evaluation of six strawberry cultivars after machine harvest

Shahrokh Khanizadeh[1,2], Jacinthe Fortin[3], Michel J. Lareau[1] & Deborah J. Buszard[2]
[1] *Agriculture Canada, Research Station, 430 Boul. Gouin, St-Jean-sur-Richelieu, Quebec, Canada J3B 3E6;* [2] *Department of Plant Science, McGill University, Ste-Anne-de-Bellevue, Quebec, Canada H9X 3V9; and* [3] *Food Research and Development Centre, Agriculture Canada, 3600, Casavant Ouest, Saint-Hyacinthe, Quebec, J2S 8E3, Canada*

Key words: Fragaria × *ananassa* Duch., flavour, texture, appearance, processing, breeding

Summary

In 1991 and 1992 four standard cultivars 'Kent', 'Glooscap', 'Bounty' and 'Midway' and the recent introductions 'Chambly' and 'Oka' were evaluated for appearance, texture and flavour attributes. 'Bounty' was the only cultivar which ranked high for sweetness and high for strawberry, honey, and melon flavors, whereas 'Oka' ranked lowest. Sweetness was not significantly different among 'Chambly', 'Glooscap', 'Kent', and 'Midway'. 'Glooscap', 'Midway', 'Kent', and 'Oka' were more acidic than 'Bounty' and 'Chambly'. Chambly' and 'Kent' had pronounced raised achenes whereas those on 'Oka' fruit were depressed which makes this cultivar unsuitable for mechanical harvesting. No pithy core was observed in 'Chambly' and 'Glooscap'. 'Chambly', 'Glooscap', 'Bounty' and 'Kent' had bright skin while 'Midway' and 'Oka' had a dull appearance. 'Oka' and 'Midway' appeared less seedy than the others. Striation and white rings were observed in all cultivars, however, these were less conspicuous in the cultivars 'Oka' and 'Midway.

Introduction

There have been many attempts to develop a mechanical strawberry harvester to pick for processing (Booster 1980, Hansen & Ledebuhr 1980). Researchers have evaluated the feasibility (Lauro & Hergert 1987) and cost of mechanically harvesting strawberries (Seecharan et al. 1988). Several harvesters have been developed and some have been commercialized (Hansen & Ledebuhr 1980, Morris et al. 1978). Small berries are better suited to machine harvesting than large berries which bruise more easily (Denisen & Buchele 1967).

Strawberry breeders are faced with very specific breeding objective when selecting for suitability for mechanical harvesting. These are: 1) brittle pedicels; 2) easy decapping, 3) concentrated ripening, 4) appearance (skin colour and brightness, seediness, achene position, achene colour, white tip, white top, shape, neck size, calyx position, flesh colour, core colour, flesh striations, flesh white ring and pithy core), 5) texture and flavour attributes (firmness, juiciness, acidity, bitterness, astringency, sweetness, melon, honey, and typical strawberry flavour). These fruit characteristics (skin and flesh colour, flavour, shape, seediness, firmness, etc.) vary considerably among cultivars (Morris et al. 1978, 1985, Gallander & Hill 1982, Khanizadeh et al. 1990, 1991, 1992a,b,c).

Many other characteristics have been evaluated including chemical constituents of fresh fruits, freezing performance and sensory characteristics of fresh or thawed fruits. The quality of frozen strawberries depends on many field and processing factors (Morris & Cawthon 1979, Sistrunk 1963). For example, climate and soil greatly influence firmness (Cochran & Webster 1931, Joslyn & Hohl 1948, Mundt 1950, Sistrunk et al. 1960). Cochran & Webster (1931) and Mundt (1950) reported that high humidity and fertilizer levels decrease soluble solids and firmness. Colour, flavour, and texture of thawed frozen strawberries are directly related to the pattern of handling, processing, and storage of the berries from the field to the processor (Sistrunk & Moore 1967).

Variable	Line Scale	
Flavor	|-----------------15cm-----------------|	
Strawberry flavour	none	strong
Sweetness	none	very
Astringency	none	strong
Honey flavour	none	strong
Melon flavour	none	strong
Bitterness	none	very
Acidity	none	very
Texture	|-----------------15cm-----------------|	
Juiciness	none	very
Firmness	soft	very firm
Appearance	|-----------------15cm-----------------|	
Skin colour	light red	dark red
Brightness	none	very
Seediness	not seedy	very seedy
Flesh colour	light red	dark red

Fig. 1. Line Scale used to evaluate flavour and texture; skin colour, brightness, seediness, and flesh colour.

The present study was designed to evaluate the sensory attributes of two recently released cultivars 'Chambly' and 'Oka' (Khanizadeh et al. 1990, 1992c) in comparison to four standard cultivars for processing as well as fresh market use. The information collected should also serve to better guide our breeding program.

Materials and methods

Climate and location

The experiments were conducted in a sandy soil at Lavaltrie, located 35 km northeast of Montreal, Quebec, approximately 45° north latitude and at 46 m altitude with continental climate and large variations in temperatures between winter (−35 °C) and summer (+30 °C).

In 1990 four standard cultivars 'Kent', 'Glooscap', 'Bounty' and 'Midway' and the recent introductions 'Chambly' and 'Oka' were evaluated in a randomized complete block design with four replicates. The field was prepared and leveled to present the fruit to the harvester on a flat plane. Plants were set 30 cm apart in single rows in 50 m long plots, spaced 120 cm apart and watered immediately with 250 ml of a fertilizer solution (10–52–10) per plant. Flowers were removed twice during the season and all runners kept and spaced within a 50 cm wide matted row. In November the plantation was mulched with 6–7 cm of straw for winter protection. In early spring of each year, the rows were rolled to flatten the ground surface. Fruits were harvested between mid June and late July.

Description of harvester

The strawberry harvester used in our trials was a tractor-drawn prototype (Bragg, Nova Scotia) equipped with a comb-brush picking mechanism, conveyers and fans to remove debris (Swinkels & Murray 1991). The comb-brush bar at the front of the machine strips the fruit from the plant, then separates it from the leaves and other foreign materials by air flow before finally being conveyed to plastic containers at the rear of the harvester. Plots were harvested at a machine: ground speed of 1.6 km h^{-1}.

Sensory evaluation

Approximately 20 kg of fruit of each cultivar was collected at each harvest to evaluate appearance, (skin colour and brightness, seediness, achene position, achene colour, white tip, white top, shape, neck size, calyx position, flesh colour, core colour, flesh striations, flesh white ring and pithy core) texture and flavour attributes (firmness, juiciness, acidity, bitterness, sweetness, melon and honey flavours, astringency and typical strawberry flavour). Specific measuring systems were used for each attribute. Flavour and texture attributes, skin colour, brightness, seediness, and flesh colour were evaluated using the line scale technique (Meilgaard et al. 1990) (Fig. 1). Fruit shape, achene colour, presence of the striations, and white ring were evaluated using the classification method (Meilgaard et al. 1990) (Fig. 2). Achene position, white tip, white top, neck size, calyx position, and presence of pithy core were evaluated using the ranking method (Meilgaard et al. 1990) (Fig. 3). All judges attended a training session before the evaluation.

Data collected at each harvest were combined. Line scale data were analyzed using an F test. Classification, and rank data were analyzed using χ^2 and Friedman's tests. The multiple comparisons of total ranks were done using $\alpha = 0.15$ as the experimentwise error rate for a conservative approach. All figures given in the tables are based on sound fruit.

Variable	Classifications
Fruit shape	Oblate, Globose, Globose-conic, Conic, Long-conic, Necked, Long-wedge, Short-wedge
Achene colour	yellow, green, light brown, brown, black
Core colour	matching, lighter, darker
Striation	present, absent
White ring	present, absent

Fig. 2. Classification method used to evaluate fruit shape, achene colour, presence of the striations, and white ring.

Variable	Rank values				
	1	2	3	4	5
White tip	absent				1/2 berry
White top	absent				1/2 berry
Pithy core	absent				present
Neck Size	See (Khanizadeh *et al.* 1990, 1991)				
Achene position	See (Khanizadeh *et al.* 1990, 1991)				
Calyx form	See (Khanizadeh *et al.* 1990, 1991)				

Fig. 3. Ranking used to evaluate the achene position, white tip, white top, neck size, calyx position, and presence of pithy core.

Results and discussion

Flavour and texture

'Bounty' was the only cultivar which ranked high for sweetness, strawberry, honey, and melon flavours, whereas, 'Oka' and 'Midway' ranked lowest. 'Chambly', 'Glooscap', and 'Midway' were considered more bitter than 'Bounty'. Sweetness was not significantly different among 'Chambly', 'Glooscap', 'Kent', and 'Midway'. Strawberry flavour was not significantly different among 'Chambly', 'Glooscap', and 'Kent'. 'Bounty' was the least astringent followed by 'Glooscap', 'Kent', 'Chambly'', 'Oka', and 'Midway' (Table 1). 'Chambly' and 'Glooscap' had similar honey and melon flavour. 'Glooscap', 'Midway', 'Kent', and 'Oka' were more acidic than 'Bounty' and 'Chambly' (Table 1). 'Bounty', 'Chambly', and 'Midway' were considered juicier than 'Kent' or 'Oka'. 'Glooscap' ranked the least juicy cultivar. 'Chambly' and 'Kent' were most firm followed by 'Glooscap', 'Bounty', and 'Oka', whereas 'Midway' was softest (Table 1).

Other fruit characteristics

Achene position, white tip, white top, raised neck, reflexed calyx and pithy core are important characteristics of strawberry cultivars used for processing. Raised achenes protect the strawberry skin during harvesting and dehulling. Low proportion of white tip, or white top improve the quality of processed berries. Raised neck and reflex calyx minimize cutting waste (Khanizadeh et al. 1990, 1992b). Pithy core not only reduces the total weight by volume, but also influences berry shape after thawing. 'Chambly' and 'Kent' had pronounced raised achenes whereas those for 'Oka' were depressed which makes this cultivar not suitable for mechanical harvesting. 'Kent' and 'Oka' had lighter tip and top compared to others cultivars which is a negative characteristic. 'Chambly' had the most pronounced neck and reflexed calyx followed by 'Glooscap'. Most of the other cultivars had either no neck or a recessed hull with calyx covering the fruit (Table 2). 'Bounty', 'Midway' and 'Oka' all exhibited pithy core but non was observed in 'Chambly' or 'Glooscap' (Table 2).

Skin, flesh, achene and core colour, brightness, seediness, striation, white ring and shape are other important quality considerations of strawberry cultivars for processing or fresh market. 'Bounty' and 'Oka' had dark red coloured skin followed by 'Chambly' and 'Glooscap'. 'Kent' and 'Midway' had light red coloured skin (Table 3). 'Chambly' and 'Glooscap' had bright skin while 'Midway' and 'Oka' had a dull skin. 'Oka' appeared less seedy than the others, due the position of the achenes which are embedded in the flesh. 'Kent' and 'Midway' were the only cultivars with light coloured flesh (Table 3). All the cultivars had greenish achenes but those on 'Kent' and 'Bounty' were darker (Fig. 4A). 'Bounty', 'Kent', 'Midway' and 'Oka' all had light cores while 'Glooscap' and 'Chambly' cores matched the flesh colour. Striation and white

Table 1. Comparison of the flavour and texture attributes of six strawberry cultivars[1].

	Flavour[2]							Texture[2]	
Cultivar	Strawberry	Sweetness	Astringency	Honey	Melon	Bitterness	Acidity	Juiciness	Firmness
Bounty	10.4 a	13.2 a	5.4 b	11.4 a	10.6 a	6.2 c	4.5 c	8.7 abc	7.4 b
Chambly	7.1 bc	8.2 bc	8.4 a	8.7 bc	8.5 ab	8.8 a	6.4 bc	9.5 ab	8.8 a
Glooscap	8.2 bc	8.6 b	6.9 ab	9.1 b	8.0 b	8.4 ab	7.5 ab	7.0 c	7.4 b
Kent	9.0 ab	6.9 bc	7.7 a	7.0 cd	7.4 b	6.6 bc	8.9 a	4.8 d	8.4 a
Midway	6.5 c	6.9 bc	8.9 a	6.8 cd	7.6 b	8.5 ab	8.7 a	10.1 a	4.2 c
Oka	6.8 c	6.4 c	8.5 a	6.2 d	6.6 b	8.0 abc	8.9 a	8.0 bc	6.7 b
LSD[3]	2.1	2.0	2.1	2.1	2.1	2.1	2.0	2.0	0.9

[1] Each value is the total rank of at least 40 determinations.
[2] See Fig. 1 for detail of scaling.
[3] Least significant difference at 5% level.

Table 2. Cumulative scores for achene position, white tip, white top, neck size, calyx position, and pithy core of six strawberry cultivars[1].

Cultivar	Achene Position[2]	White Tip[2]	White Top[2]	Neck Size[2]	Calyx Form[2]	Pithy Core[2]
Bounty	3.2	2.8	3.2	2.0	3.3	4.2
Chambly	4.4	3.2	3.2	5.0	3.9	2.4
Glooscap	3.2	3.3	3.1	3.7	3.5	3.1
Kent	4.2	4.3	4.0	3.2	3.7	3.6
Midway	3.4	3.2	3.5	3.7	2.7	4.0
Oka	1.3	4.3	4.0	3.0	2.8	3.7
Chi-square	219.7	83.2	156.3	141.0	30.3	112.0
Critical Range[3]	0.45	0.45	0.45	0.45	0.45	0.45

[1] Each value is the average rank of at least 40 determinations.
[2] See Fig. 3 for detail of scaling.
[3] Critical range for multiple comparison within columns following the Friedman's analysis of variance by ranks at $\alpha = 0.15$ experimentwise error rate.

rings were observed in all cultivars, however, these were less conspicuous in 'Oka' and 'Midway'. Fruit shape was evaluated throughout the harvest. 'Bounty' fruit were short-wedge shape whereas 'Chambly' fruits characterized as having a necked. 'Glooscap' had short-wedged shaped fruit and 'Kent' had globose-conic or conic shape fruit. 'Midway' produced fruits of different shapes during the season ranging from globose-conic to short-wedge. A high percentages of 'Oka' fruit were short-wedge (Fig. 4B).

'Kent', Glooscap', Chambly' and 'Oka' lacked strawberry and honey flavour while ranking highly in bitterness and astrigency. They were not as sweet as the late ripening cultivar 'Bounty'. Of all the cultivars tested only 'Chambly' and 'Kent' were considered firm. None of the cultivars tested were found to have ideal flavour and texture attributes and further breeding is required to produce improved cultivars suitable for machine harvesting and processing.

Table 3. Comparison of the apearance of six strawberry cultivars[1].

Cultivar	Skin Colour[2]	Brightness[2]	Seediness[2]	Flesh Colour[2]
Bounty	15.0 a	11.5 a	11.0 ab	15.0 a
Chambly	14.1 bc	12.1 a	11.8 a	15.0 a
Glooscap	14.0 b	12.2 a	0.9 cd	15.0 a
Kent	12.3 d	11.7 a	10.1 bc	13.4 b
Midway	12.8 cd	10.2 b	0.9 d	13.0 b
Oka	15.0 a	0.8 c	0.6 e	15.0 a
LSD[3]	1.3	1.1	0.9	2.2

[1] Each value is the total rank of at least 40 determinations.
[2] See Fig. 1 for detail of scaling.
[3] Least significant difference at 5% level.

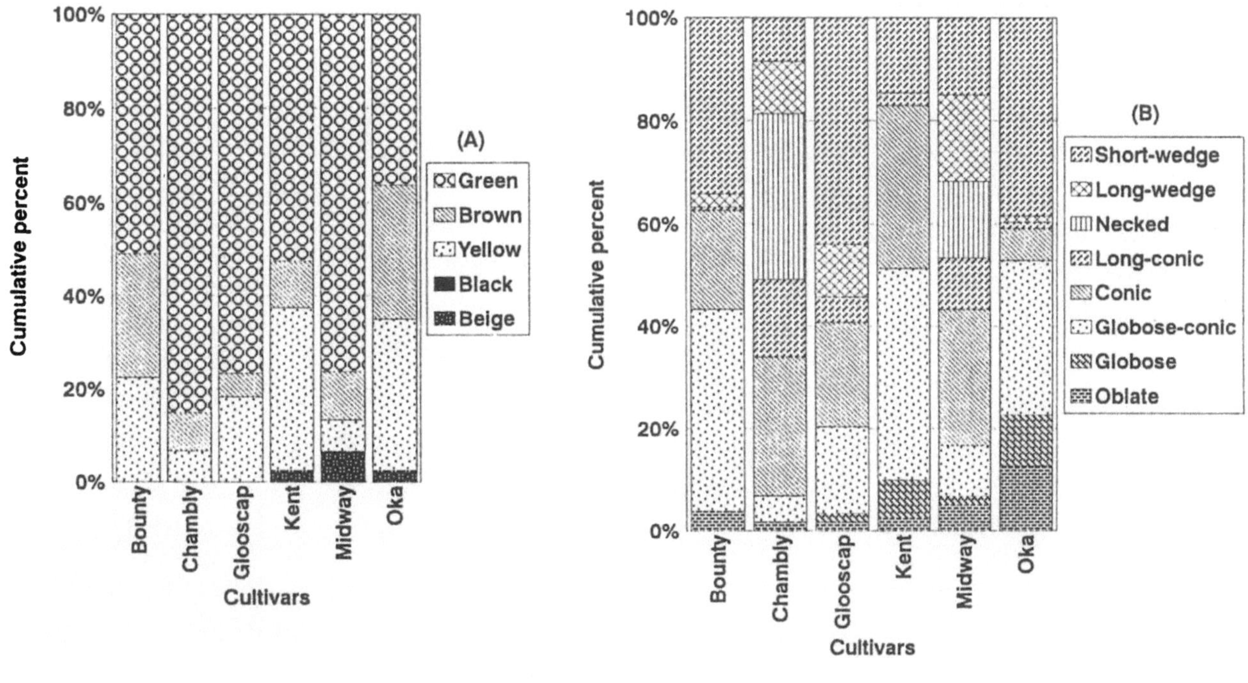

Fig. 4. Achene colour (A) and fruit shape (B) of six strawberry cultivars during the harvest period.

References

Booster, D., 1980. Mechanical harvesting of strawberries in Oregon. In: L.W. Martin & J.R. Morris (Eds) Strawberry mechanization, pp. 70–91. Agric. Exp. Stan., Oregon State University, Bull. No. 645.

Cochran, G.W. & J.E. Webster, 1931. The effect of fertilizers on the handling qualities and chemical analysis of strawberries and tomatoes. Proc. Amer. Soc. Hort. Sci. 28: 236–243.

Denisen, E.L. & W.F. Buchele, 1967. Mechanical harvesting of strawberries. Proc. Amer. Soc. Hort. Sci. 91: 267–273.

Gallander, J.F. & R.G. Hill, Jr., 1982. Effect of variety and harvest date on the quality of frozen strawberries. The Ohio State University, Wooster, Ohio, Research Circular 271, August 1982, pp. 43–46.

Hansen, C.M. & R.L. Ledebuhr, 1980. Mechanical harvesting of strawberries in Michigan. In: L.W. Martin & J.R. Morris (Eds) Strawberry mechanization, pp. 54–62. Agric. Exp. Stan., Oregon State University, Bull. No. 645.

Khanizadeh, S., D. Buszard, M.J. Lareau & D. Bagnara, 1990. 'Chambly' strawberry. HortScience 25(8): 984–985.

Khanizadeh, S., D. Buszard, M.J. Lareau & R. Pelletier, 1991. Evaluation of strawberry cultivars with different degrees of resistance to Red Stele. Fruit Var. J. 45(1): 12–17.

Khanizadeh, S., M.J. Lareau, D. Buszard & D. Bagnara, 1992c. 'Oka' Strawberry. HortScience 27(4): 374–375.

Khanizadeh, S., M.J. Lareau & D. Buszard, 1992b. Evaluation of advanced strawberry selections in Quebec. Fruit Var. J. 46(1): 53–57.

Khanizadeh, S., D. Buszard & M.J. Lareau, 1992a. Strawberry cultivar evaluation in Québec. J. Small Fruit Viticulture 1(2): 23–36.

Lauro, E.M. & G.B. Hergert, 1987. Evaluating the Technical and economic feasibility of machine harvesting strawberries. Paper No. 87–206. Presented at Can. Soc. Agr. Eng., May 18–22, 1987, MTL, QC.

Meilgaard, M., G.V. Civille & B.T. Carr, 1990. Sensory Evaluation Techniques. CRC Press, Boca Raton, Florida.

Morris, J.R. & D.L. Cawthon, 1979. Post-harvest quality of machine-harvested strawberries. J. Amer. Soc. Hort. Sci. 104(1): 138–141.

Morris, J.R., A.A. Kattan, G.S. Nelson & D.L. Cawthon, 1978. Developing a mechanized system for production, harvesting, and handling of strawberries. HortScience 13(4): 413–422.

Morris, I.R., W.A. Sistrunk, C.A. Sims, G.L. Main & E.J. Wehunt, 1985. Effects of cultivars, postharvest storage, preprocessing dip treatments and style of pack on the processing quality of strawberries. J. Amer. Soc. Hort. Sci. 110(2): 172–177.

Mundt, J.O., 1950. Microbiology of strawberries during harvest. Quick Frozen Foods 12(9): 53–56.

Seecharan, R., M. Colwell, and G. Hergert. 1988. Estimated Casts of Mechanical Strawberry harvesting. Can. J. Agri. Eng. 30(2): 221–226.

Sistrunk, W.A., 1963. Field conditions and processing practices relating to frozen strawberry quality. Proc. Amer. Soc. Hort. Sci. 83: 440–446.

Sistrunk, A.W. & J.N. Moore, 1967. Assessment of strawberry quality-fresh and frozen. Food Technol. 21: 449–453.

Sistrunk, W.A., R.F. Cain, E.K. Vaughan & H.B. Lagerstedt, 1960. Factors condtributing to the breakdown of frozen sliced strawberries. Food Technol. 14: 640–644.

Swinkels, P.M. & R.A. Murray, 1991. Development of the Bragg strawberry harvester. In: A. Dale & J.J. Luby (Eds) Strawberry in 21th Century, pp. 266–267. Timber, press, Portland, Oregon.

CHAPTER 5

Molecular markers

H. Schmidt and M. Kellerhals (Eds.), Progress in Temperate Fruit Breeding, 257–262.

DNA fingerprinting – A useful tool in fruit breeding

Hilde Nybom
Balsgård – Department of Horticultural Plant Breeding, Swedish University of Agricultural Sciences, Fjälkestadsvägen 123-1, S-29194 Kristianstad, Sweden

Key words: Cultivar identification, genetic relatedness, genome mapping, minisatellite DNA, RAPD, pedigree analysis

Summary

Identification of cultivars and selections, pedigree analysis, estimation of genetic relatedness and genome mapping – all of these objects are achieved with various methods of DNA fingerprinting which can distinguish even closely related genotypes. Initial results were obtained by a variation of the well-known RFLP-technique (Restriction Fragment Length Polymorphism): hybridization of restriction enzyme-digested DNA samples with hypervariable minisatellite DNA probes. Technical improvements, including a number of additional DNA probes and various DNA fragment detection methods, have since been published. With the recent advent of the PCR technique (Polymerase Chain Reaction) and PCR-based methods like RAPD (Random Amplified Polymorphic DNA), DNA fingerprinting has widened its applicability and attracted a growing number of scientists.

Introduction

In the first plant DNA fingerprint paper, Ryskov et al. (1988) could distinguish two barley cultivars following hybridization of restriction enzyme-digested DNA samples to a minisatellite DNA probe derived from the bacteriophage M13. According to this method, the digested DNA fragments are separated on an agarose gel and transferred to a filter through Southern blotting as in common RFLP analysis. The radioactively labelled M13 probe then hybridizes to homologous minisatellite DNA regions on the filter. These minisatellite DNA regions consist of a number of tandem repeated core sequences. Thus the hybridization and subsequent autoradiography results in a DNA fragment pattern where fragment sizes are proportional to the number of tandem repeats in each fragment. Somatic stability of DNA fragment patterns obtained with this method was reported in another paper (Rogstad et al., 1988). In a third paper appearing that year, Dallas (1988) was able to distinguish rice cultivars by hybridization to a probe derived from human DNA, i.e. the 33.6 minisatellite. Dallas also performed a genetic analysis of DNA fragment inheritance with F2 plants from a mass crossing between two rice cultivars. Fragments found only in one of the two grandparents were identified, and each F2 offspring was scored for presence or absence of these fragments. All fragments exhibited in the offspring could also be found in one or both grandparental genotypes except in one case.

Hybridization of genomic DNA samples to artificially derived oligonucleotide probes like $(GATA)_4$ have similarly yielded individual-specific DNA fragment patterns (Weising et al., 1991). The level of variation detected is dependent on which particular probe is used (Weising et al., 1992). Thus, a careful matching between probe and material enables us to focus our research at suitable levels of genetic variability.

The introduction of PCR-based methods constituted a new milestone in the field of DNA fingerprinting. DNA fragments are enzymatically amplified between pairs of DNA primers that hybridize to the sample DNA. These primers can be designed from known nucleotide sequences in the DNA sample, thereby targeting specific genes. However, primers with an arbitrary nucleotide sequence will also yield a number of DNA amplification products. Separation of these with electrophoresis and staining with ethidium bromide commonly result in DNA fragment patterns that are polymorphic between genotypes. The most widely applied PCR-based method for DNA fingerprinting is RAPD (Random Amplified Polymorphic DNA), which makes use of commercially available primers with 10 base pairs (Williams et al., 1990). Due mainly to the rapidity with which large numbers of samples can be processed, the PCR-based methods have become increasingly popular. Selected primers may yield genetic polymorphisms of the same order of magnitude as those commonly observed following hybridization of restriction enzyme-digested DNA samples to minisatellite and oligonucleotide probes.

Cultivar identification and estimation of genetic relatedness

Cultivar identification can be achieved very accurately from DNA fingerprint data, especially in materials characterized by high levels of genetic variation between cultivars and no variation within. Such examples are typically found in vegetatively propagated cultivars derived from cross-pollinating species like most fruit and berry crops. Even closely related cultivars are usually distinguished with the DNA fingerprint methods. The chance probability that two different genotypes exhibit the same DNA fingerprint pattern is usually considerably less than 10^{-3} with minisatellite DNA hybridization (Nybom, 1991). With oligonucleotide hybridization, different levels of genetic variation may be encountered depending on the probe. The sensitivity of PCR-based methods is quite variable, with some primers resulting in monomorphic DNA fragment patterns and others in relatively high levels of variation. Usually several primers are used consecutively in each investigation to obtain sufficient levels of fragment pattern variability.

Estimation of relatedness between cultivars with DNA fingerprinting is more complicated. Whereas RFLPs generally are codominantly inherited, the complex DNA fragment patterns obtained with multi-locus fingerprint probes usually makes it impossible to identify allelic fragments. Therefore we can seldom determine whether a given fragment occurs in a homozygotic or a heterozygotic state. However, by analysis of carefully chosen populations, like an F2 derived from two inbred lines, valuable fragment segregation data can nevertheless be obtained. The DNA fragments are usually found in one, or sometimes both of the parental genotypes in this type of analysis. Unfortunately most studies carried out as yet comprise very low numbers of offspring. However, Mendelian inheritance has been reported for almost all fragments analysed whereas linkage between fragments occurs relatively seldom (Dallas, 1988; Nybom & Schaal, 1990; Lavi et al., 1991; Matsuyama et al., 1992; Sharon et al., 1992).

RAPD markers are usually dominantly inherited, and allelic relationships between pairs of fragments have been reported only in a few cases. Efforts to relate staining intensity of the fragments with homozygosity vs. heterozygosity have usually been unsuccessful. Moreover, non-Mendelian segregation ratios have frequently been reported. One reason to this may be that fragments amplified by the PCR-reaction probably represent the most successful products among many more competing candidates for amplification. Overall genetic background may thus determine which candidates are actually amplified, resulting in the occurrence of 'epistatic effects' as discussed (but not actually encountered) by Heun & Helentjaris (1993).

Vegetatively propagated material. Cultivar identification has been achieved in apple with hybridization of DNA samples to the M13 probe (Nybom et al., 1990, 1992) as well as with PCR-based methods (Nybom et al., 1992; Harada et al., 1993; Koller et al., 1993; Mulcahy et al., 1993; Dunemann, 1994). Also pear cultivars can be distinguished with DNA fingerprinting (Teramoto et al., 1993). A close association was found between genetic relatedness and fragment profile similarities in a study of ornamental apple trees and their offspring (Nybom, 1990a).

Blackberry and raspberry cultivars have been distinguished with M13 hybridization, and their genetic relatedness found to be correlated with fragment profile similarities (Nybom et al., 1989; Nybom & Hall, 1991; Hoepfner et al., 1993). In a study of red and purple raspberry cultivars, DNA samples were instead hybridized to a 22-mer probe (Parent & Pagé, 1992). All cultivars investigated were again distinguished, whereas no intra-cultivar variation was encountered when analysing the effects of age of the raspberry plantation, developmental stage during the growing season, or position of the sampled leaf on the stem. These same cultivars were subsequently distinguished with RAPD markers (Parent et al., 1993). Also in this study, an investigation of possible effects of environmental factors on DNA fragment patterns turned out negative.

Cultivars have been successfully distinguished with DNA fingerprinting also in e.g. citrus fruits (Matsuyama et al., 1992, 1993; Komatsu et al., 1993), grapes

(Striem et al., 1990; Büscher et al., 1993; Collins & Symons, 1993; Gogorcena et al., 1993; Jeanjaques et al., 1993), avocado (Lavi et al., 1991), papaya (Sharon et al., 1992; Stiles et al., 1993) and banana (Kaemmer et al., 1992). A recent study on cultivar identification and genetic relatedness in blueberry cultivars makes use of DNA fragments derived by RAPD (Aruna et al., 1993).

Sports and somaclonal variation. Somatic mutations affecting plant structure and productivity give rise to so called 'sports'. Most of these deviate from the original variety only in minor characteristics and may thus be very difficult to distinguish. Minisatellite analysis was applied to a set of 15 different sports of Red Delicious apples (Nybom, 1990b). Qualitative differences in fragment profiles were not found but some minor differences in relative fragment intensity were noted, which perhaps can be explained by the chimaeric nature of many of these Red Delicious mutations. Similarly, no variation was found among sports of several other apple varieties using the RAPD method (Mulcahy et al., 1993). The authors estimated that only 0.002% of the genome was actually investigated, and suggest that attempts to obtain markers for point mutations should employ an approach which does not use single trees. Instead sibling populations segregating for the mutation of interest could be analyzed with e.g. the bulked segregation analysis, as described by Michelmore et al. (1991). Nevertheless, Kaemmer et al. (1992) were able to distinguish a mutated banana plant using RAPD markers. A 10-mer primer, either by itself or in combination with another primer, recognised this induced mutant from its parental strain. Similarly Yang & Schmidt (1994) report than one RAPD primer, out of the 40 screened, was useful in differentiating between an induced mutant in sweet cherry and the parental strain.

Tissue-culture techniques may result in a proportion of regenerants that differ from the parental type, a phenomenon called 'somaclonal variation'. However, the inability to document such variation with minisatellite DNA hybridization has been reported e.g. in studies of red raspberry, including micropropagated plantlets as well as field-grown material (Parent & Pagé, 1992), and micropropagated material from shoot tip cultures as well as callus-derived cell lines (Hoepfner et al., 1993). A similar lack of variation was encountered in protoplast-generated plants of apple (Nybom, unpublished results) as well as strawberry (Nyman, 1993), in spite of different ploidy levels and morphological variation in the latter material. Oligonucleotide DNA probe hybridization as well as RAPD markers have failed to detect somaclonal variation also in other materials like tomato (Vosman et al., 1992) and sugar beet (Schmidt et al., 1993). On the contrary, a morphologically deviating protoplast-derived rapeseed plant showed some DNA fragment variation after hybridization to the $(GATA)_4$ probe (Poulsen et al., 1993). Surprisingly high levels of variation were reported in a RAPD study of wheat (Brown et al., 1993). Some of these polymorphisms were, however, associated with the DNA isolation methodology.

Pedigree analysis

Whereas paternity determination has been one of the major applications of DNA fingerprinting in humans and many animal species, relatively few such studies have been presented in plants. One of these concerns apple seedlings derived from open pollination in a field with three cultivars (Nybom & Schaal, 1990). DNA was extracted from the obtained seedlings and then hybridized to the M13 probe. Paternity could be deduced for almost all seedlings investigated. This analysis was greatly facilitated by the identification of a multiallelic locus with five fragments. Applying RAPD markers, the paternity of apple cv Tsugaru could be determined by comparing fragment profiles of this cultivar, the known seed parent and six putative pollen parents (Harada et al., 1993).

Some other applications of pedigree analysis like monitoring of outcrossing rates, detection of gene introgression in breeding programs, and analysis of somatic hybridization, have been accomplished with DNA fingerprinting in crops like cauliflower (Boury et al., 1992), potato (Baird et al., 1992; Waugh et al., 1992; Xu et al., 1993) and rapeseed (Poulsen et al., 1994).

Genome mapping

Genetic maps comprising closely-spaced DNA markers are very useful for genome analysis. Up until recently, the molecular markers used for mapping were RFLPs derived from low copy number genomic clones or from cDNA clones. This is a time-consuming and also expensive technique. Lately RAPD markers have been used in different types of genome mapping with notable success. One problem is that RAPD markers are dominant. However, there are various ways of circumventing this problem in some crops, such as using inbred homozygous parents to generate the cross to be mapped, or to use recombinant inbred lines. Other approaches have e.g. made use of NILs (nearly isogenic lines) or bulked DNA samples. Another possible drawback is that RAPD maps may not be transferable to other crosses unless these are interspecific, with no polymorphism within species. However, RAPD markers obtained in various apple crosses appear to be homologous, suggesting that this problem is of minor importance for at least some crops (S.E. Gardiner, personal communication; N.F. Weeden, personal communication). Genome mapping with RAPDs have been successfully performed with numerous crops, e.g. legumes (Williams et al. 1990; Miklas et al. 1993), tomato (Giovanonni et al., 1991; Martin et al. 1993;) and lettuce (Michelmore et al., 1991; Paran et al., 1991). Several projects concerning fruit trees are now also on their way, e.g. in apple (Weeden et al., 1992, 1994; Gardiner et al., 1994; Hemmat et al., 1994; King, 1994), *Prunus* (Abbot et al., 1992; Chaparro et al., 1992; Grattapaglia et al., 1992a, b; Shimada et al., 1993; Dirlewanger & Bodo, 1994), grapes (Lodhi et al., 1992; Weeden et al., 1992, 1994), *Citrus* (Cai et al., 1992) and banana (Fauré et al., 1992).

Future prospects

In the future, protection of plant variety rights will probably involve one or several of the above-mentioned DNA fingerprint methods. Recently UPOV (International Union for the Protection of New Varieties of Plants) initiated an investigation of the applicability of various molecular markers in a number of model crops like citrus fruits, maize, soybean and tomato.

Marker-assisted selection, especially using the relatively easily applied RAPD markers, should enable breeders to screen large numbers of seedlings for e.g. genes affecting disease resistance, thereby reducing costs and labour involved in the growing of numerous plants in experimental orchards. Promising results on the search for RAPD markers for important genetic traits like the *Vf* gene for scab resistance in apple have been reported (Gianfranceschi et al., 1994; Yang & Krüger, 1994).

Increased knowledge of the pests and parasites involved in fruit tree and berry growing may also be achieved through DNA fingerprinting. DNA polymorphism was thus detected in *Colletotrichum gloeosporioides* by hybridization to human minisatellite probes (Braithwaite & Manners, 1989). A successful application of RAPD markers recently enabled the differentiation of various strains of apple scab, *Venturia inaequalis* (Sierotzki et al., 1994).

Several more advanced DNA fingerprint methods, which have been developed especially in human DNA research, may in the future also be applied successfully in plants. Among these methods are hybridization of Southern blots with highly polymorphic single-locus probes derived from minisatellite DNA (Wong et al., 1987) or from microsatellite DNA (= simple sequences, simple tandem repeats) (Zischler et al., 1991). Another method, which has to some extent already been used in plants, relies on the PCR-amplification of microsatellite DNA (Zhao & Kochert, 1993). A third approach, also involving the PCR-reaction, is taken by the Selective Restriction Fragment Amplification method (European Patent Office, no. 0 534 858 A1).

References

Abbott, A., L. Belthoff, S. Rajapakse, R. Ballard, W.V. Baird, R. Monet, R. Scorza, P. Morgans & A. Callahan, 1992. Development of a genetic linkage map of peach. Plant Genome I, San Diego Nov. 9–11, 1992: 44.

Aruna, M., P. Ozias-Akins, M.E. Austin & G. Kochert, 1993. Genetic relatedness among rabbiteye blueberry (*Vaccinium ashei*) cultivars determined by DNA amplification using single primers of arbitrary sequence. Genome 36: 971–977.

Baird, E., S. Cooper-Bland, R. Waugh, M. DeMaine & W. Powell, 1992. Molecular characterization of inter- and intra-specific somatic hybrids of potato using randomly amplified polymorphic DNA (RAPD) markers. Mol. Gen. Genet. 233: 469–475.

Boury, S., I. Lutz, M.-C. Gavalda, F. Guidet & A. Schlesser, 1992. Empreintes génétiques du chou-fleur par RAPD et vérification de la pureté hybride F1 d'un lot de semences. Agronomie 12: 669–681.

Braithwaite, K.S. & J.M. Manners, 1989. Human hypervariable minisatellite probes detect DNA polymorphisms in the fungus *Colletotrichum gloeosporioides*. Curr. Genet. 16: 473–475.

Brown, P.T.H., F.D. Lange, E. Kranz & H. Lörz, 1993. Analysis of single protoplasts and regenerated plants by PCR and RAPD technology. Mol. Gen. Genet. 237: 311–317.

Büscher, N., E. Zyprian & R. Blaich, 1993. Identification of grapevine cultivars by DNA analyses – Pitfalls of random amplified polymorphic DNA techniques using 10mer primers. Vitis 32: 187–188.

Cai, Q., C. Guy & G.A. Moore (1992) Genetic mapping of random amplified polymorphic DNA (RAPD) markers in *Citrus*. Plant Genome I, San Diego Nov. 9–11, 1992: 20.

Chaparro, J., D. Werner, D. O'Malley & R. Sederoff, 1992. Targetted mapping and linkage analysis in peach. Plant Genome I, San Diego Nov 9–11, 1992: 21.

Collins, G.G. & R.H. Symons, 1993. Polymorphisms in grapevine DNA detected by the RAPD PCR technique. Plant Mol. Biol. Reporter 11: 105–112.

Dallas, J.F., 1988. Detection of DNA "fingerprints" of cultivated rice by hybridization with a human minisatellite probe. Proc. Natl. Acad. Sci. USA 85: 6831–6835.

Dirlewanger, E. & C. Bodo, 1994. Molecular genetic mapping of peach. Eucarpia Fruit Breeding Sect. Meeting, Aug. 30–Sept. 3, 1993. In: H. Schmidt and M. Kellerhals (Eds.) Progress in temperate fruit breeding, pp. 309–311. Kluwer Academic Publishers, Dordrecht.

Dunemann, F., 1994. Genetic relationships in *Malus* evaluated by RAPD "fingerprinting" of cultivars and wild species. In: H. Schmidt and M. Kellerhals (Eds.) Progress in temperate fruit breeding, pp. 293–300. Kluwer Academic Publishers, Dordrecht.

European Patent Office, Publication number 0 534 858 A1. (Applicant: Keygene N.V., Wageningen, The Netherlands).

Fauré, S., J.L. Noyer, J.P. Horry, F. Bakry, C. Lanaud & D. Gonzalez de Léon, 1992. Genetic mapping of the diploid banana genome using molecular markers. Plant Genome I, San Diego Nov. 9–11, 1992: 46.

Gardiner, S.E., J.M. Zhu, H.C.M. Whitehead & C. Madie, 1994. The New Zealand apple genome mapping project. In: H. Schmidt and M. Kellerhals (Eds.) Progress in temperate fruit breeding, pp. 275–279. Kluwer Academic Publishers, Dordrecht.

Gianfranceschi, L., J.M. McDermott, N. Seglias, B. Koller, M. Kellerhals & C. Gessler, 1994. First results in developing a marker assisted breeding for resistance against scab. In: H. Schmidt and M. Kellerhals. Progress in temperate fruit breeding, pp. 291–294. Kluwer Academic Publishers, Dordrecht.

Giovanonni, J.J., R.A. Wing, M.W. Ganal & S.D. Tanksley, 1991. Isolation of molecular markers from specific chromosomal intervals using DNA pools from existing mapping populations. Nucleic Acids Res. 19: 6553–6558.

Gogorcena, Y., S. Arulsekar, A.M. Dandekar & D.E. Parfitt, 1993. Molecular markers for grape characterization. Vitis 32: 183–185.

Grattapaglia, D., J. Chaparro, P. Wilcox, S. McCord, D. Werner, H. Amerson, S. McKeand, F. Bridgewater, L. McIntyre, R. Doerge, B. Weir, R. Whetten, D. O'Malley & R. Sederoff, 1992a. RAPD mapping and tree improvement. Plant Genome I, San Diego Nov. 9–11, 1992: 14.

Grattapaglia, D., J. Chaparro, P. Wilcox, S. McCord, D. Werner, H. Amerson, S. McKeand, F. Bridgewater, R. Whetten, D. O'Malley & R. Sederoff, 1992b. Mapping in woody plants with RAPD markers: Applications to breeding in forestry and horticulture. Application of RAPD technology to plant breeding, Joint plant breeding symposia series, Minneapolis Nov. 1, 1992: 37–40.

Harada, T., K. Matsukawa, T. Sato, R. Ishikawa, M. Niizeki & K. Saito, 1993. DNA-RAPDs detect genetic variation and paternity in *Malus*. Euphytica 65: 87–91.

Hemmat M., N.F. Weeden, A.G. Manganaris & D.M. Lawson, 1994. Molecular marker linkage map for apple. J. Hered. 85: 4-11.

Heun, M. & T. Helentjaris, 1993. Inheritance of RAPDs in F1 hybrids of corn. Theor. Appl. Genet. 85: 961–968.

Hoepfner, A.-S., H. Nybom, U. Carlsson & R. Franzén, 1993. DNA fingerprinting useful for monitoring cell line identity in micropropagated raspberries. Acta Agricult. Scand., Sect. B 43: 53–57.

Jeanjaques, I., A. Defontaine & J.N. Hallet, 1993. Characterization of *Vitis vinifera* cultivars by random amplified polymorphic DNA markers. Vitis 32: 189–190.

Kaemmer, D., R. Afza, K. Weising, G. Kahl & F.J. Novak, 1992. Oligonucleotide and amplification fingerprinting of wild species and cultivars of banana (*Musa* spp.) Bio/Technology 10: 1030–1035.

King, G., 1994. The European apple genome mapping project – progress in mapping agronomic genes in apple. In: H. Schmidt and M. Kellerhals. Progress in temperate fruit breeding, pp. 263–267. Kluwer Academic Publishers, Dordrecht.

Koller, B., A. Lehmann, J.M. McDermott & C. Gessler, 1993. Identification of apple cultivars using RAPD markers. Theor. Appl. Genet. 85: 901–904.

Komatsu, A., T. Akihama, T. Hidaka & M. Omura, 1993. Identification of *Poncirus* strains by DNA fingerprinting. In: T. Hayashi, M. Omura & N.S. Scott (Eds). Techniques on Gene Diagnosis and Breeding in Fruit Trees, pp. 88–95. Fruit tree research station, Ibaraki.

Lavi, U., J. Hillel, A. Vainstein, E. Lahav & D. Sharon, 1991. Application of DNA fingerprints for identification and genetic analysis of avocado. J. Amer. Soc. Hort. Sci. 116: 1078–1081.

Lodhi, M.A., B.I. Reisch & N.F. Weeden, 1992. Molecular genetic mapping and genome size of *Vitis*. Plant Genome I, San Diego Nov. 9–11, 1992: 37.

Martin, G.B., M.C. de Vicente & S.D. Tanksley, 1993. High-resolution linkage analysis and physical characterization of the *Pto* bacterial resistance locus in tomato. Molecular Plant-Microbe Interactions 6: 26–34.

Matsuyama, T., R. Motohashi, T. Akihama & M. Omura, 1992. DNA fingerprinting in *Citrus* cultivars. Japan. J. Breed. 42: 155–159.

Matsuyama, T., M. Omura & T. Akihama, 1993. DNA fingerprinting in *Citrus* cultivars. In: T. Hayashi, M. Omura & N.S. Scott (Eds). Techniques on Gene Diagnosis and Breeding in Fruit Trees, pp. 26–30. Fruit tree research station, Ibaraki.

Michelmore, R.W., I. Paran & R.V. Kesseli, 1991. Identification of markers linked to disease-resistance genes by bulked segregant analysis: A rapid method to detect markers in specific genomic regions by using segregating populations. Proc. Natl. Acad. Sci. USA 88: 9828–9832.

Miklas, P.N., J.R. Stavely & J.D. Kelly, 1993. Identification and possible use of a molecular marker for rust resistance in common bean. Theor. Appl. Genet. 85: 745–749.

Mulcahy, D.L., M. Cresti, S. Sansavini, G.C. Douglas, H.F. Linskens, G.B. Mulcahy, R. Vignani & M. Pancaldi, 1993. The use of random amplified polymorphic DNAs to fingerprint apple genotypes. Sci. Horticult. 54: 89–96.

Nybom, H., 1990a. Genetic variation in ornamental apple trees and their seedlings (*Malus*, Rosaceae) revealed by DNA 'fingerprinting'. Hereditas 113: 17–28.

Nybom, H., 1990b. DNA fingerprints in sports of 'Red Delicious' apples. HortScience 25: 1641–1642.

Nybom, H., 1991. Applications of DNA fingerprinting in plant breeding. In: T. Burke, G. Dolf, A.J. Jeffreys & R. Wolff (Eds). DNA Fingerprinting: Approaches and Applications, pp. 294–311. Birkhäuser, Basel.

Nybom, H. & H.K. Hall, 1991. Minisatellite DNA "fingerprints" can distinguish *Rubus* cultivars and estimate their degree of relatedness. Euphytica 53: 107–114.

Nybom, H. & B.A. Schaal, 1990. DNA "fingerprints" applied to paternity analysis in apples (*Malus* × *domestica*). Theor. Appl. Genet. 79: 763–768.

Nybom, H., B.A. Schaal & S.H. Rogstad, 1989. DNA "fingerprints" can distinguish cultivars of blackberries and raspberries. 5th International Symposium on *Rubus* & *Ribes*, Acta Horticult. 262: 305–310.

Nybom, H., S.H. Rogstad & B.A. Schaal, 1990. Genetic variation detected by use of the M13 "DNA fingerprint" probe in *Malus*, *Prunus*, and *Rubus* (Rosaceae). Theor. Appl. Genet. 79: 153–156.

Nybom, H., U. Carlson & A.-S. Hoepfner, 1992. [The DNA laboratory at Balsgård]. Balsgård – Dept. Horticult. Plant Breeding, Swedish Univ. Agricult. Sci., Biennial Rept 1990–91: 146–154.

Nyman, M., 1993. Protoplast technology in strawberries. Acta Universitatis Upsaliensis, Comprehensive Summaries of Uppsala Dissertations from the Faculty of Science, No. 451.

Paran, I., R. Kesseli & R. Michelmore, 1991. Identification of restriction fragment length polymorphism and random amplified polymorphic DNA markers linked to downy mildew resistance genes in lettuce, using near-isogenic lines. Genome 34: 1021–1027.

Parent, J.-G. & D. Pagé, 1992. Identification of raspberry cultivars by nonradioactive DNA fingerprinting. HortScience 27: 1108–1110.

Parent, J.-G., M.G. Fortin & D. Pagé, 1993. Identification de cultivars de framboisier par l'analyse d'ADN polymorphe amplifié au hasard (RAPD). Can. J. Plant Sci. 73: 1115–1122.

Poulsen, G.B., G. Kahl & K. Weising, 1993. Abundance and polymorphism of simple repetitive DNA sequences in *Brassica napus* L. Theor. Appl. Genet. 85: 994–1000.

Poulsen, G.B., G. Kahl & K. Weising, 1993. Oligonucleotide fingerprinting of resynthesized *Brassica napus*. Euphytica. 70: 53–59.

Rogstad, S.H., J.C. Patton II & B.A. Schaal, 1988. M13 repeat probe detects DNA minisatellite-like sequences in gymnosperms and angiosperms. Proc. Natl. Acad. Sci. USA 85: 9176–9178.

Ryskov, A.P., A.G. Jincharadze, M.I. Prosnyak, P.L. Ivanov & S.A. Limborska, 1988. M13 phage DNA as a universal marker for DNA fingerprinting of animals, plants and microorganisms. F.E.B.s Letters 233: 388–392.

Schmidt, T., K. Boblenz, M. Metzlaff, D. Kaemmer, K. Weising & G. Kahl, 1993. DNA fingerprinting in sugar beet (*Beta vulgaris*) – Identification of double-haploid breeding lines. Theor. Appl. Genet. 85: 653–657.

Sharon, D., J. Hillel, A. Vainstein & U. Lavi, 1992. Application of DNA fingerprints for identification and genetic analysis of *Carica papaya* and other *Carica* species. Euphytica 62: 119–126.

Shimada, T., T. Haji & K. Hosaka, 1993. Classification and parent determination by RAPD in mume. In: T. Hayashi, M. Omura & N.S. Scott (Eds). Techniques on Gene Diagnosis and Breeding in Fruit Trees, pp. 77–80. Fruit tree research station, Ibaraki.

Sierotzki, H., M. Eggenschwiler, J. McDermott & C. Gessler, 1994. Specific virulence of isolates of *Venturia inaequalis* on "susceptible" apple cultivars. Eucarpia Fruit Breeding Sect. Meeting, Aug. 30–Sept. 3, 1993.

Stiles, J.I., C. Lemme, S. Sondur, M.B. Morshidi & R. Manshardt, 1993. Using randomly amplified polymorphic DNA for evaluating genetic relationships among papaya cultivars. Theor. Appl. Genet. 85: 697–701.

Striem, M.J., P. Spiegel-Roy, G. Ben-Hayyim, J. Beckmann & D. Gidoni, 1990. Genomic DNA fingerprinting of *Vitis vinifera* by the use of multi-locus probes. Vitis 29: 223–227.

Teramoto, S., Y. Kano-Murakami, M. Hori & K. Kamiyama, 1993. "DNA fingerprinting" can distinguish cultivar of Japanese pear. In: T. Hayashi, M. Omura & N.S. Scott (Eds). Techniques on Gene Diagnosis and Breeding in Fruit Trees, pp. 74–76. Fruit tree research station, Ibaraki.

Vosman, B., P. Arens, W. Rus-Kortekaas & M.J.M. Smulders, 1992. Identification of highly polymorphic DNA regions in tomato. Theor. Appl. Genet. 85: 239–244.

Waugh, R., E. Baird & W. Powell, 1992. The use of RAPD markers for the detection of gene introgression in potato. Plant Cell Reports 11: 466–469.

Weeden, N.F., G.M. Timmerman, M. Hemmat, B.E. Kneen & M.A. Lodhi, 1992. Inheritance and reliability of RAPD markers. Application of RAPD technology to plant breeding, Joint plant breeding symposia series, Minneapolis Nov. 1, 1992: 12–17.

Weeden, N.F., M. Hemmat, D.M. Lawson, M. Lohdi, R.L. Bell, A. Manganaris, B.I. Reisch, S.K. Brown & G.N. Ye, 1994. Development and application of molecular marker linkage maps in woody fruit crops. In: H. Schmidt and M. Kellerhals (Eds.) Progress in temperate fruit breeding, pp. 269–273. Kluwer Academic Publishers, Dordrecht.

Weising, K., J. Ramser, D. Kaemmer, G. Kahl & J.T. Epplen, 1991. Oligonucleotide fingerprinting in plants and fungi. In: T. Burke, G. Dolf, A.J. Jeffreys & R. Wolff (Eds). DNA Fingerprinting: Approaches and Applications, pp. 313–329. Birkhäuser, Basel.

Weising, K., D. Kaemmer, F. Weigand, J.T. Epplen & G. Kahl, 1992. Oligonucleotide fingerprinting reveals various probe-dependent levels of informativeness in chickpea (*Cicer arietinum*). Genome 35: 436–442.

Williams, J.G.K., A.R. Kubelik, K.J. Livak, J.A. Rafalski & S.V. Tingey, 1990. DNA polymorphisms amplified by arbitrary primers are useful as genetic markers. Nucl. Acids Res. 18: 6231–6235.

Wong, Z., V. Wilson, I. Patel, S. Povey & A.J. Jeffreys, 1987. Characterization of a panel of highly variable minisatellites cloned from human DNA. Ann. Hum. Genet. 51: 269–288.

Xu, Y.-S., M.S. Clark & E. Pehu, 1993. Use of RAPD markers to screen somatic hybrids between *Solanum tuberosum* and *S. brevidens*. Plant Cell Reports 12: 107–109.

Yang, H. & J. Krüger, 1994. Identification of a RAPD marker linked to the *Vf* gene for scab resistance in apples. Plant Breeding, in press.

Yang, H. & H. Schmidt, 1994. Selection of a mutant from adventitious shoots formed in X-ray treated cherry leaves and differentiation of standard and mutant with RAPDs. In: H. Schmidt and M. Kellerhals (Eds.) Progress in temperate fruit breeding, pp. 287–290. Kluwer Academic Publishers, Dordrecht.

Zhao, X.P. & G. Kochert, 1993. Phylogenetic distribution and genetic mapping of a $(GGC)_{(n)}$ microsatellite from rice (*Oryza sativa*). Plant Molec. Biol. 21: 607–614.

Zischler, H., A. Hinkkanen & R. Studer, 1991. Oligonucleotide fingerprinting with $(CAC)_5$: Nonradioactive in-gel hybridization and isolation of individual hypervariable loci. Electrophoresis 12: 141–146.

H. Schmidt and M. Kellerhals (Eds.), Progress in Temperate Fruit Breeding, 263–267.

Progress in mapping agronomic genes in apple (The European Apple Genome Mapping Project)

Graham J. King
Breeding & Genetics Department, Horticulture Research International, Wellesbourne, Warwick CV35 9EF, UK

Key words: Apple, breeding, databases, DNA markers, fruit trees, genetic mapping

Summary

The progress of the European Apple Genome Mapping Project is described. Populations segregating for a range of agronomic genes have been established in six European countries. The need for robust methods of analysis has been identified, especially with regard to the development of molecular markers. Isozyme systems, RAPDs, RFLPs and amplified genes are being used to construct a reference genetic linkage map. Standardisation and precise definition of both genotypic and phenotypic measurements has been recognised as being essential for future exploitation of genetic markers in apple breeding. Phenotypic measurements are being replicated in different geographical locations over several years. Statistical and genetic analyses are aimed at defining components of genetic variation which account for 'genes', as defined by apple breeders. A relational database is being constructed which will combine disparate sources of data relating to the genetics of apple. Comparative mapping has been identified as an efficient means of expanding genetic knowledge within and between Rosaceae genomes.

Introduction

The European Apple Genome Mapping Project was initiated in 1989 with the long term objective of improving the understanding of apple genetics (King et al., 1991). The work is based around five segregating mapping populations which have now been replicated and established in six European countries. Associated research is being carried out on the genetics and diversity of fungal resistance genes, fruit quality, and the socio-economic aspects of apple production.

A primary consideration is the need to make results available for exploitation by apple breeders. This depends on the facility with which molecular markers and the associated genetic information can be used. In this paper I will emphasise the efforts being made to ensure that methodologies adopted are standardised and robust, thus ensuring that reproducible results can be obtained in a wide range of situations from year to year.

Mapping populations

Five mapping populations have been distributed to different locations throughout Europe, one of which, 'Prima' × 'Fiesta', has been adopted as the core population for mapping genetic markers. This population of 165 trees is being screened using isozyme systems, Randomly Amplified Polymorphic DNA (RAPD) markers and Restriction Fragment Length Polymorphic (RFLP) probes. The genetic map produced will then be used as a reference for mapping genes in the other populations.

As there is incomplete replication of all five populations on six sites, where necessary leaf samples are collected and processed at one site and sent elsewhere for isozyme analysis. In addition, freeze-dried leaves are being exchanged between countries for subsequent DNA extraction.

Building a genetic map

For a molecular genetic marker to be of lasting use to plant breeders, it must be possible to generate it readily and reproducibly in a range of laboratories over a period of years. We have therefore concentrated on adopting methodologies which are both robust and portable.

Isoenzymes. At least 18 enzymes are being analysed electrophoretically to determine polymorphism at loci in the mapping populations.

DNA RAPDs. Since the first publications describing Randomly Amplified Polymorphic DNA markers (RAPDs) as a means for rapidly generating genetic markers (Williams et al., 1990), hundreds of laboratories worldwide have used the technique, many generating genetic linkage maps for a wide range of organisms. The value of such maps has been confounded by doubts about the relative lack of reproducibility of the marker systems in different laboratories. Evidence has accumulated that minor changes of experimental parameters in some circumstances alter the observed range of polymorphic fragments.

Relatively few studies have addressed the reproducibility of RAPD markers from laboratory to laboratory, which is a primary requirement for use as a universal breeding aid. For the purposes of a pan-european project it has therefore been vital to ensure that participants are able to generate reproducible and reliable experimental results. As a result we have undertaken a pilot experiment to establish the most appropriate set of experimental parameters.

A survey of existing methods established that at least five protocols had successfully been used to generate RAPD markers from apple DNA alone. Although modifications of thermal cycling conditions may result in a higher frequency of polymorphic markers (Koller et al., 1993), this can lead to problems when scoring the fragments as genetic markers in segregating populations. Simplicity has been the primary criterion in experimental design and interpretation of results, and has proved successful in establishing a robust methodology. Our experience is that primers which produce many bands (even if bright) are difficult to score accurately on segregating populations. Only primers that produce a few, very bright bands are suitable for systematic mapping work. Others, which produce variable or equivocal results are best discarded at the earliest stage, when screening primers against parental DNA samples. We have found it preferable to select primers which produce one or more non-polymorphic bands flanking the polymorphic bands in the track. These act as good internal controls of the efficiency of the amplification reaction, and assist accurate scoring.

Standardisation has been assisted by adopting identical molecular weight markers in all experiments. We have found 100 bp ladders to be particularly useful in calibrating fragment patterns. Selection of RAPD markers for genetic mapping of progeny from highly heterozygous parents, such as apple, may be divided into several stages. The first stage determines which primers yield fragments polymorphic between parents. The second stage involves selecting parents which will yield informative segregation ratios in the progeny.

Table 1 represents the possible genotypic and phenotypic segregation of a RAPD band [—] when it has been detected in both parents. In this case, 10–12 progeny are tested to detect pairs of parents, both of which are heterozygous for the locus defined by the marker band. Whilst the 'fragment absent' phenotype is useful for mapping the polymorphism, it only occurs in a quarter of the progeny, and so the information gained is limited. However, such RAPDs are particularly useful for identifying homologous linkage groups between the parents (Hemmat et al., 1994).

Table 2 shows the possible segregation of a band when it is present only in one parent. Only five or six progeny need to be tested to detect the presence of the heterozygous parent. In this case a dominant/recessive allelic system displays the same 1:1 ratio as any other polymorphic locus.

RFLPs. Initial work had suggested that polymorphic RFLPs could be detected using probes derived from a number of apple genomic libraries. Early experience with the Digoxygenin/chemiluminescent non-radioactive labelling system (Boehringer Mannheim) had demonstrated the ability to detect single copy sequences in genomic blots of plant DNA. However, we have subsequently found great variability in the sensitivity of this assay, an experience common to a significant number of laboratories worldwide. Our conclusion is that this labelling/detection system is inherently unstable and does not conform to the requirements of a robust assay suitable for routine analysis. Recent experience with the ECL non-radioactive direct labelling system (Amersham) suggests that this may meet our requirements, and allow RFLP technology to be used in many apple breeding stations.

We have subsequently detected a relatively high number of repetitive sequences in our existing apple genomic libraries and will therefore be assessing a range of other strategies for generating single copy polymorphic markers. Further genomic and DNA libraries are now being screened in order to detect

Table 1. Possible segregation of dominant RAPD fragments in progeny from a cross between two heterozygous diploids, when the fragment is detected in both parents.

Female	Male	Progeny	Ratio presence: absence
AA [—]	AA [—]	AA[—] : AA[—] : AA[—] : AA[—]	4 : 0
AA [—]	Aa [—]	AA[—] : AA[—] : Aa [—] : Aa [—]	4 : 0
Aa [—]	AA [—]	Aa [—] : Aa [—] : AA[—] : AA[—]	4 : 0
Aa [—]	Aa [—]	Aa [—] : AA[—] : Aa [—] : aa [ ]	3 : 1

Table 2. Possible segregation of dominant RAPD fragments, when the fragment is detected in only one parent.

	Female	Male	Progeny	Ratio
Fl	AA [—]	aa [ ]	Aa[—] : Aa[—] : Aa[—] : Aa[—]	4 : 0
Backcross	Aa [—]	aa [ ]	Aa[—] : Aa[—] : aa [ ] : aa [ ]	2 : 2

reliable single copy polymorphic markers. Parental genomic DNAs are being digested with a panel of five restriction enzymes (four-base and six-base cutters).

Gene maps. Isozymes provide a more direct measure of polymorphism at defined gene loci than do randomly selected RFLP or RAPD markers. They are also useful in providing a universal 'anchor' or milestone for comparing genetic linkage maps from different sources. In order to extend the range of such markers, we are now able to exploit the accumulated databases of defined gene DNA sequences available for plant species. We have already shown that it is possible to design primers for PCR DNA amplification of gene sequences from apple DNA by alignment of consensus sequences from other species (King, 1992). The amplified sequences can be labelled directly and used as RFLP probes. However, there is a danger that different sets of sequences may be amplified on different occasions, leading to conflicts in segregation data. We are therefore now cloning the PCR products, and using selected clones as probes for defined gene loci. In addition, we will sequence the ends of the amplified fragments to design primers for use in simple PCR-based assays.

Eleven *Malus* gene sequences are present in the EMBL DNA database (release 34.0), with at least ten from other members of the Rosaceae. As more DNA sequences are established for apple we will be in a position to refine the design of PCR primers. Horticulture Research International has initiated research to develop learning algorithms which would be able to predict, within defined limits of probability, the likely DNA gene sequence for one species, given the sequence from one or more other species for which the principles of sequence construction have been established.

Phenotypic assessments (what do we mean by GENES?)

The consensus view of the definition of a gene has changed dramatically over the past forty years. People in different disciplines are therefore at risk of mutual misunderstanding, with the result that they may not capitalise on their combined knowledge.

Working definitions of 'GENES':

Breeder:	any trait which appears to be inherited	*pragmatic*
Geneticist:	a locus which displays one or more alleles	↕
Molecular Biologist:	a stretch of DNA coding for a protein	*reductionist*

In order to provide breeders with useful tools to carry out design and selection of crosses, it is useful to be able to categorise the range of 'genes' for which different types of information will be available. Various facets of agronomic genes need to be considered, and may be posed as a set of questions:

Major gene or complex trait?
Dominance?
Heritability?
G × E interaction?
Linkage?
Pleiotropy?
Ease of manipulation?

Depending on the answers to the above questions, which will be established from the genetic mapping studies, decisions may be made as to whether a given trait is worth analysing further, and if so, whether it is amenable to selection with molecular markers.

Since little is known of the genetic basis of most agronomic traits currently manipulated by apple breeders, it is prudent to adopt the null hypothesis that all traits studied are complex. Precise and robust definitions are therefore being accumulated for all phenotypic measures prior to genetic analysis. It is often necessary to divide a breeder's definition of a 'major' gene into a number of readily scored components. Each component may then be subjected to linkage analysis with molecular markers. Only when two or more components converge on a genetic map is it reasonable to regard them as representing the same gene.

In order to accumulate data of sufficient quality which are amenable to such analysis, adequate replication is required. This is being achieved by replication both site to site, and from year to year. Environmental effects are then being quantified prior to genetic analysis. Evidence is accumulating from other studies that population size is probably more important in determining the likelihood of detecting 'genes', than the availability of highly saturated genetic linkage maps.

Resistance genes. These are being studied in detail, both to test the reliability of molecular markers as predictive tools, and to provide information on sources and interaction of different resistance genes. In particular, different sources of resistance to scab are being investigated to determine whether they are clustered or dispersed on the genetic map.

Habit, juvenility, winter chilling. The genetic basis of tree habit is poorly understood. In the case of the columnar phenotype, there is evidence for a single mutation at one locus (Lapins & Watkins, 1973). It is therefore reasonable to search for linkage of markers to this locus, and then study the interaction with other genetically determined components of tree habit. Initial work has included measurements of height, stem diameter, node number, and length and number of 'feathers'. Statistical analysis has involved standardisation of data followed by manipulation of measurement ratios to develop an discriminant function to identify components of the tree habit, which may then be used to detect the presence of the columnar allele in segregants. Bulked segregant analysis with up to 160 RAPD markers has to date yielded one putative genetic marker linked to the columnar phenotype.

The mapping populations are segregating for length of juvenile phase. The family segregating from a cross between Crimson Spy × 3762 displays the widest variation, with time to first flower expected to range over 10 years. After four years only 20% of this population have flowered, compared to 70% in the progeny from Fiesta × 3762.

Further characters, including winter chilling requirement, root suckering, leaf colour and burr-knots are being scored on the populations and will be subjected to genetic analysis.

Fruit. The mapping populations are clearly segregating for a range of fruit characters, including red skin colour, red flesh colour, size, sweetness, acidity and deciduous calyx. Data are being collected over a number of years.

Database development

Given the large amount of data being generated as a result of the genotypic and phenotypic analysis of the mapping populations, an adequate database retrieval and analysis system is required. A number of databases have been developed by other groups accumulating 'genome' data for a range of species. Whilst these possess a number of attractive features, none combine the management of genotypic and phenotypic data in a manner suitable for long-term breeding studies. We are therefore designing a relational database using the Oracle system. The development version of this is being constructed on a DEC-alpha workstation, and may be accessed by participants to the project via Email or Internet. In addition, smaller production releases of portions of the total database are being made available either in spreadsheet or flat database format for access on PCs. Considerable effort is being invested in the design of the database, to incorporate as much information relevant to the genetics of apple as possible. Appropriate precautions are being taken to ensure that data validation and security are such that the accumulated information will be available for use by apple breeders.

Comparative mapping

Evidence is accumulating from a range of eukaryote families that synteny of linkage groups is conserved over large evolutionary distances. In the Poaceae, linkage groups have been aligned from the large grained cereals and the rice genome (Moore et al., 1993). Evolutionary preservation of synteny is encouraging development and exploitation of genetic maps, as it enables data from one species to be extrapolated to another. Preliminary isozyme data suggest that there is some conservation of synteny within the Rosaceae. Projects developing genetic map information for *Malus* and *Prunus* should therefore benefit from sharing linkage information and molecular markers. Since the smaller *Prunus* genome is packaged into only 8 chromosomes compared to the 17 of apple, one would expect large scale duplications within the *Malus* genome, with a corresponding degree of internal synteny.

Acknowledgements

The work of the European Apple Genome Mapping Project is now funded as a shared-cost project as part of the CEC-DGVI 'AIR' programme (CT920473), with matching funding arising from government and other sources within the EC, and from the Swiss Federal Ministry of Technology. The work at Horticulture Research International is funded by the Ministry of Agriculture, Fisheries and Food and the Agricultural and Food Research Council. I would like to acknowledge and thank all contributors to this work over the last four years.

References

Hemmat, M., N.F. Weeden, A.G. Manganaris & D.M. Lawson, 1994. Molecular marker linkage map for apple. Journal of Heredity (in press).

King, G.J., F.A. Alston, I. Battle, E. Chevreau, C. Gessler, J. Janse, P. Lindhout, A.G. Manganaris, S. Sansavini, H. Schmidt & K. Tobutt, 1991. The European Apple Genome Mapping Project – developing a strategy for mapping genes coding for agronomic characters in tree species. Euphytica 56: 89–94.

King, G.J., 1992. Mapping genes in apple: the future for molecular-aided breeding. Outlook on Agriculture 21: 163–168.

Koller, B., A. Lehmann, J.M. McDermott & C. Gessler, 1993. Identification of apple cultivars using RAPD markers. Theoretical and Applied Genetics 85: 901–904.

Lapins, K.O. & R. Watkins, 1973. Genetics of compact growth habit. Report of East Malling Research Station for 1972, p. 136.

Moore, G., M.D. Gale, N. Kurata & R.B. Flavell, 1993. Molecular analysis of small grain cereal genomes: current status and prospects. Biotechnology 11: 584–589.

Williams, J.G.K., A.R. Rubelik, K.J. Livak, J.A. Rafalski & S.V. Tingey, 1990. DNA polymorphism amplified by arbitrary primers are useful as genetic markers. Nucleic Acids Research 18: 6531–6535.

H. Schmidt and M. Kellerhals (Eds.), *Progress in Temperate Fruit Breeding*, 269–273.

Development and application of molecular marker linkage maps in woody fruit crops

N.F. Weeden[1], M. Hemmat[1], D.M. Lawson[1], M. Lodhi[1], R.L. Bell[2], A.G. Manganaris[3], B.I. Reisch[1], S.K. Brown[1] & G.-N. Ye[1]

[1] *Department of Horticultural Sciences, Cornell University, Geneva, NY, U.S.A.;* [2] *USDA, ARS, Appalachian Fruit Research Station, Kearneysville, WV, U.S.A.;* [3] *National Agricultural Research Foundation, Pomology Institute, 59200 Naoussa, Greece*

Key words: Apple, grape, marker assisted selection, pear, RAPDs, *Pyrus communis*

Summary

Extensive linkage maps, consisting primarily of molecular markers, are being developed for apple, pear, and grape varieties. The intrinsically high heterozygosity of outcrossing perennial species is utilized to produce segregating populations directly from a cross between varieties. Nearly complete linkage maps have been generated for the apple varieties 'Rome Beauty' and 'White Angel'. The map for 'Rome Beauty' contains 161 molecular markers, while that for 'White Angel' has 251 markers. Maps for the pear varieties, 'Bartlett' and NY10353, also are being developed. Linkages conserved between the pear and apple genomes have been identified. In grapes, maps for four varieties are available, the most extensive being those for 'Cayuga White' and 'Aurore'. The apple maps have been used to investigate the genetic basis of morphological and physiological characters. A gene controlling the presence of anthocyanins in the skin of the fruit is located on linkage group 3. Genes controlling early bud break, branching habit, and production of root suckers have also been identified and mapped.

Introduction

Marker assisted selection has proven to be a useful tool in many annual crops. However, the technique has not been used extensively in woody perennials despite its advantage in eliminating undesirable genotypes before planting in the field. A dearth of genetic markers in perennial fruit crops is one important reason this approach has not been applied. Very few monogenic morphological polymorphisms have been identified in any perennial species. Isozyme loci, although excellent markers (Manganaris & Alston, 1987, 1988; Manganaris, 1989) are not numerous enough to satisfy the demands of the breeder/geneticist. Thus, the development of DNA marker technology has been followed with considerable interest by fruit breeders.

A major goal of ours has been the identification of genetic markers in apple, grape, and pear. We believe these markers will benefit the breeding programs directly through marker assisted selection and indirectly by facilitating a better understanding of some characters being manipulated. We are particularly interested in a "user friendly" approach. That is, a method that is relatively easy to apply, permitting breeders to incorporate the techniques into their own programs without the need to collaborate with a molecular biology laboratory. In the following sections we describe our approach, present genetic tools being developed, and report how these tools have been used to elucidate, at least partially, the genetic basis of specific traits.

The exploitation of highly heterozygous genomes

Early speculations that apple was highly heterozygous have been confirmed convincingly by isozyme studies (Chevreau et al., 1985; Weeden & Lamb, 1985, 1987; Manganaris, 1989). This high level of heterozygosity has greatly facilitated genetic studies in apple because each variety becomes, in effect, an F_1, and the progeny produced from an open pollinated tree will segregate for all the loci heterozygous in that tree.

For genetic studies it is convenient to make a controlled cross between two varieties, thereby specifying and limiting the number of alleles segregating in the following generation. Genetically, this type of arrangement may be referred to as a "double pseudotestcross". The terminology is awkward, but the cross is familiar to all fruit breeders, being the most commonly used cross in most breeding programs. In such crosses, alleles segregate in a 1:1 ratio unless both parents are heterozygous at a locus in which case a 3 : 1, 1 : 2 : 1 or 1 : 1 : 1 : 1 ratio is expected. As in any test cross, dominant markers are just as valuable as codominant markers because for a 1 : 1 segregation, phenotype equals genotype. This last consideration is particularly important in our approach, for we rely heavily on dominant DNA polymorphisms generated by amplification with arbitrary primers (RAPDs in the terminology of Williams et al., 1990). Although the dominance of RAPD markers does not appear to be a significant disadvantage in the double pseudotestcross format, other factors such as degree of polymorphism, distribution in the genome, or poor reproducibility may limit their value. Early in our mapping experiments we attempted to establish the merits and limitations of RAPDs relative to other commonly used markers.

Our studies in apple, grape and pear indicate that a large number of RAPDs can be identified quickly among varieties and conveniently scored (Table 1). On average, three heterozygous markers were identified in each of the varieties examined. As these varieties represent a wide selection of germplasm, the results suggest that 150 to 300 polymorphic markers can be identified in most varieties using relatively few (50 to 100) short primers. If the RAPDs are randomly distributed across the genome, this number of markers should be sufficient to permit the tagging of important characters segregating in any cross.

To confirm the genetic basis of these polymorphisms and to examine their distribution on the nuclear genome we have analyzed RAPD segregation patterns in four double pseudotestcross populations, and compared these segregation patterns with those of segregating allozyme and RFLPs. The crosses investigated included one apple ('Rome Beauty' × 'White Angel'), one pear ('Bartlett' × NY10353) and two grape ('Aurore' × 'Cayuga White' and 'Horizon' × Ill 547–1) (Table 2). Extensive linkage maps are being developed for each of the varieties. The most thoroughly analyzed population is that from 'Rome Beauty' × 'White Angel'. The maps assembled for these two varieties appear to cover the majority of the nuclear genome, for nearly all the RAPD markers, as well as all the segregating isozyme loci and RFLPs, could be included on just over 20 linkage groups (Hemmat et al., 1994).

Using several analytical techniques, we established that nearly all the RAPDs scored reflected true genetic polymorphism and that our error rate in scoring RAPD segregation averaged about 3%, ranging from undetectable for the intensely staining fragments to 8% for faint fragments (Weeden et al., 1992). Markers heterozygous in both parents were used to identify homologous linkage groups in the two parents. Fourteen of the linkage groups identified in 'Rome Beauty' could be paired with their counterparts in 'White Angel'.

Our results in pear and grape are very similar to those in apple, except that an extensive error analysis has not been performed for these data and that, in pear, the number of linkage groups has yet to coalesce to a figure close to the haploid chromosome number. Although RAPDs have yet to prove useful for identification of linkages conserved between genera, results on isozyme loci indicate that the linkage between *Aat-p* and *Gpi-c* has been maintained in both apple and pear.

Use of markers in genetics and breeding of apple

Once an extensive set of genetic markers is available for a variety, the tagging of any heterozygous locus becomes relatively trivial. We were able to find RAPD markers within 10 cM of 27 of the 31 known genes (isozymes, known DNA probes, and powdery mildew resistance) segregating in the 'Rome Beauty' × 'White Angel' population. Further saturation of the maps of 'Rome Beauty' and 'White Angel' is feasible but may be only of academic interest because neither variety is being used currently in our breeding effort. However, the other three crosses are integral parts of grape and pear breeding programs and continue to be investigated intensely.

The identification of markers near genes involved in the expression of more complex characters such as fruit quality, tree habit, and flowering time is of great potential for breeding programs. The set of segregating markers available in the 'Rome Beauty' × 'White Angel' populations were used to help elucidate the genetic basis of certain of these complex characters (Table 3). The approach used to identify the major

Table 1. Genetic polymorphism in apple, pear, and grape as revealed by arbitrary primers

Variety	No. primers tested	No. RAPDs detected	Polymorphisms per primer
Apple			
White Angel	62	227	3.66
Rome Beauty	62	162	2.61
Pear			
Bartlett	70	237	3.39
NY 10353	70	221	3.15
Grape			
Aurore	105	274	2.60
Cayuga White	105	270	2.57
Horizon	81	276	3.41
Ill 547–1	81	290	3.58

Table 2. Number of segregating markers identified in single varieties and number of linkage groups identified by joint segregation analysis

Variety	Number of segregating markers	Number of linkage groups
Apple		
White Angel	253	24
Rome Beauty	156	21
Pear		
Bartlett	252	–
NY 10353	232	–
Grape		
Aurore	299	24
Cayuga White	299	22
Horizon	276	20
Ill 547-1	290	25

genes was basically the same for each character and can be exemplified by our analysis of fruit skin color.

The fruit skin color of both 'Rome Beauty' and 'White Angel' is red, with the former variety exhibiting stripes and the latter a solid, uniform red. In the progeny we observed a variety of color patterns on the fruit surface, including a yellow skinned fruit totally lacking red pigmentation. These skin color patterns were grouped in various ways and for each group performed a joint segregation analysis with alleles at each of the marker 'loci'. Although the pattern or hue of red pigmentation did not appear to cosegregate with any of the markers, the red/yellow polymorphism displayed a clear correlation with several markers. Approximately one fourth of the trees produced yellow fruit (65 trees producing some red pigment in fruit skin, 24 trees lacking red pigment in the fruit skin). This ratio suggested the action of two genes or a single locus heterozygous in both parents. In either case, absence of red pigmentation would be the genetically defined (homozygous recessive) phenotype, whereas presence of red pigmentation could be due to one of several genotypes. Thus the segregation at marker loci initially was examined only within the trees producing yellow fruit. Two RAPD markers displayed a major distortion in their segregation ratio within the yellow-fruiting group: S27c from White Angel (17 "–", 1 "+") and S45b from Rome Beauty (13 "–", 1 "+"). Both markers gave segregation ratios near 1 : 1 for the entire population. These results suggested that both 'Rome Beauty' and 'White Angel' possessed genes affecting the red/yellow dimorphism. Consulting the

Table 3. Markers displaying linkage with major genes controlling morphological or physiological characters in apple

Character	Marker	Linkage group	Gene symbol
Terminal bearing	P27e	5	*Tb*
Initial bud break	Prx-c	6	*Rbb*
Leaf expansion	P27e	5	*Tb*[a]
Bloom time	no correlation with a P < 0.001		
Rootsucker formation	124e	1	*Rs*

[a] Time of leaf expansion appears to be a pleiotropic effect of *Tb*.

Table 4. Linkage between *Idh-2* and fruit skin color in a 'Rome Beauty' × 'White Angel' progeny

Idh-2 genotype of tree	No. of trees with red pigment in fruit	No. of trees lacking red pigment in fruit
bb	2[a]	22
ab	44	1
aa	23	0

[a] Red coloration very light blush.

maps for each parent, we found that both markers had been placed on the respective linkage group 3 and that a tightly linked isozyme locus, *Idh-2*, was heterozygous in both parents. Joint segregation analysis of *Idh-2* and fruit skin color is presented in Table 4. The high correlation between the two segregation patterns ($\chi^2 = 77.0$, $P < 0.00001$) indicates that in the population examined, the presence of red pigmentatin in the skin of the fruit is controlled by a single locus, here designated R_f to conform to the terminology used by Alston & Watkins (1973). Both 'Rome Beauty' and 'White Angel' are heterozygous at this locus and therefore display red pigmentation in the fruit skin. The hue and distribution of red pigmentation appear to be controlled by factors unlinked to *Idh-2*. At present, these results can be applied only to the two parents; however, the results of Schmidt (1988) suggest that R_f is the primary locus governing fruit skin color in many apple varieties.

Similar analyses demonstrated that branching habit, timing of initial bud break, and production of rootsuckers were primarily controlled by single loci, each heterozygous in White Angel (D.M. Lawson, M. Hemmat, and N.F. Weeden, 1995). Although many other characters were investigated for which the genetic basis could not be elucidated (Hagens, 1992), it is clear that marker driven genetic analysis represents a powerful tool for finding tags for apparently complex traits. We currently are applying this approach to the pear and grape populations and continuing our analyses in apple. Present results indicate that this approach has excellent prospects for permitting breeders of perennial fruit to use marker assisted selection in their programs.

Acknowledgements

We thank R.C. Lamb, R.D. Way, J.N. Cummins, and H.S. Aldwinckle for their advice, interest, and support on our studies in apple. The work in grape was funded in part by a United States-Israel BARD Project (US 1888–90) and that in pear by a USDA Specific Cooperative Agreement (# 58–1931–2–046).

References

Alston, F.H. & R. Watkins, 1973. Apple breeding at East Malling. Proc. Eucarpia Fruit Section, Canterbury: 14–29.

Chevreau, E., Y. Lespinasse & M. Gallet, 1985. Inheritance of pollen enzymes and polyploid origin of apple (*Malus* × *domestica* Borkh.) Theor. Appl. Genet. 71: 268–277.

Hagens, D.M, 1992. Genetic studies in *Malus*: Integration of isozymes, DNA markers and morphological traits. MS Thesis, Cornell University. 55 pp.

Hemmat, M., N.F. Weeden, A.G. Manganaris & D.M. Lawson, 1994. Molecular marker linkage map in *Malus*. J. Hered. 85: 4-11.

Lawson, D.M., M. Hemmat & N.F. Weeden, 1995. The use of molecular markers to analyze the inheritance of morphological and developmental traits in apple. J. Amer. Soc. Hort. Dci., *in press*.

Manganaris, A.G., 1989. Isoenzymes as genetic markers in apple breeding. Ph.D. Dissertation, University of London. 430 pp.

Manganaris, A.G. & F.H. Alston, 1987. Inheritance and linkage relationships of glutamate oxaloacetate transaminase isoenzymes in apple 1. The gene *Got-1*, a marker for the *S* incompatibility locus. Theor. Appl. Genet. 74: 154–161.

Manganaris, A.G. & F.H. Alston, 1988. The acid phosphatase gene *ACP-1* and its linkage with the endopeptidase gene *ENP-1* and the pale green lethal gene *l* in apple. Acta Hort. 224: 177–184.

Schmidt, H., 1988. The inheritance of anthocyanin in apple fruit skin. Acta Hort. 224: 89–97.

Weeden, N.F. & R.C. Lamb, 1985. Identification of apple cultivars by isozyme phenotypes. J. Amer. Soc. Hort. Sci. 110: 509–515.

Weeden, N.F. & R.C. Lamb, 1987. Genetics and linkage analysis of 19 isozyme loci in apple. J. Amer. Soc. Hort. Sci. 112: 865–872.

Weeden, N.F., G.M. Timmerman, M. Hemmat, B.E. Kneen & M.A. Lodhi, 1992. Inheritance and reliability of RAPD markers. In: Applications of RAPD Technology to Plant Breeding, pp. 12–17. Proc Joint Plant Breeding Symposia, Minneapolis, MN. Crop Science Soc Amer, Madison, WI.

Williams, J., A.R. Kubelik, D.J. Livak, J.A. Rafalski & S.V. Tingey, 1990. DNA polymorphisms amplified by arbitrary primers are useful as genetic markers. Nucleic Acids Res. 18: 6531–6535.

H. Schmidt and M. Kellerhals (Eds.), Progress in Temperate Fruit Breeding, 275–279.

The New Zealand apple genome mapping project

A progress report

Susan E. Gardiner, Ji Mei Zhu, Heather C.M. Whitehead & Charlotte Madie
The Horticulture and Food Research Institute of New Zealand Ltd (HortResearch), Private Bag 11 030, Palmerston North, New Zealand

Key words: Apple, chromosomes, gene mapping, *Malus*, RAPD, RFLP

Summary

This project was initiated three years ago to support the New Zealand apple breeding programme at HortResearch and is funded by both the New Zealand Foundation for Research, Science and Technology and the New Zealand Apple and Pear Marketing Board. Our initial goal is to construct a linkage map comprising RFLP, RAPD and isoenzyme markers as well as characters of importance to our breeders and this will enable the breeders to use marker assisted selection to identify the most promising seedlings within progeny populations at an early age. Characters of interest in the shorter term include resistances to scab, powdery mildew, woolly apple aphid and silver leaf, and in the longer term more complex traits such as early and late fruit maturity, fruit quality characters, low temperature tolerance and rootstock influences such as dwarfing and precocity are under consideration. We are currently developing long-term plans for isolation of apple genes in association with other molecular biologists in HortResearch. We have used a cDNA library from apple flesh as a prolific source of RFLP probes and detect their hybridization to Southern blots using a chemiluminescent method. We have screened our mapping line of seedlings for more than 160 markers so far (1/3 RFLP, 2/3 RAPD). Detection of in situ hybridization of key RFLP probes to metaphase chromosome preparations for the purpose of associating marker linkage groups with physical chromosomes is by a digoxygenin or biotin linked immunoassay. The project has collaborative links to apple genome mapping projects in America and Europe.

Introduction

The New Zealand Apple Genome Mapping Project was initiated during 1990 in order to support the New Zealand apple industry by assisting this country's apple breeding programme to produce new cultivars which are tailored to fit specific market niches and meet the constantly changing demands of international markets. Such new cultivars must not only possess the quality attributes of colour, flavour and texture desired by the consumer in the country of consumption and fit a particular time slot in the market, but should also have genetically determined resistances to pests and diseases in order to reduce the requirement for application of chemical sprays during fruit production and meet growing environmental concerns.

An increased understanding of apple genetics and the ability to use molecular markers to identify seedlings with the best genetic potential will enable use to be made of a much wider base of germplasm in breeding new cultivars. Most current commercial cultivars are derived from a very narrow genetic base, leading to a high level of inbreeding in apple breeding programmes worldwide, but there is a vast resource of other germplasm available in older, non-commercial cultivars and wild species of apple (Noiton & Shelbourne, 1992). Such material has previously been little used in apple breeding because of the difficulty of selecting from progeny lines derived from outcrosses the individual seedlings which contain the desired character, but which have a minimum of other exotic genes. Marker assisted selection has the potential to markedly reduce this effect of linkage drag during gene introgression as molecular markers can be used to detect the small fraction of individuals in a large progeny population which have experienced recombination near the gene of interest (Tanksley et al., 1989).

A genome map with molecular markers which cosegregate with major genes of particular interest will enable breeders to assess the genetic potential of seedling populations at a very early age and to select the small fraction of elite individuals for planting out and later field observation. Such early progeny screening would result in an overall reduction in the requirement for land and maintenance in the breeding programme and a decrease in costs and in the time required to produce new cultivars. Breeders will not only be able to obtain information on the presence of genes which are not expressed because of the juvenile state of the plant, but also on those which are not expressed in the heterozygous state, or in the absence of the inducing pathogen.

Identification of plants with different loci coding for resistance to the same organism will enable breeders to design crosses to aggregate such loci as they require, and detect those individual progeny seedlings possessing the desired combinations of loci. Similar benefits allowing more informed decisions about further crosses will result from an understanding of the makeup of the component loci constituting a particular quantitative or polygenic trait and their interactions. The distribution of these quantitative loci amongst the individuals in a progeny population will only be able to be determined by molecular markers.

The availability of a gross structure map for apple will enable initiation of work on map-based cloning of genes coding for an unknown product, such as genes for pathogen resistance. Methods involving scoring of polymorphisms in DNA bulked according to phenotype (Michelmore et al., 1991; Williams et al., 1993) now offer the rapid identification of tightly linked markers near the gene of interest and will enhance the chances of a successful chromosome walk and ultimate gene isolation. Current improved transformation techniques for apple being developed in our Institute should allow the successful transfer of such a gene into a given cultivar which possesses a number of desirable characters but is lacking in a particular pathogen resistance.

Genetic traits and plant material

The seedling progeny line of 90 plants being used to generate our genome map resulted from the pollination in 1984 by HortResearch apple breeders of 'Scijoy' with pollen obtained from Rutgers University, U.S.A., which was believed at the time to be from 'Redfree' (A. White, personal communication). This particular progeny population was selected as being particularly suitable as a mapping line because it was predicted that it would display segregation between individuals for both molecular and agronomic traits including the Vf gene and fruit characters and because it had not been subjected to preselection for disease resistance at the seedling stage. Molecular marker data collected since the beginning of the project indicate that it is unlikely that 'Redfree' is the pollen parent, but the plants are obviously segregating for a number of fruit characters and a resistance to apple scab.

Other existing breeding populations have been designated for locating genes which may not be segregating in the mapping line and further crosses are planned for spring 1994. Insect resistances include those for woolly apple aphid (Er1, Er2 and a resistance out of *M. sieboldii*), rosy apple aphid (Sm_h), and rosy leaf curling aphid (Sd_3). Resistances against disease include powdery mildew (Pl1 and Pl2) and apple scab (Vf), as well as a possible resistance against silverleaf. Plant growth attributes under consideration are burrknot formation (bu_1, bu_2), suckering (from Robusta 5), dwarfing (from M9), leaf size (M26 $\times$ Robusta 5), blossom type (M26 $\times$ Robusta 5). Physiological characters which are predicted to segregate in our crosses for mapping include duration of the juvenile period, winter chilling requirement, cold and heat tolerance of the fruit and timing of fruit maturity. In addition, the breeding population of 500 genotypes in the apple quantitative genetics programme at HortResearch (Noiton & Shelbourne, 1992) will provide the long term resource for a range of characters not found in the current cultivar breeding programme.

Map construction

Our aim is to produce a map constructed with a backbone of RFLP and isoenzyme markers and fleshed out with RAPDs. This map will be able to be used in the shorter term to provide markers for marker assisted selection by breeders, and in the long term will enable map-based cloning of a specific gene. Mapping of individual loci comprising quantitative traits will also provide a tool to increase the understanding of some aspects of apple physiology.

The isoenzyme and RFLP markers will provide the anchor points to correlate our map with maps being constructed in the USA and Europe (Hemmat et al., 1993; King et al., 1991). RAPDs should be useful genetic tags for use by breeders in marker assisted selection and also for the fine mapping leading to gene cloning. We plan to use bulk segregant analysis (Michelmore et al., 1991; Williams et al., 1993) to identify RAPD markers which are linked to genes of particular interest.

Anonymous RFLPs are being generated by probing Southern blots of DNA which has been isolated from expanding leaf tissue and digested with a range of restriction enzymes. Our probes are clones from a cDNA library from ripe apple cortical tissue (Ross et al., 1992). Probes for genes with known function and gene specific PCR primers have been made available to us by other collaborators (R. Atkinson, M. Lay-Yee, G. King, G. Ross, C. Simon). We have developed the use of the Amersham ECL non-radioactive detection system for routine detection of hybridization of high copy number (Nybom et al., 1992) and low copy number probes in the apple system. We have found that the system not only has the advantage of safety, but that it also offers a quick (overnight) turn-around for blots.

The RAPD technique (Williams et al., 1990) which uses the PCR reaction and single random 10 base primers to generate amplified fragments of DNA from a genomic template has proved to be a rich source of markers in apple for us and others (Hemmat et al., 1993). We found that technique required initial careful optimization, but after this it is quick and relatively simple to perform on a routine basis. About 50% of primers tested reveal one or more polymorphisms segregating amongst the individuals in our mapping line.

Protocols involving electrophoresis of leaf extracts on both starch and polyacrylamide gels are now established at the HortResearch Havelock North Research Orchard (D. Noiton) for 10 isoenzyme systems and the mapping line will be screened for polymorphisms using these initially.

Marker linkage groups will be assigned to physical chromosomes by in situ hybridization of key RFLP probes from each marker linkage group to metaphase chromosome spreads.

Gene isolation

We are currently initiating a programme for map-based cloning of disease resistance genes from apple which could be used to introduce these characters into commercial cultivars by transformation. If we obtain this funding, we then will use our initial gross structure map as a starting point for construction of a fine structure map around the designated gene. Techniques will be developed for cloning large (> 250 kb) apple chromosome fragments into YACs so as to reduce to a minimum the number of walking steps required to isolate the genes. We envisage that we will need to generate a library of approximately 10,000 clones to cover 3n to 5n apple genomes.

Progress

Map construction

A major effort is being made to accumulate marker segregation data in the mapping line of seedlings. These 90 seedlings have so far been screened for the segregation of more than 160 molecular markers. Approximately a third of these are RFLPs and the remainder are RAPDs. Initial joint segregation analysis using LINKAGE-1 has been made for a third of the data set.

In situ hybridization

Apple rootstocks have been used as model systems to develop protocols for in situ hybridization of low and high copy number probes to metaphase chromosomes prepared from root tip squashes. Non-radioactive labelling of probes coupled with colour detection systems has proved a reliable and rapid technique for determining the sites of hybridization of probes of high and low copy number to the very small (1.0–3.5 μm) chromosomes of apple. The methods developed will be used for probing chromosome preparations from 'Scijoy' as soon as we have a ready source of self-rooted plants of this cultivar.

Hybridization of an apple rDNA probe labelled with digoxygenin and comprising the 18S coding plus spacer sequences has been localised to the telomeric regions of the short arms of chromosomes 6 and 14 of 'MM106'. These in situ results were confirmed both by silver staining for NORs and also for nucleoli in interphase nuclei (Zhu & Gardiner, in preparation).

Table 1. Chromosomal assignment in 'A106' (*Malus sieboldii*) of three genes associated with apple ripening

Probe (no. of replicates)	Chromosome assignment		Arm	Position
	> 80% spreads	50% spreads		
1. ACC synthase (25)	9		short	near telomere
	10		short	near telomere
	12		short	near telomere
		17	short	near telomere
2. ACC oxidase (16)	1		long	near centromere
	5		short	near telomere
	11		short	near telomere
		13	long	near centromere
3. Endopolygalacturonase (18)	6		long	near centromere
	11		long	near centromere
	14		long	near centromere
		3	short	near telomere

A total of three probes related to apple fruit ripening (gifts of M. Lay Yee and G. King) have been tested for hybridization to metaphase chromosomes of 'A106' (*M. sieboldii*) using biotin labelling and a streptavidin-alkaline phosphatase linked colour detection system. Our analysis indicates that both the genes for ACC synthase and ACC oxidase which are required for ethylene biosynthesis, and endopolygalacturonase, which is involved in cell wall softening (Dilley & Wilson, 1992; Ross et al., 1992), are located on independent chromosome sites in apple and are not clustered (see Table 1). These results are consistent with our RFLP data for these probes. For each probe, hybridization was detected to three different chromosomes in at least 80% of the 16–25 metaphase chromosome spreads analyzed, indicating a relatively close homology between probe and binding site for these chromosomes. The hybridization detected to a single chromosome for each probe in 50% of spreads suggests a divergence in sequence for each gene on one chromosome.

Conclusion

The New Zealand Apple Genome Mapping Project is making steady progress towards producing a primary genetic map of apple with molecular marker linkage groups to be correlated with the physical chromosomes. Traits of importance to breeders which are planned to be included on this map include a variety of pathogen resistances, plant growth attributes and physiological characters. The project was developed in association with the HortResearch apple breeders, for the benefit of the New Zealand apple industry and there are strong links with the other apple molecular biology programmes within HortResearch. The existing collaboration with the two other apple gene mapping projects internationally (that of N. Weeden and colleagues at Geneva, N.Y., U.S.A., and the European project coordinated by G. King at H.R.I., Wellesbourne, England) will be a key factor in developing systems for using molecular biology to aid apple breeding.

Acknowledgements

This research is funded both by the New Zealand Foundation for Research, Science and Technology (Contract Number CO6209) and also by the New Zealand Apple and Pear Marketing Board. We sincerely thank Allan White, Dominique Noiton, Vincent Bus and Michael Malone for their ongoing input into the project.

References

Dilley, D.R. & I.D. Wilson, 1992. Molecular biological investigations of gene expression attending fruit ripening: Current status and future prospects. HortTechnology 2: 294–301.

Hemmat, M., N.F. Weeden, A.G. Manganaris & D.M. Lawson, 1993. Molecular marker map for apple. J. Heredity, in press.

King, G.J., F.H. Alston, I. Battle, E. Chevreau, C. Gessler, J. Janse, P. Lindhout, A.G. Manganaris, S. Sansavini, H. Schmidt & K. Tobutt, 1991. The 'European Apple Genome Mapping Project' – Developing a strategy for mapping genes coding for agronomic characters in tree species. Euphytica 56: 89–94.

Michelmore, R.W., I. Paran & R.V. Kesseli, 1991. Identification of markers linked to disease-resistance genes by bulked segregant analysis: A rapid method to detect markers in specific genomic regions by using segregating populations. Proc. Natl. Acad. Sci. USA 88: 9828–9832.

Noiton, D. & C.G.A. Shelbourne, 1992. Quantitative genetics in an apple breeding strategy. Euphytica 60: 213–219.

Nybom, H., S. Gardiner & C.J. Simon, 1992. RFLPs obtained from an rDNA probe and detected with enhanced chemiluminescence in apples. HortScience 27: 355–356.

Ross, G.S., M.L. Knighton & M. Lay-Yee, 1992. An ethylene-related cDNA from ripening apples. Plant Mol. Biol. 19: 231–238.

Tanksley, S.D., N.D. Young, A.H. Paterson & M.W. Bonierbale, 1989. RFLP mapping in plant breeding: New tools for an old science. Bio/Technology 7: 257–264.

Williams, J.G.K., A.R. Kubelik, K.J. Livak, J.A. Rafalski & S.V. Tingey, 1990. DNA polymorphisms amplified by arbitrary primers are useful as genetic markers. Nucleic Acid Research 18: 6531–6535.

Williams, J.G.K., R.S. Reiter, R.S. Young & P.A. Scolnik, 1993. Genetic mapping of mutations using phenotypic pools and mapped RAPD markers. Nucleic Acids Research 21: 2697–2702.

Zhu, J.M. & S.E. Gardiner, 1993. Chromosomal localization of ribosomal DNA sequences in an apple (*Malus*) rootstock cultivar using a non-radioactive detection system. In preparation.

H. Schmidt and M. Kellerhals (Eds.), Progress in Temperate Fruit Breeding, 281–285.

Identification of a RAPD marker linked to the *Vf* gene for scab resistance in apples

Hongyu Yang & Jutta Krüger
Federal Centre for Breeding Research on Cultivated Plants, Institute for Breeding of Ornamental Species, Bornkampsweg 31, D-22926 Ahrensburg, Germany

Key words: Apple scab, DNA marker, *Malus floribunda*, *Venturia inaequalis*, *Vf* gene

Summary

Apple scab, caused by *Venturia inaequalis* (Cke.) Wint., is the most serious disease of apple trees in many areas of the world. Resistance to *V. inaequalis*, derived from the small fruited species *Malus floribunda* 821, is determined by a major dominant gene *Vf*. Our objective was focused on the identification of RAPD (random amplified polymorphic DNA) markers linked to the *Vf* gene. The approach in this paper is based on the introgression of the *Vf* gene from *M. floribunda* into commercial varieties. Comparing RAPD patterns of a genomic DNA sample of *M. floribunda* with a pooled DNA sample of resistant individuals and that of 10 susceptible commercial apple varieties, we identified fragments which are derived from *M. floribunda*. One of them, *OPD20/600*, proved to be linked to the *Vf* gene by cosegregation analysis with two backcross progenies. It is the first DNA marker so far for the scab resistance.

Introduction

The resistance to *V. inaequalis*, derived from the small fruited species *Malus floribunda* 821, is determined by a major dominant gene assigned as *Vf*. The *Vf* gene was most widely used in many breeding programmes, because it is the most stable one (Hough, 1944; Lespinasse et al., 1979). Since 1978 apple breeding for scab resistance is done at Ahrensburg, using varieties or clones carrying the resistance source from *M. floribunda* 821. In backcross progenies with one parent possessing the *Vf* gene a segregation of 1 : 1 for resistance was observed (Krüger, 1983), indicating that *Vf* segregates as a single locus trait. An isozyme locus, *Pgm1*, was found to be linked to the apple scab by Manganaris (1989). In this paper identification of DNA sequence linked to the *Vf* gene is described.

Material and methods

Assessment of scab infection

The scoring of the level of resistance to scab was performed according to the following classification: 1. no sporulating spots; 2. small sporulating spots on single leaves; 3. larger or numerous small sporulating spots on several leaves; 4. a large part of the leaves is covered with sporulating spots. Plants in classes 1 and 2 with only occasional sporulation were considered to be resistant and to have the *Vf* gene.

Bulk composition and RAPD analysis

Two bulks were composed for RAPD analysis according to Michelmore et al. (1991). One bulk consisted of DNA samples from 5 resistant individuals of the segregating progeny '81/10' derived from a cross between 'Prima' and a mildew (*Podosphaera leucotricha*) resistant clone 'A142/5' from East Malling was used to identify identical segments. 'Prima' was selected from a *M. floribunda* 821 backcross progeny in the 4th generation (BC4) and contains the *Vf* gene. 'A142/5' is derived from the cross between 'Jonathan' and *Malus robusta* and possesses the gene *Pl1* for mildew resistance. The other bulk was composed of DNA samples of 10 susceptible varieties: 'Golden Delicious', 'Cox's Orange Pippin', 'Gravensteiner', 'Jonathan', 'Idared', 'James Grieve', 'McIntosh', 'Melba', 'Boskoop', and 'Granny Smith'. Some of these had been used as recurrent parents in backcrosses. The original clone 821 of *M. floribunda* was not available. In place of it a scab resistant *M. floribunda* clone from the Pillnitz collection was taken for the following analysis.

DNA was prepared from fresh leaves by using the simple isolation method described by Colosi and Schaal (1993). The samples were assayed in comparison with a standard after gel electrophoresis and contained approximately 20 ng/μl DNA. The PCR procedure described by Williams et al. (1990) was followed with minor modifications.

Cosegregation analysis

To check potential cosegregation of DNA markers and disease phenotype, primers which generated an identical fragment both in *M. floribunda* and the resistant bulk were analysed on individuals of known resistance phenotypes from the progeny '90/15' and that of a subset of the progeny '93/22'. '90/15' consisted of 28 individuals derived from a BC6 between the scab resistant selection '81/19–35' with the *Vf* gene and the variety 'Margol'. '93/22' is derived from a cross between '81/19–35' and the mildew resistant seedling '87/7–10' with the gene Pl_2 from *M. zumi*. This progeny has just passed the first assessment for scab resistance in the greenhouse. Of 209 plants, 48 heavily scabed ones were analysed in comparison to a subset of 110 plants randomly chosen from the remaining progeny.

Measurements of linkage were computered according to Mather (1938). Chi-square tests were used to confirm the linkage of the DNA marker identified to scab resistance. The recombination frequency between the RAPD marker and the resistant phenotype was calculated as:

$$p = \frac{(r_1 + r_2)}{n}$$

where p = recombination frequency, r_1, r_2, and n are the number of recombinants resistant/no marker, recombinants susceptible/having marker, and total individuals respectively. The standard error was estimated as:

$$S_p = \sqrt{p(1-p)/n}$$

Results

Screening RAPD fragments originating from M. floribunda

In the resistant bulk approximately 423 discernible fragments, ranging from 0.4 to 4.0 kilo basepairs, were amplified by 131 out of 170 primers screened (average of 3.2 products per primer). No fragments were produced by 29 primers at the given amplification conditions. Besides fragments that are present in all samples (48.9%), a large part (45.4%) of the products was identical in both, resistant and susceptible varieties bulk. As expected, only few fragments that appeared in both *M. floribunda* and the resistant bulk but not in the varieties bulk, were observed (1.4%). These products could reflect the presence of chromosomal segments of the donor parent in the resistant plants. Few fragments were found exclusively in the resistant bulk (4.3%, Table 1). This leads to the suggestion of a linkage between the *Vf* gene and the specific fragments.

Cosegregation analysis

To confirm the putative linkage between the fragments originating from *M. floribunda* and the *Vf* gene, the progeny '90/15' which segregated for scab resistance, was analysed. Of the RAPD fragments derived from *M. floribunda*, most segregated independently to scab resistance, but one appeared to be linked to the *Vf* gene (Tables 2 and 3). Figure 1 shows the RAPD fragments amplified by *OPD20*. The fragment of approximately 600 bp cosegregating to scab resistance is marked and designed *OPD20/600*. The amplification of OPD20/600 was highly repeatable, enabling clear visualization on agarose gels.

Confirmation of the cosegregation

According to the reaction to *V. inaequalis* after the first infection in the greenhouse 77 plants of the progeny '93/22' were scored as susceptible, the others remained for further scab tests. Forty-eight out of the 77 plants infected by *V. inaequalis* were screened with the primer *OPD20*. Except for 8, all susceptible plants screened did not show the RAPD fragment *OPD20/600* found in the resistant bulk. The screening of the randomly chosen 110 remaining plants resulted in a segregation of 88 positive to 22 negative (Table 4).

Table 1. Comparison of fragments amplified in *M. floribunda*, resistant bulk, and varieties bulk

Fragments present in all three samples		Fragments present in both res. -and var. bulk		Fragments present both in res. bulk and *M. floribunda*		Fragments present only in res. bulk	
n	%	n	%	n	%	n	%
192	45.4	207	48.9	6	1.4	18	4.3

Table 2. Segregation of scab resistance and the DNA fragment, originating from *M. floribunda* in the progeny '90/15'

Seedling	Reaction to scab	*OPD20/600*	Seedling	Reaction to scab	*OPD20/600*
1	R	–	9	S	+
2	R	+	11	S	–
3	R	+	15	S	–
4	R	+	18	S	–
5	R	+	21	S	–
6	R	+	22	S	–
7	R	+	23	S	–
8	R	–	24	S	–
10	R	+	25	S	–
12	R	+	26	S	–
13	R	–	27	S	–
14	R	–	28	S	–
16	R	–			
17	R	+			
19	R	–			
20	R	+			

Discussion

The approach to tag the DNA sequences linked to the *Vf* gene in this paper relies on the availability of resistant material from backcross progenies, which have been developed by introgression at Ahrensburg. They have been found to segregate nearly 1 : 1 for the *Vf* gene from *M. floribunda* (Krüger, 1983). Introgression was accomplished by repeatedly backcrossing the resistant lines carrying the *Vf* gene to susceptible varieties (recurrent parent) having otherwise desirable properties. After each cross, resistant plants were selected that were supposed to possess the *Vf* gene for scab resistance. This process resulted in lines that probably only carry a small part of DNA sequences from the donor parent *M. floribunda* 821. Because the commercial apple varieties (*M.* × *domestica*) are sufficiently different from *M. floribunda* 821 in the genetic background, it should be possible to detect few identical segments in both, donor parent *M. floribunda* 821 and the resistant selections from the backcross progenies. Markers that detect such identical segments, which are absent in the recurrent parents, will likely be linked to the target gene, if most of the resistant individuals carry the same DNA segment.

An application of this strategy resulted in identifying 6 significant RAPD fragments which appeared both in the resistant bulk and *M. floribunda*. One of them has been shown to be linked to the *Vf* gene. The high recombination rate of 25% in the progeny '90/15' may due to the small amount of plants analysed. This strategy could also be used in the identification of molecular markers linked to other important characters which are introgressed from wild species to commercial apples, e.g. mildew resistance and anthocyanin production.

Although the scab resistance is a qualitative trait, screening a new progeny for the presence of the *Vf* gene based on the phenotype requires several years, since

Table 3. Linkage analysis of scab resistance and the RAPD fragment originating from *M. floribunda* in the progeny '90/15'

Locus		Phenotype*							Recombination
A	B	AB	Ab	aB	ab	χ^2_A	χ^2_B	χ^2_{AB}	frequency (%)
Vf	*OPD20/600*	10	6	1	11	0.57	1.28	7.00**	25.0 ± 8.2

* Phenotypes: AB: scab resistant and has *OPD20/600*; Ab: scab resistant without *OPD20/600*; aB: scab susceptible and shows *OPD20/600*; ab: scab susceptible without *OPD20/600*.
** Significant at the 0.01 probability level.

Table 4. Cosegregation for resistance and RAPD marker in the progeny '93/22'

subset of	total of plants	*OPD20/600*		
		present	absent	Recombination frequency (%)
susceptible plants	48	8	40	20.0 ± 3.2
selected plants*	110	88	22	

* In the selected plants susceptible individuals may be present, so the true recombination frequency may be lower.

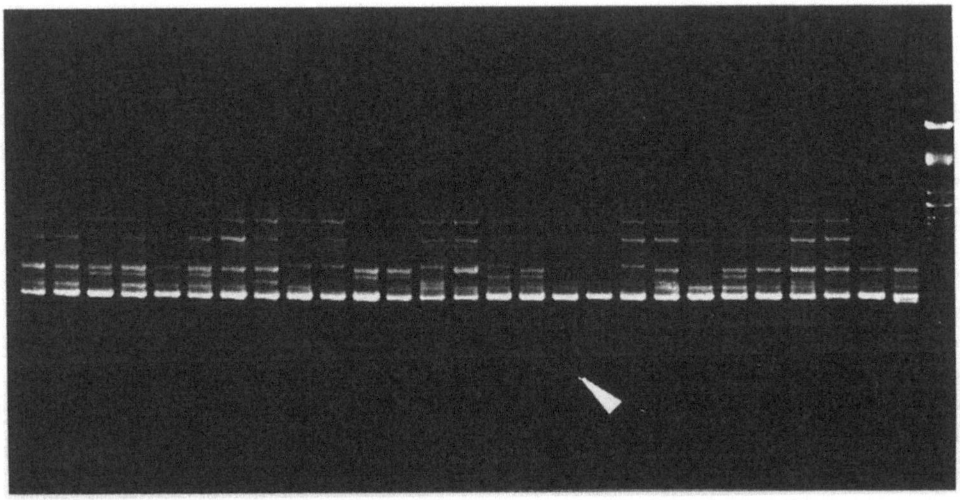

Fig. 1. RAPD fragments amplified by *OPD20*, the marked one having 600 bp cosegregates scab resistance.

the expression of the resistant/susceptible gene can be influenced by the enviroment. Some individuals may escape infection even when susceptible. In this case, *OPD20/600* may be a valuable initial screening tool for segregation analysis to confirm a putative linkage of an other marker to the target gene, because it is relatively easy to score.

References

Colosi, J.C. & B.A. Schaal, 1993. Tissue grinding with ball bearing and vortex mixer for DNA extraction. Nucleic Acids Research 21(4): 1051–1052.

Hough, L.F., 1944. A survey of the scab resistance of the foliage on seedlings in selected apple progenies. Proc. Amer. Soc. Hort. Sci. 44: 260–271.

Krüger, J., 1983. Sources of resistance to apple scab (*Venturia inaequalis*) used at the Federal Research Center for Horticultural Plant Breeding at Ahrensburg. WPRS Bull. III/4: 123–128.

Lespinasse, Y., J.M. Olivier & M. Godicheau, 1979. Études entreprisés dans le cadre de la résistance à la tavelure du pommier, Proc. Eucarpia Fruit Section Symposium Tree Fruit Breeding Angers, Sept. 3.–7, 1979: 97–105.

Manganaris, A.G., 1989. Isozymes as genetic markers in apple breeding. PhD thesis, University of London.

Mather, K., 1938. The Measurement of Linkage in Heredity. Methuen Co. Ltd, London.

Michelmore, R.W., I. Paran & R.V. Kesseli, 1991. Identification of markers linked to disease resistance genes by bulked segregant analysis: A rapid method to detect markers in specific genome regions by using segregating populations. Proc. Natl. Acad. Sci. USA 88: 9828–9832.

Williams, J.G.K., A.R. Kubelik, K.J. Livak, A. Rafalsky & S.V. Tingey, 1990. DNA polymorphisms amplified by arbitary primers are useful as genetic markers. Nucleic Acids Res. 18: 6531–6535.

H. Schmidt and M. Kellerhals (Eds.), Progress in Temperate Fruit Breeding, 287–290.

Selection of a mutant from adventitious shoots formed in X ray treated cherry leaves and differentiation of standard and mutant with RAPDs

Hongyu Yang & Hanna Schmidt
Federal Centre for Breeding Research on Cultivated Plants, Institute for Ornamental Plant Breeding, Bornkampsweg 31, D-22926 Ahrensburg, Germany

Key words: adventitious shoots, cherry, chimeras, mutation breeding, *Prunus cerasus*, RAPD marker, X ray

Summary

An obstacle when using scions or *in vitro* shoots for mutation induction is the occurrence of chimeras. When adventitious shoots are formed from irradiated material these usually are derived from single cells, this leading to homohistont mutants. Since *Prunus avium* regenerates adventitious shoots from leaves at a low rate only (Yang & Schmidt, 1992), leaves of the interspecific cherry rootstock '209/1' (*P. cerasus* × *P. canescens*) were irradiated. '209/1' regenerates adventitious shoots readily. Dosages applied were 5, 10, 20 and 40 Gy. Shoot production following 5 Gy irradiation was similar to the control. The application of 40 Gy resulted in strong damage with only few leaves regenerating. Among the adventitious shoots from leaves irradiated with 20 Gy one shoot was evident already *in vitro* with thicker and smaller leaves having a serrate margin. It was cloned as '209/1-20m'. The clone stayed stable since 1990 *in vitro*, in the greenhouse and the field. Compared with standard '209/1' the mutant is very dwarf.

Research to differentiate between standard '209/1' and '209/1-20m' was done using RAPDs. Among the decamer primer kits D, J, and T from Operon Technologies, Calif. Only primer OPJ05 (5'CTCCATGGGG3') differentiated between '209/1' and '209/1-20m'. Rootstock '209/1' showed one band of 2 kb additionally. This band is missing in the mutant.

Introduction

For mutation breeding the production of non-chimeric mutants is highly desirable. Unstable mericlinal chimeras automatically occur when plant parts with multi-cellular apices are irradiated. Moreover, the so called intrasomatic selection in a multi-cellular apex often prevents the appearance of mutated cells. Chimeras can be converted into stable solid mutants by repeated pruning. Vegetative propagation is often necessary.

The possibility to regenerate plants from single cells would automatically result in a high percentage of solid mutants. The adventitious bud technique has successfully been used for mutation breeding of vegetatively propagated horticultural crops (Broertjes & van Harten, 1988). This method involves irradiation of, and adventitious shoot regeneration from detached leaves. The basis for the method is the origin of adventitious shoots from single cells.

Since adventitious shoots could be regenerated from *in vitro* leaves of cherry rootstock '209/1' (Yang et al., 1991), this genotype was used as model plant to investigate the potential significance of the adventitious shoot technique for mutation breeding of woody fruit plants. At first attention was paid to assay the radiosensitivity of *in vitro* leaves, because such informations are prerequisites for a mutation induction experiment. An induced mutant with changed leaves can be identified during the stages of *in vitro* shoot regeneration. This mutant was investigated morphologically and by means of PCR techniques.

Table 1. Adventitious shoot regeneration from leaves of '209/1' in relation to irradiation dosages, n = 60

Dose (Gy)	% of leaves forming shoots	Number of adventitious shoots/leaf
0	52a	2,0a
5	55a	1,9a
10	32b	1,1b
20	28b	0,6b
40	2	0

Values having different letters are significantly different at $\alpha = 5\%$.

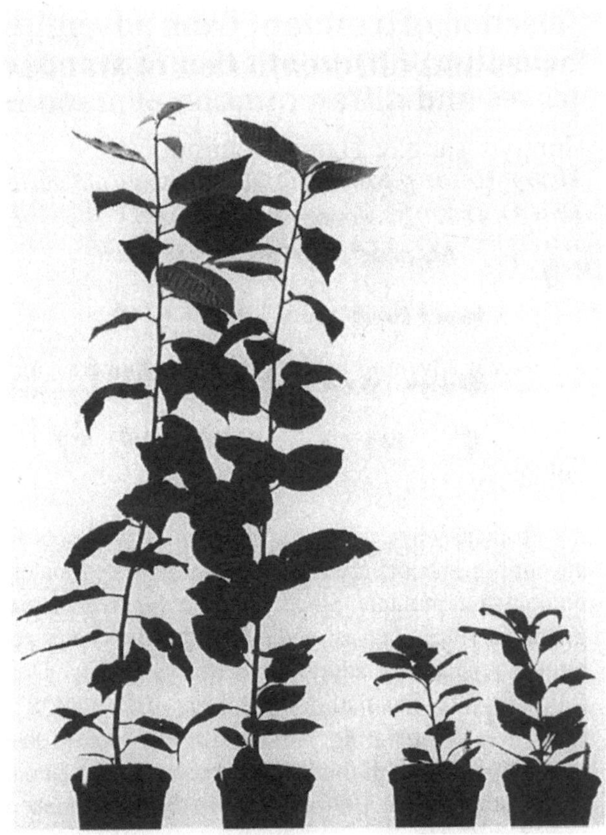

Fig. 1. Normal plant (l.) and mutant (r.) derived from irradiated leaf of '209/1' using 20 Gy.

Material and methods

The cherry rootstock '209/1', an interspecific hybrid derived from the crossing between *Prunus cerasus* × *P. canescens*, was used in all experiments.

Adventitious shoot regeneration from detached leaves
All leaves used are derived from *in vitro* propagated shoots. The media and environmental conditions previously described (Yang, 1992) were used.

Irradiation procedure
Immediately after plating, the detached leaves were irradiated with doses between 5, 10, 20 and 40 Gy, using an X-ray machine. The technical data were: 12 mA, 150 kV, 1.7 mm Al-filter; dose rate at a distance of 55 cm from the focus to the petridishes: 0.9 Gy min^{-1}. For the measurement of the radiosensitivity of the *in vitro* leaves the mean number of adventitious shoots per explant and the percentages of leaves forming shoots were estimated.

Identification of the mutant with RAPDs
DNA was prepared from fresh leaves by using the isolation method described by Colosi and Schaal (1993). The samples were assayed in comparison with a standard after gel electrophoresis and contained approximately 20 ng/μl DNA. Two microliter DNA sample containing app. 40 ng of genomic DNA were used for the PCR. The PCR procedure described by Williams et al. (1990) was followed with minor modifications (Yang & Krueger, 1993). The decamer primers of Kits D, J and T from Operon Technologies, Calif. were screened.

Results and discussion

Radiosensitivity of in vitro *leaves*

X-ray treated leaves were cultivated under optimal conditions for adventitious shoot regeneration. Differences were observed four weeks after the irradiation with regard to the mean number of shoots per explant and the percentage of leaves forming shoots (Table 1). The treatment by 5 Gy does not show significant irradiation damages, the treated leaves produced as many shoots as does the control. LD50 was calculated as nearly 20 Gy. Since only one out of fifty leaves treated with 40 Gy formed shoots this dose is regarded too high for *in vitro* leaves of '209/1'.

Mutant selection and identification with RAPDs

Among the adventitious shoots from leaves irradiated with 20 Gy one shoot was evident already *in vitro*

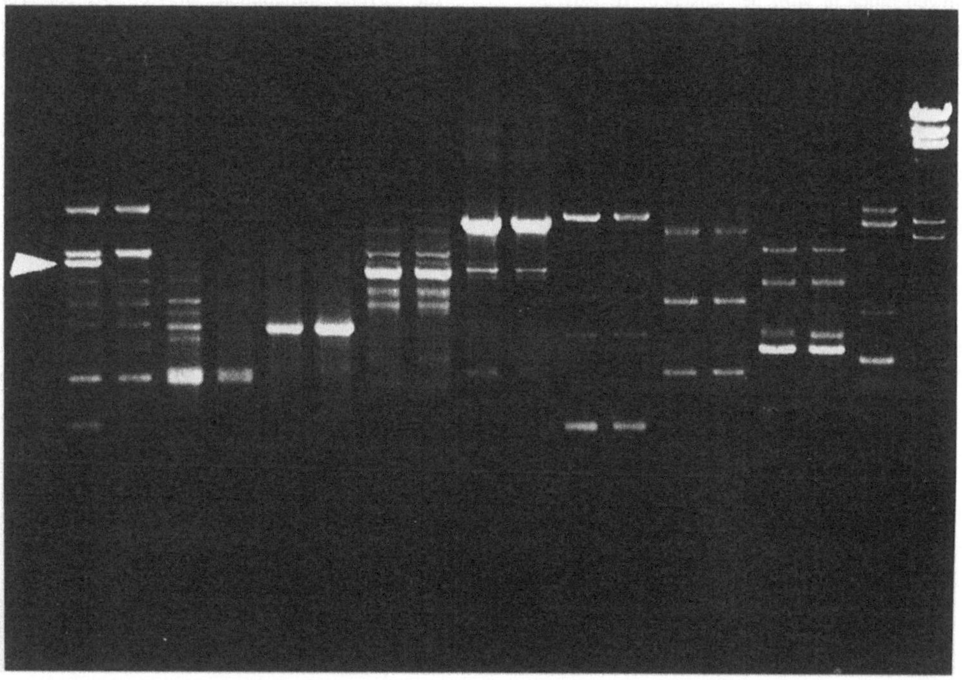

Fig. 2. Comparison of RAPD fragment patterns of '209/1' (l.) and its X ray mutant '209/1-20m' (r.), produced by different decamer primers, the lane 1 and 2 were amplified using primer OPJ05.

with thicker and smaller leaves having a serrate margin. Compared with standard '209/1' the mutant is very dwarf (Fig. 1). The mutant was cloned *in vitro*; several hundred propagated plants proved to be identical, indicating that '209/1-20m' is a homohistont mutant.

The screening of 60 decamer primers of Kits D, J and T resulted in the selection of primer OPJ05 which differentiated between '209/1' and '209/1-20m' as is shown in Fig. 2. The standard has one band of 2 kb additionally; this band is missing in the mutant, indicating that '209/1-20m' is a solid mutant. The band of two kb must be also present in the mutant when being a chimera.

A prerequisite for the induction of solid mutants using the adventitious bud technique is the single cell origin of shoots. Because regeneration from leaves is an unpredictable event, it was quite difficult to observe the patterns of shoot formation. The results presented in this paper give an indirect evidence for the origin of the adventitious shoot from single cells.

If it is possible to achieve a reasonably high number of leaves regenerating the adventitious bud technique could be of great potential value for mutation breeding of woody fruit plants as was shown in the herbacious plants (Broertjes et al., 1976; Broertjes, 1982).

References

Broertjes, C., 1982. Significance of *in vitro* adventitious bud techniques for mutation breeding of vegetatively propagated crops. In: IAEA, Induced Mutations in Vegetatively Propagated Plants II. Vienna.

Broertjes, C. & A.M. van Harten, 1988. Fruit Crops: In: C. Broertjes & A.M. van Harten (Eds) Applied mutation breeding for vegetatively propagated crops. Elsevier Scientific Publ. Co. Amsterdam, Oxford, New York, Tokyo.

Broertjes, C., S. Roest & G.S. Bokelmann, 1976. Mutation breeding in *Chrysanthemum morifolium* Ram. using *in vivo* and *in vitro* adventitious bud techniques. Euphytica 25(1): 11–19.

Colosi, J.C. & B.A. Schaal, 1993. Tissue grinding with ball bearing and vortex mixer for DNA extraction. Nucleic Acids Research 21(4): 1051–1052.

Williams, J.G.K., A.R. Kubelik, K.J. Livak, A. Rafalsky & S.V. Tingey, 1990. DNA polymorphisms amplified by arbitary primers are useful as genetic markers. Nucleic Acids Res. 18: 6531–6535.

Yang, H.Y., 1992. Untersuchungen zur *in vitro*-Regeneration, *in vitro*-Mutationsinduktion und Selektion bei Süßkirschen (*Prunus avium* L.), Ph. D. thesis, University of Hannover.

Yang, H.Y. & J. Krueger, 1993. Identification of a RAPD Marker linked to the *Vf* gene for scab resistance in apples, Plant Breeding, in press.

Yang, H.Y. & H. Schmidt, 1992. Untersuchungen zur Adventivsproß-regeneration *in vitro* bei Kirschen, II. Adventivsproßbildung an *in vitro*-Blättern verschiedener *Prunus avium*-Idiotypen. Gartenbauwiss. 57(1): 7–10.

Yang, H.Y., H. Schmidt & U. Kett, 1991. Untersuchungen zur Adventivsproßbildung *in vitro* bei Kirschen, I. Adventivsproßbildung an *in vitro* Blättern der Kirschunterlage 209/1. Gartenbauwiss. 56(5): 210–213.

H. Schmidt and M. Kellerhals (Eds.), Progress in Temperate Fruit Breeding, 291–294.

Towards a marker assisted breeding for resistance against apple scab

L. Gianfranceschi,[1] J.M. McDermott,[1] N. Seglias,[1] B. Koller,[1] M. Kellerhals[2] & C. Gessler[1]
[1] *Pathology group, Institute of Plant Sciences, Swiss Federal Institute of Technology, CH-8092 Zürich, Switzerland;*
[2] *Swiss Federal Research Station, CH-8820 Wädenswil, Switzerland*

Key words: Venturia inaequalis, apple scab, molecular markers

Summary

We present the first attempts to find molecular markers (RAPDs) in the apple genome linked to the *Vf* gene conferring resistance to scab. The availability of genetic markers will allow selection of individuals with resistance genes in progeny. Moreover genetic markers allow us to recognise individuals with minimal linkage drag in back-crossing experiments with wild species. The feasibility of the bulk-segregant analysis to identify linked markers has been demonstrated on several occasions (Martin et al., 1991; Giovannoni et al., 1991; Michelmore et al., 1991). In this paper we report the preliminary results proving that the method could also be applied successfully in allogamous plants.

Introduction

Globally today's agricultural production systems are either highly productive due to sophisticated technology and high industrial inputs or of low productivity with little technological input. High quantity and quality production systems, even if necessary for human nutrition, are now widely seen as not sustainable and are increasingly questioned for environmental and consumer acceptability. Crop protection by intensive use of pesticides is one of the most questionable aspects. Alternative strategies and biocontrol have brought advances in pest control and in some cases also in weed control. Fungal disease control still relies on fungicides to a major extent, although the development of new fungicides has dramatically reduced toxicity towards non-target organisms and chemicals are being used in a more controlled condition (e.g. IP-strategies). Further improvement towards safer fungal pathogen control can be achieved through resistance breeding and eventually plant genetic engineering. In apple breeding, for example, the introduction of resistance against the major disease, scab caused by *Venturia inaequalis*, mostly from wild *Malus* sources, has a long tradition (Alston, 1989; Kellerhals, 1989a; Lespinasse, 1989), with many cultivars available, although most have resistance from *M. floribunda* 821 (*Vf*, a dominant major gene) (Crosby et al., 1992). However, this resistance has recently been overcome by a specific race (Parisi et al., 1993). The additional use of ephemeral resistance genes in breeding strategies to render the resistances more durable has already been discussed (Gessler, 1993; Gessler & Blaise, 1994).

A major requirement for pyramiding resistance genes was to recognise functionally different resistances in order to select progeny accordingly. In most cases, vertical resistance (Vanderplank, 1982) confers resistance to its carrier against some isolates of the pathogen while others can easily attack the plant. This type of resistance is often controlled by a single gene (often called R 'major gene' or R-gene). By using the corresponding pathogen genotypes, different resistances can be distinguished.

Other resistance genes derived from wild *Malus* species (*Vf, Va, Vm, Vr, Vbj*) have been recognised and it has also been shown that old cultivars, as well as particular modern cultivars, carry functionally different resistances (Gessler, 1994). These resistances could be used in apple breeding, especially when complemented with appropriate planting strategies (Gessler & Blaise, 1994).

To recognise functionally different resistances, selection should be based on serial testing with various races, but this is impractical. A more powerful approach than selecting resistance in the greenhouse or field is to identify and mark the genome segment carrying the resistance and to select only those individuals in the progeny carrying that particular DNA segment. The success of such a method is based on the reasonable assumption that resistances at different loci are also functionally different.

In recent years molecular biology has provided the tools suitable for rapid and detailed genetic analysis. Perhaps the most fundamental of these are the molecular markers, and especially the DNA markers. There are mainly two different approaches that could be used. One is the construction of a linkage map of molecular markers to survey the whole genome in order to detect co-segregation between the gene of interest and the molecular markers linked to it. The second is the new and powerful approach of Bulked Segregant Analysis (Giovannoni et al., 1991; Michelmore et al., 1991). This technique is based on the comparison between two DNA bulks, one made from resistant and the other from susceptible plants of a segregating population. Due to the selection, the bulks have completely random genotype all over the genome except in the region around the gene conferring the resistance. Because we select the plants for their resistance and because we know that the resistance is inherited as a single gene, the presence of polymorphism between the amplification pattern of the two bulks is expected only for those bands that are genetically linked to the resistance gene (Giovannoni et al., 1991). Because each band represents a specific amplified genomic DNA segment, markers differentiating the bulks are probably linked to the disease resistance genes. Such markers can then be analysed in a larger progeny for segregation with the trait and used to find further markers (Michelmore et al., 1991). The traits and the linked markers will segregate together and genetic distance can be calculated by appropriate methods from the number of cases of dissociation.

Another approach to genetic enhancement of host resistance is the direct insertion of the gene or genes of interest via biotechnology. For the moment this is not feasible on a large scale both because the technique still needs to be improved and second because there is no precise information about the R-loci. None of them has been cloned and characterised. The first step for the introduction of resistance genes either via marker assisted selection or via biotechnology is their identification and genomic localisation. For the biotechnological approach a further step is required that is the cloning and more accurate characterisation of the regulation and expression of those genes (Newbury, 1992).

In this paper we present the first results obtained in our laboratory applying the bulk-segregant analysis to detect markers for the *Vf*-resistance in apple.

Material and methods

The markers used in this work are the Random Amplified Polymorphic DNA. Those markers are generated by PCR amplification of random DNA segments with single primers of arbitrary nucleotide sequence (Williams et al., 1993). At the suitable annealing temperature, if the single primer binds to sites on opposite strands of the genomic DNA, that are within the amplifiable distance of each other, a discrete DNA segment is produced. The presence or absence of this specific product will be diagnostic for the oligonucleotide-binding site on the genomic DNA. Although the number of bands produced during a normal RAPD reaction vary depending on the genome complexity, generally an average of at least five bands is seen. The fact that the assay is non radioactive, requires only nanograms of DNA and is relatively fast, compared to RFLP, makes it suitable for large scale screenings.

The following crosses have been analysed: Redfree (*Vf*) × Pinova, Idared × *M. floribunda* 821 (*Vf*) and P22R24A8(*Vf*) × K1R11A26(*Vf*).

The *Vf*-resistance gene inheritance was identified, first in the greenhouse after infection with a spore suspension (Swiss Federal Research Station, Kellerhals, 1989b) and classification according to the reaction type from 0 to 4 proposed by Chevalier et al. (1991). Secondly, the progenies were both macro and microscopically re-tested in a greenhouse with environmental controlled conditions following a published protocol (Gessler, 1992).

Besides the resistance assessment we also carried out analysis of the electrophoretic pattern of the Phosphoglucomutase (PGM) isoenzyme (protocols kindly provided by Dr. Elisabeth Chevreau INRA, Angers, France) as preliminary results obtained by Dr. Manganaris indicated that the PGM-1 locus is linked (about 6 cM) to the *Vf*-locus (Manganaris, personal communication).

Table 1. Classification of the progeny of two crosses after macroscopic assessment as proposed by Chevalier et al. (1991)

Crosses	Classes 0	1	2	3a	3b	4
P22R24A8(*Vf*) × K1R11A26(*Vf*)	65	20	160	53	18	31
Idared × *M. floribunda* 821 (*Vf*)	120	17	122	29	13	55

0 = resistant to 4 = fully susceptible.

Results

After microscopic assessment of the Redfree × Pinova progeny, we found 16 resistant plants. Unfortunately, most of them had red leaves and were smaller than normal. For this reason we decided not to continue the analysis on the cross since some other factor might influence the expression of the *Vf* gene (e.g. interference of some other molecule present in the leaf with the *Vf* expression or with ontogenic resistance since the plants were smaller then normal). We then analysed the two other crosses. The results of the first greenhouse test are presented in Table 1.

Because it is very important for the bulk-segregant analysis that none of the plants composing the bulks is mis-classified a second analysis was performed following the fungal stages of development microscopically. For this purpose all the plants classified as '0' and '1' in the previous test were chosen. After microscopic assessment only about half of the plants classified as resistant in the first test could be considered the same also after the second test.

We used the second and more restrictive evaluation method to choose the resistant plants to construct the bulk. The PGM isozyme analysis was also performed. About 20 resistant and 20 susceptible plants from each cross have been tested. It was not possible to use the isozyme system in the Idared × *M. floribunda* 821 cross as the two parental lines have the same allelic variant. This makes it impossible to evaluate in the progeny from which parental line the allele derives and therefore if it is the one that is linked to *Vf*.

In the second cross, between P22R24A8 × K1R11A26, we saw a strong distortion in the segregation of PGM-1 both among the resistant and the susceptible plants. This confirms that the PGM-1 locus can be used as a linked marker.

We then constructed the bulks according to both resistance classification and PGM-1 pattern.

Primary screening of the bulks is currently in progress. We have preliminary results showing a polymorphic band between the bulks from the cross P22R24A8 × K1R11A26. The band is amplified, as expected, only in the resistant bulks. However at the moment we cannot be more precise since the analysis of the linkage to the *Vf* gene is not completed.

Discussion

The bulked-segregant analysis has been used successfully in e.g. lettuce (Michelmore et al., 1991), tomato (Martin et al., 1991) and bean (Haley et al., 1993), to find markers linked to genes for resistance. Although all of these examples were in autogamous plants, there is no reason why the same encouraging results could not be obtained also in apple and other woody plants even though the genetic analysis of heterogamous plants is more complex.

In apple, the availability of markers for several differential resistance genes, including the unrecognised resistance genes of older cultivars, should allow selection of individuals with several resistance genes in an appropriate progeny.

Recognition of ephemeral resistance genes in commercial cultivars would be greatly facilitated by a set of DNA-probes able to detect those differential resistance genes. It would be also possible to provide each cultivar with a key denoting its resistances which in turn would facilitate the choice of optimum partners for a cultivar mixture at a particular site. Such information is fundamental to the use of apple cultivar mixtures proposed by Gessler & Blaise (1994).

Moreover, breeding perennials pose several additional problems, not least that back-crossing is not true back- crossing because the recurrent parent changes from generation to generation (Wolfe & Gessler, 1992). A genetic marker method that allows us to recognise progeny individuals with minimal linked segments can reduce considerably the number of progeny selected at an early stage in the breeding programme.

Acknowledgements

We thank all the European Apple Project components as well as Dr. N.F. Weeden and Dr. S.E. Gardiner for fruitful discussions and comments. Plant material was kindly provided by the Swiss Federal Research Station Wädenswil. We thank Martin Wolfe for critical reading of the manuscript. The authors acknowledge the support by the Swiss National Foundation for Scientific Research grants 31-29928.90 & 31-36271.92, grant NF-Biotechnology Module 6 5002-034614, Swiss Federal Office for Education and Science grant no. A6010.

References

Alston, F.H., 1989. Breeding pome fruits with stable resistance to diseases. In: C. Gessler, D. Butt & B. Koller (Eds) Integrated control of pome fruit diseases II, pp. 90–99. OILB-WPRS Bulletin XII/6.

Chevalier, M., Y. Lespinasse & S. Renaudin, 1991. A microscopic study of the different classes of symptoms coded by the *Vf* gene in apple for resistance to scab (*Venturia inaequalis*). Plant Pathology 40: 249–256.

Crosby, J.A., J. Janick, P.C. Pecknold, S.S. Korban, P.A. O'Connor, S.M. Ries, J. Goffreda & A. Voordeckers, 1992. Breeding Apples for Scab resistance: 1945–1990. Fruit Varieties Journal 46: 145–166.

Gessler, C., 1992. Scab resistance in apple: the minor genes in the *Vf*-resistance. Acta Phytopathol. Entomol. Hungaricae 27: 257–263.

Gessler, C., 1994. Biology and biotechnology in strategies to control apple scab. Norwegian Journal of Agricultural Sciences. In press.

Gessler, C. & Ph. Blaise, 1994. Differential resistance in apple against scab and its use in breeding and in orchard planting strategies to control the disease. In: H. Schmidt and M. Kellerhals (Eds.) Progress in temperate fruit breeding, pp. 99–104. Kluwer Academic Publishers, Dordrecht.

Giovannoni, J.J., R.A. Wing, M.W. Ganal & S.D. Tanksley, 1991. Isolation of molecular markers from specific chromosomal intervals using DNA pools from existing mapping population. Nucleic Acid Research 19: 6553–6558.

Haley, S.D., P.N. Miklas, J.R. Stavely, J. Byrum & J.D. Kelly, 1993. Identification of RAPD markers linked to a major rust resistance gene block in common bean. Theoretical and Applied Genetics 86: 505–512.

Kellerhals, M., 1989a. Breeding of pome fruits with stable resistance to diseases. In: C. Gessler, D. Butt & B. Koller (Eds) Integrated control of pome fruit diseases II, pp. 116–129. OILB-WPRS Bulletin XII/6.

Kellerhals, M., 1989b. Breeding disease resistant apple cultivars in Switzerland. In: C. Gessler, D. Butt & B. Koller (Eds) Integrated control of pome fruit diseases II, pp. 130–136. OILB-WPRS Bulletin XII/6.

Lespinasse, Y., 1989. Breeding of pome fruits with stable resistance to diseases: Genes, resistance mechanisms, present work and prospects. In: C. Gessler, D. Butt & B. Koller (Eds) Integrated control of pome fruit diseases II, pp. 100–115. OILB-WPRS Bulletin XII/6.

Martin, G.B., J.G.K. Williams & S.D. Tanksley, 1991. Rapid identification of markers linked to a Pseudomonas resistance gene in tomato by using random primer and near-isogenic lines. Proc. Natl. Acad. Sci. USA 88: 2336–2340.

Michelmore, R.W., I. Paran & R.V. Kesseli, 1991. Identification of markers linked to disease-resistance genes by bulked segregant analysis: A rapid method to detect markers in a specific genomic region by using segregating population. Proc. Natl. Acad. Sci. USA 88: 9828–9832.

Newbury, H.J., 1992. Fungal resistance: The isolation of a plant R gene by transposon tagging. In: Gatehouse, Hilder & Boulter (Eds) Plant Genetic Manipulation for Crop Protection, pp: 109–134, CAB. Biotechnology in agriculture No. 7: 266.

Parisi, L., Y. Lespinasse, J. Guillaumes & J. Krüger, 1993. A new race of *Venturia inaequalis* virulent to apples with resistance due to the *Vf* gene. Phytopathology 83: 533–537.

Vanderplank, J.E., 1982. Host-Pathogen Interaction in Plant Disease. Academic Press, New York, 207 pp.

Williams, J.G.K., A.R. Kubelik, K.J. Livak, J.A. Rafalski & S.V. Tingey, 1990. DNA polymorphisms amplified by arbitrary primers are useful as genetic markers. Nucleic Acid Research. 18: 6531–5435.

Wolfe, M.S. & C. Gessler, 1992. Resistance genes in breeding: epidemiology: In: T. Boller & F. Meins (Eds) Plant gene research. Genes Involved in Plant Defense, pp. 3–23, Springer Verlag.

H. Schmidt and M. Kellerhals (Eds.), Progress in Temperate Fruit Breeding, 295–300.

Molecular classification of *Malus* with RAPD markers

F. Dunemann
Institute for Ornamental Plant Breeding, Federal Centre for Breeding Research on Cultivated Plants, Bornkampsweg 31, D-22926 Ahrensburg, Germany

Key words: Malus, RAPD markers, cultivar identification, genetic similarity, genetic relationships, taxonomy

Summary

The potential use of RAPD markers for taxonomic studies in *Malus* was investigated using sets of genotypes consisting of wild species and apple cultivars. DNA-'fingerprints' generated with preselected random primers were analysed for polymorphic amplification fragments, and coefficients estimating genetic similarity were calculated on the basis of about 50 polymorphic RAPD loci in each set of genotypes. Cluster analysis by an unweighted pair-group method with arithmetic averages (UPGMA) revealed that, in cultivars, the molecular classification was in good agreement with the known lineage. A dendrogram generated for wild species gave relationships that were, in principle, in accordance with the known phylogenetic information. Closely related species from section I were clearly distinguishable from those of sections III and IV. A high degree of genetic diversity was found on the molecular level among both different apple cultivars and wild species of the genus *Malus*. The results gave additional evidence for the hypothesis, that *M. pumila* and *M. sylvestris* were involved in the origin of the cultivated apples.

Introduction

RAPD markers have been used over the past three years for the efficient development of genetic maps, marker-assisted selection in breeding, DNA 'fingerprinting' for the assessment of individual identity, and for taxonomic studies (Tingey & Del Tufo 1993). The usage of the RAPD approach in apple has been described (Koller et al. 1993, Harada et al. 1993). With regard to the present taxonomic classification of *Malus*, many ambiguities still exist both within the cultivated apples and wild species of the genus. Many apple cultivars are of unknown or at least unclear origin and the genetic diversity even between some widely grown cultivars is not fully known. Furthermore, it can be assumed, that *Malus* × *domestica* is a complex interspecific hybrid with several, not yet all definitively known species involved in the process of domestication (Korban & Skirvin 1984, Way et al. 1990). Therefore, the genetic basis of all cultivated apples should be relatively broad.

The objective of the study presented here was to evaluate the potential taxonomic use of RAPD markers in apple cultivars and wild species. Considering known genetic relationships, the usefulness of genetic similarity data analysis based on polymorphic RAPD loci is investigated in three different sets of genotypes consisting of cultivars, wild species, and a combination of both, respectively.

Materials and methods

A set of 27 apple cultivars consisting of both economically important and old domestic cultivars was used in this study (set C, Table 1). 20 accessions of *Malus* wild species representing sections *Malus* (I), *Eriolobus* (III) and *Chloromeles* (IV) (set W, Table 2) were studied in a second experiment. In order to investigate relationships between cultivars and wild species in a common experiment, 9 cultivars from set C and 13 wild species from set W were combined in a third set designated as CW (see Fig. 2c). With the exception of *M. sieboldii* and *M. hupehensis*, clones from all wild accessions have been

Table 1. List of apple cultivars combined in set C and known lineage for cultivars originated under a contribution of cultivars involved in this study.

Cultivar	Abbreviations Used in Fig. 2 a–c	Known Lineage (as far as relevant for this study)
Ingrid Marie	IngMarie	possibly Cox Orange × unknown
Margol		Ingrid Marie × Golden Delicious
Elstar		Golden Delicious × Ingrid Marie
Golden Delicious	GoldDeli	
Jonagold		Golden Delicious × Jonathan
Jonathan		
Karmijn		Cox Orange × Jonathan
Kent		Cox Orange × Jonathan
Cox Orange	CoxOrang	
Fiesta		Cox Orange × Idared
Idared		Jonathan × Wagenerapfel
James Grieve	JamesGri	possibly Potts Seedling × Cox Orange
Jamba		Melba × James Grieve
Melba		McIntosh × unknown
Boskoop		
Gravensteiner	Gravenst	
McIntosh		
Granny Smith	GrannySm	
Prima		
Goldparmäne	Goldparm	
Glockenapfel	Gloapfel	
Gelber Edelapfel	GelbEdel	
Jacob Lebel	JacLebel	
Pirja		Huvitus × Melba
Gascoynes	Gascoyne	
Zuccalmaglio	Zuccalma	
S 13/1–30		Golden Delicious (Douds) × unknown

obtained from the genebank 'Obst' in Dresden-Pillnitz (Germany).

Total plant DNA for PCR was isolated from leaves using a modification of the CTAB protocols from King (HRI Wellesbourne pers. comm.) and Doyle & Doyle (1990). DNA amplification was performed using 29 random decamer primers purchased from Operon Technologies (Alameda, USA) (OP primer) and the Nucleic Acid-Protein Service Unit (University Vancouver, Canada) (UBC primer). The PCR was based on the protocol of Williams et al. (1990) with minor modifications. The reaction mix for a 25 ul reaction comprised 25 ng template DNA, 0.4 uM primer, 100 uM each of dNTPs and 0.5 units Taq polymerase. Amplification was performed in a Hybaid Omnigene thermocycler programmed as follows: cycle no. 1: 5 min 94 °C/1 min 35 °C/2 min 72 °C, cycles no. 2–39: 30 s 94 °C/1 min 35 °C/2 min (+ 2 s time increment/cycle) 72 °C, cycle no. 40: 30 s 94 °C/1 min 35 °C/9 min 72 °C. Amplified DNA fragments were separated in 1.3% agarose gels, stained with ethidium bromide and photographed under UV light with Polaroid films.

The amplification patterns were visually scored. A number was assigned to each scorable polymorphism and presence or absence of a band was coded by 1 or 0, respectively. Coefficients of genetic similarity, either simple matching (SM) or Jaccard's (J) were calculated using the specific programmes of the Numerical Taxonomy and Multivariate Analysis System program package for PC (NTSYS-pc, version 1.70, Rohlf 1993). The same software was used for cluster analysis of the SM and J coefficients, based on the UPGMA method, and dendrograms have been constructed.

Table 2. List of *Malus* wild species combined in set W and their sectional classification or lineage according to Way et al. (1990), section I: *Malus* with subsections A: *Pumilae*, B: *Sieboldianae*, C: *Kansuensis*, section III: Eriolobus, section IV: Chloromeles.

Wild species	Accession No.	Abbreviations Used in Fig. 2b + c	Section/Subsection	Lineage
M. pumila	188	pumila	I (A)	
M. hupehensis 'Rosea'	?	hupehens	I (A)	
M. sylvestris	84	sylvestr	I (A)	
M. prunifolia	175	prunifol	I (A)	
M. micromalus	158	micromal	I (A)	
M. baccata jackii	419	baccjack	I (A)	
M. baccata sibirica	349	baccsibi	I (A)	
M. baccata mandshurica	420	baccmand	I (A)	
M. sikkimensis	409	sikkimen	I (A)	
M. sieboldii	?	sieboldi	I (B)	
M. yunnanensis	258	yunnanen	I (C)	
M. trilobata	428	trilobat	III	
M. ioensis	368	ioensis	IV	
M. baccata flavescens	104	baccflav		= M. pumila
M. rockii	352	rockii		interspecific hybrid?
M. dawsoniana	121	dawsonia		M. fusca × M. pumila
M. robusta persicifolia	205	robusta		M. baccata × M. prunifolia
M. floribunda	54	floribun		unknown interspecific hybrid
M. toringo	27	toringo		= M. sieboldii? (Synonym?)
M. zumi calocarpa	274	zumi		M. baccata mandshurica × M. sieboldii

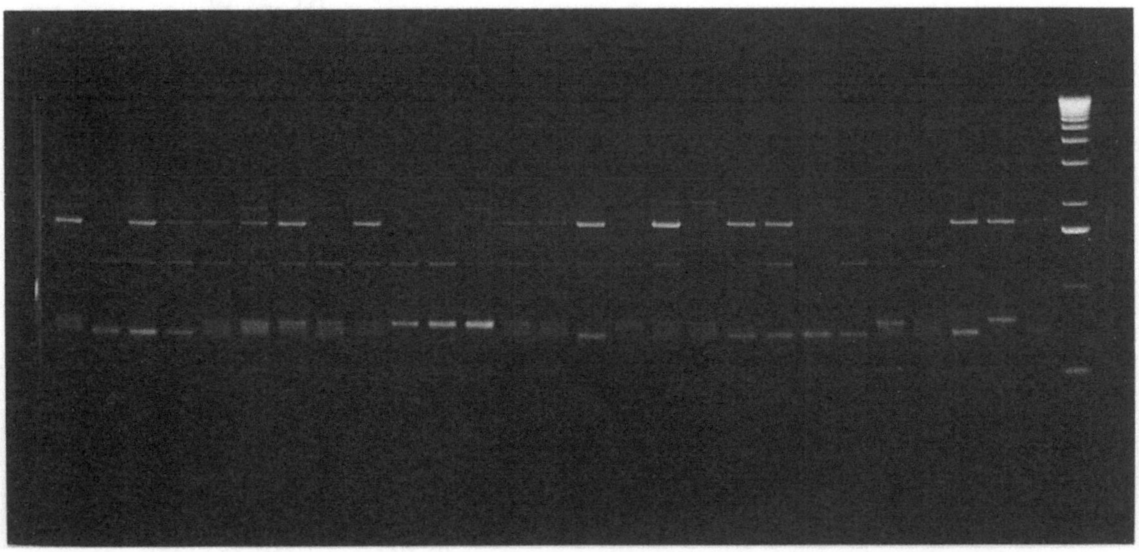

Fig. 1. RAPD amplification patterns of 27 apple cultivars generated with primer UBC 227 (from left to right: Ingrid Marie – S13/1–30 according to Table 1; last lane: 1 Kb-ladder).

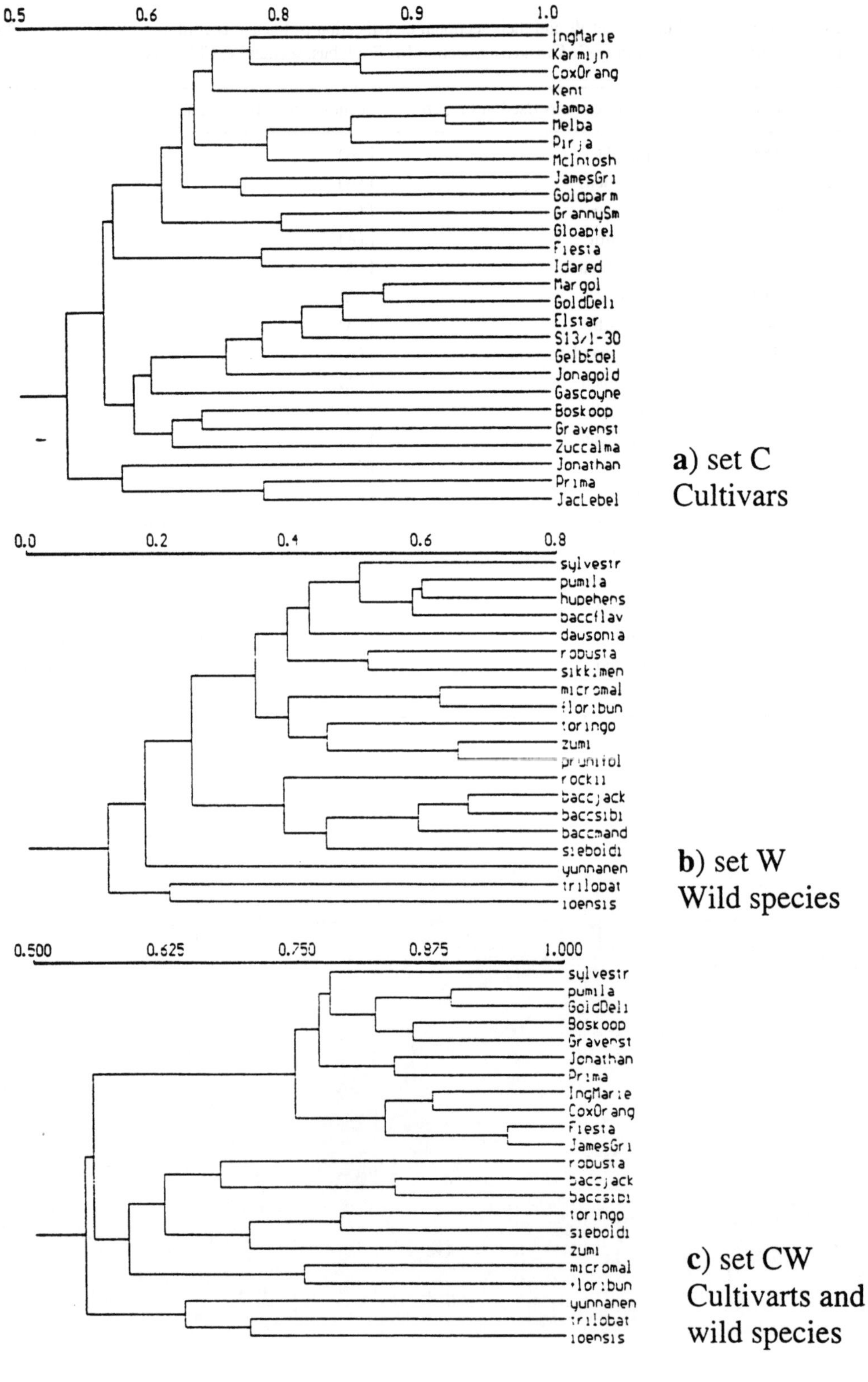

Fig. 2A-C. Dendrograms of apple cultivars and wild species from sets C, W and CW derived from UPGMA analysis performed on the basis of SM (sets C, CW) or J (set W) coefficients for genetic similarity.

Results

A variety of 29 preselected informative random primers were applied to the different sets of *Malus* genotypes. High levels of polymorphisms were observed both in cultivars and species. A demonstration of the amplification patterns obtained with set C is given in Fig. 1. A total of 49 to 57 polymorphic band positions was selected in the three sets for statistical analysis. Genetic similarity among cultivars estimated by SM coefficients ranged from only 42% for 'Fiesta' – 'Golden Delicious' up to 90% for the 'Jamba' – 'Melba' comparison demonstrating the close relationship of the latter two cultivars. Other closely related cultivars showing SM values higher than 80% are 'Karmijn' and 'Cox Orange' or 'Golden Delicious' and 'Margol', respectively. Corresponding data for relatedness in wild species ranged from 30% (*M. bacc. mandshurica* – *M. sylvestris*) to 84% (*M. bacc. jackii* – *M. bacc. sibirica*) indicating the broader genetic background compared with the cultivars of *M.* × *domestica*.

Dendrograms constructed using UPGMA clustering are shown in Fig. 2a–c. Generally, the results obtained from statistical analysis of the cultivar set are in good agreement with the known lineage. For example, cultivars 'Karmijn' and 'Kent', originating both from the cross 'Cox Orange' × 'Jonathan', are placed beside their maternal parent. 'Fiesta', originating from the cross 'Cox Orange' × 'Idared', grouped closely with the paternal parent (Fig. 2a). The dendrogram generated using common band information in wild species gave relationships that were in accordance with known phylogenetic information in most cases. More closely related species such as *M. sylvestris* and *M. pumila* grouped together and were separated with maximal distance from the individuals of sections III (*M. trilobata*) and IV (*M. ioensis*) (Fig. 2b). The dendrogram from the combined set CW is shown in Fig. 2c. It was found that *M. sylvestris* and *M. pumila* clustered rather with apple cultivars than with the other wild accessions. As in the analysis of set W, *M. ioensis* was characterized as the most distantly related species.

Discussion

The results presented here indicate that the RAPD 'fingerprinting' technique can be applied successfully to a wide spectrum of apple genotypes for the estimation of genetic relationships. The potential use of RAPD markers for molecular classification could especially be demonstrated for a set of cultivars. On an intraspecific level a considerable degree of genetic variation seems to exist still in apple cultivars. However, revealing genetic relatedness of cultivars with completely known lineage it was found, that at least one parent was classified nearly beside the hybrid cultivar. The hypothesis that 'Cox Orange' was involved as maternal parent in the origin of 'Ingrid Marie' (Götz & Silbereisen 1989) could be supported by this way. Some cultivars with an unclear origin seem to be relatively closely related, as for example 'Boskoop' and 'Gravensteiner', with a genetic similarity of about 70%.

With respect to the classification system of the genus *Malus*, many obscurities still exist. RAPD markers appear to have the potential to complement 'classical' taxonomic studies. Way et al. (1990) pointed out that *M.* × *domestica* is a complex interspecific hybrid with several species involved. Wild species from the Altai region (*M. sieversii*) and from the Caucasus region (*M. orientalis*) were supposed as potential ancestors (Götz & Silbereisen 1989), but a participation of the European species *M. sylvestris* and *M. pumila* was also taken into consideration (Rheder 1940, Korban & Skirvin 1984, Roach 1988). The dendrogram for the set CW (Fig. 2c) shows a remarkably small genetic distance between both *M. sylvestris* and *M. pumila*, and some apple cultivars. A participation of these two species in the domestication process of apple seems therefore to be very likely. *M. baccata flavescens*, which was grouped in the vicinity of *M. pumila* is not a member of *M. baccata* but could be classified by conventional methods as *M. pumila* (Büttner pers. comm.). The other three subspecies of *M. baccata* (*jackii*, *sibirica*, *mandshurica*) proved to have a common ancestor.

Acknowledgements

The valuable technical assistance of R. Kahnau is gratefully acknowledged. I wish to thank Dr R. Büttner and Prof. Dr. H. Schmidt for helpful discussions.

References

Doyle, J.J. & J.L. Doyle, 1990. Focus 12: 13–15.

Götz G. & R. Silbereisen, 1989. Obstsorten-Atlas, E. Ulmer, Stuttgart.

Harada, T., K. Matsukawa, T. Sato, R. Ishikawa, M. Niizeki & K. Saito, 1993. Euphytica 65: 87–91.

Koller, B., A. Lehmann, J.M. McDermott & C. Gessler, 1993. Theor. Appl. Genet. 85: 901–904.
Korban, S.S. & R.M. Skirvin, 1984. HortScience 19: 177–180.
Rheder, A., 1940. Manual of Cultivated Trees and Shrubs, 2nd Edition. MacMillan, New York.
Roach, F.A., 1988. HortScience 23: 51–55.
Rohlf, F.J., 1993. Numerical Taxonomy and Multivariate Analysis System, Version 1.70. Exeter Software, Setauket.
Tingey, S.V. & J.P. del Tufo, 1993. Plant Physiology 101: 349–352.
Way, R.D., H.S. Aldwinckle, R.C. Lamb, A. Rejman, S. Sansavini, T. Shen, R. Watkins, M.N. Westwood & Y. Yoshida, 1990. Apples (*Malus*). In: J.N. Moore & J.R. Ballington Jr. (Eds) Genetic Resources of Temperate Fruit and Nut Crops I, pp. 3–62. Int. Soc. for Hort. Sci., Wageningen.
Williams, J.G.K., A.R. Kubelik, K.J. Livak, J.A. Rafalski & S.V. Tingey, 1990. Nucleic Acids Research 18: 6531–6535.

H. Schmidt and M. Kellerhals (Eds.), Progress in Temperate Fruit Breeding, : 301–303, 1994.

Inheritance of leaf isozymes in apple (*Malus domestica* Borkh. and *Malus floribunda* Van Houtte)

F. Štampar, J. Smole, B. Javornik, A. Solar, M. Viršček-Marn
Biotechnical Faculty, Agronomy Department, University of Ljubljana, 61000 Ljubljana, Slovenia

Key words: Isozymes, *Malus domestica*, *Malus floribunda*, inheritance

Summary

Eight progenies from controlled crosses (*Malus domestica* and *Malus floribunda* Van Houtte) were examined for isozyme phenotype using starch gel electrophoresis. Inheritance of eight enzyme systems were analyzed. Young leaves were used for enzyme extraction. Twenty eight loci were defined and nine of them displayed polymorphism (PX-1; two alleles, PX-2; two alleles, PX-3; two alleles, PX-4, three alleles, 6-PGD-1; four alleles, 6-PGD-2; two alleles, ME-1; two alleles; MDH-4; two alleles and TPI-1; two alleles). In the majority of progenies, the inheritance of enzymes for each of these loci was consistent with Mendelian principles. The linkage between polymorphic loci was not statistically significant.

Introduction

Isozyme polymorphism has been examined in the majority of plant species (Soltis & Soltis 1989). Considerable polymorphism has been revealed in isozyme of apple scion and rootstock varieties (Torres 1989). The study of polymorph enzyme systems in apple can generate information on linkage groups present in the genome and the relationship between agronomical characters and isozyme polymorphism (Chevreau et al. 1985, Chevreau & Laurens 1987, Weeden & Lamb 1987, Manganaris & Alston 1987, 1988, 1992a,b,c, King et al. 1991). The purpose of this work was to study isozymes polymorphism in progenies of *Malus domestica* Borkh. and *Malus floribunda*.

Material and methods

Eight progenies from controlled crosses involving (*Malus domestica* Borkh., *Malus floribunda* Van Houtte and *Malus purpurea*; ('Elstar' × 'Granny Smith'), ('Elstar' × 'Idared'), ('Elstar' × *Malus floribunda*), ('Elstar' × *Malus floribunda* 'Hillieri'), ('Elstar' × *Malus purpurea* 'Aldenhamensis'), ('Jonagold' × 'Granny Smith'), ('Jonagold' × *Malus floribunda*), ('Jonagold' × *Malus floribunda* 'Hillieri') were examined for isozymic phenotypes using horizontal starch gel electrophoresis. Young leaves were used for enzyme extraction. Extracts from leaf tissue were prepared as described by Weeden & Lamb (1985). Starch gel electrophoresis was performed as described by Wendel & Weeden (1989). Two buffer systems were used for the electrophoresis (Weeden & Lamb 1985). Slices from the Tris citrate/lithium borate gel were assayed for peroxidase (PX), aspartate aminotransferase (AAT), triosephosphate isomerase (TPI) and glucosephosphate isomerase (GPI). Slices from the Histidine gel were assayed for malate dehydrogenase (MDH), malic enzyme (ME), 6-phosphogluconate dehydrogenase (6-PGD) and phosphoglucomutase (PGM). Staining solutions used for enzyme identification were those described by Vallejos (1983). Inheritance of isozymes and linkage between polymorphic loci were analysed by chi-square test.

Results

Extracts from young leaves gave clear activity bands. The segregation of phenotypes for the eight polymorphism identified is presented in Table 1. Each polymor-

Table 1. Single locus segregation and chi-square analysis for polymorphic isozyme loci.

Isozyme[1] (Lokus)	Cross[2]	Parental Genotypes	Progeny Genotypes	Expected Ratio	Chi Square
PX-2	3	aa × bb	6aa:3bb	1:1	1.00
	4	aa × bb	5aa:3bb	1:1	0.50
	5	aa × bb	4aa:4bb	1:1	0.00
PX-4	2	aa × bb	6aa:5bb	1:1	0.09
	6	aa × bb	4aa:5bb	1:1	0.11
	7	aa × bb	4aa:5bb	1:1	0.11
	8	aa × bb	6aa:3bb	1:1	1.00
6-PGD-1	1	bd × ad	3ab:5ad:2bd:0dd	1:1:1:1	4.81
	2	bd × bd	4bb:5bd:2dd	1:2:1	1.37
	3	bd × aa	3ab:6ad	1:1	1.00
	4	bd × aa	4ab:4ad	1:1	0.00
	5	bd × aa	4ab:4ad	1:1	0.00
	6	bd × ad	2ab:5ad:2bd:0dd	1:1:1:1	6.66
	7	bd × aa	5ab:4ad	1:1	0.11
	8	bd × aa	4ab:5ad	1:1	0.11
6-PGD-2	4	aa × ab	2aa:6ab	1:1	2.00
	8	aa × ab	3aa:6ab	1:1	1.00
ME-1	3	aa × ab	4aa:5ab	1:1	0.11
	4	aa × ab	3aa:5ab	1:1	0.50
	5	aa × ab	3aa:5ab	1:1	0.50
	7	aa × ab	6aa:3ab	1:1	1.00
	8	aa × ab	6aa:3ab	1:1	1.00
MDH-4	3	aa × ab	4aa:5ab	1:1	1.00
	4	aa × ab	0aa:8ab	1:1	8.00
	5	aa × ab	5aa:3ab	1:1	0.50
	7	aa × ab	5aa:4ab	1:1	0.11
	8	aa × ab	5aa:4ab	1:1	0.11
TPI-1	3	aa × ab	5aa:4ab	1:1	0.11
	4	aa × ab	5aa:3ab	1:1	0.50
	5	aa × ab	4aa:4ab	1:1	0.00
	7	aa × ab	6aa:3ab	1:1	1.00
	8	aa × ab	4aa:5ab	1:1	0.11

[1] Isozyme designations: Peroxidase (PX), 6-phosphogluconate dehydrogenase (6-PGD), malic enzyme (ME), malate dehydrogenase (MDH), triosephosphate isomerase (TPI).
[2] Cross designation: 1 – ('Elstar' × 'Granny Smith'), 2 – ('Elstar' × 'Idared'), 3 – ('Elstar' × *Malus floribunda*), 4 – ('Elstar' × *Malus floribunda* 'Hillieri'), 5 – ('Elstar' × *Malus purpurea* 'Aldenhamensis'), 6 – ('Jonagold' × 'Granny Smith'), 7 – ('Jonagold' × *Malus floribunda*), 8 – ('Jonagold' × *Malus floribunda* 'Hillieri').

phism will be discussed under the respective isozyme system.

Peroxidase Five major sets of PX activity bands were present in apple leaf extracts (PX-1, PX-2, PX-3, PX-4 and PX-5). Polymorphism was displayed on PX-1, PX-2, PX-3 and PX-4.

Aspartate aminotransferase Two sets of AAT activity bands were present in progenies from controled crosses ('Elstar' × 'Granny Smith'), ('Elstar' × 'Idared') and ('Jonagold' × 'Granny Smith'), and three sets of AAT activity bands were present in progenies from controlled crosses ('Elstar' × *Malus floribunda*), ('Elstar' × *Malus floribunda* 'Hillieri'), ('Elstar' × *Malus purpurea* 'Aldenhamensis'), ('Jonagold' ×

Malus floribunda), ('Jonagold' × *Malus floribunda* 'Hillieri').

Triosephosphate isomerase Four monomorphic regions and one polymorphic regions of TPI activity were observed in leaf extracts.

Glucosephosphate isomerase Two monomorphic regions of GPI activity were observed in leaf extracts.

Malate dehydrogenase Five major sets of MDH activity were present in the apple leaf extracts. Polymorphism was displayed only by MDH-4, which showed a single fast band in progenies from controlled crosses ('Elstar' × 'Granny Smith'), ('Elstar' × 'Idared') and ('Jonagold' × 'Granny Smith'), or a tripling of bands in progenies from controled crosses ('Elstar' × *Malus floribunda*), ('Elstar' × *Malus floribunda* 'Hillieri'), ('Elstar' × *Malus purpurea* 'Aldenhamensis'), ('Jonagold' × *Malus floribunda*), ('Jonagold' × *Malus floribunda* 'Hillieri').

Malic enzyme One polymorphic set of ME activity bands were present in leaf extracts.

6–Phosphogluconate dehydrogenase Two polymorphic set of 6-PGD actvity were present in apple leaf extracts.

Phosphoglucomutase Five sets of PGM activity bands were observed in apple leaf extracts. Three of them were monomorphic and PGM-1 and PGM-2 were polymorphic.

28 loci were defined and 9 of them displayed polymorphism (PX-1; two alleles, PX-2; two alleles , PX-3; two alleles , PX-4; three alleles, 6-PGD-1; four alleles, 6-PGD-2; two alleles, ME-1; two alleles, MDH-4; two alleles and TPI-1 two alleles).

In the majority of progenies (Table 1), the inheritance of enzymes for each of these loci was consistent with Mendelian principles.

The linkage between polymorphic loci was not statistically significant.

Discussion

Our result demonstrate that the polymorphism presented in eight enzyme systems is produced by allelic variants at 9 distinct loci.

Linkage groups were not detected due to small number of offsprings and cross combination.

Our studies and those of Chevreau et al. (1985, 1987), Vinterhalter & James (1982), Manganaris & Alston (1987, 1988, 1992a,b,c) and Weeden & Lamb (1985, 1987) demonstrate that allozyme variation can be an important genetic tool in apple breeding. The high level of allozyme polymorphism in *Malus* has already been used to distinguish cultivars and could be used to construct linkage maps and mark commercially important genes. Such information about the apple genome could be used in a wide range of breeding and genetic studies.

References

Chevreau, E., Y. Lespinase & M. Gallet, 1985. Inheritance of pollen enzymes and polyploid origin of apple (*Malus domestica* Borkh). Theor. Appl. Genet. 71: 268–277.

Chevreau, E. & F. Laurens, 1987. The pattern of inheritance in apple (*Malus* × *domestica* Borkh.): further results from leaf isozyme analysis. Theor. Appl. Genet. 75: 90–95.

King, G.J., F.A. Alston, I. Battle, E. Chevreau, C. Gessler, J. Janse, P. Lindhout, A.G. Manganaris, S. Sansavini, H. Schmidt & K. Tobutt, 1991. The European Apple Genome Mapping Project – Developing a strategy for mapping genes coding for agronomic characters in tree species. Euphytica 56: 89–94.

Manganaris, A.G. & F.H. Alston, 1987. Inheritance and linkage relationships of glutamate oxaloacetate transaminase isoenzymes in apple. Theor. Appl. Genet. 74: 154–161.

Manganaris, A.G. & F.H. Alston, 1988. Inheritance and linkage relationships of glutamate oxaloacetate transaminase isoenzymes in apple. Theor. Appl. Genet. 76: 449–454.

Manganaris, A.G. & F.H. Alston, 1992a. Genetics of leucine aminopeptidase in apple. Theor. Appl. Genet. 83: 345–352.

Manganaris, A.G. & F.H. Alston, 1992b. Inheritance and linkage relationships of peroxidase isoenzymes in apple. Theor. Appl. Genet. 83: 392–399.

Manganaris, A.G. & F.H. Alston, 1992c. Genetics of esterase isozymes in *Malus*. Theor. Appl. Genet. 83: 467–475.

Soltis, D.E. & P.S. Soltis, 1989. Isozymes in Plant Biology. Dioscorides Press, Portland.

Torres, A.M., 1989. Isozyme analysis of tree fruits. In: Isozymes in Plant Biology, pp. 192–250. Dioscorides Press Portland.

Vallejos, C.E., 1983. Enzyme activity staining. In: Isozymes in Plant Genetics and Breeding, Part A, pp. 469–516. Elsevier, Amsterdam.

Vinterhalter, D.V. & D.J. James, 1982. The use of peroxidase polymorphism in the identification of apple scion cultivars. Scentia Horticulturae. 18: 253–261.

Weeden, N.F. & R.C. Lamb, 1985. Identification of apple cultivars by isozyme phenotypes. J. Amer. Soc.Hort. Sci. 110: 509–515.

Weeden, N.F. & R.C. Lamb, 1987. Genetic and linkage analysis of 19 isozyme loci in apple. J. Amer. Soc. Hort. Sci. 112: 865–872.

Wendel, J.F. & N.F. Weeden, 1989. Visualization and Interpretation of Plant Isozymes. Isozymes in Plant Biology. Dioscorides Press, Portland.

H. Schmidt and M. Kellerhals (Eds.), Progress in Temperate Fruit Breeding, 305–308.

The European *Prunus* mapping project

Progress in the almond linkage map

P. Arús[1], R. Messeguer[1], M. Viruel[1], K. Tobutt [2], E. Dirlewanger[3], F. Santi[4], R. Quarta[5] & E. Ritter[6]

[1] *IRTA, Centre de Cabrils, Departament de Genètica Vegetal, 08348 Cabrils, Spain;* [2] *Horticultural Research International, East Malling, Kent ME19 6BJ, U.K.;* [3] *INRA, Station de Recherches Fruitières, Centre de Recherches de Bordeaux, BP81, 33883 Villenave d'Ornon Cedex, France;* [4] *INRA, Station d'Amélioration des Arbres Forestiers, Centre de Recherches d'Orléans, F-45160 Olivet, France;* [5] *Istituto Sperimentale per la Frutticoltura, Via Fioranello, 52. 00040, Ciampino Aeroporto, Roma, Italy;* [6] *IKT-CIMA, Granja Modelo, Apartado 46, 01080 Vitoria-Gasteiz, Spain*

Key words: Almond, isozymes, RFLP, linkage, *Prunus amygdalus*, genetic map

Summary

Six European research groups are collaborating to develop genetic markers and linkage maps for use in *Prunus* breeding programmes. A basic map with 200 RFLPs and 50 more markers including isozymes and RAPDs will be constructed using two highly segregating populations: an interspecific peach × almond F_2 and a cherry F_2. Then, the parents of eleven almond, cherry, peach or plum breeding progenies segregating for target characters will be screened for polymorphisms at the marker loci, and a set of reduced maps, one per progeny, will be constructed with markers spaced 20–30 cM and covering the whole genome. Cosegregation analysis of markers and characters of interest will allow us to find linkages between markers and major genes or quantitative trait loci responsible for the expression of these traits. A map with 72 markers, 7 isozymes and 65 RFLPs, has been developed at the IRTA-Cabrils laboratory using an intraspecific almond progeny, 'Ferragnes' × 'Tuono'. Probes for the analysis of RFLPs were obtained from almond genomic and cDNA libraries. The level of polymorphism for RFLPs and the distribution of markers in the chromosomes of almond are discussed.

Introduction

A considerable amount of information on isozyme markers of fruit tree crops has been gathered during the last decade (Torres, 1990). Isozymes have been used for such applications as genotype identification, gene tagging, and the evaluation of the variation of different species. The development of DNA markers, like RFLPs or RAPDs, is very recent in fruit crops (Eldredge et al., 1992; Koller et al., 1993). Their potential as tools for breeding and evolutionary studies of this group of species is high, since they can rapidly improve the generally poor knowledge of the fruit tree genetics, and can be of immediate use for marker-assisted selection, reducing the long time span needed for the development of new varieties and allowing the use of more efficient breeding schemes.

In this paper we report on the objectives of the European *Prunus* Mapping Group, which has recently obtained funds from the EC AIR Programme, and on the progress by one of the partners of this group, IRTA, in constructing an almond marker map.

The European *Prunus* mapping group

The European *Prunus* Mapping Group consists of six research groups: IRTA (Cabrils, Mas Bové and Barcelona) and CIMA (Arkaute) from Spain, HRI (East Malling) from the U.K., INRA (Bordeaux and

Orléans) from France, and ISF (Roma) from Italy. All of them have *Prunus* breeding projects in progress: IRTA in almond and apricot, HRI and INRA (Orléans) in cherry, INRA (Bordeaux) and ISF in peach, and CIMA in plum. Moreover, most of the partners have experience on the development of isozyme markers in *Prunus* crops, like peach (Messeguer et al., 1987; Monet & Gibault, 1991), almond (Cerezo et al., 1989; Arús et al., 1994) and cherry (Santi & Lemoine, 1990; Boskovic & Tobutt, 1994). The joint initiative is a four year project, starting during autumn 1993, which consists of two parts.

First, a densely populated map with 250 markers (200 RFLPs and 50 isozymes or RAPDs) will be constructed with the collaboration of all partners (four of them will develop 50 polymorphic probes for RFLPs each) for two F_2 populations: the progenies from selfing an individual of the cross almond 'Texas' × peach 'Earlygold', and from selfing cherry 'Stella'. This map is the tool needed for further applications of markers, and will allow comparison of the alignment of homologous markers between the linkage groups of two of the most divergent genomes of the genus, peach/almond and cherry.

In the second part, the basic map will be used for detecting useful linkages between markers and major genes or quantitative trait loci responsible for characters that are the breeding objectives of each partner. Eleven progenies, 4 of peach, 3 of cherry, 2 of almond and 2 of plum, segregating for agronomic characters have been chosen. After the analysis of polymorphism in the parents of each of these progenies, reduced maps (i.e., maps with 30–40 segregating markers spaced 20–30 cM apart and covering the entire genome) will be elaborated for each progeny, and the co-segregations between agronomic characters and the markers of these reduced maps will be analysed. Twenty-six agronomic characters will be studied including monogenic and polygenic traits. Five are pest or disease resistances, 8 are quality aspects of fruits or kernels, 3 concern product diversification, 5 concern production season and stability, and the remaining 5 relate to other vegetative characters.

In addition, the level of genetic variation in the five *Prunus* crops will be evaluated on the basis of the marker variation observed in a set of genotypes of each species; such information could be useful for conservation purposes. And the potential of marker loci for cultivar identification, of interest to nurserymen and plant variety rights bureaux will be assessed.

A molecular marker map for almond

Materials and methods

The parents and unselected progeny (60 individuals) of the almond intraspecific cross 'Ferragnes' × 'Tuono' were provided by F. Vargas from IRTA (Mas Bové). Nuclear DNA was extracted from young leaves according to the method of Bernatzky & Tanksley (1986) with the modification that the DNA was resuspended in TE after isopropanol precipitation and further purified through a CsCl gradient. Probes used for RFLPs were obtained from a *Pst*I genomic library in pBluescript (Stratagene) and from two mRNA derived libraries, one from seedling roots and the other from developing seeds. Parental DNA was digested with five enzymes (*Bam*HI, *Dra*I, *Eco*RI, *Hin*dIII, and *Mva*I) according to suppliers' indications, separated in 0.9% agarose gels and Southern blotted into Hybond N^+, Hybond N (Amersham) or Nytran (Schleicher and Schuell) filters. Filters were probed with ^{32}P-dATP radiolabeled DNA (Amersham) by the random hexamer method (Sambrook et al., 1989), or with dig-dUTP (Boehringer Mannheim) by insert amplification. Dig-UTP labeled probes were detected according to the protocol supplied by Boehringer Mannheim. Combinations of enzymes and probes that proved polymorphic were likewise studied in the progeny.

Extraction, electrophoresis procedures and staining methods for isozymes were as described by Arús et al. (1994).

Linkage groups were established using LINKAGE-1 software (Suiter et al., 1983). Two maps were elaborated (Fig. 1), one for loci heterozygous in 'Ferragnes' ('FER') and one for 'Tuono' ('TUO'). Loci heterozygous in both parents, i.e., those segregating 1 : 2 : 1 and 1 : 1 : 1 : 1 were included in both maps and used as bridges between them. Four class type segregations were converted into two backcross ratios, one segregating in 'Ferragnes' and the other in 'Tuono'.

Results and discussion

Seventy-two loci were analyzed in the progeny of 'Ferragnes' × 'Tuono'. Seven of them (*Pgm-2, Gpi-2, Aat-1, Aat-2, Lap-1, Sdh-1,* and *Idh-2*) were isozyme genes, and the remaining 65 were RFLPs. Eleven RFLPs were obtained from cDNA clones (11 clones), and 54 from genomic clones (39). Two of the cDNA clones corresponded to characterized genes of almond (Garcia-Mas

GI 'TUO'	GI 'FER'	GII 'TUO'	GII 'FER'	GIII 'TUO'	GIII 'FER'
AG23B*	AG25C	AG24C	AC10*	*Idh-2*	AG7
AG5	AG29A	AC10*	AC6*	*Aat-1*	AG12A
AG22A*	AG23B*	AG21A*	AG26B+	AG28A*	AG28A*
AG36A	AG9	AC6*	AG21A*	AG3A*	AG1A
Pgm-2+	AG22A*	AG26B+	AG20A*		AG3A*
AC3A	*Pgm-2+*	AG20A*	*Ole1*		AG32A
Lap-1+	AC7A	AG35A			AG37A
Gpi-2	AG30A				
AG38A	*Lap-1+*				
	Aat-2				
	AG11A				

GIV 'TUO'	GIV 'FER'	GV 'TUO'	GV 'FER'	GVI 'TUO'	GVI 'FER'
AG23A*	AG24B	AG21B+	AG16A*	AG13*	*Pglm1*
AG6	AC1	AG16A*	AG25A+	*Sdh-1+*	AG13*
AG18A	AG23A*	AG25A+	AG21B+	AG27B	AG32D
AG31A	AG39A	AG25B	AG33*	AG19A*	*Sdh-1+*
AG12B	AG28B	AG15A			AC8A
		AG33*			AG18B
					AG26A
					AG19A*

GVII 'TUO'	GVII 'FER'	GVIII 'TUO'	GVIII 'FER'	UNLINKED 'TUO'	UNLINKED 'FER'
AG10A+	AG8A	AG4A*	AG24A	AG30B	-
	AG27C		AG2A	AG8B	
	AG10A+		AG4A*		
	AG17A		AG32B		
	AG34A		AG14A		
			AC5A		
			AC12		
			AC11A		

Fig. 1. Marker loci included in the linkage groups for almond cultivars 'Tuono' ('TUO') and 'Ferragnes' ('FER'). Common loci to both maps are denoted with an asterisk (1:1:1:1 segregations) or with a plus sign (1:2:1 segragations) after the locus name.

et al., unpublished) coding for oleosine (*Ole1*) and for phosphoglyceromutase (*Pglm1*).

Most of the loci studied, 51, segregated 1 : 1, whereas 14 segregated 1 : 1 : 1 : 1 and 7 segregated 1 : 2 : 1. RFLP analysis revealed also 7 loci segregating 3 : 1, but these have not yet been included in the map. From the 86 segregations examined (the 14 1 : 1 : 1 : 1 progenies were analyzed as two backcross progenies), 12 (14%) were significantly different from the expected Mendelian ratios.

The 'FER' map contained 54 loci, which could be assigned to eight linkage groups (GI to GVIII) with a maximum of 11 loci in GI and a minimum of 4 loci in GV. Six groups were constructed with the 39 loci of the 'TUO' map, the two smallest with 4 loci each, GIII and GVI, and the largest with 9 loci, GI. Four loci of the 'TUO' map were unlinked, but two of them, AG10A and AG4A, were heterozygous in both parents and could be assigned to GVII and GVIII respectively. Five of the loci with distorted segregations were located in GIII of the 'FER' map, whereas no departures from expected were observed in loci located at similar positions of the 'TUO' map. These data suggest that a selected allele may be located at this region of the 'Ferragnes' genome. It is too early to say if the eight linkage groups found in the 'FER' map correspond to the eight chromosomes of almond. Some of them, like GII, GIII, GIV and GVII are short, and as new markers are added to the map it may be that some groups merge to form larger ones and that new groups arise.

References

Arús, P., C. Olarte, M. Romero & F. Vargas, 1994. Linkage analysis of ten isozyme genes in F_1 segregating progenies of almond. J. Amer. Soc. Hort. Sci. 119(2): 339–344.

Bernatzky, R., & S.D. Tanksley, 1986. Genetics of actin-related sequences in tomato. Theor. Appl. Genet. 72: 314–321.

Boskovic, R. & K.R. Tobutt, 1994. Inheritance and linkage of isoenzymes in two interspecific cherry progenies. Acta Horticulturae, in press.

Cerezo, M., R. Socias i Company & P. Arús, 1989. Identification of almond cultivars by pollen isoenzymes. J. Amer. Soc. Hort. Sci. 114: 164–169.

Eldredge, L., R. Ballard, W.V. Baird, A. Abbott, P. Morgens, A. Callahan, R. Scorza & R. Monet, 1992. Application of RFLP analysis to genetic linkage mapping in peaches. HortScience 27: 160–163.

Koller, B., A. Lehmann, J.M. McDermott & C. Gessler, 1993. Identification of apple cultivars using RAPD markers. Theor. Appl. Genet. 85: 901–904.

Messeguer, R., P. Arús and M. Carrera, 1987. Identification of peach cultivars with pollen isozymes. Scientia Hortic. 31: 107–117.

Monet, R. & B. Gibault, 1991. Polymorphisme de l'α-amylase chez le pêcher. Étude génétique. Agronomie 11: 353–358.

Sambrook, J., E.F. Fritsch & T. Maniatis, 1989. Molecular Cloning. A Laboratory Manual. Cold Spring Harbor Laboratory Press, New York.

Santi, F. & M. Lemoine, 1990. Genetic markers for *Prunus avium* L.: Inheritance and linkage of isozyme loci. Ann. Sci. For. 47: 131–139.

Suiter, K.A., J.F. Wendel & J.S. Case, 1983. LINKAGE-1: A PASCAL computer program for the detection and analysis of genetic linkage. J. Hered. 74: 203–204.

Torres, A.M., 1990. Isozyme analysis in tree fruits. In: D.E. Soltis & P.S. Soltis (Eds). Isozymes in Plant Biology, pp. 192–205. Chapman and Hall, London.

H. Schmidt and M. Kellerhals (Eds.), Progress in Temperate Fruit Breeding, 309–311.

Molecular genetic mapping of peach

E. Dirlewanger & C. Bodo
INRA, Centre de Recherches de Bordeaux, Station de Recherches Fruitières, BP 81, 33883 Villenave d'Ornon Cedex, France

Key words: Prunus persica, genetic map, molecular markers, RAPD

Summary

A project to develop a linkage map of the peach (*Prunus persica*) genome is underway using an F2 population segregating for several morphological characters and pest resistance e.g., nectarine (*g*), weeping shape (*pl*) and aphid resistance (*Rm1*). The RAPD technique was used to analyse 270 plants. Linkage analysis of the F2 population was performed using the MAPMAKER software. Eight linkage groups were established and RAPD markers flanking the *pl* gene were found.

Introduction

Although *Prunus* is an important genus including several important fruit species and cultivars, no genetic linkage map has been published. Peach, (*Prunus persica* L. Batsch), is diploid with 2n = 16. Since it is considered to be the best genetically characterized species of the genus, we have used it as a model for other *Prunus* species. Molecular markers linked to valuable agronomic characters would permit earlier selection of such characters, particularly those relating to the fruit.

Plant material

An F2 population composed of 270 individuals was provided by R. Monet (INRA, Station de Recherches Fruitières, Centre de Recherches de Bordeaux, France). This population derived from a cross between two heterozygous parents (1161 : 12 × 2678 : 47) 1 : 55, a weeping clone, and the cultivar Early Sungrand. The two parents differed for peach aphid (*Myzus persicae*) resistance and for morphological characters e.g., nectarine and weeping shape. The peach aphid resistance is controled by a single dominant gene (*Rm1*) (Massonie et al., 1984), while the nectarine and the weeping shape are each controled by a single recessive gene, respectively *pl* (Monet et al., 1988) and *g* (Bailey & French, 1949). The peach aphid induces leaf fall and transmits the Sharka virus. As the F2 individuals were planted in 1993, the nectarine character and the aphid resistance have not been analysed. Aphid resistance will be tested in 1994 and the nectarine character will be examined as soon as the trees begin fruiting. Only the weeping shape was examined in the present paper. A weeping shape would enable an easier fruit harvest. Young leaves were collected from orchards and were used for linkage map construction.

Markers

Five hundred and twenty-four primers from Operon Technologies (Alameda, CA, U.S.A.) were tested on the two parents and on nine individuals of the F2 population. As the two parents were heterozygous we selected only the primers amplifying the bands segregating in the nine F2 individuals.

The DNA was extracted according to the method of Saghai-Maroof et al. (1984).

For RAPD analysis, the procedure described by Williams et al. (1990) was used. Thirty ng of genomic DNA per sample were used. The samples were held at 95 °C for 1 min and were then subjected to 40 PCR cycles of melting at 94 °C for 10 s, annealing at 37 °C for 15 s, extending at 72 °C for 75 s using

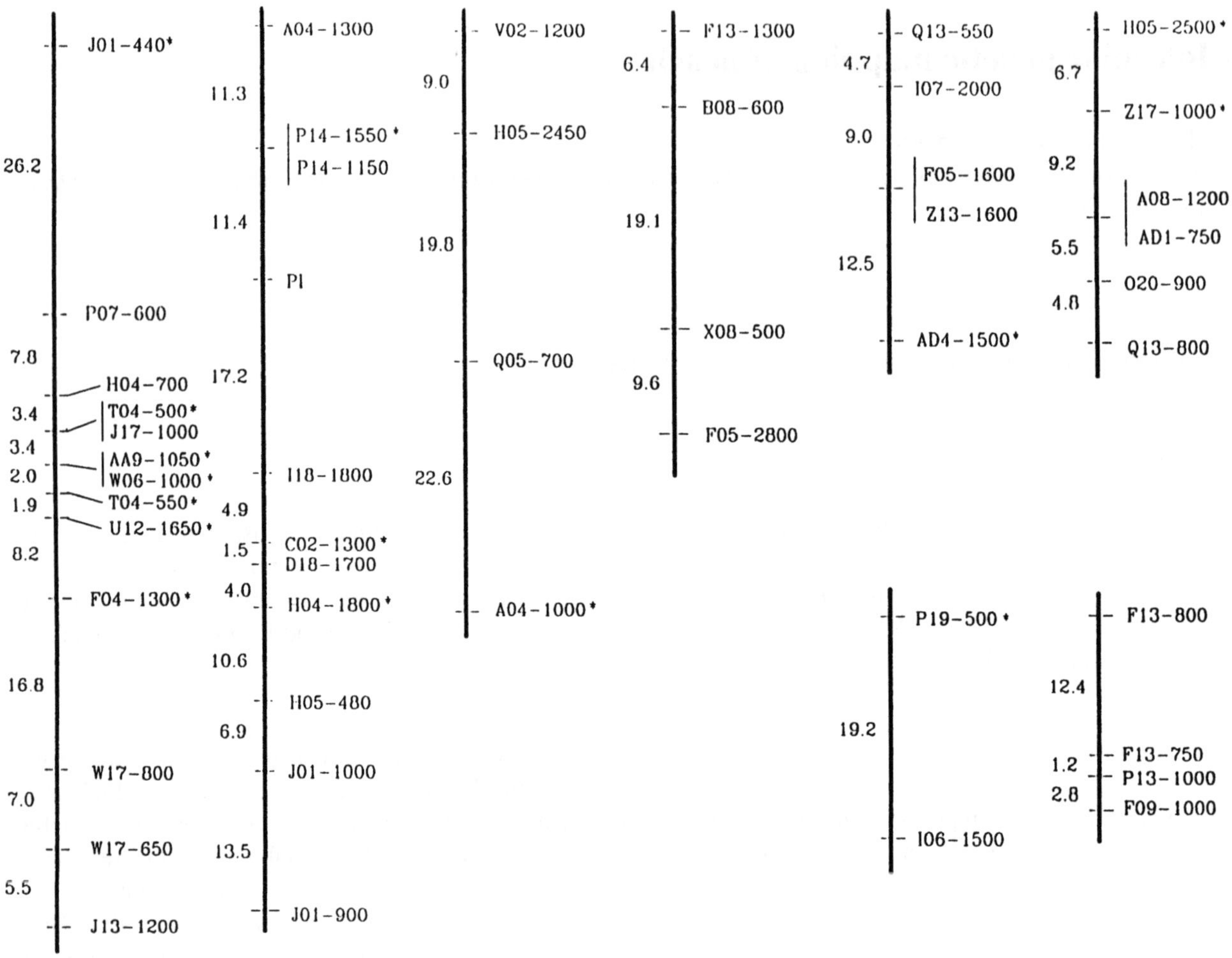

Fig. 1. Partial genetic linkage map of peach constructed with MAPMAKER. Lod > 4, recombination rate < 0.3. The genes and the markers, with the size of the band in base pair, are on the right side of the chromosomes. Map distances are given in centimorgans on the left side of the chromosomes (Kosambi). Markers V17–750, Z3–700, Z13–1600 and AC11–1500 and remain independant. * RAPD markers presenting a deviation from the 1 : 3 ratio.

the Perkin Elmer GeneAmp PCR System 9600. The products were separated on 1.5 % agarose gels.

Results

Polymorphism

Thirty-eight out of the 524 primers (i.e., 7%) segregated in the F2 population. Only eleven primers gave several polymorphic bands: eight primers gave two polymorphic bands and three primers gave three polymorphic bands. Thus, 52 RAPD markers were analysed in the F2 population. Six out of the 52 RAPD markers did not reveal polymorphism between the two parents though segregated in the F2 population.

Segregation data

For the weeping shape, 56 F2 plants were weeping and 210 were normal. This segregation ratio fits the expected 1 : 3 ratio for a single dominant gene (X^2 = 2.2). For 15 out of the 52 RAPD markers a significant deviation from the 1 : 3 ratio was observed.

Genetic map

Linkage analysis of the F2 population was performed on a Vax workstation using the MAPMAKER software (Lander et al., 1987). Map distances (centiMorgan, cM) were then estimated using the Kosambi mapping function (Kosambi, 1944). The linkage groups were obtained by choosing 0.3 as the maximum recombinant fraction and 4 as the minimum LOD score value. The genetic linkage map consisted of eight linkage groups covering more than 350 cM (Fig. 1). The lengths of the linkage groups were 82.2 cM, 81.3 cM, 51.4 cM, 35.1 cM, 26.2 cM, 26.2 cM, 19.2 cM and 16.4 cM. This map included the *pl* gene conferring the weeping shape (11.4 (±0.9) cM from P14–1150, P14–1550 and 17.2 (±1.2) cM from I18–1800) and 52 RAPD markers. Four RAPD markers remained independent (V17–750, Z3–700, Z13–1600, AC11–1500). By using 3 as the minimum LOD score value, these RAPD markers remained independent. In the biggest linkage group, six RAPD markers spanning in 15.5 cM presented a deviation from the 1 : 3 ratio. For five of them, spanning in 12 cM, an excess of the weeping clone recessive allele was observed. These data suggest that a selected allele may be located at this region.

Conclusion

It is too early to say if the eight linkage groups found correspond to the eight chromosomes of peach. However, if we consider that the genome of peach is only slightly larger than that of *Arabidopsis* (Baird et al., 1992) we would be able to saturate the map with a few additional markers (i.e., map with segregating markers covering the entire genome). The different characters of the fruits will be included as soon as the trees begin fruiting. Future work will investigate QTLs in relation to fruit quality, yield and precocity.

References

Bailey, J.S. & A.P. French, 1949. The inheritance of certain fruit and foliage characters in the peach. Mass. Agr. Expt. Sta. Bul.: 452.

Baird, V., L. Belthoff, R. Ballard, R. Scorza, A. Callahan, R. Monet & A. Abbott, 1992. Analysis of the nuclear genome, and construction of a genetic map for peach (*Prunus persica* L. Batch). HortScience 27(6): 580.

Kosambi, D.D., 1944. The estimation of map distance from recombination values. Ann. Eugen. 12: 172–175.

Lander, E.S., P. Green, J. Abrahamson, A. Barlow, M.J. Daly, S.E. Lincoln & L. Newburg, 1987. MAPMAKER: An interactive computer package for constructing primary genetic maps of experimental and natural populations. Genomics: 174–181.

Massonie, G., R. Monet, Y. Bastard & P. Maison, 1984. Hérédité de la réaction d'hypersensibilité du pêcher *Prunus persica* (L) Batsch à *Myzus persicae* Sulzer. Bulletin OILB/SROP 7: 69.

Monet, R., Y. Bastard & B. Gibault, 1988. Étude génétique du caractère "port pleureur" chez le pêcher. Agronomie 8(2): 127–132.

Saghai-Maroof, M.A., K.M. Soliman, R.A. Jorgensen & R.W. Allard, 1984. Ribosomal DNA spacer-length polymorphisms in barley: Mendelian inheritance chromosomal location and populations dynamics. Proc. Nat. Acad. Sci (USA) 81: 8014–8018.

Williams, J.G.K., A.R. Kubelik, K.J. Livak, A. Kafalski & S.V. Tingey, 1990. DNA polymorphisms amplified by arbitrary primers are useful as genetic markers. Nucleic Acids Res. 18: 6531–6535.

H. Schmidt and M. Kellerhals (Eds.), Progress in Temperate Fruit Breeding, 313–320.

Characterization of isozyme variation in walnut (*Juglans regia L.*)

A. Solar, J. Smole, F. Štampar, & M. Viršček-Marn
Biotechnical Faculty, Agronomy Department, University of Ljubljana, 61000 Ljubljana, Slovenia

Key words: Cultivar, identification, isozyme, variation, walnut, *Juglans regia* L.

Summary

Four leaf enzymes malate dehydrogenase (MDH), 6-phosphogluconate dehydrogenase (6PGD), peroxidase (PX) and aspartate aminotransferase (AAT) of 17 walnut cultivars and two pollen enzymes malate dehydrogenase (MDH) and 6-phosphogluconate dehydrogenase (6PGD) of 15 walnut cultivars were analysed using horizontal starch gel electrophoresis. Walnut cultivars of different origin exhibited different numbers of electrophoretic bands and also different relative mobility. Different activity levels and phenotypic groups were detected in studied enzyme systems. Pollen enzymes revealed higher variability than enzymes extracted from the leaves. 15 walnut cultivars were classified into 10 malate dehydrogenase phenotypic groups and 14 6-phosphogluconate dehydrogenase phenotypic groups based on pollen analyses. 17 cultivars were classified into 9 peroxidase phenotypic groups and 7 6-phosphogluconate dehydrogenase phenotypic groups based on analyses of leaves. All of the 15 walnut cultivars could be identified and distinguished with electrophoretic analyses of MDH and 6PGD from the pollen while only 10 cultivars were distinguishable with analyses of 6PGD and PRX from the leaves. No variability useful for cultivar identification was observed in MDH and AAT from the leaves.

Introduction

The first isozymic genes identified in walnut species were for glucose phosphate isomerase (GPI), aspartate aminotransferase (AAT), phosphoglucomutase (PGM) and esterase (EST) (Arulsekar et al. (1986). McGranahan et al. (1986) identified intergeneric hybrids between *Pterocarya* and *Juglans regia* based on isozyme variation. Cheng and Yang (1987) examined the origin of different walnut types in southern China and Tibet using isozyme analysis. Arulsekar et al. (1985) and Germain et al. (1991) studied genetic differences between some of *Juglans* species and interspecific hybrids using isozyme banding patterns. Malvolti et al. (1991) reported on electrophoretic polymorphism in walnut cultivars. Aleta et al. (1990) studied eight enzymatic systems and found that four of them, 6PGD, MDH, PGM and SDH were sufficient to group 29 walnut cultivars into 25 phenotypic groups. Solar et al. (1993) found differences in number and intensity of electrophoretic bands among 15 walnut cultivars while studying the variation of MDH and 6PGD in pollen. Aleta et al. (1993) reported on the inheritance of 6PGD, MDH, PGM and SDH in eight progenies of *Juglans regia*.

Electrophoretic polymorphism of two pollen enzymes, MDH and 6PGD and four leaf enzymes, MDH, 6PGD, AAT and PRX was studied here in order to determine the most variable enzymatic system useful for identification and distinguishing of walnut cultivars.

Material and methods

Fifteen cultivars, part of the walnut germplasm collection in Slovenia, were studied for isozymic variation of pollen enzymes and 17 cultivars from the same collection were studied for isozymic variation of leaf enzymes. Pollen was collected at the time of full bloom of catkins, phenophase Fm 2 (Germain et al. 1981). Leaf extracts were prepared from young growing leaves. The extraction buffer was that described by Weeden and Lamb (1985). Electrophoresis was performed in horizontal starch gels (12 histidine/citrate, pH 6,5 and tris citrate/lithium borate, pH 8,3 were used). Staining solutions used for enzyme identification were those described by Vallejos (1983).

Results and discussion

Isozymes of all the systems were detected on histidine/citrate buffer except AAT from leaves where the second buffer, tris citrate/lithium borate gave better results.

Results of relative mobility

Relative mobility of electrophoretic bands was detected as Kuhns and Fretz, (1978) as follows:

$$Rm = \frac{distance\ enzyme\ band\ migrated}{distance\ marker\ dye\ migrated}$$

The number of bands and their relative mobility were used to produce tables and figures which represented electrophoretic polymorphism of isozymes in different walnut cultivars.

Malate dehydrogenase (pollen). Between seven and nine electrophoretic bands were detected. Eight cultivars had seven bands, six cvs. had eight bands and one cv. had nine electrophoretic bands. The second, the third, the seventh and the nineth band migrated equally in all cultivars. The fifth electrophoretic band was the most variable. It was detected only in cultivars 'G-139', 'Hartley', 'Mayette', 'Šampion', 'Vujan', 'Elit' and 'MB-24'.

Variation of the first, the fourth and the sixth electrophoretic bands was very low. The first band was detected in all cultivars except 'Hartley'. The eighth band appeared only in the cultivars 'Hartley', 'Concord' and 'Mayette'.

6-phosphogluconate dehydrogenase (pollen).
Three or four electrophoretic bands were detected in all included cultivars except in cv. 'G-26' which had only two. The first band stained very strong in all cultivars except in cv. 'Concord'. Intensity of the other bands was different. In cv. 'Hartley' all of the bands were very strongly expressed while the expression of the bands in cv. 'G-26' was very weak. Thirteen cultivars were quite different in respect of variation of the first and the second electrophoretic band. In cvs. 'Concord' and 'Mayette' the second band was not detected. The third and the fourth electrophoretic band was detected in all 15 cultivars. Their variation was a little lower.

Peroxidase (leaves).
The separation of electrophoretic bands was poor. The bands and their relative mobility could not be detected in seven cultivars ('Jupiter', 'G-139', 'G-26', 'Drenovo', 'Elit', 'Petovio' and 'MB-24'). In the other cultivars between one and four electrophoretic bands were detected. Cultivars 'Šampion' and 'Vujan' which are both of Serbian origin had only one band. Six cultivars ('G-120', 'Franquette', 'Mayette', 'Parisienne', 'Hartley' and 'Rače-866') had three bands and two cvs. ('Šejnovo' and 'Esterhazy-II') had four electrophoretic bands.

6-phosphogluconate dehydrogenase (leaves).
Five electrophoretic bands of 6PGD were detected in all of the 17 cultivars. The first, the second, the fourth and the fifth bands had the same relative mobility in all cultivars. Relative mobility of the third electrophoretic band was more variable: from 0.15 ('Hartley') to 0.21 ('G-139'). The bands with the same relative mobility were less intense than the third band which was very strongly expressed in all cultivars except 'G-120' and 'Petovio'.

Malate dehydrogenase (leaves).
Six electrophoretic bands were detected in all cultivars except 'G-139'. The second, the third, the fourth and the fifth bands had the same relative mobility in all cultivars. Variation of the first electrophoretic band was very low. The sixth band was less intense and not clear. It relative mobility was found to vary between 0.41 and 0.45 and it could not be detect precisely. With cv. 'G-139' the zymogram was not clear. We were able to detect only the first and the sixth electrophoretic band in this cultivar.

Aspartate aminotransferase (leaves).
Very low variation was established. Four electrophoretic bands with nearly the same relative mobility were detected in the leaf extracts of all cultivars.

Description of the zymograms

MDH (pollen): The bands were assigned to eight zones: MDH-1,..., MDH-8. Five of them (MDH-2, MDH-3, MDH-4, MDH-5 and MDH-8) were polymorphic and the other were monomorphic (table 1,

Table 1. Relative mobility of MDH and 6PGD from the pollen extract in 15 walnut cultivars.

cultivar	relative mobility															
	malate dehydrogenase (MDH) / activity level													6 PGD / activity level		
	MDH-1	MDH-2		MDH-3			MDH-4		MDH-5	MDH-6	MDH-7	MDH-8		6PGD-1	6PGD-2	6PGD-3	6PGD-4
JUPITER	0.49		0.36		0.30			0.24		0.15	0.09	0.045		0.39	0.32	0.27	0.17
G-120	0.49		0.36		0.30			0.24		0.15	0.09	0.045		0.38	0.32		0.16
G-139	0.49		0.36	0.31		0.285		0.24		0.15	0.09	0.045		0.385	0.32	0.25	0.14
G-26	0.49		0.36		0.30			0.24		0.15	0.09	0.045		0.38	0.32		
G-1247	0.49		0.36		0.30			0.24		0.15	0.09	0.045		0.38	0.32	0.25	0.145
HARTLEY	0.49	0.39	0.36		0.30			0.24	0.19	0.15	0.09			0.38	0.31	0.25	0.145
CONCORD	0.49	0.39	0.36	0.31				0.24		0.15	0.09	0.045		0.38	0.32		0.14
ŠAMPION	0.49		0.36			0.29	0.26	0.24		0.15	0.09	0.05		0.38	0.305	0.25	0.145
VUJAN	0.49		0.36			0.29	0.26			0.15	0.09	0.05	0.38		0.32	0.235	0.14
ELIT	0.49		0.36		0.30		0.26		0.21	0.15	0.09	0.05		0.38	0.30	0.23	0.14
PETOVIO	0.49		0.36		0.30			0.24		0.15	0.09	0.05		0.38	0.32	0.23	0.14
MB-24	0.49		0.36		0.30		0.26	0.24		0.15	0.09	0.05		0.38	0.33	0.28	0.14
ESTERHAZY	0.49		0.36		0.30			0.24		0.15	0.09	0.05		0.39	0.33	0.27	0.13
MAYETTE	0.49	0.40	0.36		0.30			0.24	0.20	0.15	0.09	0.05		0.38	0.32		0.14
KILLINGER	0.49		0.36		0.30			0.24		0.15	0.09	0.045		0.39	0.32	0.25	0.145

Table 2. Relative mobility of PRX and 6PGD from the leaf extract in 17 walnut cultivars.

cultivar	relative mobility											
	peroxidase (PRX) / activity level							6 PGD / activity level				
	PRX-1	PRX-2					PRX-3	6PGD-1		6PGD-2	6PGD-3	
JUPITER								0.31	0.28	0.18	0.12	0.04
G-120		0.35				0.31	0.23	0.31	0.28	0.17	0.12	0.04
G-139								0.31	0.28	0.21	0.12	0.04
G-26								0.31	0.28	0.19	0.12	0.04
FRANQUETTE	0.39	0.35		0.33				0.31	0.28	0.17	0.12	0.04
MAYETTE	0.38	0.35			0.32			0.31	0.28	0.16	0.12	0.04
PARISIENNE	0.39		0.34		0.32			0.31	0.28	0.16	0.12	0.04
HARTLEY	0.38			0.33		0.31		0.31	0.28	0.15	0.12	0.04
RAČE-866	0.37			0.33		0.31		0.31	0.28	0.19	0.12	0.04
DRENOVO								0.31	0.28	0.20	0.12	0.04
ŠAMPION	0.37							0.31	0.28	0.20	0.12	0.04
VUJAN	0.37							0.31	0.28	0.16	0.12	0.04
ELIT								0.31	0.28	0.17	0.12	0.04
PETOVIO								0.31	0.28	0.16	0.12	0.04
MB-24								0.31	0.28	0.20	0.12	0.04
ŠEJNOVO	0.37		0.34		0.32		0.24	0.31	0.28	0.16	0.12	0.04
ESTERHAZY-II	0.38		0.34		0.32		0.24	0.31	0.28	0.16	0.12	0.04

Table 3. Relative mobility of MDH and AAT from the leaf extract in 17 walnut cultivars.

cultivar	relative mobility										
	malate dehydrogenase (MDH) / activity level							AAT / activity level			
	MDH-1	MDH-2			MDH-3			AAT-1	AAT-2	AAT-3	AAT-4
JUPITER	0.43 - 0.41	0.37	0.34	0.30	0.25		0.21	0.61	0.55	0.48	0.37
G-120	0.43 - 0.41	0.37	0.34	0.30	0.25		0.21	0.61	0.55	0.49	0.37
G-139	0.45 - 0.41						0.21	0.61	0.56	0.49	0.37
G-26	0.45 - 0.41	0.37	0.34	0.30	0.25		0.21	0.61	0.55	0.49	0.37
FRANQUETTE	0.45 - 0.41	0.37	0.34	0.30	0.25		0.21	0.62	0.56	0.50	0.38
MAYETTE	0.43 - 0.41	0.37	0.34	0.30	0.25	0.23		0.62	0.56	0.50	0.38
PARISIENNE	0.43 - 0.41	0.37	0.34	0.30	0.25		0.21	0.62	0.56	0.50	0.38
HARTLEY	0.43 - 0.41	0.37	0.34	0.30	0.25	0.23		0.62	0.56	0.50	0.38
RAČE-866	0.43 - 0.41	0.37	0.34	0.30	0.25		0.21	0.62	0.56	0.51	0.38
DRENOVO	0.43 - 0.41	0.37	0.34	0.30	0.25		0.21	0.62	0.56	0.51	0.39
ŠAMPION	0.43 - 0.41	0.37	0.34	0.30	0.25		0.21	0.62	0.56	0.51	0.39
VUJAN	0.43 - 0.41	0.37	0.34	0.30	0.25		0.21	0.62	0.57	0.51	0.39
ELIT	0.43 - 0.41	0.37	0.34	0.30	0.25		0.21	0.62	0.56	0.51	0.39
PETOVIO	0.45 - 0.41	0.37	0.34	0.30	0.25		0.21	0.62	0.56	0.51	0.38
MB-24	0.45 - 0.41	0.37	0.34	0.30	0.25		0.21	0.62	0.57	0.51	0.37
ŠEJNOVO	0.43 - 0.41	0.37	0.34	0.30	0.25		0.21	0.61	0.57	0.50	0.37
ESTERHAZY-II	0.43 - 0.41	0.37	0.34	0.30	0.25		0.21	0.61	0.56	0.50	0.37

Table 4. Isozyme phenotypes of MDH, 6PGD and PRX in 20 walnut cultivars.

cultivar	origin	isozyme phenotypes												
		MDH (pollen)					6 PGD (pollen)				PRX (leaves)			6PGD(l.)
		MDH-2	MDH-3	MDH-4	MDH-5	MDH-8	6PGD-1	6PGD-2	6PGD-3	6PGD-4	PRX-1	PRX-2	PRX-3	6PGD-2
HARTLEY	U.S.A.	bc	b	b	c	-	c	c	c	c	A_2	B_3B_5		B_7
CONCORD	U.S.A.	bc	a	b	-	b	c	b	-	d	x	x	x	x
KILLINGER	U.S.A.	c	b	b	-	b	a	b	c	c	x	x	x	x
FRANQUETTE	France	x	x	x	x	x	x	x	x	x	A_1	B_1B_3		B_5
MAYETTE	France	ac	b	b	b	a	c	b	-	d	A_2	B_1B_4		B_6
PARISIENNE	France	x	x	x	x	x	x	x	x	x	A_1	B_2B_4		B_6
G - 120	Germany	c	b	b	-	b	c	b	-	b	-	B_1B_5	C_2	B_5
G - 139	Germany	c	ad	b	-	b	b	b	c	d	-	-	-	B_1
G - 26	Germany	c	b	b	-	b	c	b	-	-	-	-	-	B_3
G - 1247	Germany	c	b	b	-	b	c	b	c	c	x	x	x	x
ELIT	Slovenia	c	b	a	a	a	c	e	e	d	-	-	-	B_5
PETOVIO	Slovenia	c	b	b	-	a	c	b	e	d	-	-	-	B_6
MB - 24	Slovenia	c	b	ab	-	a	c	a	a	d	-	-	-	B_2
RAČE - 866	Slovenia	x	x	x	x	x	x	x	x	x	A_3	B_3B_5		B_3
ESTERHAZY-II	Hungary	c	b	b	-	a	a	a	b	e	A_2	B_2B_4	C_1	B_6
ŠEJNOVO	Bulgaria	x	x	x	x	x	x	x	x	x	A_3	B_2B_4	C_1	B_6
DRENOVO	Bulgaria	x	x	x	x	x	x	x	x	x	-	-	-	B_2
JUPITER	Czech R.	c	b	b	-	b	a	b	b	a	-	-	-	B_4
ŠAMPION	Yugoslavia	c	c	ab	-	a	c	d	c	c	A_3			B_2
VUJAN	Yugoslavia	c	c	a	-	a	c	b	d	d	A_3			B_6

x - not analysed

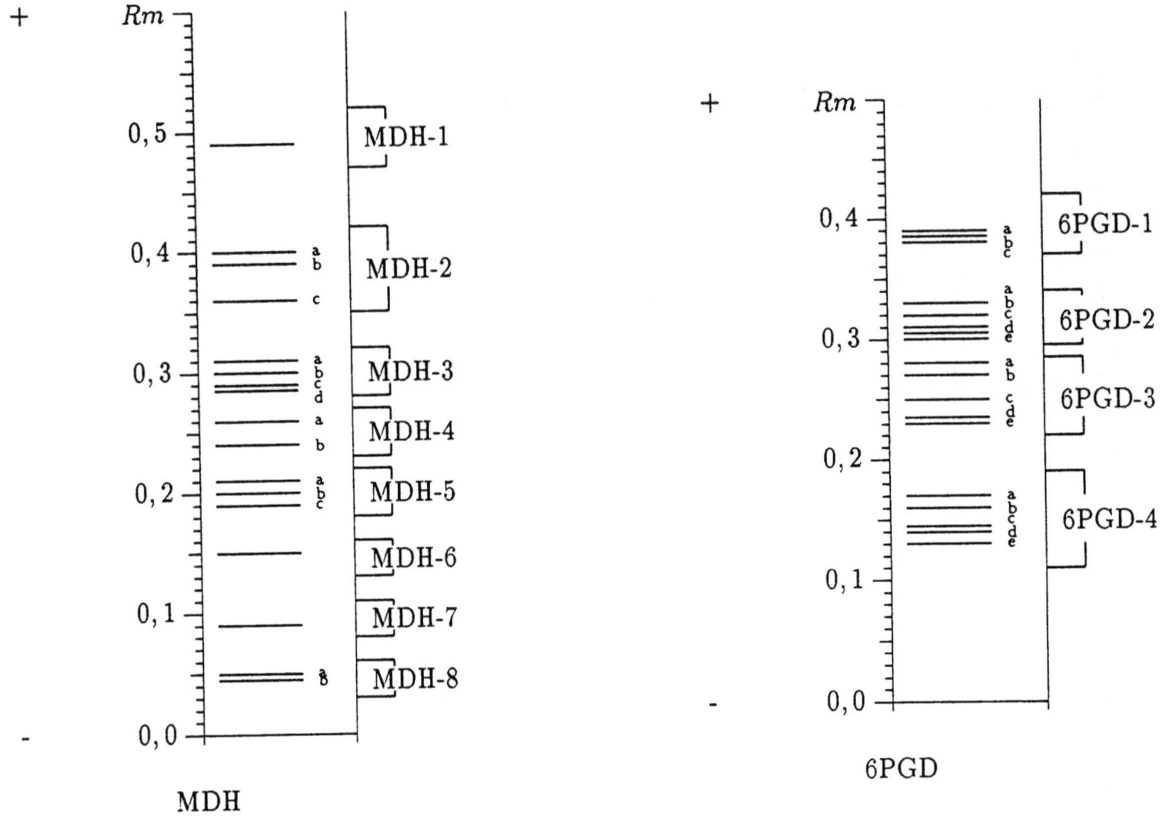

Fig. 1. Schematic representation of banding phenotypes in the MDH and 6PGD from the pollen.

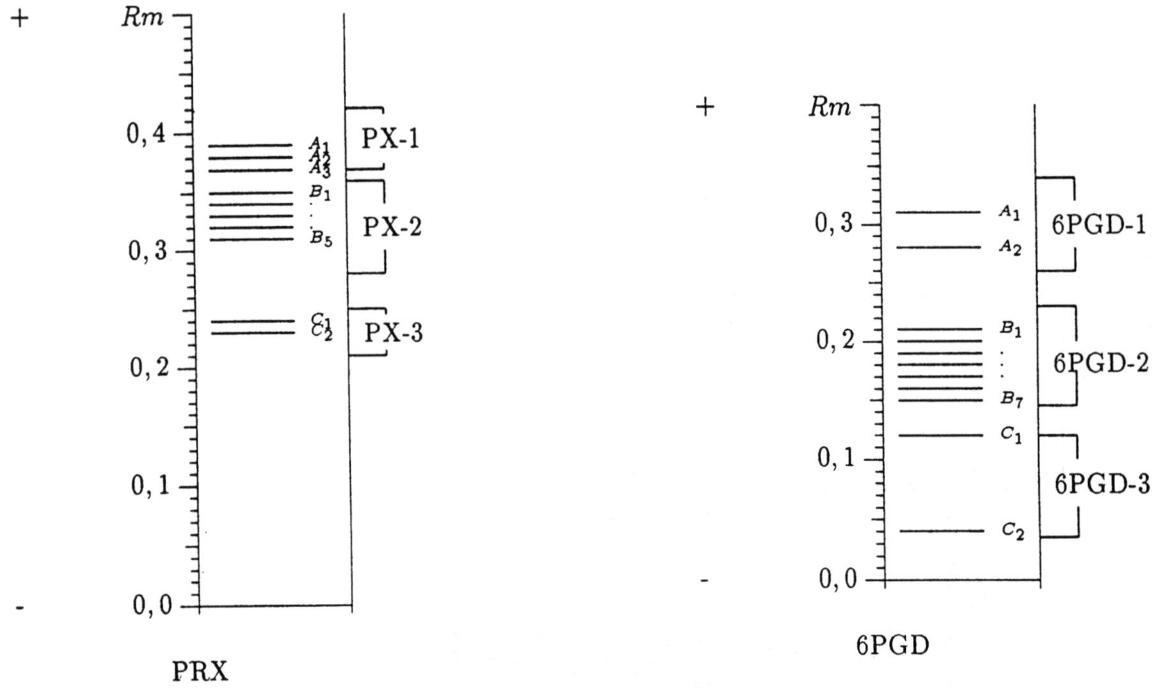

Fig. 2. Schematic representation of banding phenotypes in the PRX and 6PGD from the leaves.

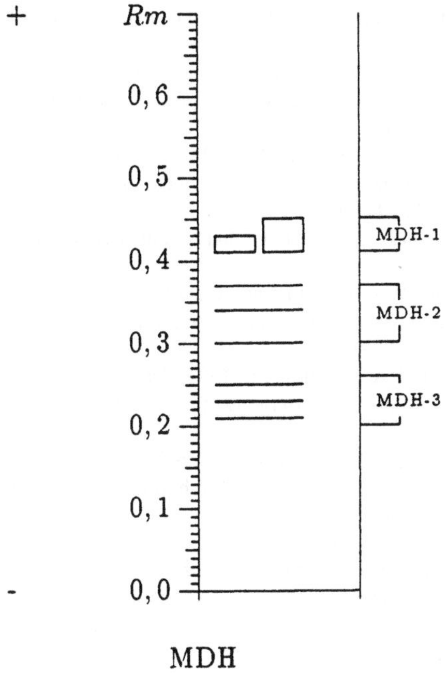

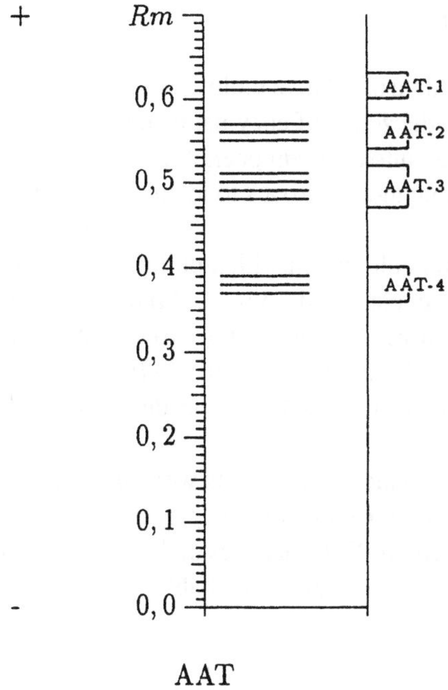

Fig. 3. Schematic representation of banding phenotypes in the MHD and AAT from the leaves.

figure 1).

6PGD (pollen): Four sets of 6PGD (6PGD-1, ..., 6PGD-4) were present. All of them were polymorphic (table 1, figure 1).

PRX (leaves): Three polymorphic sets of PRX activity were present in the leaf extracts (table 2, figure 2).

6PGD (leaves): Three regions of 6PGD activity were observed. Polymorphism was displayed only in 6PGD-2 (table 2, figure 2). Aleta et al. (1990) observed only two activity regions of 6PGD: the first was monomorphic and the second was polymorphic. She did not detect the third region because of the low separability in the most cathodal part of the zymogram.

MDH (leaves): Three sets of MDH activity were present (table 3, figure 3). It was not possible to distinguish bands in the first zone of activity. Aleta et al. (1990) observed from six to nine electrophoretic bands in different cultivars. We detected only six bands in zones MDH-2 and MDH-3. It is possible that some of cultivars had more electrophoretic bands in MDH-1 zone which we could not distinguish.

AAT (leaves): Four zones of AAT activity were observed (table 3, figure 3). Variation was not good enough for cultivar distinction. This result is in agreement with those obtained by Arulsekar et al. (1985) and Aleta et al. (1990).

The most intense electrophoretic bands were detected in MDH in the leaf extract, what is in agreement with Arulsekar and Parfitt (1986). The same authors cited that young, actively growing leaves are the most convenient tissue for isozyme analyses. In our case, electrophoresis of pollen enzymes brought better results in comparison with leaf enzymes.

We compared the isozyme patterns of cultivars of the same origin. German cultivars 'G-120', 'G-26' and 'G-1247' have the same phenotype of pollen MDH, French cultivars 'Mayette' and 'Parisienne' have the same isozyme pattern of leaf 6PGD and Yugoslav cultivars 'Šampion' and 'Vujan' are the same as far as PRX phenotypes are concerned.

The same pattern of pollen 6PGD was observed in the American cultivar 'Concord' and the French cultivar, 'Mayette'. 'Mayette' had the same leaf isozyme pattern for MDH as the American cultivar 'Hartley'. Aleta et al. (1990) found the same MDH phenotype in 'Mayette' and 'Hartley', too. *Juglans regia* L. also known as the English walnut was introduced from Europe to U.S.A in the early 1800s (Chandler, 1947; Arulsekar et al., 1986; Korać, 1986 etc.). Therefore some American and French cultivars could have the same ancestors which might explain similarity of isozyme patterns in the above mentioned cultivars.

Isozymic maps

Based on the data of relative mobility of electrophoretic bands and the zones of enzymatic activity cultivars were classified into different phenotypic groups (Table 4). They enable most cultivars to identified and distinguished.

Fifteen walnut cultivars in which pollen extracts were analysed were classified into 10 MDH phenotypic groups and 14 6PGD phenotypic groups. With electrophoretic analyses of MDH and 6PGD from pollen all of the cultivars included could be identified and distinguished.

17 cultivars in which leaf extracts were analysed were classified into 9 PRX phenotypic groups and 7 6PGD phenotypic groups. With analyses of these two systems only 10 cultivars were distinguishable.

Conclusions

Malate dehydrogenase and 6-phosphogluconate dehydrogenase extracted from the pollen are very useful enzyme systems for identification walnut cultivars. All of the 15 included cultivars were identificated and distinguished with electrophoresis of these two systems. Among the enzyme systems, extracted from the leaves, 6PGD and PRX were the most variable. We could identify and distinguish 10 walnut cultivars out of 17 using electrophoresis of 6PGD and PRX from the leaves.

References

Aleta, N., Olarte, C., Truco, M. J. & Arus, P. (1990). Identification of walnut cultivars by isozyme analysis. Acta Horticulturae 284: 91-96.

Aleta, N., Rovira, M., Ninot, A. & Arus, P. (1991). Inheritance of four isozymes in walnut. Acta Horticulturae 311:62-65.

Arulsekar, S., Parfitt, D. E. & McGranahan, G. H. (1985). Isozyme gene markers in *Juglans* species. The Journal of Heredity 76:103-106.

Arulsekar, S., and Parfitt, D. E. (1986). Isozyme analysis procedures for stone fruits, almond, grape, walnut, pistachio and fig. Hort Science 21(4):928-933.

Arulsekar, S., McGranahan, G. H. & Parfitt, D. E. (1986). Inheritance of phosphoglucomutase and esterase isozymes in Persian walnut. The Journal of Heredity 77:220-221.

Chandler, W.H. (1947). Decidious Orchards. Philadelphia, ZDA, 388.

Cheng, S. and Yang, W. (1987). Taxonomic studies of ten species of the genus *Juglans* based on isozymic zymograms. Acta Horticulturae Sinica 14(2):90-96.

Germain, E., Jalinat, J. & Marchou, M. (1981). Divers aspects de la biologie florale du noyer. In: Bergougnoux, F. & Grospierre, P.: Le Noyer, 13-27.

Germain, E., Hanquier, I. & Monet, R. (1993). Identification of eight *Juglans* sp. and some inter-specific hybrids by isoenzyme patterns. Acta Horticulturae 311:73-81.

Korać, M. (1986). Orah. Nolit, Beograd, 7,8.

Kuhns, L.J. and Fretz, T.A. (1978). Distinguishing rose cultivars by polyacrylamide gel electrophoresis. I. Extraction and storage of protein and active enzymes from rose leaves. J. Amer. Soc. Hort. Sci. 103(4):503–508.

Malvolti, M.E., Paciucci, M., Cannata, F. & Fineschi, S. (1993). Genetic variation in Italian populations of *Juglans regia* L. Acta Horticulturae 311:86-91.

Manganaris, A.G. and F.H. Alston (1992). Inheritance and linkage relationships of peroxidase isoenzymes in apple. Theor. Appl. Genet. 83:392-399.

McGranahan, G.H., Tulecke, W., Arulsekar, S. & Hansen, J.J. (1986). Intergeneric Hybridization in the Juglandaceae: *Pterocarya* sp. x *Juglans regia*. J.Amer. Soc.Hort.Sci. 114:627-630.

Peirce, L.C. & Brewbaker, J.L. (1973). Applications of Isozyme Analysis in Horticultural Science. Hort Science 8(1):17-22.

Scandalios, J.G. (1969). Genetic Control of Multiple Molecular Forms of Enzymes in Plants. A review. Biochem. Genet. 3:37-79.

Shields, C.R., Orton, T.J. & Stuber, C.W. (1983). An outline of general resource needs and procedures for the electrophoretic separation of active enzymes from plant tissue. In: Tanksley, S. D., Orton, T. J.: Isozymes in Plant Genetics and Breeding. Part A, Elsevier, Amsterdam, 443-468.

Solar, A., Smole, J. Štampar, F. (1993). Identification of Walnut Cultivars by Pollen Isozymes. Acta Horticulturae 311:95-100.

Vallejos, C.E. (1983). Enzyme activity staining. In: Tanksley, S. D., Orton, T.J.: Isozymes in plant genetics and breeding. Part A, Elsevier, Amsterdam, 469-516.

Weeden, N.F. & Lamb, R.C. (1985). Identification of Apple Cultivars by Isozyme Phenotypes. J. Amer. Soc. Hort. Sci. 110:509-515.

H. Schmidt and M. Kellerhals (Eds.), Progress in Temperate Fruit Breeding, 321–325.

The use of isoenzyme polymorphisms as an aid for cultivar identification in strawberry

J.A. Bell & D.W. Simpson
Horticulture Research International, East Malling, West Malling, Kent, ME19 6BJ, U.K.

Key words: Cultivar identification, isoenzymes, strawberry, *fragaria* × *ananassa*

Summary

The study was intended to uniquely classify 32 strawberry cultivars and two breeders' selections, mostly of Northern European origin, using four polymorphic enzyme systems: phosphoglucoisomerase (PGI), phosphoglucomutase (PGM), esterase (EST) and leucine amino peptidase (LAP). Three of these (PGI, PGM and LAP) had been used previously for this purpose by workers in North America. The techniques of both starch (SGE) and polyacrylamide gel electrophoresis (PAGE) were used. Consistent banding patterns were obtained for PGI, PGM and LAP while EST produced a large number of sharp bands but the patterns were inconsistent. The three stable enzymes only were thus used for cultivar characterisation. Banding patterns were divided phenotypically to produce three regions for PGI and two for PGM and LAP. Nine different banding categories were observed for PGI, 13 for PGM and 10 for LAP. In combination these three enzyme systems permitted 30 of the 34 clones examined to be uniquely characterised. Although much of the material is closely related, e.g. nine of the cultivars have Gorella as one parent, the technique offers a high level of discrimination. In the two cases where it was not possible to separate cultivars no common ancestors for at least three generations exist. The procedure will assist greatly in cases of misidentification where a particular clone is known to be one of a small number of alternatives. Also, the banding patterns could be used as an additional criterion for identification, alongside various morphological characteristics, when publishing cultivar descriptions.

Introduction

In any germplasm collection it is extremely important to maintain varietal purity. Morphological differences between genotypes are often small, particularly if they are closely related, and may also be inconsistent, so a method of distinguishing between genotypes using reliable genetic markers has considerable advantages. Isoenzyme variation has been used to identify cultivars of a wide range, including seed and vegetatively propagated species. Twenty-nine North American strawberry cultivars have been classified using three enzyme systems (Bringhurst et al., 1981; Nehra, et al., 1991), although in several cases not uniquely.

Production of strawberry nuclear stocks for the United Kingdom is done at Horticulture Research International (HRI), East Malling, on behalf of the Nuclear Stock Association (NSA). These plants, which are produced under the Plant Health Propagation Scheme (PHPS), are free from all known pests and diseases and are ultimately the source of all certified strawberry plants propagated in the UK, so must be guaranteed 'true to type'. The PHPS maintains healthy plants of existing cultivars and produces pest and disease-free material of new cultivars and breeders' selections from the UK and overseas breeding programmes.

The intention of this study was to classify uniquely 32 cultivars and two breeders' selections, mainly from the NSA collection, using four polymorphic enzyme systems: phosphoglucoisomerase (PGI), phosphoglucomutase (PGM), esterase (EST) and leucine amino peptidase (LAP). Three of these (PGI, PGM and LAP) had also been used by Bringhurst et al. (1981) and Nehra et al. (1991). The techniques of both starch

(SGE) and polyacrylamide gel electrophoresis (PAGE) were used.

Materials and methods

The 32 cultivars studied were; Aromel, Cambridge Favourite, Cambridge Vigour, Harvester, Hedley, Pandora, Pantagruella, Pegasus, Providence, Redgauntlet, Royal Sovereign, Silver Jubilee, Tantallon and Troubadour from the U.K.; Bogota, Elsanta, Elvira, Gorella, Karina, Ostara, Rapella and Tamella from the Netherlands; Domanil and Hapil from Belgium; Gea from Italy; Annapolis, Honeoye, Rainier, Selva, Tioga, Totem and Tyee from North America; and the two HRI breeders' selections LA 467 and LA 957. All plants were conventionally propagated from runners and maintained in a glasshouse at 20 °C with daylength extended to 16 hours during the winter to maintain active growth.

Newly opened leaves for PGI, PGM and EST, and older leaves (dark green) for LAP were collected and placed in a refrigerator at 4 °C for short term storage. Samples were prepared by adding liquid N_2 to 100 mg of leaf tissue in a mortar and then grinding to a fine powder. When the N_2 had evaporated 3ml of extraction buffer was added and thawed samples were then centrifuged for 20 minutes at 18,000 rpm and 4 °C; the supernatant was collected. Samples for PAGE (EST and LAP) then had 30 μl of bromophenol blue added. Samples were kept cool or on ice at all times. Different extraction buffers were used for each enzyme studied:

PGI – 0.001M DTT, 0.02M tris HCl pH7.5, 1% PVP (soluble).

PGM – 0.01M mercaptoethanol, 0.01M Na_2EDTA, 0.1M KCl, 0.1M $MgCl_2$, 8% PVP (soluble), 0.1M Tris HCl pH7.5.

LAP – 0.05M phosphate pH7, 0.01M mercaptoethanol, 10% PVPP, 15% sucrose.

EST – 0.1M tris HCl pH7.5, 15% sucrose, 10% PVPP, 0.001M DTT, 0.05% Triton X-100.

For PGI and PGM, starch gels were prepared following the method of Smithies (1955). For PGI a tris citrate – lithium borate buffer – system at pH 7.6 was used (Simpson, 1981), while for PGM a histidine – EDTA – tris citrate system at pH 7.0 (Florence, 1981) was found to give the best resolution of the bands. PGM gels had a tracker dye (bromophenol blue) loaded on a filter paper wick at either end of the samples; a visible front is produced with the discontinuous buffer system employed for PGI. The current for both gels was initially set at 50 mA for 30 minutes, the wicks were then removed and current reset to 5 mA (PGM) and 10 mA (PGI) and run overnight (approximately 16 hours). Gels were stopped when the front had moved 8–10 cm from the origin, then sliced and stained.

The following recipes, modified from those of Gottlieb (1972) and Yang (1971) were used to stain for enzyme activity:

PGI – fructose-6-phosphate 20 mg, NADP 10 mg, PMS 1 mg, MTT 10 mg, 1M $MgCl_2$ 0.5 ml, 0.05M tris HCl pH8 50 ml, glucose-6-phosphate dehydrogenase 50 units.

PGM – glucose-1-phosphate 150 mg, NADP 15 mg, MTT 20 mg, PMS 2 mg, 1M $MgCl_2$ 1 ml, 0.04M tris HCl pH7 50 ml, glucose-6-phosphate dehydrogenase 50 units.

Gels were incubated in the dark at 37 °C.

Polyacrylamide gels were prepared and run (with slight modifications) following the method of Manganaris & Alston (1987) with 70 μl of sample introduced into each well. For EST the gel was run until the tracking dye reached the end of the cassette, while for LAP a further 2 hours was necessary after this point.

Stain recipes were modified from those of Devonshire (1975) and Scandalios (1969).

LAP – 0.2M phosphate buffer pH6 50 ml, black K salt 100 mg, L-leucine-β-naphthylamide 25 mg (dissolved in 1 ml methanol).

EST – 0.2M phosphate buffer pH6 50 ml, fast blue RR salt 100 mg, α-naphthyl acetate 5 mg (dissolved in 1 ml acetone).

Results

Consistent banding patterns were obtained with three of the enzymes studied: PGI, PGM and LAP. The fourth enzyme system, EST, produced between 3 and 10 sharp bands but the patterns were inconsistent. Nehra et al. (1991) reported that EST in strawberry had unstable phenotypes when comparing leaves from glasshouse and in vitro plants. The results of this study showed that even when comparable leaves from glasshouse plants were used the phenotypes were still unstable, probably being influenced by the physiological state of the plants and factors such as daylength. Only the three stable enzymes were used for cultivar characterisation.

Bringhurst et al. (1981) and Nehra et al. (1991) each used a different system of classification for their

Table 1. Classification of strawberry varieties using banding patterns of three enzyme systems, phosphoglucoisomerase, phosphoglucomutase and leucine amino peptidase

Class	PGI	PGM	LAP	
1	1,2	1,1	2,1	Honeoye
2	1,2	3,1	3,2*	Tioga
3	1,2	5,2	2,3*	Rapella
4	2,2	3,1	1,1	Ostara
5	2,2	3,4*	2,1	Annapolis
6	3,1*	5,1*	1,3	Bogota
7	3,2	1,1	1,1	Elsanta
8	3,2	1,2	4,3	Tamella
9	3,2	1,3	1,1	LA 467
10	3,2	3,1	4,1	Tyee
11	3,2	3,3	1,1	Selva
12	3,2	3,3	4,1	Totem, Troubadour
13	3,2	4,1	1,1	Hedley
14	3,2	5,2	1,1	Gorella
15	3,2	5,3	5,3*	Elvira
16	4,2	2,1	1,1	Hapil
17	4,2	3,1	1,1	Cambridge Favourite
18	4,2	3,1	4,1	Harvester
19	4,2	3,5*	1,1	LA 957
20	4,2	4,1	1,1	Pegasus
21	5,1	1,2	1,3	Gea
22	5,1	1,3	2,1	Rainier
23	5,1	3,2	4,1	Domanil
24	5,2	2,1	1,3	Aromel, Providence
25	5,2	2,1	4,2	Tantallon
26	5,2	3,1	1,1	Redgauntlet
27	5,2	3,2	5,1*	Karina
28	5,2	5,3	4,2	Silver Jubilee
29	6,1*	5,3	1,1	Pantagruella
30	6,2	2,1	1,1	Royal Sovereign
31	6,2	3,3	4,3	Cambridge Vigour
32	6,2	4,1	1,1	Pandora

* Unique classification with one enzyme.

observed banding patterns. These results revealed a wider range of phenotypes than either of the two previous studies so, to simplify the scoring procedure, the banding patterns were divided phenotypically to produce three regions for PGI (Fig. 1) and two for PGM (Fig. 2) and LAP (Fig. 3). The gels were scored for the observed pattern within each region and then each phenotype allocated a banding category (Table 1).

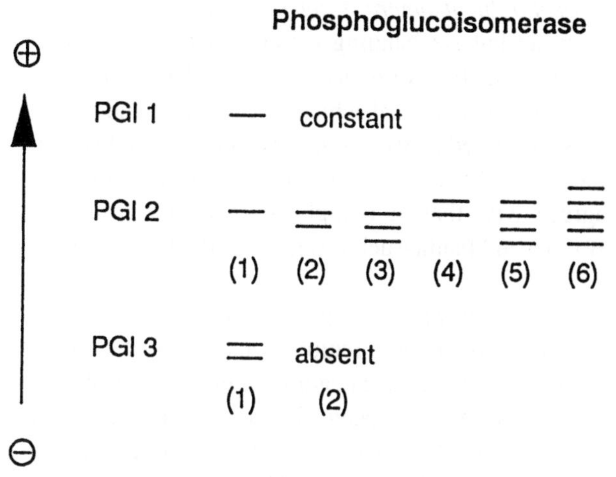

Fig. 1. Observed banding patterns for PGI.

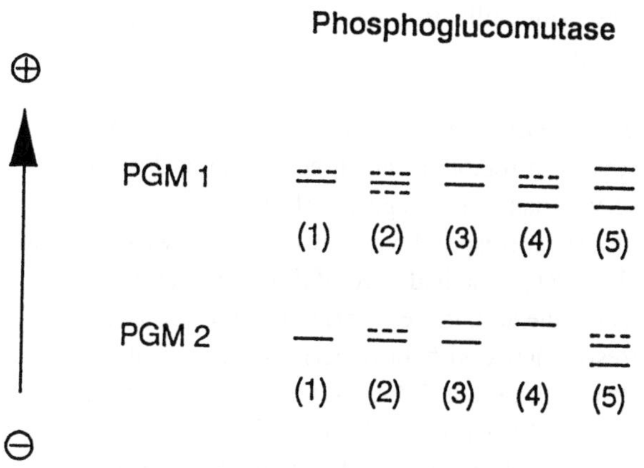

Fig. 2. Observed banding patterns for PGM.

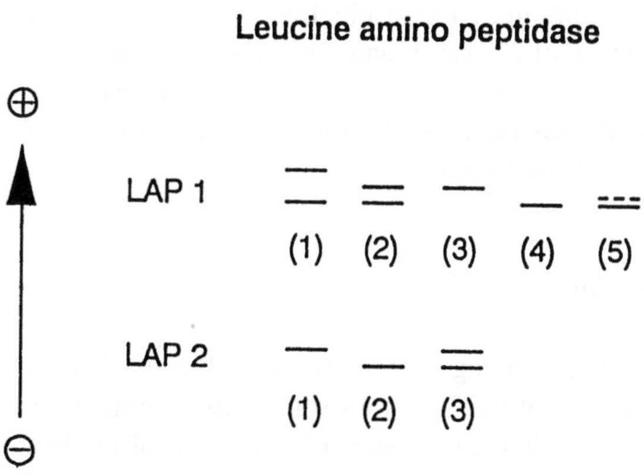

Fig. 3. Observed banding patterns for LAP.

Phosphoglucoisomerase (pgi). Region 1 was monomorphic but six banding patterns were observed for region 2 and two for region 3 (Fig. 1). There were nine banding categories of which three corresponded with those observed by Bringhurst et al. (1981) and two with those of Nehra et al. (1991). Our classification of Tioga agreed with that of Bringhurst et al. Two cultivars, Bogota and Pantagruella, were classified uniquely.

Phosphoglucomutase (pgm). Five banding patterns were observed in both regions 1 and 2 giving a total of 13 observed banding categories (Fig. 2). Four of these categories had also been observed by Bringhurst et al. (1981) but there was no clear correspondence with the patterns of Nehra et al. (1991), who had used the PAGE system. Bringhurst et al. had classified Tioga as C3, corresponding to 5,1, whereas here 3 bands were observed rather than 4 and it was scored 3,1. Further work will be required to resolve this anomaly. Three clones, Annapolis, Bogota and LA 957 were classified uniquely.

Leucine amino peptidase (lap). Five patterns were observed in region 1 and three in region 2 giving 10 observed banding categories (Fig. 3). These did not obviously correspond with any observed by Bringhurst et al. (1981), who had used SGE rather than PAGE for this enzyme and, more surprisingly, there was no clear correspondence with the patterns of Nehra et al. (1991). Nehra et al. used a PAGE system with 5% acrylamide compared to 9% here. Elvira, Karina, Rapella and Tioga were uniquely classified using LAP alone.

Using two enzyme systems increased the number of clones that could be classified uniquely. The best two-enzyme combination was PGI with PGM which gave 29 classes and unique classification for 26 out of 34 clones. Combining the results from all three enzymes gave a total of 32 classes and it was possible to classify uniquely 30 of the 34 clones (Table 1). It was still not possible to distinguish between Totem and Troubadour or Aromel and Providence.

Discussion

It was encouraging that it was possible to classify uniquely 30 of the 34 clones studied using only three enzymes. This is rather surprising in view of the fact that much of the material is closely related, e.g. nine of the cultivars have Gorella as one parent. Interestingly, in the two cases where it was not possible to separate cultivars they are not closely related. There are no common ancestors between Providence and Aromel for at least four generations, while Totem and Troubadour have Howard 17 as a common great great grandparent but are otherwise unrelated.

While a good level of separation between clones has been achieved using the three enzyme systems PGI, PGM and LAP it is obvious that these systems alone would not be sufficient to identify uniquely each clone in a large germplasm collection such as that maintained for a breeding programme or plant variety rights testing. They would, however, assist greatly in cases of misidentification where a particular clone was known to be one of a small number of alternatives. Furthermore the banding patterns could be used as an additional criterion for identification, alongside various morphological characteristics, when publishing cultivar descriptions.

The technique would be enhanced by including additional enzymes but EST is not suitable for this purpose as the bands were found to be inconsistent. Preliminary studies with several other enzymes have suggested that acid phosphatase, glutamate oxaloacetate transaminase, glycerate-2-dehydrogenase, malate dehydrogenase and malic enzyme are all polymorphic in the cultivated strawberry but experimental procedures need to be refined to produce consistently clear distinct bands. Other enzymes examined were aldolase, fumarase, glucose-6-phosphate dehydrogenase, glutamate dehydrogenase, α-glycerophosphate dehydrogenase, isocitrate dehydrogenase, 6-phosphogluconate dehydrogenase and peroxidase but no polymorphism was observed for any of these.

A further enhancement of the system using just PGI, PGM and LAP would be to consider also the intensity of the bands. Arulsekar et al. (1981) suggested this would be possible with PGI as the differences in intensities are a consequence of octoploidy in the cultivated strawberry. Therefore, using the banding patterns on a purely qualitative basis means that different genotypes can result in the same phenotype. Differences in band intensity were observed to be consistent although this was not measured. Objective recording of this could be attempted using a scanning densitometer to record the relative intensities.

Future work will also involve investigating the inheritance of isoenzyme banding patterns in strawberry, particularly for PGM, which has not been studied before.

Acknowledgements

This work was carried out as part of a project commissioned by the Ministry of Agriculture, Fisheries and Food (Project number HH0113JFV).

References

Arulsekar, S., R.S. Bringhurst & V. Voth, 1981. Inheritance of PGI and LAP isozymes in octoploid cultivated strawberries. Journal of the American Society for Horticultural Science 106(5): 679–683.

Bringhurst, R.S., S. Arulsekar, J.F. Hancock & V. Voth, 1981. Electrophoretic characterisation of strawberry cultivars. Journal of the American Society for Horticultural Science 106(5): 684–687.

Devonshire, A.L., 1975. Studies of the carboxylesterases of *Myzus persicae* resistant and susceptible to organophosphorus insecticides. In: Proceedings of the 8th British Insecticide Fungicide Conference 1, pp. 67–73.

Florence, L.Z., 1981. Allozyme variation and population structure in loblolly pine. Ph.D. dissertation. Stephen F. Austin, State University, Nacogdoches, Texas.

Gottlieb, L.D., 1972. Enzyme differentiation and physiology in *Clarkia franciscana*, *C. rubicunda* and *C. amoena*. Evolution 27: 205–214.

Manganaris, A.G. & F.H. Alston, 1987. Inheritance and linkage relationships of glutamate oxaloacetate transaminase isoenzymes in apple. 1. The gene GOT-1, a marker for the *S* incompatibility locus. Theoretical and Applied Genetics 74: 154–161.

Nehra, N.S., K.K. Kartha & C. Stushnoff, 1991. Isozymes as markers for identification of tissue culture and greenhouse-grown strawberry cultivars. Canadian Journal of Plant Science 71: 1195–1201.

Scandalios, J.G., 1969. Genetic control of multiple molecular forms of enzymes in plants: A review. Biochemical Genetics 3: 37–79.

Simpson, D.W., 1981. Variation in natural populations of three *Ranunculus* species. PhD thesis. University of Birmingham, U.K.

Smithies, O., 1955. Zone electrophoresis in starch gels: Group variations in serum proteins of normal human adults. Biochemical Journal 61: 629–641.

Yang, S.Y., 1971. Studies and genetics VI. An appendix. University of Texas publication 7103: 85–90.

CHAPTER 6

In vitro methods and biotechnology

H. Schmidt and M. Kellerhals (Eds.), Progress in Temperate Fruit Breeding, 329–331.

Stable gene expression in transgenic apple tree tissues and segregation of transgenes in the progeny – preliminary evidence

D.J. James, A.J. Passey & S.A. Baker
Horticulture Research International, East Malling, West Malling, ME19 6BJ, U.K.

Key words: Apple, inheritance, stable gene expression, transgenes, trees, *Malus pumila*

Summary

We report here, for the first time, the stable expression and Mendelian segregation of transgenes in a tree species. So far we have evidence for a 1 : 1 segregation of the *nos* gene in the R1 of transgenic apple progeny. In addition we present evidence for stable gene expression of both *nos* and the co-transferred gene *nptII* in the fruit flesh of the apple fruit some 7 years after the initial transformations.

abbreviations: nos – nopaline synthase, *nptII* – neomycin phosphotransferase, ACC – 1-aminocyclopropane carboxylic acid

Introduction

Transgenic apple clones (cv Greensleeves) were first produced in 1986 (James et al., 1989) using the disarmed Ti-binary vector pBIN6 (Bevan, 1984) in an Agrobacterium-mediated leaf disc transformation procedure (Horsch et al., 1985). The plants were transgenic for a reporter gene nopaline synthase (*nos*) and a selectable marker neomycin phosphotransferase (*nptII*) conferring resistance to kanamycin.

Procedure

In 1987 six transgenic clones of *in vitro* rooted plants of pBIN6 apple clones were transferred to soil and grown under glass for 2 years before being grafted onto M27 dwarfing rootstock to reduce size and encourage early flowering. Annual tests, since 1987, for expression of *nos* in the leaves of all clones have always been positive.

All clones appeared phenotypically normal except one clone, B, which showed some tendency towards premature leaf fall and senescence. This clone had previously shown reduced rooting ability *in vitro* (James et al., 1989). In 1992 two clones flowered under growth room conditions and in 1993 the other 4 did so.

In 1992 controlled pollinations were performed on the first 2 clones using pollen of the apple cv Baskatong which carries a dominant, homozygous gene for anthocyanin pigmentation (R) (Church & Williams, 1978). This separated hybrid zygotic embryos from any selfed or apomictic embryos. Fruit set was normal on all pollinated trees.

To expedite segregation analyses immature embryos were excised aseptically from fruit 30–40 days post anthesis and micro-propagated by standard procedures (James & Dandekar, 1991) to provide early material for biochemical and molecular analyses.

Gene expression assays of apple fruit

nopaline synthase (nos)

Gene expression assays for *nos* were carried out using paper electrophoresis (Otten & Schilperoort, 1978). Samples of apple 'flesh' were excised directly from mature fruit raised under greenhouse conditions. Freshly excised 'flesh' tissue always gave a positive response for *nos* from several different apple fruits as

Table 1. ELISA assays for the NPTII protein performed on transgenic and control apple tissues

Clone	ng NPTII/mg protein	
	Callus from fruit 'flesh'	Leaf extract
Untransformed tissue Replicate 1	0	0
Untransformed tissue Replicate 2	0	0
Transformed tissue Clone A Replicate 1	not available	1.4
Transformed tissue Clone A Replicate 2	not available	1.0
Transformed tissue Clone E Replicate 1	7.8	not done
Transformed tissue Clone E Replicate 2	11.2	not done

did the callus tissues derived from flesh tissue after 5 weeks growth on a callus-inducing medium containing 50μg/ul of the antibiotic kanamycin.

neomycin phosphotransferase (nptii)

A double anti-body sandwich ELISA method for the NPTII protein was used for biochemical assays (Nagel et al., 1992). NPTII ELISA assays were performed on different tissues of transformed and non-transformed Greensleeves apple plants. Transgenic fruit flesh was able to form callus on callus-induction media containing either 0 or 50 μg/ul of kanamycin during a 5 week growth period, whereas control flesh from untransformed fruit of the same cultivar and grown under the same conditions failed to grow in the presence of the antibiotic. The NPTII protein was detectable in both fruit 'flesh' callus (8–11 ng/mg protein) and leaf material (1–1.5 ng/mg protein) from transgenic plants but not in equivalent material from control untransformed plants (Table 1).

Table 2. Segregation of the *nos* gene in two independent transgenic apple clones

Transgenic clone	*nos* positive embryos	*nos* negative embryos
Clone B	7	5
Clone E	27	25

Inheritance data

Micropropagated immature embryos, exhibiting the anthocyanin pigmentation, and derived from two independent transgenic apple clones, B and E, were assessed for segregation and expression of the *nos* gene. Data from electrophoresis assays showed that the gene segregated 1 : 1 according to a normal Mendelian ratio, where clone E showed 27 'seedlings' positive for expression of nopaline synthase and 25 negative from a population of 52, whilst clone B showed 7 positive and 5 negative from a population of 12 'seedlings' (Table 2).

Future work and prospects

Currently molecular evidence is being sought to confirm the inheritance of the transgenes. The data presented here suggest that long-lived stable gene expression in both vegetative and reproductive parts of the apple plant should in the future permit the introduction of novel genes that control such processes as fruit ripening. Already two genes that code for key enzymes involved in the ripening of climacteric fruit by modulating endogenous levels of ACC, and thereby ethylene biosynthesis (see review by Theologis, 1992), have been isolated i.e. ACC synthase and ACC oxidase or the ethylene forming enzyme (EFE). By inverting these 'ripening genes' with respect to their promoter, i.e. making the 'anti-sense form' of the genes, it has already proven possible to reduce ethylene biosynthesis markedly and alter tomato ripening for improved fruit quality (Picton et al., 1993). Since the corresponding genes have also been isolated, cloned and sequenced from ripening apple fruit (Dong et al., 1991, 1992) it should be possible to apply the same anti-sense strategy to control apple ripening. At the same time it should also permit the identification of fruit-specific genes and their corresponding promoters for the targeting of gene expression directly to the fruit for improvements in both fruit quality and enhanced pest and disease resistance.

References

Bevan, M.W., 1984. Binary *Agrobacterium* vectors for plant transformation. Nuc. Acids Res. 12: 8711–8721.

Church, R. & R.R. Williams, 1978. Pollination of pome fruits. Report of the Long Ashton Research Station for 1977, pp. 21–22.

Dong, J.G., W.T Kim, W.K. Yip, G.A. Thompson, L. Li, A.B. Bennett & S.F. Yang, 1991. Cloning of a cDNA encoding 1-aminocyclopropane-1-carboxylate synthase and expression of its mRNA in ripening apple fruit. Planta 185: 38–45.

Dong, J.G., D. Olson, A. Silverstone & S.F. Yang, 1992. Sequence for a 1-aminocyclopropane-1-carboxylate oxidase from apple fruit. Plant Physiol. 98: 1530–1531.

Horsch, R.B., J.E. Fry, N.L. Hoffman, D. Eicholtz, S.G. Rogers & R.T. Fraley, 1985. A simple and generalised method of transferring genes into plants. Science 227: 1229–1231.

James, D.J. & A.M. Dandekar, 1991. Regeneration and transformation of apple (*Malus pumila* Mill.). Plant Tissue Culture Manual B8: 1–18.

James, D.J., A.J. Passey, D.J. Barbara & M.W.Bevan, 1989. Genetic transformation of apple (*Malus pumila* Mill.) using a disarmed Ti-binary vector. Plant Cell Rep. 7: 658–661.

Nagel, R.J., J.M. Manners & R.G.B irch, 1992. Evaluation of an ELISA assay for rapid detection and quantitation of neomycin phosphotransferase II in transgenic plants. Plant Molec. Biol. Rep. 10: 263–272.

Otten, L.A.B.M. & R.A. Schilperoort, 1978. A rapid microscale method for the detection of lysopine and nopaline dehydrogenase activities, Biochim. et Biophys. Acta 527: 494–500.

Picton, S., S.L. Barton, M. Bouzayen, A.J. Hamilton & D. Grierson, 1993. Altered fruit ripening and leaf senescence in tomatoes expressing an antisense ethylene-forming enzyme transgene. The Plant J. 3: 469–482.

Theologis, A. 1992. One rotten apple spoils the whole bushel: The role of ethylene in fruit ripening. Cell 70: 181–184.

H. Schmidt and M. Kellerhals (Eds.), Progress in Temperate Fruit Breeding, 333–338.

Transgenic 'Malling 26' apple expressing the attacin E gene has increased resistance to *Erwinia amylovora*

John L. Norelli,[1] Herb S. Aldwinckle,[1] Luis Destéfano-Beltrán[2,3] & Jesse M. Jaynes[2,4]
[1] *Department of Plant Pathology, Cornell University, New York State Agricultural Experiment Station, Geneva, NY 14456-0462, USA;* [2] *Department of Biochemistry, Louisiana State University, Baton Rouge, LA 70803, USA;* [3] *present address: Centro Internationalde Agricultura Tropical, P.O. Box 6713 Cali, Colombia;* [4] *present address: Demeter Biotechnologies, Limited, P.O. Drawer 14388, Research Triangle Park, NC 27709, USA*

Key words: apple, attacin, transformation, resistance, fire blight, *Erwinia amylovora, Malus domestica*

Summary

Apple (*Malus domestica*) transgenic T1 was obtained by *Agrobacterium tumefaciens*-mediated transformation of Malling 26 rootstock using the plasmid binary vector pLDB15. pLDB15 contains within its T-DNA a gene encoding the lytic protein attacin E. The integration of the attacin E gene into the apple genome was confirmed by Southern analysis. Northern analysis indicated the presence of an attacin E mRNA in plants inoculated with *Erwinia amylovora*. After inoculation of *in vitro* grown plants of T1, Malling 26, and Malling 7 (resistant control) with *E. amylovora*, the $\log_{10}$ of the inoculum concentration lethal to 50% of the plants was 5.4, 4.4, and 5.6, respectively. In greenhouse trials for resistance to fire blight, T1 was significantly more resistant than 'Malling 26'.

Abbreviations: DNA – deoxyribonucleic acid; GUS – beta-glucuronidase; M.7 – Malling 7; M.26 – Malling 26; MU – 4-methylumbelliferone; MUG – 4-methylumbelliferyl-beta-D-glucuronide; *npt*II – neomycin phosphotransferase gene; RNA – ribonucleic acid; X-gluc – 5-Bromo-4-chloro-3-indolyl-beta-D-glucuronic acid

Introduction

Insects have an inducible resistance response to bacterial infection known as cell free or humoral immunity (Boman & Hultmark, 1987). When diapausing pupae of the Giant Silk Moth, *Hyalophora cecropia*, are injected with bacteria, they respond by producing a set of approximately 15 proteins, including attacins and cecropins (Boman & Hultmark, 1987; Hultmark et al., 1980). Attacin E and the synthetic cecropin-like peptides SB-37 and Shiva-1 possess substantial lytic activity against many important plant pathogenic bacteria (Destéfano-Beltrán, 1991; Destéfano-Beltrán et al., 1990; Jaynes et al., 1993). Chimeric genes comprised of the lytic protein genes driven by plant promoters have been cloned into *Agrobacterium tumefaciens* binary plasmid vectors (Destéfano-Beltrán, 1991; Destéfano-Beltrán et al., 1990; Jaynes et al., 1993). Transgenic tobacco lines containing Shiva-1 lytic protein have increased resistance to infection by *Pseudomonas solanacearum* (Destéfano-Beltrán, 1991; Jaynes et al., 1993). Fire blight, caused by the bacterium *Erwinia amylovora*, is a serious and difficult-to-control disease of apple scion and rootstock cultivars (Van der Zwet & Beer, 1992). Our goal was to obtain lytic protein-transgenic apple in order to evaluate the protein's effect on the susceptibility of apple to *E. amylovora*.

Materials and methods

Binary plasmid vector pLDB15. Plasmid pCP521 (Kockum et al., 1984) contains the complete cDNA sequence of the mature form of attacin E. Four modifications were made to this attacin E clone to allow for its transfer to and expression in plants (Destéfano-Beltrán, 1991). First, a plant consensus AACA*ATG* initiation codon was placed immediately upstream of the first codon of the gene. Second, much of the animal transcription termination signal was removed from the 3′ end. Third, the attacin E gene was placed between the promoter and terminator of the proteinase inhibitor gene from potato (Keil et al., 1989). Fourth, it was cloned into the *A. tumefaciens* binary plasmid vector, pBI121 (Clontech Labs, Inc., Palo Alto, CA).

Plant tissue culture and transformation. The apple cultivar used for transformation was the rootstock Malling 26. Methods and media used for shoot tip proliferation, rooted-plant culture, and plant regeneration from leaf pieces were as previously described (Norelli & Aldwinckle, 1993; Norelli et al., 1988) except that light conditions for plant regeneration were 40 μEinsteins/m^2/sec at room temperature (21°–30° C). The disarmed *A. tumefaciens* strain LBA4404 (Hoekema et al., 1983) was grown in Kado 523 broth (Kado & Heskett, 1970) for 18 hours at 28° C, resuspended in 0.5 × Murashige-Skoog micro- and macro-elements (Murashige & Skoog, 1962), pH 5.4, containing 100 μM acetosyringone, and adjusted to a density of 2 × 10^9 cfu/ml by measuring absorbance at 600 nm. Leaves were cut from rooted *in vitro* plant cultures, sliced transversely into segments 3–5 mm wide, inoculated with *A. tumefaciens,* and placed abaxial side up on regeneration medium without antibiotics. After 48 hours incubation in the dark at room temperature, leaf segments were transferred to regeneration medium containing 10 μg/ml paromomycin and 250 μg/ml cefotaxime. Regenerating cultures were divided into pieces containing 1 or a few meristems and were placed on proliferation medium containing 50 μg/ml kanamycin. Shoot tips that remained green on kanamycin medium after 6 wk were screened for GUS activity.

GUS assays. The presence of GUS in putative transgenic plants was determined using a fluorometric assay based on the cleavage of MUG to MU as described by Jefferson et al. (1987). Quantitative GUS assays were conducted as previously described (Norelli & Aldwinckle, 1993).

*Neomycin phosphotransferase gene (*nptII*) activity assay.* *npt*II activity was assayed by evaluating the ability of *in vitro* grown shoot tips to root in the presence of 25 or 50 μg/ml kanamycin. A single 114 ml (4-oz) baby food jar of rooting medium containing 5 shoot tips was the unit of replication; there were 25 replicates per treatment.

Southern analysis. Plant DNA was isolated using the Gawel & Jarret (1991) procedure for DNA extraction from *Musa* except that 1) the leaf tissue-extraction buffer mixture was incubated at 37° C for 45 min, and 2) following treatment of DNA with RNAse, the DNA was treated with Proteinase K (1.5 mg/ml at 55° C for 90 min. DNA was analyzed by standard Southern procedures (Sambrook et al., 1989). The attacin probe was the 2.2 kb *Hin*dIII fragment of pLDB15 and consisted of the 5′ and 3′ region of the proteinase inhibitor II gene from potato and the attacin gene. The GUS probe was the ca. 2.1 kb *Bam*HI-*Eco*RI fragment of pBI121 and consisted of the GUS gene and the nopaline synthase terminator from *A. tumefaciens.* The *npt*II probe was the ca. 1.6 kb *Pst*I fragment of pBI121 and consisted of most of the *npt*II coding region and the nopaline synthase terminator.

Northern analysis. Source plants for RNA isolation were inoculated with *E. amylovora* 72 hours prior to leaf harvest to induce expression of the attacin E gene. Total RNA was isolated from leaf tissue using the protocol of Napoli et al. (1990). Poly (A) RNA was purified from the bulk of cellular RNA by affinity chromatography on oligo(dT)-cellulose and analyzed by standard northern procedures (Sambrook et al., 1989). The attacin probe was the 2.2 kb attacin E *Hin*dIII fragment of pLDB15 (Fig. 1A).

In vitro *evaluation of disease resistance. In vitro* grown rooted plants were inoculated with *E. amylovora* as previously described (Norelli et al., 1988) except that inoculum was prepared at 6 or more concentrations ranging from 1 × 10^4 to 1 × 10^7 cfu/ml. *E. amylovora* strain Ea273 was used as inoculum. SAS (SAS Institute Inc., Cary, NC) probit procedure was used to calculate the inoculum dose necessary for 50% of the plants to become infected (LD_{50}). Evaluations were repeated three times and LD_{50} values were averaged for the three evaluations. A single 114 ml baby food jar of rooting medium containing 5 shoot tips was the unit of replication for inoculations; there were means of 2.3, 2.3, and 6.9 replicates per inoculum dose × cultivar combination in the three trials.

Greenhouse evaluation of disease resistance. In vitro propagated plants of T1, M.26, and M. 71 were adapted to growth in the greenhouse, and then grown as single shoot plants. Vigorously growing plants with 14.5 to 26 cm of shoot growth (mean 19.9 cm) were inoculated with *E. amylovora* strain Ea273 as previously described (Aldwinckle & Preczewski, 1976) except that 1) inoculum concentration was 5 × 10^6 cfu/ml and 2) inoculum was prepared in 0.01 M phosphate buffer, pH 6.5. The percent of the shoot length that developed fire blight symptoms after inoculation with *E. amylovora* was observed every 2 to 3 days until all lesions had ceased extension into healthy tissue. Seven replicates of each cultivar were inoculated 23 April 1993. Disease progress was modeled to an exponential regression curve described by $y = a + br^x$ using Genstat 5 (1987) where: y is the proportion of the shoot length blighted, x is the number of days after inoculation, a is the y asympote, $(a + b)$ is the origin, and r is the non-linear 'rate' parameter. Differences between regression lines were tested by an F test comparing the regression variance obtained when the data were fitted to a common regression line for both cultivars with that obtained for separate lines for each cultivar.

Results

Recovery of transgenic plant. From 648 leaf segments inoculated with *A. tumefaciens* strain LBA4404 containing plasmid binary vectors, 36 regenerants were recovered on medium with 10 μg/ml paromomycin. Among these was an *npt*II transgenic (T1) from an M.26 leaf slice inoculated with LBA4404 (pLDB15). T1 possessed *npt*II and beta-glucuronidase (GUS) activity which non-transgenic M.26 did not. In an *in vitro* test, 76%, 0%, and 0% of M.26 plants, and 48%, 60%, and 45% of T1 plants rooted in medium containing 0, 25, and 50 μg/ml kanamycin, respectively. The rate of MUG to MU conversion (nanomoles/min/mg fresh weight) in M.26 and T1 tissue was 0.6 and 81.3, respectively.

Southern analysis. Since the attacin E gene is contained in a 2.2 kb *Hind*III fragment within the T-DNA transferred to T1 (Fig. 1A), digestion of either plasmid DNA or transgenic genomic plant DNA should result in a 2.2 kb fragment that hybridized with the attacin E gene probe. As predicted, a ca. 2.2 kb fragment from pLDB15 (Fig. 1B, lane 2) and T1 (lane 4) but not from pBI121 (lane 1) or M.26 (lane 3) hybridized with the attacin gene probe. Since the GUS gene and *npt*II are flanked by T-DNA borders and a *Hind*III site (Fig. 1A), hybridization of these probes to plasmid DNA and transgenic genome DNA digested with *Hind*III should result in hybridizing fragments of different sizes. The pLDB15 DNA fragment that hybridizes with the probes should be the same size as the pBI121 hybridizing fragment (Fig. 1A). The size of the hybridizing fragment in transgenic genomic DNA will be of unknown size because it will include both

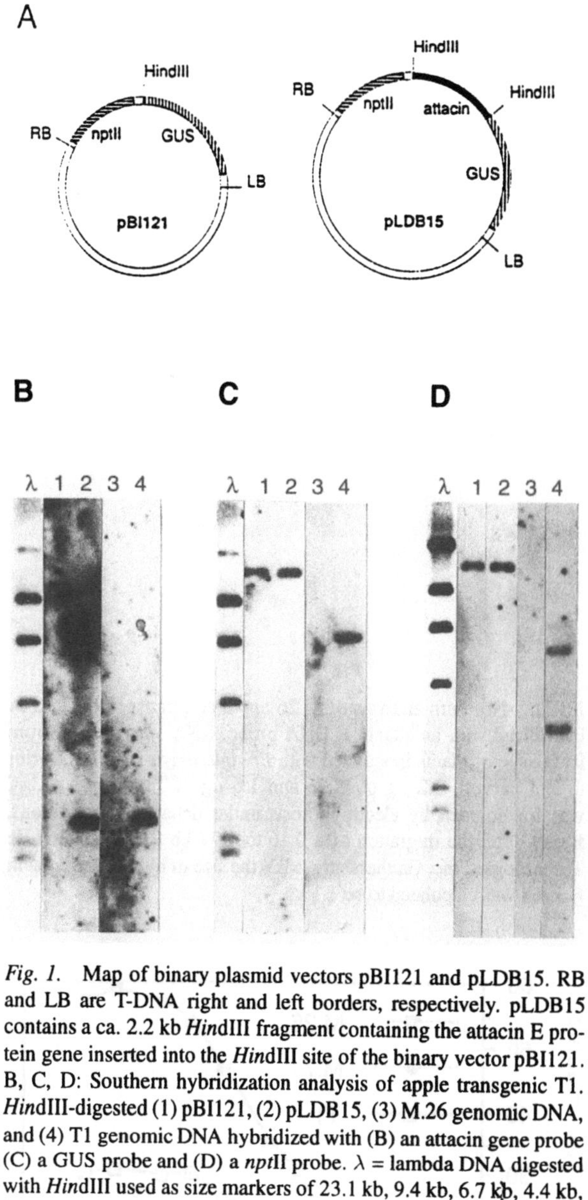

Fig. 1. Map of binary plasmid vectors pBI121 and pLDB15. RB and LB are T-DNA right and left borders, respectively. pLDB15 contains a ca. 2.2 kb *Hind*III fragment containing the attacin E protein gene inserted into the *Hind*III site of the binary vector pBI121. B, C, D: Southern hybridization analysis of apple transgenic T1. *Hind*III-digested (1) pBI121, (2) pLDB15, (3) M.26 genomic DNA, and (4) T1 genomic DNA hybridized with (B) an attacin gene probe (C) a GUS probe and (D) a *npt*II probe. λ = lambda DNA digested with *Hind*III used as size markers of 23.1 kb, 9.4 kb, 6.7 kb, 4.4 kb, 2.3 kb, and 2.0 kb.

T-DNA (*Hind*III site to T-DNA border) and plant DNA (site of integration to next plant *Hind*III site). Since the GUS gene and *npt*II of pLDB15 are separated by *Hind*III sites, the DNA fragments that hybridize with the probes should be of different sizes. As predicted, hybridization of *Hind*III-digested pBI121 and pLDB15 indicated that a fragment with the approximate size of pBI121 hybridized with the GUS gene probe (Fig. 1C, lanes 1 and 2) and the *npt*II probe (Fig. 1D, lanes 1 and 2). M.26 DNA did not hybridize with either the GUS gene probe (Fig. 1C, lane 3) or the *npt*II probe (Fig. 1D, lane 3). Hybridization of T1 genomic DNA indi-

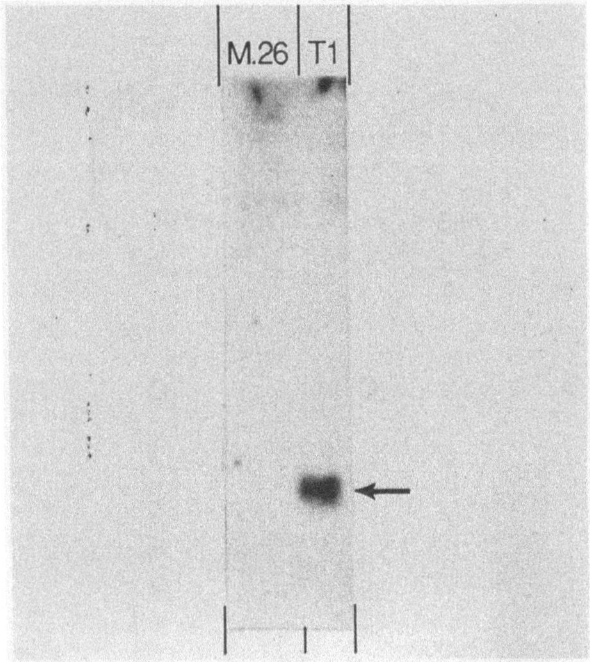

Fig. 2. Northern analysis of M.26 apple and transgenic T1 RNA hybridized with an attacin E DNA probe. RNA was isolated from leaf tissue of plants inoculated with *Erwinia amylovora* 72 hr prior to leaf harvest. 3.2 μg of M.26 and 1.6 μg of T1 poly (A) RNA was fractionated by electrophoresis under denaturing conditions. Based upon the migration of a 0.16 to 1.77 kb RNA ladder (Life Technologies, Inc, Gaithersburg, MD) the size of hybridizing attacin E RNA was calculated to be 1.1 kb.

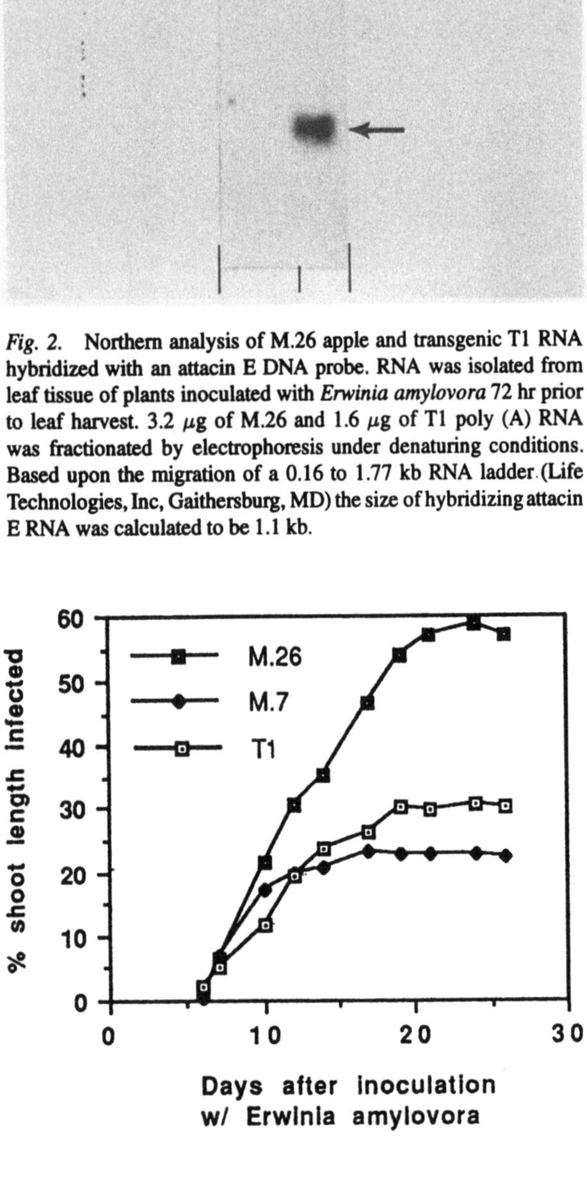

Fig. 3. The development of fire blight in greenhouse plants following inoculation with *E. amylovora*. T1 is an attacin E transgenic of the fire blight susceptible apple rootstock M.26. Liberty is a disease-resistant scion cultivar with moderate resistance to fire blight.

cated that a fragment of unique size (approximately 7.5 kb) hybridized with the GUS probe (Fig. 1C, lane 4). Hybridization of T1 genomic DNA with an *npt*II probe resulted in 2 fragments of unique size (approximately 5.9 and 3.4 kb).

Northern analysis. Northern analysis indicated the presence of an attacin E m-RNA in T1 plants 72 hours after inoculation with *E. amylovora* (Fig. 2). That mRNA was not present in M.26 plants. Since twice as much M.26 RNA as T1 RNA was loaded onto the gel (Fig. 2), these results cannot be due to non-specific hybridization. Based upon a 0.16 to 1.7 kb RNA size marker, the size of the mRNA that hybridized with the attacin probe was estimated to be 1.1 kb, which is very close to the predicted size of the attacin E m-RNA (ca. 1.0 kb, including ca. 20 bp leader, 640 bp coding region, ca. 200 bp 3′ untranslated region, and ca. 100 bp poly (A) tail).

Evaluation of disease resistance. When *in vitro* grown plants of T1, M.26, and (resistant control cultivar) were inoculated with *E. amylovora*, the mean $\log_{10}$ of the inoculum concentration necessary to kill 50% of the plants (LD_{50}) was 5.4, 4.4, and 5.6, respectively. The LD_{50} values obtained for T1 were less than those obtained for M.26 in all three of the independent inoculation trials. The resistance of T1 was compared with that of M.26 and M.7 by evaluating the amount of fire blight that developed over time when greenhouse plants were inoculated with *E. amylovora* (Fig. 3). The use of the exponential regression curve $y = a + br^x$ to model disease progress accounted for 99% of the variance. Disease progress in M.26 was significantly different from that observed in T1 ($P < 0.0001$). In M.26, 57% of the shoot length was necrotic 21 days after inoculation with *E. amylovora* compared with 30% in T1. The amount of fire blight that occurred in T1 and M.7 was similar 7 and 12 days after inoculation (Fig. 3), although, by 21 days there was a significant difference between the disease progress of T1 and M.7 ($P < 0.0001$).

Discussion

Recent studies have indicated that selection with 25 to 63 μg/ml paromomycin is optimal to select for *npt*II transgenic M.26 cells on agar solidified medium (Norelli & Aldwinckle, 1993). The low rate of *npt*II transgenic selection was probably due to the low concentration of paromomycin. Paromomycin was used to select *npt*II transgenics because the frequency of regeneration from *npt*II-transgenic M.26 leaf tissue on agar solidified media containing paromomycin is higher than on media containing kanamycin (Norelli & Aldwinckle, 1993).

The presence in T1 genomic DNA of a unique fragment that hybridized with the GUS probe is proof of T1 transformation. Transformation of T1 is also supported by hybridization with the *npt*II probe, although, multiple hybridizing fragments were not expected. Hybridization of multiple fragments could result from partial digestion of DNA or multiple T-DNA insertions. Since aliquots of the same *Hin*dIII-digested DNA were used for Figs 1B, 1C, and 1D and there is no evidence of incomplete digestion in Figs 1B or 1C, the multiple bands cannot be explained by incomplete digestion resulting from DNA preparation or digestion. Incomplete digestion of specific, but not all, restriction sites of a particular restriction endonuclease have been reported for plant genomic DNA including apple (Simon & Weeden, 1992). If the multiple fragments were due to partial digestion of the *npt*II-*Hin*dIII site, then digestion with another enzyme should result in a single hybridizing fragment. However, when T1 genomic DNA was digested with *Eco*RI and hybridized with the *npt*II probe, multiple fragments hybridized with the probe (data not shown). This suggests that the multiple hybridizing fragments are due to multiple T-DNA insertions. Since hybridization with the GUS and attacin probe resulted in a single hybridizing fragment, the hybridization results could be explained by a single full-length T-DNA insertion and at least one insertion of an incomplete T-DNA fragment consisting of T-DNA right border and at least part of *npt*II.

This work demonstrates that an M.26 transgenic containing and expressing the attacin E gene has increased resistance to infection by *E. amylovora* when compared with the parent cultivar M.26. Confirmation of the genetic basis of the increased fire blight resistance of T1 must await flowering of T1 plants and subsequent progeny analysis.

The fire blight resistance of T1 in *in vitro* and greenhouse evaluations was more similar to that of M.7 than of M.26. Van der Zwet and Keil (1979) have summarized fire blight susceptibility data for the most common Malling and Malling-Merton rootstocks evaluated by several other researchers. The mean blight ratings for natural infection plus artificial inoculation of ungrafted rootstocks were placed in four general categories. M.26 was placed in the most susceptible 'very severe' category, while M.7 was placed in the most resistant 'light' category.

Apple and other perennial fruits are beset by many bacterial diseases, some of which are uncontrollable by current technology, and others controllable only by excessive use of chemical pesticides. Although the use of resistant cultivars has been a very effective means of controlling the bacterial diseases of annual crops, its success with perennial crops has been limited due to unavailability of resistant cultivars, strong market demands for specific susceptible cultivars, and difficulties in developing resistant cultivars by classical breeding methods. The introduction of bacterial disease resistance genes into existing apple cultivars of economic importance by biotechnology could be of great benefit to the control of bacterial diseases on perennial crops.

Acknowledgements

We thank Scott Demmin and Karin Kroeger, USDA-ARS, Plant Genetic Resources Unit, Cornell University, Geneva, NY for advice on the isolation of apple genomic DNA; and John Barnard, Computer Services, Cornell University, Geneva, NY for statistical advice. We also thank Leah Abraham, Herb Gustafson, and Laurie Jensen for technical assistance. This research was funded in part by the New York Apple Research Association Inc., the New York State Apple Research and Development Program, and the New York State Integrated Pest Management Program.

References

Aldwinckle, H.S. & J.L. Preczewski, 1976. Reaction of terminal shoots of apple cultivars to invasion by *Erwinia amylovora*. Phytopathology 66: 1439–1444.

Boman, H.G. & D. Hultmark, 1987. Cell-free immunity in insects. Ann. Rev. Microbiol. 41: 103–126.

Destéfano-Beltrán, L.J.C., 1991. The introduction into tobacco plants of genes which encode some of the natural components of the humoral immune response of *Hyalophora cecropia*. Ph.D. Dissertation Louisiana State University, Baton Rouge, LA.

Destéfano-Beltrán, L, P.G. Nagpala, M. Selim Cetiner, J.H. Dodds & J.M. Jaynes, 1990. Enhancing bacterial and fungal disease resistance in plants: application to potato. In: M.E. Vayda & W.D. Park (Eds) The molecular and cellular biology of the potato, pp. 205–221. CAB International, Wallingford, UK.

Gawel, N.J. & R.L. Jarret, 1991. A modified CTAB DNA extraction procedure for *Musa* and *Ipomoea*. Plant Mol. Biol. Rep. 9: 262–266.

Genstat 5 Committee, 1987. Genstat 5 Reference Manual. Clarendon Press, Oxford, UK.

Hoekema, A., P.R. Hirsch, P.J.J. Hooykaas & R.A. Schilperoort, 1983. A binary vector strategy based on separation of *vir*- and

T-region of the *Agrobacterium tumefaciens* T1-plasmid. Nature 303: 179–180.

Hultmark, D., H. Steiner, T. Rasmuson & H.G. Boman, 1980. Purification and properties of three inducible bactericidal proteins from the hemolymph of immunized pupae of *Hyalophora cecropia*. Eur. J. Biochem. 106: 7–16.

Jaynes, J.M., P. Nagpala, L. Destéfano-Beltrán, J.H. Huang, J.H. Kim, T. Denney & S. Cetiner, 1993. Expression of a Cecropin B lytic peptide analog in transgenic tobacco confers enhanced resistance to bacterial wilt caused by *Pseudomonas solanacearum*. Plant Science 89: 43–53.

Jefferson, R.A., T.A. Kavanagh & M.W. Bevan, 1987. GUS fusions: β-glucuronidase as a sensitive and versatile gene fusion marker in higher plants. EMBO J. 6: 3901–3907.

Kado, C.I. & M.G. Heskett, 1970. Selective media for isolation of *Agrobacterium, Corynebacterium, Erwinia, Pseudomonas*, and *Xanthomonas*. Phytopathology 60: 969–976.

Keil, M., J.J. Sanchez-Serrano & L. Willmitzer, 1989. Both wound-inducible and tuber-specific expression are mediated by the promoter of a single member of the potato proteinase inhibitor II gene family. EMBO J. 8: 1323–1330.

Kockum, K., I. Faye, P.V. Hofsten, J.-Y. Lee, K.G. Xanthopoulos & H.G. Boman, 1984. Insect immunity. Isolation and sequence of two cDNA clones corresponding to acidic and basic attacins from *Hyalophora cecropia*. EMBO J. 3: 2071–2075.

Murashige, T. & F. Skoog, 1962. A revised medium for rapid growth and bioassay with tobacco tissue culture. Physiol. Plant 15: 473–497.

Napoli, C., C. Lemieux & R. Jorgensen, 1990. Introduction of a chimeric chalcone synthase gene into petunia results in reversible co-suppression of homologous genes *in trans*. The Plant Cell 2: 279–289.

Norelli, J.L. & H.S. Aldwinckle, 1993. The role of aminoglycoside antibiotics in the regeneration and selection of neomycin phosphotransferase-transgenic apple tissue. J. Amer. Soc. Hort. Sci. 118: 311–316.

Norelli, J.L., H.S. Aldwinckle & S.V. Beer, 1988. Virulence of *Erwinia amylovora* strains to *Malus* spp. Novole plants grown *in vitro* and in the greenhouse. Phytopathology 78: 1292–1297.

Sambrook, J., E.F. Fritsch & T. Maniatis, 1989. Molecular Cloning: a laboratory manual, 2nd ed. Cold Spring Harbor Laboratory Press, New York.

Simon, C.J. & N.F. Weeden, 1992. Molecular analysis and cloning of *Malus* ribosomal DNA. J. Amer. Soc. Hort. Sci. 117: 164–168.

Van der Zwet, T. & H.L. Keil, 1979. Fire Blight: A Bacterial Disease of Rosaceous Plants. U.S. Dep. Agric. Handb. 510: 200pp.

Van der Zwet, T. & S.V. Beer, 1992. Fire blight – its nature prevention and control, a practical guide to integrated orchard management. Agric. Info Bull. 631, USDA.

H. Schmidt and M. Kellerhals (Eds.), Progress in Temperate Fruit Breeding, 339–342.

Isolation of apple genomic clones containing putative β-1,3-glucanase gene(s)

Jyothi Thimmapuram & Schuyler S. Korban
Department of Horticulture, University of Illinois, 310 Plant & Animal Biotechnology Laboratory Building, 1201 W. Gregory, Urbana, IL 61801, USA

Key words: Malus × domestica, apple, plant defense gene(s), β-1,3-glucanase, disease resistance, gene cloning, restriction mapping

Summary

An apple genomic library was constructed using Sau3A digested 'Golden Delicious' DNA cloned into a Charon 35 Lambda cloning vector. At screening time, the titer of the library was 3.4 × 10^3 plaque forming units per microliter (pfu/μl). This library was used to screen for β-1,3-glucanase gene(s) (EC3.2.1.39); (a) plant defense gene(s) which encode(s) for a hydrolytic enzyme that catalyzes β-1,3-glucans already present in fungal cell walls. This enzyme may also act indirectly by degrading polysaccharides to oligosaccharides which in turn act to elicit the plant's defense system. The objective of this study is to isolate and characterize β-1,3-glucanase gene(s) in apple. The apple genomic library was screened using a β-1,3-glucanase cDNA clone from maize as a probe. After four screenings, 8 positive clones were isolated, 2 of which appear to be different from each other. A restriction enzyme map of one clone was constructed. A 2.1 Kb EcoRI/HindIII fragment has shown repeated hybridization in several Southern blots suggesting that this fragment contains a DNA sequence homologous to the maize β-1,3-glucanase cDNA clone.

Introduction

One of the many plant responses to infection by a pathogen involves the induction of hydrolytic enzymes such as β-1,3-glucanase and chitinase. β-1,3-glucanase, along with chitinase, has been shown to inhibit fungal growth on agar plates (Mauch et al., 1988a). This enzyme may also act indirectly by degrading polysaccharides to oligosaccharides from fungal cell walls which in turn act to elicit the plant's defense response (Ham et al., 1991).

In plants, it is often observed that a hypersensitive response to a single pathogen induces resistance to various other pathogens (Collinge & Slusarenko, 1987). This type of resistance may be due to induction of various proteins such as pathogenesis-related (PR)-proteins. β-1,3-glucanase, one of the PR-proteins, has been investigated both at the physiological and molecular level in herbaceous plants (Edington et al., 1991; Xu et al., 1992; Chang et al., 1993). There are two isoforms of β-1,3-glucanases reported in plants, a vacuolar form which is basic, and a cell wall- localized form which is acidic (Linthorst, 1991). β-1,3-glucanase appears to be coordinately expressed along with chitinase, following pathogen infection, wounding, ethylene treatment, or chemical stress (Simmons et al., 1992). Expression of glucanase and chitinase have been reported to correlate with resistance reactions in plants (Mauch et al., 1988b). Identical isoforms are induced in response to various treatments listed above suggesting that induction of β-1,3-glucanase is a non-specific form of response by the host. β-1,3-glucanase has been reported to play a role in the signaling leading to a hypersensitive response by releasing elicitors from cell walls of plant pathogens (Mauch & Staehelin, 1989; Lamb et al., 1989). Isolation and characterization of cDNA and genomic clones from different plant species showed similarities between the acidic and basic isoforms of β-1,3-glucanases (Linthorst et al., 1990; Hoj et al., 1989). Both isoforms are encoded by a small gene family, with higher level of similarity observed among members of each group than between groups (Linthorst, 1991).

Woody plants differ from herbaceous plants not only in growth patterns, longevity, size, and morphology, but also in biochemistry, physiology, and development. Therefore, it is possible that these differences may extend to the molecular level. The apple, *Malus* × *domestica* Borkh., is an important fruit crop in the world. Only a few genes from apple have been isolated and characterized (Korban & Chen, 1992). β-1,3-glucanase genes, which are important in the overall plant defense response have not been studied in woody plants as far as we know. Investigating the structure and expression of β-1,3-glucanase gene(s) from apple provides insights into the biology of defense-related genes in woody plants. Hence, we are interested in isolating and characterizing this gene(s) from apple.

Materials and methods

Screening an apple genomic DNA library

An apple genomic DNA library was constructed in our laboratory using Sau3A-digested 'Golden Delicious' DNA cloned into a BamHI-digested Charon 35 Lambda cloning vector (Chen et al., 1990). The library was titered and propagated in *Escherichia coli* strain LE392.

To screen the library, duplicate lifts of plaques were hybridized with a maize β-1,3-glucanase cDNA clone used as a probe. The cDNA probe was labeled with digoxigenin-11-dUTP using the random primed method. Labeling and detection were conducted using the Genius system (Boehringer Mannheim Corporation, Indianapolis, Indiana), a non-radioactive labeling and detection system. Hybridization was done in 5 × SSC, 1.0% blocking reagent, 0.1% N-lauroylsarcosine, 0.02% SDS, and with 10 ng/ml of probe, overnight at 65° C. Membranes were washed in 3 × SSC and 0.1% SDS twice at room temperature, 5 min per wash, and an additional two times at 65° C, 15 min per wash. Membranes were treated with a blocking reagent and then with anti-digoxigenin Fab fragments conjugated with alkaline phosphatase. The detection was done using Lumi-Phos 530 and exposing the X-ray film for 15 min at room temperature. Positive plaques were picked, titered, and screened for four rounds. When phage stocks were homogeneous, liquid lysates were prepared from each of the recombinant clones and stored at 4° C.

Restriction enzyme mapping of a recombinant clone

DNA purified from homogeneous phage stock was digested with one or more enzymes, and fragments were separated by electrophoresis on an 0.8% agarose gel. The gel was stained with ethidium bromide and photographed. Based on the size and number of fragments in different digestions, a restriction enzyme map was constructed. To determine the DNA fragment containing sequences similar to β-1,3-glucanase gene and localize the gene in the insert, DNA fragments were blotted onto a nylon membrane and hybridized to the maize β-1,3-glucanase cDNA probe. Labeling and detection were conducted as mentioned in the above section.

Subcloning of the DNA fragment containing putative apple β-1,3-glucanase gene(s) into a plasmid vector

A 2.1 Kb EcoRI/HindIII fragment from a Lambda clone was subcloned into the plasmid vector pMOB (Strathmann et al., 1991). The DNA from the vector and the selected recombinant clone was digested with EcoRI and HindIII, and electrophoresed on a low-melting point agarose gel. A 2.1 Kb fragment from the Lambda clone was subcloned into a 1.8 Kb plasmid DNA. Both fragments were excised from the gel and placed in the same microfuge tube. The agarose was dissolved and DNA was purified using a GENECLEAN kit (BIO 101, Inc., LaJolla, California) following manufacturer's directions. A ligation buffer (consisting of 250 mM Tri/HCl (pH 7.6), 50 mM $MgCl_2$, 5 mM ATP, 5 mM DTT, and 25% (w/v) PEG-8000) and ligase were added to the eluted DNA mixture, and ligation was conducted for 3 hr at 15° C. Frozen competent *E. coli* strain DH5α cells were transformed using the complete ligation mix (15 μl) and plated onto LB plates containing ampicillin. Transformants were picked and grown in overnight cultures. The DNA from these cultures was extracted and digested with EcoRI and HindIII and analyzed on an agarose gel along with the original plasmid pMOB DNA which was digested with EcoRI and HindIII.

Results and discussion

The titer of the apple genomic DNA library at the time of screening was to be about 3.4 × 10^3 pfu/μl. Initial screening of approximately 150,000 pfu resulted in a very high number of positives. Only 40 plaques were

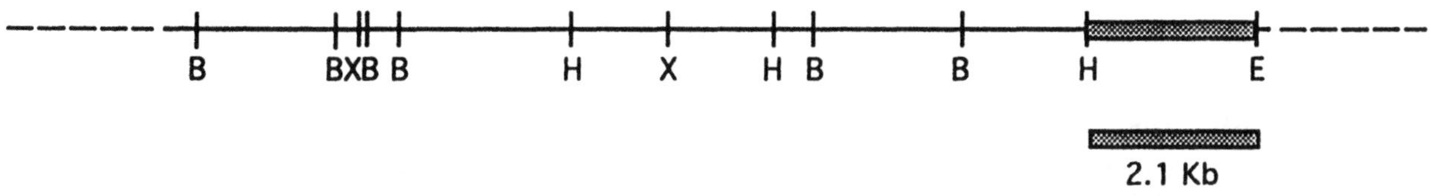

Fig. 1. Restriction enzyme map of an apple genomic clone, λ-AG7142. The stippled box corresponds to the β-1,3-glucanase gene identified by hydridization using a maize β-1,3-glucanase cDNA probe. Restriction sites: B, BamHI; E, EcoRI; H, HindIII; S, SalI; Ss, SstI; X, XhoI. Dotted lines correspond to right and left arms of the pMOB plasmid vector.

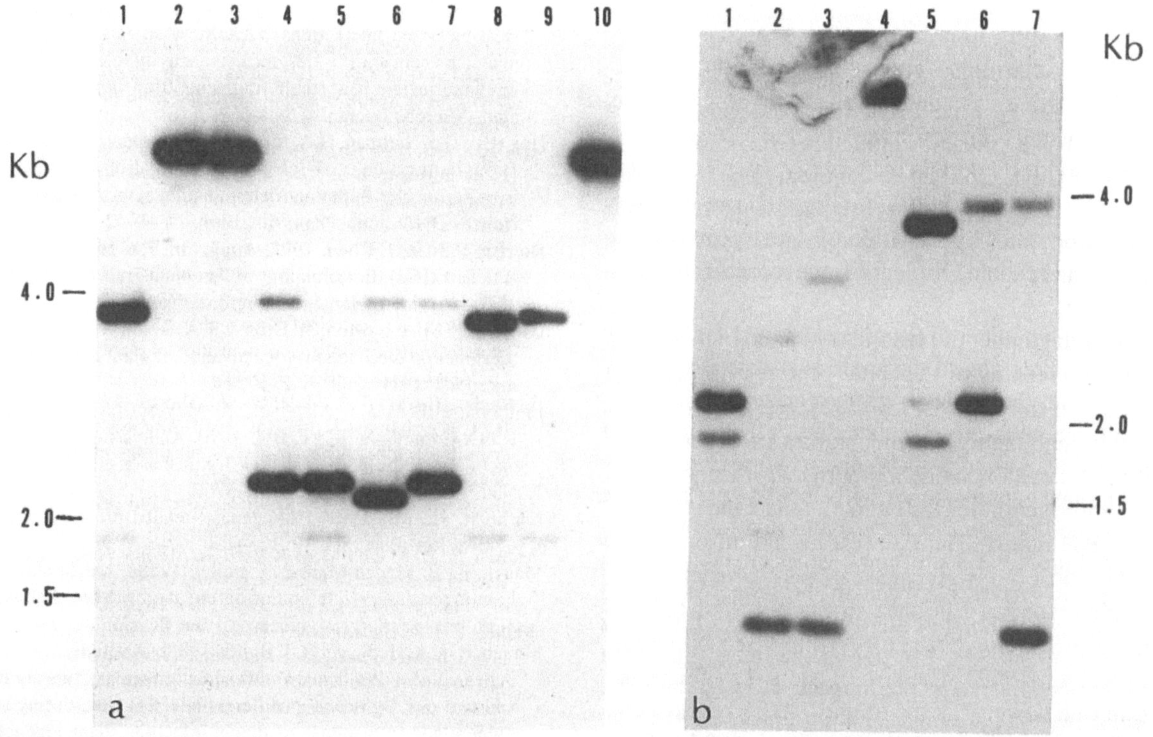

Fig. 2. Southern blot analysis of an apple genomic clone, λ-AG7142. The DNA from this clone was digested with different restriction enzymes. (a) Lanes 1, 2, 3, 4, 5, 6, 7, 8, 9, and 10 correspond to BamHI, EcoRI; SalI; HindIIII; HindIII & BamHI; HindII & EcoRI; HindIII & SalI; BamHI & EcoRI; BamHI & SalI; and EcoRI & SalI, respectively. (b) Lanes 1, 2, 3, 4, 5, 6, and 7 correspond to EcoRI, BamHI & HindIII; EcoRI, BamHI & SstI; EcoRI, HindIII & SstI; Eco RI & XhoI; EcoRI, BamHI & XhoI; EcoRI, HindIII & XhoI; and EcoRI, SstI & XhoI, respectively.

picked after the first round of screening, and some of these were used in subsequent screenings. After four rounds of screening, 15 clones were slotblotted, and hybridized to the β-1,3-glucanase cDNA probe from maize. Eight clones showed clear hybridization after 15 min of exposure. DNA from four of these clones was digested with BamHI, EcoRI, and SalI, electrophoresed, blotted, and hybridized with the β-1,3-glucanase cDNA probe. The autoradiograph showed that the hybridizing fragment in one clone was different from those fragments found in the other three clones. This indicates that there are at least two different clones containing sequences similar to the β-1,3-glucanase gene.

One of the recombinant clones (λAG7142) was screened further for yet another round and then characterized by restriction enzyme mapping (Fig. 1). The approximate size of the insert in the lambda clone was estimated equal to 13.5 Kb. Mapping was done using BamHI, EcoRI, HindIII, SalI, SstI, and XhoI enzymes either alone or in various combinations. There were no SalI restriction sites within the insert. There were seven SstI sites and hence these could not be mapped accurately. BamHI, EcoRI, HindIII and XhoI restriction sites are clearly designated on the map (Fig. 1).

A 2.1 Kb EcoRI/HindIII fragment strongly hybridized to the β-glucanase probe used (Fig. 2). This fragment was subcloned into EcoRI/HindIII sites in the polylinker region of the plasmid pMOB. Transformation was conducted using *E. coli* strain DH5α competent cells. The DNA from putative colonies was analyzed by restriction digestion. Out of nine colonies tested, three were identified as the desired recombinant pMOB plasmids. The cloned plasmid will be used to generate nested deletions; the gene will be sequenced using transposon site-specific primers using the TN1000 kit (Gold BioTechnology, Inc., St. Louis, Missouri). The non-radioactive digoxigenin labeling of the probe and hybridization system proved to be sensitive and reliable for identifying recombinant fragments.

The high number of positives obtained in this study following screening of the apple genomic library may indicate that the apple β-1,3-glucanase gene belongs to a multigene family. Further studies involving other positive clones along with Southern hybridization analysis of apple genomic DNA will elucidate the structure of this gene family. The size (exons and introns combined) of most β-1,3-glucanase genes characterized in other plant species has been in the range of 659 bp in tobacco (Linthorst et al., 1990) to 2385 bp in pea (Chang et al., 1993). The size of the β-1,3-glucanase gene in rice however is the longest, 4151 bp, and this is attributed to its unusually long intron of 3145 bp (Simmons et al., 1992). Therefore, we assume that the 2.1 Kb EcoRI/HindIII fragment characterized in this study may contain the complete, or most of the, apple β-1,3-glucanase gene.

Acknowledgements

This work was support in part by a grant received from Gerber Products Company and the Michigan Apple Research Committee. The authors wish to acknowledge Drs. Alan Kriz and Shenchuan Wu for providing us with the maize cDNA probe used in this study.

References

Chang, M-M, D.E. Culley & L.A. Hadwiger, 1993. Nucleotide sequence of a pea (*Pisum sativum* L.) β-1,3-glucanase gene. Plant Physiol. 101: 1121–1122.

Chen, H., S.S. Korban & D.E. Buetow, 1990. Nucleotide sequence of the gene encoding the light harvesting chlorophyll a/b polypeptide of photosystem I from apple. Nucl. Acids Res. 18: 679.

Collinge, D.B. & A.J. Slusarenko, 1987. Plant gene expression in response to pathogens. Plant Mol. Biol. 9: 389–410.

Edington, B.V., C.J. Lamb & R.A. Dixon, 1991. cDNA cloning and characterization of a putative β-1,3-β-D-glucanase transcript induced by fungal elicitor in bean cell suspension cultures. Plant Mol. Biol. 16: 81–94.

Ham, K-S, S. Kauffmann, P. Albersheim & A.G. Darvill, 1991. Host-pathogen interactions XXXIX. A soybean pathogenesis-related protein with β-1,3-glucanase activity releases phytoalexin elicitor-active heat stable fragments from fungal walls. Mol. Plant-Microbe Interac. 4: 545–552.

Hoj, P.B., D.J. Hartman, N.A. Morrice, D.N.P. Doan & G.B. Fincher, 1989. Purification of β-1,3-glucan endohydrolase isoenzyme II from germinated barley and determination of its primary structure from a cDNA clone. Plant Mol. Biol. 13: 31–42.

Korban, S.S. & H. Chen, 1992. Apple. In: F.A. Hammerschlag & R.E. Litz (Eds) Biotechnology of Perennial Fruit Crops, pp. 203–227. CAB International, Wallingford, Oxon, U.K.

Lamb, C.J., M.A. Lawton, M. Dron & R.A. Dixon, 1989. Signals and transduction mechanisms for activation of plant defenses against microbial attack. Cell 56: 215–224.

Linthorst, H.J.M., L.S. Melchers, A. Mayers, J.S.C. van Roekel, B.J.C. Cornelissen & J.F. Bol, 1990. Analysis of gene families encoding acidic and basic β-1,3-glucanases of tobacco. Proc. Natl. Acad. Sci. USA 87: 8756–8760.

Linthorst, H.J.M., 1991. Pathogenesis-related proteins of plants. Critical Rev. Plant Sci. 10: 123–150.

Mauch, F., B. Mauch-Mani & T. Boller, 1988a. Antifungal hydrolases in pea tissue. II. Inhibition of fungal growth by combinations of chitinase and β-1,3-glucanase. Plant Physiol. 88: 936–942.

Mauch, F., L.A. Hadwiger & T. Boller, 1988b. Antifungal hydrolases in pea tissue. I. Purification and characterization of two chitinases and two β-1,3-glucanases differentially regulated during development and in response to fungal infection. Plant Physiol. 87: 325–333.

Mauch, F. & L.A. Staehelin, 1989. Functional implications of the subcellular localization of ethylene-induced chitinase and β-1,3-glucanase in bean leaves. Plant Cell 1: 447–457.

Simmons, C.A., J.C. Litts, N. Huang & R.L. Rodriguez, 1992. Structure of a rice β-glucanase gene regulated by ethylene, cytokinin, wounding, salicylic acid and fungal elicitors. Plant Mol. Biol. 18: 33–45.

Strathmann, M., B.A. Hamilton, C.A. Mayeda, M.I. Simon, E.M. Meyerowitz & M.J. Palazzole, 1991. Transposon-facilitated DNA sequencing. Proc. Natl. Acad. Sci. USA 88: 1247–1250.

Xu, P., J. Wang & G.B. Fincher, 1992. Evolution and differential expression of the (1,3)-β-glucan endohydrolase-encoding gene family in barley, *Hordeum vulgare*. Gene 120: 157–165.

H. Schmidt and M. Kellerhals (Eds.), Progress in Temperate Fruit Breeding, 343–347.

Agrobacterium-mediated transformation of the apple cultivar Granny Smith

Adelina Trifonova, Desislava Savova & Kristina Ivanova
Institute of Genetic Engineering, 2232 Kostinbrod, Bulgaria

Key words: Apple, genetic transformation, *ipt* gene, *malus*, regeneration, rooting-anticytokinin

An efficient adventitious shoot regeneration system from leaf segments of the apple cultivar Granny Smith was developed. Regenerants in sufficient frequency were obtained under the optimal conditions in presence of 3 mg/l BA, 2 mg/l 2iP and 0,2 mg/l NAA. Putative transgenic plants were regenerated from leaf segments that were co-cultivated with disarmed C58 *Agrobacterium* strain containing either of the following binary plasmids: *pGV2449* or *pGV2492*. The chimeric marker gene for neomycin phosphotransferase II (*nptII*) and *ipt* genes – encoding for isopentenyl transferase, the first enzyme in cytokinin biosynthetic pathway were integrated in both plasmid derivatives. Seven putative transgenic plants were obtained on the selective medium containing 50 μg/ml kanamycin after transformation with *pGV 2449*. The expression and integration of *nptII* marker gene was detected in leaves of the plants. Rooting of the propagated plants was only achieved in presence of anticytokinin substance – 4-substituted-triazolo [4,5,d] pyrimidine and 0.5 mg/l IBA.

Introduction

A basic prerequisite for the application of gene transfer techniques in apple improvement is the ability to regenerate plants from cultured tissues. Several reports have illustrated that the regeneration of apple plants from in vitro derived leaves has been successfully used in genetic transformation experiments (Dandekar et al. 1993). Phytohormones are involved in all developmental processes during the plant life cycle. From a breeding point of view the plant hormone level manipulation could influence such characteristics as tree growth habit, branching and apical dominance. Hormone production – encoding genes have been isolated from *Agrobacterium* strains. One of them, the *ipt* gene codes for an isopentenyl transferase that is involved in cytokinin biosynthetic pathway. Transgenic plants expressing the *ipt* gene demonstrate reduced apical dominance and failure to root (Beinsberger et al. 1992).

The present study deals with the regeneration and transformation of the apple cultivar Granny Smith by *Agrobacterium tumefaciens* strains carrying plasmids harbouring the *ipt* cytokinin gene and *nptII* marker gene.

Materials and methods

Well-developed young leaves of *in vitro* propagated plants from apple cultivar Granny Smith were used in regeneration and transformation experiments.

Three nutrition media based on salt-vitamins mixture MS, with 3% sucrose solidified with 0.8% agar, pH 5.4, supplemented with growth regulators in following combinations: MS1: 3 mg/l 6-benzylaminopurine (BA) and 0.2 mg/l α-naphtalene acetic acid (NAA); MS2: 3 mg/l BA, 2 mg/l 6-(τ,τ-dimethylallylamino) Purine (2iP) and 0.2 mg/l NAA; MS3: 3 mg/l BA, 2 mg/l Kinetin Riboside and 0.2 mg/l NAA were tested. The explants were cultured at 23 ° C for 30 days in dark.

The *Agrobacterium tumefaciens* constructs used to perform leaf section transformation were: *pGV2449* plasmid, containing the *ipt* gene isolated from *pTiB653*, and *pGV2492* plasmid, containing the *ipt* gene isolated from *pTIC58*. The T-DNA region of both vector includes the chimeric marker gene coding for neomycin phosphotransferase II (*nptII* gene). Plasmids were mobilized in *A. tumefaciens* strain C58C1 in the Ti plasmid vector *pGV2260*.

Table 1. Effect of the nutrition medium on the regeneration of leaf segment explants from apple cultivar Granny Smith.

Media	MS1	MS2	MS3
regenerated explants (%)	35 ± 0.98 a	69 ± 1.36 b	43.3±0.73 c
mean no. shoots per explant	3.1 ± 1.5 a	5.1 ± 0.9 b	3.6 ± 1.2 a

One way ANOVA. Detailed comparisons of mean were made with a Student-Newman-Keuls criteria ($P < 0.05$).

Leaf segments were co-cultivated in 10 ml MS medium with 3 mg/l BA in the presence of 0.5 ml overnight bacterial suspension (OD_{560} – 0.8) for 24 h in dark on a rotary shaker. The explants were washed and placed on selective media upside down. Putative transgenic shoots were selected on 50 mg/l kanamycin (Km) and 200 mg/l cefotaxime containing medium. The plants which passed first selection were excised from explant tissue and callus and micropropagated following standard procedure.

The putative transformants obtained after second selection were tested for *nptII* gene expression according to McDonnell et al. (1987). The integration of *nptII* gene was confirmed by PCR assay (Saiki 1990).

Root formation was induced on medium containing 10^{-6} M anticytokinin substance (4-substituted-triazolo [4,5,d] pyrimidine), (Karanov et al. 1993) and $2.5 \cdot 10^{-6}$ M Indole-3-Butyric Acid (IBA) in presence of 50 μg/ml Km.

Results and discussion

Adventitious primordia formation was observed after 20 days of culturing on the regeneration medium. Adventitious buds occur more frequently near the main vein on the cut site of the basal segments (Fig. 1). Three different regeneration media were compared in order to optimize the conditions for adventitious bud induction. Data presented in Table 1 shows significant differences between the percentage of explants regenerating. In five experiments regenerants in sufficient frequency were obtained under the optimal conditions in presence of 3mg/l BA, 2 mg/l 2iP and 0.2 mg/l NAA, where regeneration averaged 69% ± 1.36 with 5.1 ± 0.9 shoots per regenerating explant.

Green plants were regenerated with different frequency after co-cultivation with an *A. tumefaciens* strain containing either of the plasmids *pGV 2449* and *pGV 2492*. Thirty plants were obtained after first selection when the *pGV 2449* plasmid was used. Lower number of green regenerants (19) was observed when the transfection was carried out by the *pGV2492*.

The second selection test showed that seven clones out of 30 putative transformants exhibit resistance to 50 μg/ml kanamycin. The plants passed through selection tests were phenotypically normal. The expression of *nptII* marker gene was detected in leaves of the obtained plants (Fig. 2), as a positive control transgenic tobacco plants were used.

Rooting of the shoots obtained was not successful by using 1/2 MS medium supplemented with IBA in several concentrations, following standard procedure (Ivanova 1991). The application of anticytokinins recovered the rooting ability of the shoots (Fig. 3). Those plants were successfully rooted on selection medium containing 50 μg/ml Km in the presence of anticytokinin substance 10^{-6} M and $2.5 \cdot 10^{-6}$ M IBA.

The polymerase chain reaction (PCR) was used to amplify a 553 bp fragment of *nptII* coding sequence in the transgenic plants. The results are shown in Fig. 4.

Plant regeneration and transformation of only some apple cultivars have been described (James et al. 1989, 1991, Maheswaran et al. 1992). Our work has shown that apple cultivar Granny Smith has the potential to form adventitious buds with sufficient frequency after genetic transformation experiments. Combination of BA, 2iP and NAA proved to be optimal for shoot regeneration compared to the media with BA only and with BA + kinetin.

The *Agrobacterium* construct *pGV2492* has been used to obtain tobacco transgenic plants expressing the *ipt* gene that show different phenotypic alterations (Beinsberger et al. 1992, Schmulling et al. 1992). Our study suggests that *pGV 2449* gives positive results in apple transformation. Almost all of the apple shoots obtained were morphologically normal except two which show 'bushy' phenotype.

The transformation and selection procedure we have used has resulted in a low transformation frequency (0.07).The fact that after the second selection test fewer plants survived may indicate that mixed population of cells take part in regeneration, and transformed cells deplete the level of selective agent in the surrounding medium.

The failure of putative apple transformants to root has also been observed by James et al. (1989). It

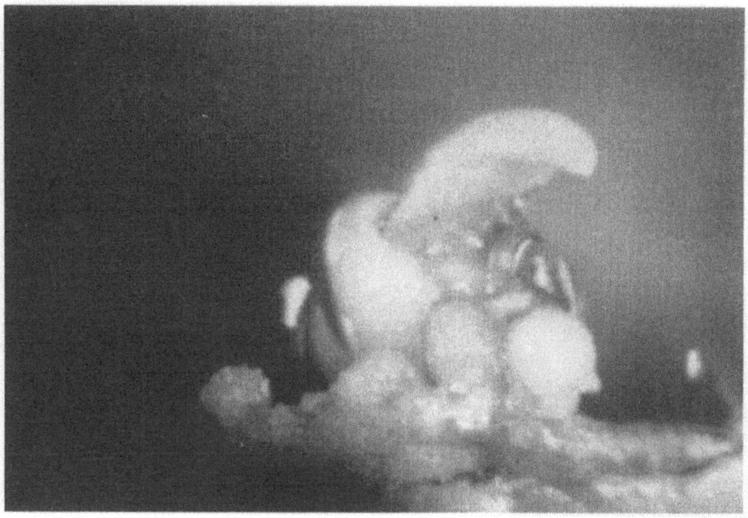

Fig. 1. Adventitious bud primordia formation from leaf explant.

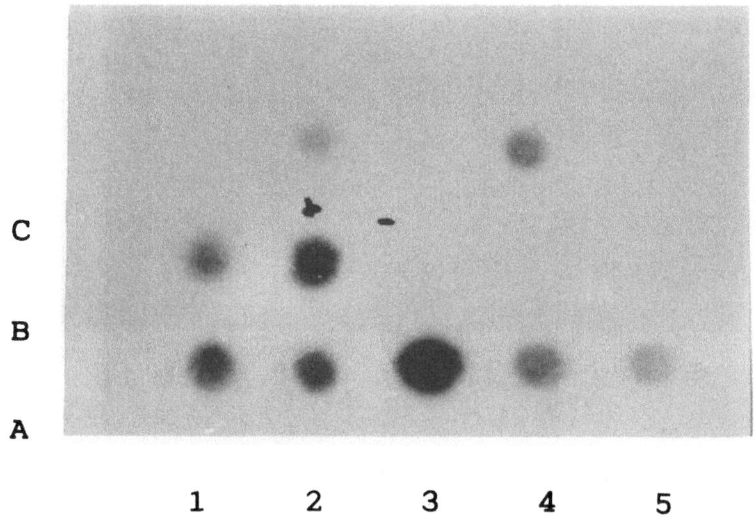

Fig. 2. Detection of *nptII* activity in apple transgenic plants. 2B – positive transgenic tobacco control, 3B – negative apple control plant.

could be due to the changes in endogenously produced cytokinins (Medford et al. 1989). The application of anticytokinin in root induction medium in combination with auxin is a new approach. It might be that the anticytokinins suppress autotrophic cytokinin synthesis and the presence of the auxin stimulates the root formation.

Fig. 3. Rooted transformed Granny Smith apple plants on a medium containing 10^{-6} M anticytokinin substance, $2.5 \cdot 10^{-6}$ M IBA and 50 mg/l kanamycin.

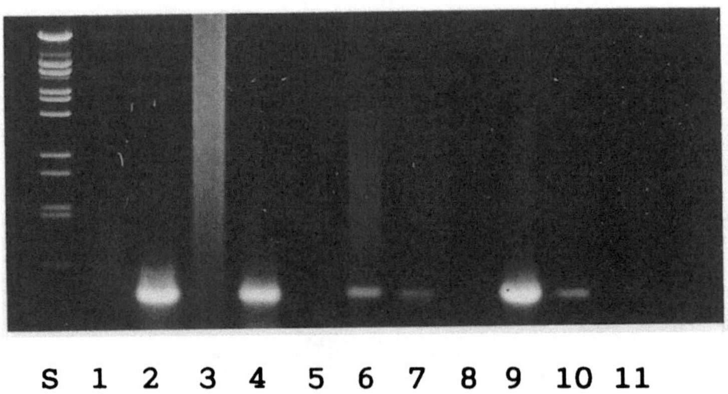

Fig. 4. Results of PCR using primers for *nptII* coding sequence. Line S-1kb DNA standard; line 1 – negative control, apple plant; lines 2, 3, 4, 6, 7, 9 – genomic DNA from leaves of transgenic apple plants; line 5 – negative control, tobacco plant; line 8 – negative control apple plant; line 10 – positive control, tobacco plant; line 11 – all buffers.

Acknowledgements

This work was supported by National Foundation of Agricultural Science, Bulgaria.

We are grateful to Dr Dirk Inze, Ghent University, Belgium, for kindly providing the *Agrobacterium* constructs.

We are also grateful to Prof E. Karanov, Institute of Plant Physiology, Bulgaria, for kindly providing the anticytokinin substance.

References

Beinsberger, S.E., R.L.M. Valcke, H.M.M. Clijsters, J.A. De Greef & H.A. Van Onckelen, 1992. Effects of enhanced cytokinin levels in ipt transgenic tobacco. In: M. Kaminik, D.W.S. Mok & E. Zazimalova (Eds) Physiology and biochemistry of cytokinins in plants, pp. 77–82. Proceeding of the Int. Symp. on Physiol. and Bioch. of Cytokinins in Plants, Liblice, Czechoslovakia, September 10–14, 1990.

Dandekar, A.M., G.H. McGranahan & D.J. James, 1993. Transgenic Woody Plants. In: Shain-dow King & Ray Wu (Eds) Transgenic plants, Vol. 1, pp. 129–151. Academic Press, Inc, CA San Diego.

Ivanova, K., 1991. Clonal microprogation of apple cultivars. Rast. nauki. 3(6): 61–65.

James D.J., A.J. Passey, D.J. Barbara & M. Bevan, 1989. Genetic transformation of apple (*Malus pumila* Mill.) using a disarmed Ti-bynary vector. Plant Cell. Rep. 7: 658–661.

James, D.J., A.M. Dandekar, 1991. Regeneration and transformation of apple (*Malus pumila* Mill). In: K. Lindsey (Ed.) Plant Tissue Culture Manual: Fundamentals and Applications, pp. B8/1–B8/18. Kluwer Academic Publishers, Dordrecht.

Karanov, E., V. Alexieva, E. Golovinski & M. Haimova, 1993. Cytokinin and anticytokinin activity of some 4-substituted 1H-pyrazoles and 8-aza analogues of adenine. Plant Growth Regul. 13: 7–11.

Maheswaran, G., M. Welander, J. Hutchinson, M. Graham, D. Richards. 1992. Transformation of apple rootstock M26 with Agrobacterium tumefaciens. J. Plant Physiol. vol. 139. p. 560–568.

McDonnell R., R. Clark, W.A. Smith & M. Hinchee, 1987. A simplified method for the detection of neomycin phosphotransferaseII activity in transformed plant tissue. Plant Mol. Biol. Rep. 5: 380–386.

Medford, J.I., R. Horgan, Z. El-Sawi & H.J. Klee, 1989. Alterations of endogenous cytokinins in transgenic plants using a chimeric isopentenyl transferase gene. Plant Tissue Cell Cult. 1: 403–413.

Schmulling T., J. Schell & A. Spena, 1992. Construction of cytokinin overproducing mutants in higher plants. In: M. Kaminik, D.W.S. Mok & E. Zazimalova (Eds) Physiology and biochemistry of cytokinins in plants, pp. 77–82. Proceeding of the Int. Symp. on Physiol. and Bioch. of Cytokinins in Plants, Liblice, Czechoslovakia, September 10–14, 1990: 87–94.

Saiki, R., 1990. Amplification of genome DNA. In PCR protocols: A guide to Methods and Applications, M.A. Innia, D.H. Gelfand, J.J. Sninsky and T.J. White eds.; San Diego CA, Academic Press, Inc. p. 13–20.

H. Schmidt and M. Kellerhals (Eds.), Progress in Temperate Fruit Breeding, 349–354.

Coat protein-mediated protection against plum pox virus in herbaceous model plants and transformation of apricot and plum

Artur da Câmara Machado, Hermann Katinger & Margit Laimer da Câmara Machado
Institute of Applied Microbiology, University of Agriculture and Forestry, Nußdorfer Lände 11, 1190 Vienna, Austria

Key words: Coat protein-mediated protection, Plum Pox Virus, transformation, *Prunus armeniaca*, *Prunus domestica*, apricot, plum

Summary

The expression of the viral coat protein gene in transgenic plants has been shown to induce tolerance against virus infection (Beachy et al., 1990). Transgenic plants of *Nicotiana clevelandii* and *Nicotiana benthamiana* – herbaceous host plants for PPV – transformed with *Agrobacterium* strain LBA 4404 containing the plasmid pBinPPVm, regenerated on selection media containing kanamycin were tested for the expression of the PPV-coat protein gene by ELISA and immuno western blot. After rooting and acclimatisation plants were tested for the protection against PPV. Following the inoculation plants were investigated for symptom development and virus accumulation. Different lines were identified, according to the different reaction to the mechanical inoculation, ranging from a complete absence to a strong reduction of symptoms. There have not been many reports on transformation of trees in general, and in fruit trees particularly. It is obvious that the major obstacle is the regeneration of transformed plantlets. Attempts to improve crop plants by genetic engineering techniques will always depend very strongly on the availability of reliable protocols for transformation, selection and regeneration (Laimer et al., 1989, 1990). Different systems involving juvenile and adult plant material have been developed allowing the transfer of foreign genes into apricot and plum cultivars. We report the transformation and regeneration of *Prunus armeniaca* and *Prunus domestica* plants with *Agrobacterium tumefaciens* strain LBA 4404 containing various binary plasmids, pBinGUSint, carrying the marker gene β-glucuronidase (GUS) and pBinPPVm, carrying the coat protein gene of Plum Pox Virus (PPV), the causal agent of Sharka disease. The marker gene *GUS* was used for the optical evaluation of the efficiency of different transformation systems involving cotyledons of immature embryos as well as somatic embryos and leaf discs. The coat protein gene of PPV was used to introduce the coat protein mediated resistance against one of the most important pathogens of stone fruit trees in Europe and the whole Mediterranean area.

Introduction

In recent years a viral pathogen became the major threat to stone fruit cultivations in large areas of Central and Southern Europe and other mediterranean countries, being responsible for considerable economic losses and reduction of production areas. Plum Pox Virus (PPV), the causal agent of Sharka disease and member of the potyvirus family, was classified by US and EC plant quarantine agencies as the most important pathogen in apricots, plums and peaches (the only plant pathogen for which an APHIS plan exists, Scorza, 1991). Sharka infection data from Spain, Greece, France and Italy clearly demonstrate the economic consequences of this threat (COST 88), as apricot orchards are increasingly being replaced. Apricot appears to be the most sensitive stone fruit towards infection with PPV. Apricot production was practically erased in the late 70s in some valleys of Northern Italy, e.g. Vintschgau, and is seriously threatened in some Austrian valleys, e.g. Wachau (Pieber, personal communication).

Considering the severity of the disease, the difficulty to control its spread, and the lack of resistant cultivars, the necessity of resistant cultivars is evident and a strait-forward strategy is required. Biotechnology offers completely new approaches through the isolation and transfer of "resistance" genes.

"Cross protection" was originally described as the phenomenon of protection of a plant against the invasion of a severe disease-causing virus due to prior inoculation of the plant with an attenuated virus strain (McKinney, 1929). Hamilton postulated in 1980 that the expression of sequences from the viral genome, if expressed in transgenic plants, could possibly cause a protection against viruses. In fact by the expression of the viral coat protein gene in transgenic plants similar effects could be obtained, and it was therefore distinguished as coat protein mediated protection (Beachy et al., 1990).

Due to their susceptibility and high virus accumulation *Nicotiana benthamiana* and *Nicotiana clevelandii* are the ideal model plants to prove the CP mediated protection. We transformed leaf discs and stem microcuttings of *Nicotiana benthamiana* and *Nicotiana clevelandii* with *Agrobacterium tumefaciens* carrying pBinPPVm and identified the coat protein expressing plants by immuno western blot (Laimer et al., 1990). These plants were used in infection studies to gain information about virus accumulation, suppression or inhibition. Finally, the results about PPV coat protein mediated protection obtained in these experiments with herbaceous plants were inserted in a major programme with the final goal to achieve stone fruit trees resistant to PPV.

Regeneration of plants from single cells and complex explants, either by organogenesis or by somatic embryogenesis is to be considered a prerequisite for any programme aimed at improving fruit tree quality by transformation with Agrobacterium (James, 1987; Laimer et al., 1989). Attempts to improve fruit trees by genetic engineering will always depend very strongly on the availability of reliable protocols for transformation, selection and regeneration (McGranahan et al., 1988; Laimer et al., 1992). Obviously the main obstacle for transformation of fruit tree species is the regeneration of transformed plantlets from the desired cultivars. Due to their higher degree of differentiation tissues of perennial species are much more difficult to induce to undergo a process of dedifferentiation and to enter a new differentiation step. The regeneration of shoots from a number of explants which we considered useful for transformation with *Agrobacterium* (Laimer et al., 1989) could be induced, but due to the low number and the high sensitivity of the regenerants to kanamycin as selecting agent it was not possible to obtain any transformant. The second problem appearing was that regeneration took place from more than one cell, thus giving rise to transgenic regenerants, which proved to be chimera in histological analyses (unpublished data).

In the case of apricot the difficulties existed already at the tissue culture level. It was necessary to optimize, based on the few works existing at that time (Snir, 1984) a micropropagation protocol and a system for adventitious regeneration that would allow successful transformation (Weiss et al., in preparation). This still continues to be a difficult task among fruit trees, as can easily be seen from the limited number of successful reports (James, 1987; McGranahan et al., 1988; Laimer et al., 1992; Oliveira & Pais, 1992; Mante et al., 1991).

The transformation of apricots and plums with the PPV coat protein gene will bring new perspectives to the control of Sharka disease.

Materials and methods

For construction of the plant expression vector pBinPPVm the coat protein gene of PPV was mutagenized to introduce a start codon and connected to the 35S promoter and terminator from CaMV. Finally the entire recombinant gene was cloned into Bin19 (Regner et al., 1992).

Leaf discs and stem cuttings of 100 μm (Laimer et al., 1990) of *Nicotiana clevelandii* and *Nicotiana benthamiana* were cocultivated with *Agrobacterium*-suspension ($OD_{600} = 0.6$) diluted 1 : 50 in MS culture medium for 48 hours and plantlets were regenerated under culture conditions as described (Laimer et al., 1990). After one month green shoots were screened for transformation by NPT II-assay and immuno western blot (da Câmara Machado et al., 1991).

Challenge infection on acclimatized *Nicotiana clevelandii* plants and *Nicotiana benthamiana* CP+ cuttings was carried out with PPV NAT purified from *Nicotiana clevelandii* and plants were scored daily for the appearance of symptoms after inoculation (da Câmara Machado et al., 1991). Levels of transgenic coat protein were determined by ELISA at day zero and the virus accumulation at day 7, day 20 and day 40 after challenge infection (da Câmara Machado et al. 1991).

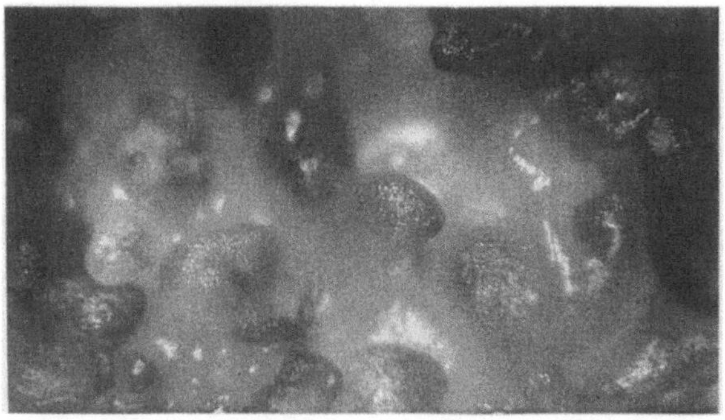

Fig. 1. Regeneration of shoots from cotyledons during the optimal stage of development (days 68 to 89).

To establish the appropriate time frame where the physiological stage of the explants would allow for highest regeneration capacity, open pollinated immature embryos of *Prunus armeniaca* cv. Kecskemeter were collected from day 49 till day 111 after full bloom at weekly intervals and taken into culture on modified culture media (Laimer da Câmara Machado et al., 1992).

To obtain transformed plantlets from plant material of adult origin, an embryogenic line (da Câmara Machado et al., 1993, in preparation) and leaf discs of *Prunus* (Cossio & Bassi, 1991) have been used to evaluate their applicability for transformation. It appears clearly that a good transformation system for somatic embryos could finally yield the high number of transformants needed to thoroughly study the phenomenon of cross protection in respect to questions like pattern and level of expression, importance of site of insertion, etc.

In order to avoid background problems with GUS-activity caused by agrobacteria persisting in the tissues due to incomplete elimination by the antibiotic treatment, we decided to use a chimeric gene construct which automatically would eliminate appearance of false positives. Therefore we worked with pBin-GUSint, which contains a plant intron, thus restricting the expression of the gene to plant cells (Vancanneyt et al., 1990).

Results

We obtained 14 independently regenerated *Nicotiana benthamiana* plants expressing in a correct manner the CPgene. After challenge infection none of the plants showed disease symptoms, even if virus titer increased (da Câmara Machado et al., 1991).

We could select 17 CP+ *Nicotiana clevelandii* plants. Plants of the Rl lines 55 (22 plants: A-V) and 58 (20 plants: 1-20) were chosen for the detailed presentation of the data. Plants of line 55 were characterized by a non systemic spread of disease symptoms. Six plants showed a similar development of symptoms as the control, most probably due to the Mendelian segregation of the trait. In the other plants, even if we could observe a different accumulation of virus, virus titer decreased at day 20 post infection, with none of the new leaves showing symptoms, and old lesions partly disappeared (da Câmara Machado et al., 1991).

Line 58 differs in a clear manner from line 55 as virus challenge caused neither disease symptoms nor virus accumulation (da Câmara Machado et al., 1991).

The regeneration rate from cotyledons of immature embryos was highest between day 68 and 89 after full bloom, not only concerning the number of cotyledons forming shoots, but also the number of shoots which could be regenerated per cotyledon. Cotyledons taken into culture during the early stages of embryo development (day 49–61) produced shoots not only in reduced number, but also showing a slower development. From explants between day 68 and 89 on one cotyledon up to 21 shoot primordia have been indentified after 3 weeks in culture (Fig. 1). Herewith the conditions for a gene transfer were established (Laimer da Câmara Machado et al. 1992).

For the embryogenic line of *Prunus* a procedure was worked out allowing to take initially advantage of the repetitive embrogenesis until the point of transformation, but then interrupting the cycle for the regeneration of normal plantlets. Results obtained in the transformation experiment with GUS are shown in Fig. 2. The evenly stained plantlets confirm that the method involves single transformed cells from the embryo surface and does not yield chimeric plants (da Câmara Machado et al., in preparation).

For leaf disc explants of *Prunus* a series of media were compared for leaves of different ages, using different manipulations. Preliminary results indicate that a regeneration rate high enough for transformation studies can be obtained.

Due to the high regeneration rate obtained and due to careful selection procedures from cotyledons, after 7 months it was possible from our transformation experiments with pBinPPVm to isolate 41 candidates out of 716, which we expected to carry the PPV-coat protein gene. PCR was the first method we used to verify

Fig. 2. Transgenic plant expressing GUS regenerated through embryogenesis. →

Fig. 4. Transgenic plant of *Prunus armeniaca cv. Kecskemeter* expressing the coat protein gene of Plum Pox Virus. ↓

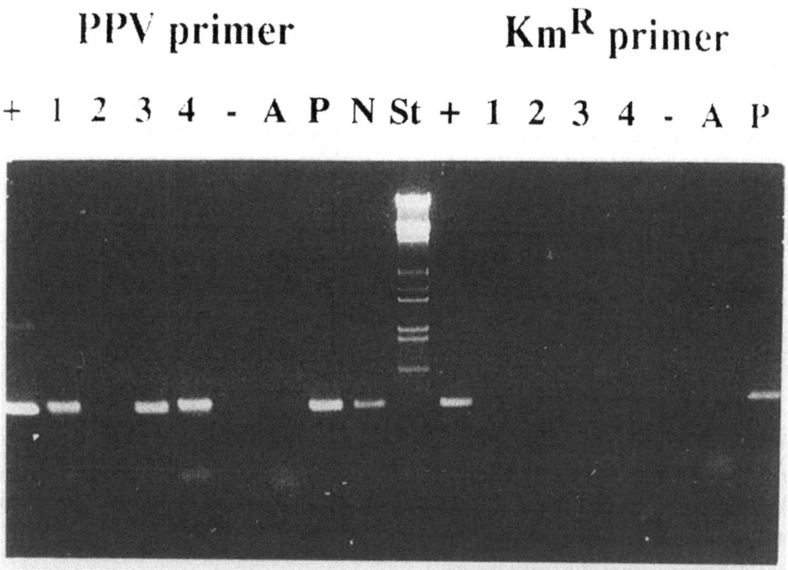

Fig. 3. Ethidium bromide stained electrophoresis of PCR-products to prove the integration of the Plum Pox Virus-coat protein gene into apricot plant genome. Primers for samples left of standard (ST) are within the PPV coat protein gene; primers for samples right of the ST are within the bacterial kanamycin resistance gene, which is located outside the border regions of the T-DNA and should help to detect contamination of agrobacteria. Besides +/- controls, a preparation of the plasmid pBinPPVm (P) was used as a further positive control. Transgenic *Nicotiana benthamiana* (N) carrying the PPV-coat protein gene was used as a positive control, to detect possible differences between woody and herbaceous tissues following this procedure. As further control tissue of apricot (A) sprayed with *Agrobacterium* LBA 4404 carrying pBinPPVm was included. Sample preparation was the same as for the transgenic plants. To make sure that any amplification of agrobacterial DNA is avoided.

the integration of the foreign DNA (Laimer da Câmara Machado et al., 1992).

Figure 3 shows the results of the PCR analyses of our regenerated apricot plants. Due to the power of the PCR technique it is important to include proper controls in plant transformation experiments, because a very low level of agrobacterial contamination persisting in cultures could give false positive results. Therefore we constructed primers for the amplification of the bacterial kanamycin resistance gene located outside the T-DNA. Any expression of this gene is an indication for the undesired presence of *Agrobacteria*.

Plants 1, 3 and 4 show a band corresponding to the relevant sequence within the coat protein gene. The same is evident for the positive control, the plasmid preparation (P) and the transgenic *Nicotiana benthamiana* (N). In plant 2 apparently no integration of the gene has occured.

Transgenic *Prunus armeniaca* plants were acclimatized to greenhouse conditions (Fig. 4) and subjected to challenge infection experiments by graft-transmission, aphid transmission and mechanical inoculation.

Discussion

In our experiments with *Nicotiana benthamiana* we could not correlate the level of transgenic coat protein with the degree of protection. These findings are in agreement with Stark & Beachy (1989) for other potyviruses.

With the use of cuttings (through vegetative propagation) we intended to preserve the originally regenerated genotype. This allows not only to test the degree of protection in the R0 generation, but also to try to understand the mechanism of cross protection without the influence of gene segregation. It is essential to gain information about the behaviour which we might expect from our transgenic *Prunus armeniaca* plants (Laimer et al., 1992), as in a fruit tree cultivar it is not possible to work with the following generations. In transgenic *Nicotiana clevelandii* plants we could observe considerable differences in reaction to virus challenge between the different lines. Data from line 55 support one of the possible explanations of the mechanism of protection, which is the inhibition of virus movement from infection sites to the upper leaves (Wiesnieski et al., 1990; Clark et al., 1990). In contrast, in the protected plants of line 58 no virus accumulation occured throughout the period of observation. Data of line 58 indicate that in an early event of virus infection the uncoating of the virus is blocked (Register & Beachy, 1988; Wu et al., 1990).

Conditions of gene transfer in fruit trees were elaborated, evaluating the efficiency of the method by GUS visualization. The presented results of transformation of *Prunus armeniaca* with the PPV coat protein gene, a major step towards coat protein-mediated protection, is a further contribution to resistance breeding in fruit trees.

Encouraged by results of coat protein-mediated protection obtained with herbaceous hosts of PPV (da Câmara Machado et al., 1991; Regner et al., 1991) we are now conducting similar challenge infection experiments to investigate resistance phenomena with transgenic *Prunus armeniaca* and *domestica* acclimatized to greenhouse conditions.

The use of adult material for transformation experiments – which would yield the great advantage of offsprings genetically identical to the mother plant and differing only in the newly introduced characteristics – was persued by experiments using leaf disc explants of *Prunus*, where only preliminary results could be obtained, and of an embryogenic line of *Prunus*, where evenly transformed plantlets confirm that the method involves single transformed cells from the embryo surface.

References

Beachy, R.N., L.S. Loesch-Fries & N.E. Tumer, 1990. Coat protein-mediated resistance against virus infection. Annual Rev. of Phytopath.: 451–474.

Clark, W.G., J.C. Register, A. Nejidat, D.A. Eichholtz, P.R. Sanders, R.T. Fraley & R.N. Beachy, 1990. Virology 179: 640–647.

Cossio, F. & G. Bassi, 1991. Ricerche sulla organogenesi di alcuni *Prunus*: Rigenerazione in vitro di germogli di susino (Prunus domestica L.) cultivar "Bluefre". Rivista di Frutticoltura 3: 77–81.

COST 88, 1993. Potyvirus Subgroup: Plum Pox Virus Workshop. Valencia, June 1993.

da Câmara Machado, A., F. Regner, H. Steinkellner, D. Mattanovich, V. Hanzer, H. Weiss, M. Laimer da Câmara Machado & H. Katinger, 1991. Coat protein-mediated protection against Plum Pox Virus. Acta Hort. 309: 203–210.

da Câmara Machado, A., M. Puschmann, H. Katinger & M. Laimer da Câmara Machado, 1993. Regeneration of transgenic plantlets of Prunus sp. via somatic embryogenesis. Vienna Cost 87 Meeting, May 1993.

da Câmara Machado, A., M. Puschmann, H. Katinger & M. Laimer da Câmara Machado. *Agrobacterium*-mediated transformation of somatic embryos of *Prunus subhirtella* and regeneration of transgenic plants, in preparation.

James, D.J., 1987. Cell and tissue culture technology for the genetic manipulation of temperate fruit trees. In: Biotechnology and genetic engineering reviews 5, pp. 33–79. Intercept Ltd, Dorset, U.K.

Laimer, M., A. da Câmara Machado, V. Hanzer, G. Himmler, D. Mattanovich & H.W.D. Katinger, 1989. Acta Hort. 235: 85–92.

Laimer, M., A. da Câmara Machado, D. Mattanovich, F. Regner, V. Hanzer, H. Steinkellner, B. Durniok, G. Himmler & H. Katinger, 1990. Acta Hort. 280: 577–580.

Laimer da Câmara Machado, M., A. da Câmara Machado, V. Hanzer, H. Weiss, F. Regner, H. Steinkellner, D. Mattanovich, R. Plail, E. Knapp, B. Kalthoff & H. Katinger, 1992. Regeneration of transgenic plants of *Prunus armeniaca* containing the coat protein gene of Plum Pox Virus. Plant Cell Rep. 11: 25–29.

Mante, S., P.H. Morgens, R. Scorza, J.M. Cordts & A.M. Callahan, 1991. *Agrobacterium*-mediated transformation of plum (*Prunus domestica*) hypocotyl slices and regeneration of transgenic plants. Bio/Techn. 9: 853–857.

McGranahan, G.H., C.A. Leslie, S.I. Uratsu, L.A. Martin & A.M. Dandekar, 1988. *Agrobacterium*-mediated transformation of walnut somatic embryos and regeneration of transgenic plants. Biotechnology 6:800–804.

McKinney, H.H., 1929. Mosaic disease in the Canary Islands, West Africa, and Gibraltar. Journal of Agricultural Research 39: 557–578.

Oliveira, M.M. & M.S. Pais, 1992. Somatic embryogenesis in leaves and leaf-derived protoplasts of *Actinidia deliciosa* var. *deliciosa* cv. Hayward. Plant Cell Rep. 11: 314–317.

Register, J. & R.N. Beachy, 1988. Resistance to TMV in transgenic plants results from interference with an early event in infection. Virology 166: 524–532.

Regner, F., A. da Câmara Machado, M. Laimer da Câmara Machado, H. Steinkellner, D. Mattanovich, V. Hanzer, H. Weiss & H. Katinger, 1992. Coat protein mediated resistance to Plum Pox Virus in *Nicotiana clevelandii* and *benthamiana*. Plant Cell Rep. 11: 30–33

Scorza, R., 1991. Gene transfer for the genetic improvement of perennial fruit and nut crops. HortScience 26: 1033–1035.

Snir, I., 1984. In vitro propagation of "Canino" apricot. HortScience 19: 229–230.

Stark, D.M. & R.N. Beachy, 1989. Protection against Potyvirus infection in transgenic plants: Evidence for a broad spectrum resistance. Bio/Technology 7: 1257–1262.

Vancanneyt, G., R. Schmidt, A. O'Connor Sanchez, L. Willmitzer & M. Rocha Sosa, 1990. Construction of an intron-containing marker gene. Mol. Gen. Genet. 220: 245–250

Weiss, H., A. da Câmara Machado, V. Hanzer, H. Katinger & M. Laimer da Câmara Machado. In vitro culture of different austrian selections of apricot (*Prunus armeniaca*), in preparation.

Wiesnieski, L., A.P. Powell, R. Nelson & R.N. Beachy, 1990. Local and systemic movement of tobacco mosaic virus in tobacco plants that express the TMV coat protein gene. Plant Cell 2:559 –567

Wu, X., R.N. Beachy, M. Wilson & J. Shaw, 1990. Inhibition of uncoating of tobacco mosaic virus particles in protoplasts from transgenic tobacco plants that express the viral coat protein gene. Virology 179: 893–895.

H. Schmidt and M. Kellerhals (Eds.), Progress in Temperate Fruit Breeding, 355–359.

Effects of explant size, pretreatment, and light intensity on shoot regeneration from *in vitro*-grown apple leaves

Richard E. Durham* & Schuyler S. Korban
*Department of Horticulture, University of Illinois, 310 Plant & Animal Biotechnology Laboratory, 1201 W. Gregory Dr., Urbana, IL 61801, USA; * Present address: Department of Plant and Soil Science, Texas Tech University, and Texas Agricultural Experiment Station, Texas A&M University, Box 42122, Lubbock, TX 79409, USA*

Key words: Malus × domestica, apple, organogenesis, adventitious shoots, tissue culture

Summary

The effects of explant size, pretreatment, and light intensity on shoot regeneration of apple (*Malus × domestica* Borkh.) were investigated using the rootstock genotype Malling 7a. Young leaves were harvested from *in vitro* proliferating shoot cultures, sliced into sections or minced into small pieces, dipped in either sterile distilled water or liquid regeneration medium, and placed onto a medium consisting of half-strength Murashige and Skoog salts, Staba vitamins, 20 μm thidiazuron, and 2.7 μM α-naphthaleneacetic acid. During the first four weeks of culture, explants were exposed to various light treatments: continuous darkness, two weeks of darkness followed by two weeks of low-light (15 μmol m^{-2} s^{-1}), continuous low-light, two weeks of low-light followed by two weeks of high-light (100 μmol m^{-2} s^{-1}), or continuous high-light. The regeneration frequency observed increased with decreasing light intensity and light duration. Mincing leaves and dipping explants in a liquid regeneration medium prior to culture also resulted in an increase in the number of shoots produced per leaf but did not significantly influence the regeneration frequency. The highest regeneration frequency (92%) was obtained when explants were sectioned, dipped in a liquid regeneration medium, and cultured in darkness for 4 weeks. However, the highest number of shoots (59.8 per leaf) was observed on minced explants cultured under the same conditions. A significantly higher regeneration frequency was obtained for five apple genotypes tested when leaves were minced rather than sectioned.

Introduction

Highly efficient and reliable shoot regeneration systems are necessary for exploitation of *in vitro* genetic manipulation techniques such as the production of somaclonal variants and/or transgenic plants. Shoot organogenesis from leaf tissue of apple was first reported by Liu et al. (1983). This report and others since investigated several factors that influence organogenesis including the orientation of the leaf surface in contact with the medium, maturity of explant, wounding, dark incubation, and medium composition (see Korban & Chen 1992, for review). None of these studies however investigated the effect of liquid pretreatment on adventitious shoot induction and/or regeneration frequency. Such a pretreatment is necessary for inoculating explants with *Agrobacterium tumefaciens* when conducting transformation experiments. In this study, we investigate the combined effects of liquid pretreatment, light intensity and duration, and explant size on shoot regeneration from *in vitro*-grown leaves of the apple rootstock Malling 7a (M7a). In addition, we report on an increased regeneration frequency for several apple genotypes.

Materials and methods

In vitro-grown leaves of M7a were harvested from proliferating shoot cultures (Korban et al. 1992). Leaves were either sliced transversely into 2–3 mm wide sections (6–7 sections per leaf) or minced into pieces 2–4 mm^2 ($\approx$ 15 pieces per leaf) with a scalpel, and soaked for 5 min in either sterile distilled water or liquid regeneration medium. Leaf sections/pieces were placed with the adaxial surface of the leaf in contact with the regeneration medium in 100 × 15 mm petri dishes (Baxter Healthcare Scientific Products, IL). The regeneration medium used was as described earlier by Korban et al. (1992) except 20 μM thidiazuron and 4.4 μM α-napthyleneacetic acid concentrations were used instead. Cultures were exposed to various light treatments during the initial four weeks of culture: continuous darkness, two weeks of darkness followed by two weeks of low-light (15 μmol m^{-2} s^{-1}), continuous low-light, two weeks of low-light followed by two weeks of high-light (100 μmol m^{-2} s^{-1}), or continuous high-light. A 2 × 2 × 5 factorial experiment was conducted with two leaves per treatment and five replications.

A second experiment was conducted to test further the effects of explant size and dipping pretreatment on shoot regeneration. Sectioned versus minced M7a leaves were either dipped in sterile distilled water or liquid regeneration medium or not dipped at all (control) prior to plating on the solidified regeneration medium. Each treatment consisted of two leaves with five replications. Cultures were incubated in darkness for four weeks.

In a third experiment, leaves from *in vitro* grown shoots of five apple genotypes including 'Dayton', 'Gala', 'Golden Delicious', 'McIntosh', and 'Royal Gala' were either sectioned or minced, dipped in a liquid regeneration medium, and then plated on a solidified regeneration medium. Each treatment consisted of two leaves with five replications per treatment. Cultures were incubated in darkness for four weeks.

The number of leaf sections/pieces developing shoots and the number of shoots produced on a per whole leaf basis were recorded for each experiment four weeks after culture initiation. Analysis of variance and mean separation by LSD at the 5% level was conducted using SAS statistical package (SAS Institute 1985); an arcsine transformation of percentage data was performed prior to analysis. For presentation of results, non-transformed data are used.

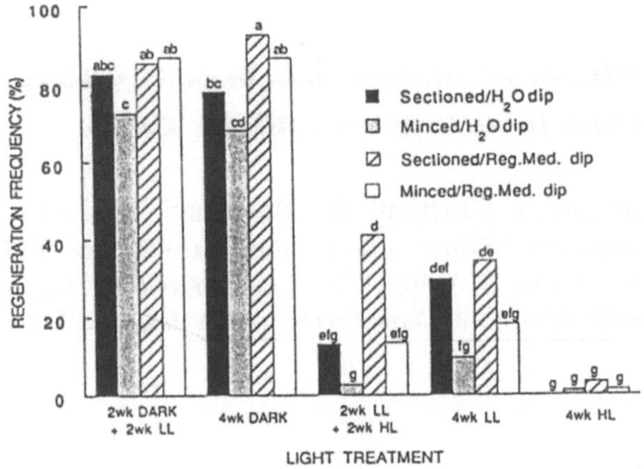

Fig. 1. Effects of light intensity, dipping pretreatment, and explant size on shoot regeneration frequency of *in vitro*-grown Malling 7a leaf explants. Bars with the same letter are not significantly different at the 5% level. LL – low-light, HL – high-light.

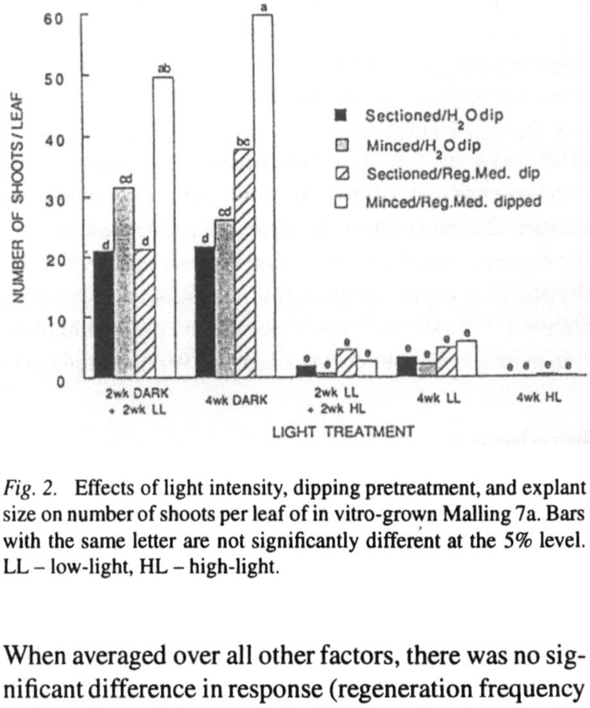

Fig. 2. Effects of light intensity, dipping pretreatment, and explant size on number of shoots per leaf of in vitro-grown Malling 7a. Bars with the same letter are not significantly different at the 5% level. LL – low-light, HL – high-light.

Results and discussion

Almost all treatment combinations resulted in shoot regeneration of M7a leaf explants; however, the highest regeneration frequency and the highest number of shoots/leaf were obtained when the treatment included 2 or 4 weeks of dark incubation (Figs. 1 and 2). When averaged over all other factors, there was no significant difference in response (regeneration frequency and number of shoots per leaf) for explants grown for 4 wks of continuous darkness or for 2 wks of darkness followed by 2 weeks under low-light. This is in agreement with previous studies which indicated that a 1–4 week dark treatment is beneficial although not necessary for shoot induction (Liu et al. 1983, Welander 1988, Fasolo et al. 1989, Predieri & Fasolo Fabbri Malavasi 1989, Dufour 1990, Korban et al. 1992). In contrast to some previous studies, shoot development

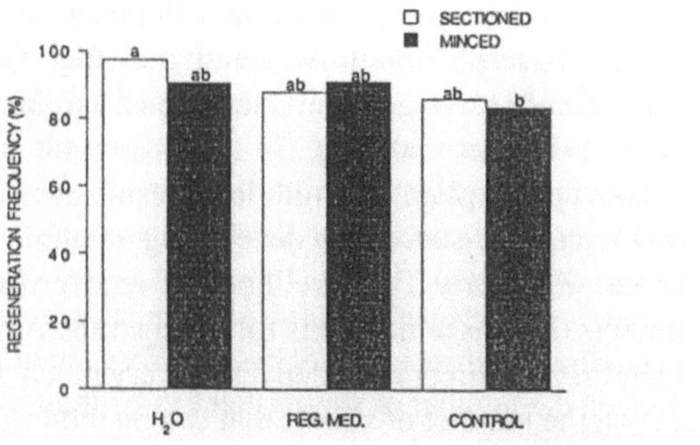

Fig. 3. Effects of three dipping pretreatments and explant size on shoot regeneration frequency from *in vitro*-grown leaves of Malling 7a. Bars having the same letter are not significantly different at the 5% level. Control – explants not dipped prior to plating.

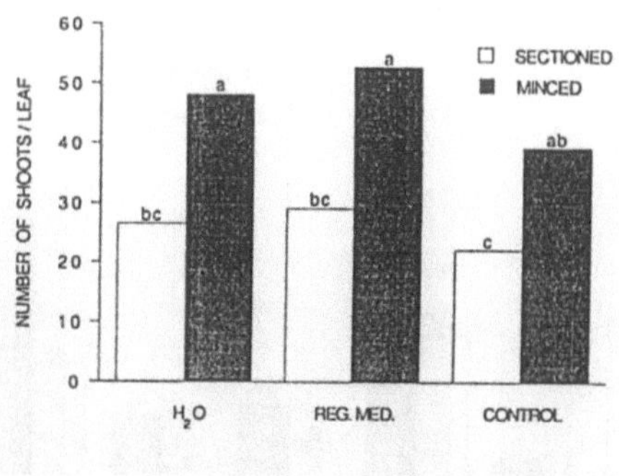

Fig. 4. Effects of three dipping pretreatments and explant size on number of shoots per leaf of in vitro grown Malling 7a. Bars with the same letter are not significantly different at the 5% level. Control – explants not dipped prior to plating.

was also observed when cultures were initially placed under low-light or even under high-light conditions (Figs. 1 and 2). Intermediate regeneration frequencies were observed under both continuous low-light conditions (22.9%, averaged over all other factors) or under 2 weeks of low-light followed by 2 weeks of high-light (17.6%, averaged over all other factors) (Fig. 1). The number of shoots developed per leaf was significantly reduced under low-light conditions with an average of 4.0 shoots/leaf for cultures grown under continuous low-light conditions and 2.4 shoots/leaf for cultures exposed to 2 weeks of low-light followed by 2 weeks of high-light (Fig. 2). The regeneration response dropped even further under continuous high-light conditions with an average of 1.5% regeneration frequency and 0.18 shoots/leaf (Figs. 1 and 2). A similar observation was reported for *Rubus* cotyledons grown under continuous high-light conditions (Fiola et al. 1990). Our finding that shoot regeneration from apple leaves can be induced under various light conditions may have useful application in gene expression studies involving light responsive promoters.

Regeneration frequency and number of shoots developed on a per leaf basis were also influenced by dipping pretreatment and explant size (Figs. 1–4). When minced or sectioned M7a leaf explants were dipped in liquid regeneration medium, an average of 46.2% regeneration frequency and 18.7 shoots per leaf were obtained; while a 35.8% regeneration frequency and 10.8 shoots per leaf were obtained when explants were dipped in sterile water (Figs. 1 & 2). Although no significant differences among dipping pretreatments were observed in the second study (Figs. 3 & 4), nevertheless, dipping explants in liquid regeneration medium prior to plating resulted in an increase in shoot number per leaf (Fig. 4). Liquid pretreatments are necessary in transformation studies involving co-culture of explants with *Agrobacterium tumefaciens*. Bacterial cells are usually resuspended in liquid prior to inoculation of plant tissue. This study suggests that using a sterile liquid regeneration medium may be better than water for resuspension and co-culture of bacterial cells and apple leaf tissues.

Explant size also exhibited a significant effect on the number of shoots produced and the regeneration frequency observed (Figs. 1–4). Overall, minced versus sectioned explants produced an average of 36% regeneration frequency and 17.8 shoots per leaf and 46% regeneration frequency and 11.7 shoots per leaf, respectively (Figs. 1 and 2). There was a significant difference in the number of shoots produced for minced versus sectioned leaves (46.7 versus 25.9), but there was no significant difference in the regeneration frequency among the types of explants. James et al. (1988) reported that cutting *in vitro* grown leaves in transverse or longitudinal strips resulted in significantly higher regeneration frequency and number of shoots per leaf than excising leaves with a 7 mm cork borer into leaf disks. This was attributed to the fact that there was less wounded surface per leaf area for disks than for thinly cut strips. Our results indicate that small explants (minced) with a higher ratio of wounded surface to overall leaf area produce more shoots on a per leaf basis than sectioned leaves, however the regeneration frequency was only equal to or less than that obtained from sectioned explants. The optimal conditions for shoot regeneration from *in vitro*-grown leaves of M7a

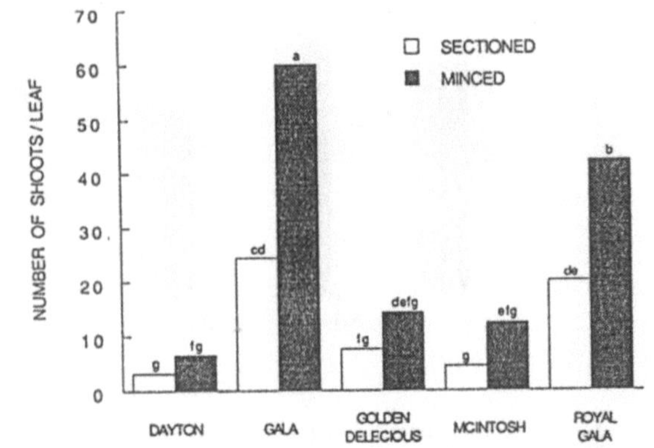

Fig. 5. Regeneration frequency of five apple genotypes; explants were either sectioned or minced prior to culture. Bars with the same letter are not significantly different at the 5% level.

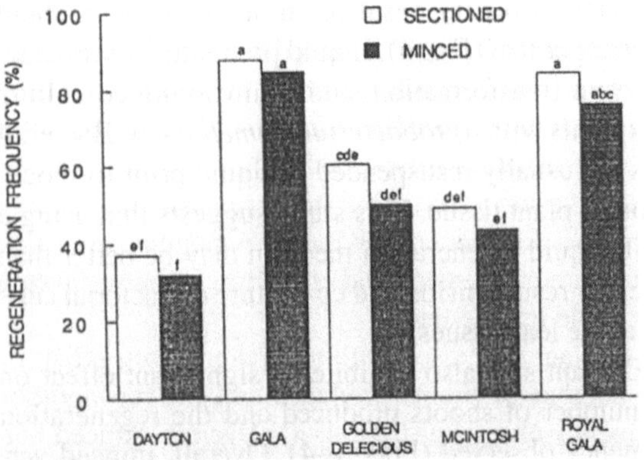

Fig. 6. Shoot number per leaf for five apple genotypes; explants were either sectioned or minced prior to culture. Bars with the same letter are not significantly different at the 5% level.

included mincing of leaves, dipping explants into the liquid regeneration medium, and dark incubation for four weeks. This resulted in an average of 59.8 shoots per leaf and a regeneration frequency of 86.6%.

When five other apple genotypes were induced to develop adventitious shoots, significant differences were observed among genotypes for both regeneration frequency and number of shoots per leaf (Figs. 5 and 6). This is similar to our observations in an earlier report (Korban et al. 1992). Overall, an increased regeneration response was observed in this study for the same genotypes tested earlier. This higher amenability to regeneration may be attributed to a number of factors including dipping pretreatment, longer dark incubation period, changes in medium composition, and explant size. Comparing sectioned versus minced explants did not show significant differences in regeneration frequency for all five genotypes (Fig. 5); however, minced leaves generally developed a higher number of shoots per leaf (Fig. 6). Increasing the number of shoots per explant is a significant result since it may allow increased success in developing an apple transformation system. The likelihood of recovering high numbers of transgenic plants following transformation via indirect or direct gene delivery is very much dependent on the number of shoots that can be induced from an explant.

Shoot regeneration under the above described conditions appeared to arise by both direct and indirect organogenesis. Shoots were first evident from the basal portion of leaf sections (proximal to the petiole), however leaf orientation in minced explants was not important. Callus was observed along the cut surface of the explant and shoots later developed from the callus within 2 to 3 weeks after culture initiation. Shoot and callus production first appeared in proximity to veins but quickly spread to interveinal areas as well. All leaf sections responded by producing callus and shoots without regard as to whether they were taken from the basal or apical portion of the leaf. Polarity effects on regeneration have been previously reported for apple cotyledons (Kouider et al. 1984) and leaves (Welander 1988). The phenomenon has been linked to the fact that vascular bundles converge near the base of cotyledons (Kouider et al. 1984); this seems to concur with our current findings in which shoot regeneration first appears associated with leaf veins (i.e., vascular bundles). In contrast to the report of Welander (1988), all portions of the leaf appeared to be responsive; however, initial callus and shoot formation was observed along the basal portion of the leaf. Proximity to vascular bundles appeared to be more important than relative maturity of leaf sections (Welander 1988) in our study. Polarity effects diminished when leaves were minced into small pieces, although mincing itself made it impossible to discern the original orientation of individual explants.

The information presented in this study should prove useful to researchers pursuing gene transfer into apple. The increased ratio of wounded surface to overall surface area of minced explants should make this source of explant a superior tissue for *Agrobacterium*-mediated transformation. Moreover, callus development from this type of cut surface will provide greater responsive surface area for bombardment with DNA-coated microprojectiles.

Acknowledgements

This work was supported in part by the Illinois Agriculture Experiment Station project no. 65–0325.

References

Dufour, M., 1990. Improving yield of adventitious shoots in apple. Acta Hort. 280: 51–60.

Fasolo, F., R.H. Zimmerman & I. Fordham, 1989. Adventitious shoot formation on excised leaves of in vitro grown shoots of apple cultivars. Plant Cell Tiss. Org. Cult. 16: 75–87.

Fiola, J.A., M.A. Hassan, H.J. Swartz, T.H. Bors & T. McNicols, 1990. Effect of thidiazuron, light fluence rates and kanamycin on in vitro shoot organogenesis from excised *Rubus* cotyledons and leaves. Plant Cell Tiss. Org. Cult. 20: 223–228.

James, D.J., A.J. Passey & E. Rugini, 1988. Factors affecting high frequency plant regeneration from apple tissues cultured *in vitro*. J. Plant Physiol. 132: 148–154.

Korban, S.S. & H. Chen, 1992. Apple. In: F.A. Hammerschlag & R.E. Litz (Eds) Biotechnology of perennial fruit crops, pp. 203–227. C.A.B. International, Wallingford, Oxon.

Korban, S.S., P.A. O'Connor & A. Elobeidy, 1992. Effects of thidiazuron, naphthaleneacetic acid, dark incubation and genotype on shoot organogenesis from *Malus* leaves. J. Hort. Sci. 67: 341–349.

Kouider, M., S.S. Korban, R.M. Skirvin & M.C. Chu, 1984. Influence of embryonic dominance and polarity on adventitious shoot formation from apple cotyledons *in vitro*. J. Am. Soc. Hort. Sci. 109: 381–385.

Liu, J.R., K.C. Sink & F.G. Dennis, 1983. Plant regeneration from apple seedling explants and callus cultures. Plant Cell Tiss. Org. Cult. 2: 293–304.

Predieri, S. & F. Fasolo Fabbri Malavasi, 1989. High-frequency shoot regeneration from leaves of the apple rootstock M26 (*Malus pumila* Mill.). Plant Cell Tiss. Org. Cult. 17: 133–142.

SAS Institute, Inc., 1985. SAS User's Guide: Statistics. Cary, North Carolina.

Welander, M., 1988. Plant regeneration from leaf and stem segments of shoots raised *in vitro* from mature apple trees. J. Plant Physiol. 132: 738–744.

H. Schmidt and M. Kellerhals (Eds.), Progress in Temperate Fruit Breeding, 361–363.

Simplified protocol for *in vitro* shoot regeneration from leaves of *Prunus domestica* L. (cv 'Susina di Dro')

Gino Bassi & Ferdinando Cossio
Istituto Sperimentale di Frutticoltura, Provincia di Verona, Italy

Key words: Prune cultivar, shoot regeneration, tissue culture

Summary

A simplified protocol for shoot regeneration by means of direct organogenesis of *in vitro* leaves of prune cv 'Susina di Dro' is described. Young and fully expanded leaves were taken from 2-week-old micropropagated shoots and transferred to Petri dishes containing a regeneration medium. The efficiency of shoot regeneration is improved when Thidiazuron is added instead of Benzyladenine, when white light is employed instead of dark and when the leaf abaxial blade is put on the medium. After culture and subculture which lasted 25 days, the average shoot regeneration per leaf reached more than 80%. The application of this protocol opens the way to further studies on genetic transformation of cultivated varieties of important stone fruits.

Abbreviations: BA – Benzyladenine, NAA – 1-naphthyl-acetic acid, TDZ – Thidiazuron (N-phenyl-N′-1,2,3-thiadiazol-5-ylurea), 2,4-D – 2,4-dichlorophenoxyacetic acid

Introduction

As recently stated, progress is needed to develop regeneration protocols from mature tissues of stone fruit species (Scorza & Hammerschlag 1992); with regard to prune cultivars there are few reports on *in vitro* shoot regeneration (Cossio & Bassi 1991, Bassi & Cossio 1991).

Our preliminary investigations on the regeneration of prunes from *in vitro* leaves gave encouraging results: the protocol included an induction phase with auxin, derived from the experiences on *Prunus canescens* (Antonelli & Druart 1990), which was demonstrated to be effective in shoot regeneration in the cv 'Bluefre' and, to a smaller extent, in the cv 'Susina di Dro' (Bassi & Cossio l.c.). In that work the cv 'Susina di Dro' showed a regeneration rate from 2.2 to 11.1% after an induction phase (with 2,4-D or NAA respectively), and 16.6% of the organogenesis from the control without an induction phase.

Some organogenetic capacity of that cultivar was therefore demonstrated in those experiments, but the development of a more efficient and simple regeneration method was foreseen. Further experiments were carried out to assess a simplified protocol.

Materials and methods

Plant material and micropropagation media. Axillary shoot cultures of mature trees of prune cv 'Susina di Dro' (*Prunus domestica* L.) had been established 3–4 years prior to the regeneration experiments and the propagules were grown on a multiplication medium consisting of macronutrients of Quoirin & Lepoivre (1977), micronutrients and iron of Murashige & Skoog (1962), 1 mg/l thiamine HCl, 100 mg/l inositol, 0.5 mg/l nicotinic acid, 2 mg/l glycine, 0.5 mg/l pyridoxine, 10 mg/l citric and ascorbic acid, 0.1 or 0.5 mg/l BA, 0.1 mg/l NAA, 30 g/l sucrose, 5 g/l agar (Biolife), pH 5.75.

The 500 ml culture jars were incubated under a 16-h photoperiod with white light (provided by Philips TLD 36W/33 lamps with intensity of 23 $\mu Em^{-2} \cdot s^{-1}$) at 22 ± 1 °C; the cultures were transferred every 3 weeks to fresh medium.

Table 1. Effects of cytokinins, light and leaf position on shoot regeneration from leaves of cultivar 'Susina di Dro'.

	Leaves forming shoots (%)	No. of shoots per leaf	Average shoot length (cm)
Cytokinins Effect (white light, abaxial surface on the medium)			
1 mg/l BAP	9.4 ± 5.4	1.5 ± 0.40	0.4 ± 0.07
1 mg/l TDZ	87.5 ± 12.5	3.1 ± 0.62	0.8±0.20
Light Effect (TDZ, abaxial surface on the medium)			
white light	84.3 ± 10.4	2.5 ± 0.74	1.3±0.32
red light	78.1 ± 13.6	1.5 ± 0.26	0.9 ± 0.22
dark	28.1 ± 16.2	1.0 ± 0.35	1.5±0.21
Leaf Position (with TDZ, white light)			
abaxial surface	90.6 ± 5.4	2.5 ± 0.48	1.5 ± 0.31
adaxial surface	25.0 ± 25.8	1.5 ± 1.54	1.4 ± 0.24

Preparing explants and regeneration media. Entire leaves with petiole were taken from 2-week-old subcultured shoots (3 to 6 per shoot), choosing those with a similar developmental stage and situated in the terminal half of the shoot, excluding apical leaves. The leaves 1–2 cm long were cut 4–6 times across the midrib (keeping the leaves intact) and quickly transferred to 100-mm-diameter Petri dishes containing 25 ml regeneration medium consisting of half macronutrients of Quoirin & Lepoivre (1977), micronutrients and iron of Murashige & Skoog (1962), 1 mg/l thiamine HCl, 100 mg/l inositol, 0.1 mg/l NAA, 20 g/l sucrose, 5 g/l agar (Biolife), pH 5.75; with regard to the cytokinins, BA was compared with TDZ, both at 1 mg/l.

After 25 days, the leaves were transferred onto fresh medium and incubated under the same conditions as the micropropagated shoots.

The standard white light conditions were compared with:

- red light (Philips 36W/15; 5 $\mu Em^{-2} \cdot s^{-1}$) throughout 50 days;
- full darkness (first 25 days) followed by 16/24 h white light during the subculture.

The effect of putting the leaf blades with the abaxial or adaxial surface on the medium was tested.

Experimental procedure. Each treatment consisted of 4 replications with 8 leaves per replicate; leaves forming shoots and the number of shoots per leaf were counted after 50 days culture.

Results and discussion

In general the experiments confirmed that the preconditioning method is important for establishing morphogenic competence in leaf explants (Litz & Gray 1992): young and fully expanded leaves from fast growing *in vitro* shoots are required in order to obtain shoot regeneration. In fact regeneration is reduced when old and/or basal leaves are employed.

For this reason it is essential to have a good multiplication medium for each specific cultivar; in addition it can be useful to vary BA levels or to shorten subculture intervals to assure good leaf morphology and to avoid the elongation stage before explanting the leaves.

In our experiments the highest number of shoots originated from the cuts of the leaf midrib, particularly in the proximal zone and from the petiole. In the early stages the shootlets are often visible under the leaf. Meanwhile the leaf areas increase, sometimes redoubling.

The recorded regeneration probably involved shoot organogenesis without a callus phase. In fact very little callus was observed.

The number of shoots per regenerated leaf ranged from 1 to 3.1 and was sometimes improved by waiting 2–3 weeks more; the shoot length varied from 0.4 to 1.5 cm.

The first shoot growth was detected at the end of the first period of culture. After subculture, the regenerated shootlets could be easily transplanted on multiplication medium (with only 0.1 mg/l BA). Later they were easily transferred to a rooting medium and then acclimatized in a greenhouse.

Cytokinin effect. Although shoot regeneration occurred in the presence of BA (9.4%), TDZ gave much better results, from 84.3 to 90.6% (Table 1); this confirmed the usefulness of this compound, which has been used to obtain morphogenesis in some *Prunus* cotyledons (Mante et al. 1989).

Light effect. In general the presence of light had a positive effect on regeneration; keeping the Petri dishes in the dark during the first 25 days reduced the regeneration percentage to 28.1% (Table 1). Between the light types, white light gave the highest numbers of shoots per leaf.

Leaf position effect. The data demonstrated the utility of putting the abaxial surface on the medium (Table 1). This appeared to be related to the greater contact of the midrib with the medium obtained in this way. Otherwise the leaves tended to lift from the medium.

Conclusion

The adjusted regeneration protocol gave consistently efficient results and is simpler than that previously described for prune cv 'Bluefre' (Cossio & Bassi l.c., Bassi & Cossio l.c.), because it avoids the elongation stage during micropropagation and excludes the auxin-induction phase before regeneration. Moreover it includes the use of 1 mg/l TDZ, white light, and placing leaf blades in an abaxial position for culture and subculture.

Using this protocol consistently high shoot regeneration rates (> 80% of leaves) were repeatedly obtained in successive experiments (unpublished data). Good results were also obtained with leaf explants from regenerated shoots of 'Susina di Dro'.

This protocol opens the way to studies on genetic transformation of cultivated varieties of important stone fruits (Dandekar 1992).

This paper was supported by M.A.F. Rome, National Project: 'Advanced Technologies Applied to Plants'.

References

Antonelli, M & Ph. Druart, 1990. The use of a brief 2,4-D pretreatment to induce leaf regeneration on *Prunus canescens* Bois. Acta Horticulturae 280: 45–50.

Bassi, G. & F. Cossio, 1991. *In vitro* shoot regeneration of 'Bluefre' and 'Susina di Dro' prune cultivars (*Prunus domestica* L.). Acta Horticulturae 289: 81–82.

Cossio, F. & G. Bassi, 1991. Ricerche sulla organogenesi di alcuni Prunus: rigenerazione '*in vitro*' di germogli di susino (*Prunus domestica* L.) cultivar 'Bluefre'. Frutticoltura 3: 77–81.

Dandekar, A.M., 1992. Transformation. In: F.A. Hammerschlag & R.E. Litz (Eds) Biotechnology of perennial fruit crops, pp. 141–168. C.A.B. International, Wallingford, Oxon.

Litz, R.E. & D.J. Gray, 1992. Organogenesis and Somatic Embryogenesis. In: F.A. Hammerschlag & R.E. Litz (Eds) Biotechnology of perennial fruit crops, pp. 3–34. C.A.B. International, Wallingford, Oxon.

Mante, S., R. Scorza & J.M. Cordts, 1989. Plant regeneration from cotyledons of *Prunus persica*, *Prunus domestica*, and *Prunus cerasus*. Plant Cell Tissue Organ Culture 19: 1–11.

Murashige, T. & F. Skoog, 1962. A revised medium for rapid growth and bio-assays with tobacco tissue cultures. Physiol. Plant. 15: 473–497.

Quoirin, M. & P. Lepoivre, 1977. Improved media for '*in vitro*' culture of *Prunus* sp. Acta Horticulturae 78: 437–442.

Scorza, R. & F.A. Hammerschlag, 1992. Stone Fruits. In: F.A. Hammerschlag & R.E. Litz (Eds) Biotechnology of perennial fruit crops, pp. 277–301. C.A.B. International, Wallingford, Oxon.

H. Schmidt and M. Kellerhals (Eds.), Progress in Temperate Fruit Breeding, 365–369.

Plant regeneration from tissue cultures of *Malus* species: The importance of the genotype

Viola Hanke
Federal Centre of Breeding Research on Cultivated Plants, Institute for Fruit Breeding Dresden-Pillnitz; Pillnitzer Platz 2, D-01326 Dresden, Germany

Key words: Adventitious shoot, apple, axillary shoot, *Malus*, protoplast, regeneration

Summary

A prerequisite for the application of biotechnological methods in fruit breeding is an efficient recovery of viable plants from *in vitro* cultures. The main components with large influence on the regeneration process are the genetic base and the cultural conditions *in vitro*. These components are discussed in different tissue culture techniques: 1) the micropropagation system; 2) the regeneration system from complex leaf explants; and 3) the protoplast system. Results are presented for several commercial grown apple cultivars, including cultivars with multiple resistance to powdery mildew disease and to apple scab, for new dwarf and semidwarf apple rootstocks and for an apple wild type with great importance in any breeding programme for disease resistance.

Introduction

Cell and tissue culture techniques offer a new way for genetic improvement of top fruit tree species to overcome the limitations of the sexual system, which can play a crucial role in a conventional breeding strategy, and to establish genotypes with discrete genetic changes on the base of existing cultivars. A prerequisite for the application of any biotechnological method is an efficient recovery of viable plants from *in vitro* cultures (James 1987, Janick 1992).

This paper deals with the main components with large influence on plant regeneration in apple, the genetic base and the cultural conditions, using different *in vitro* culture techniques: the axillary shoot propagation system, the adventitious shoot formation system from leaf explants and the isolation of protoplast from mesophyll tissue. Results are presented for several new German commercial grown apple cultivars and rootstocks bred in Dresden-Pillnitz, including cultivars with multiple resistance to powdery mildew disease and to apple scab, dwarf and semidwarf rootstocks.

Material and methods

Axillary shoot formation from shoot cultures

13 different apple genotypes (Table 1) were established *in vitro* on a Murashige & Skoog proliferation medium supplemented with BA 0.5; IBA 0.2; GA_3 0.5 mg/l and 30 g/l sucrose which has been used for several years in our laboratory as the standard propagation medium for various fruit tree species. In order to improve the yield of regenerated axillary shoots, and to optimize the growth conditions for every genotype the established shootlets were transferred to various MS-based media compositions containing a different hormone balance and if not otherwise stated 30 g/l sucrose (Jones et al. 1977, Pua et al. 1983, Pua & Chong 1985, Doughty & Power 1988, Laimer et al. 1988, Patat-Ochatt et al. 1988, Scheewe 1990). The shoot cultures were maintained at 24 °C under a 16 h light photoperiod at 2000 lux and subcultured monthly. Culture responses were assessed during three subcultures.

Table 1. Genotypes used for various tissue culture techniques.

	Genotype	Technique
Rootstocks:	Pi-AU 51–11	(1, 3)
	Pi-AU 56–83	(1, 3)
	Pi-AU 36–2	(2)
	Pi-AU 9–82	(1, 2, 3)
	Pi-AU 7–33	(1, 3)
	Pi-AU 9–16	(1, 3)
	M9	(2)
Scions:	Alkmene	(2)
	Clivia	(2)
	Golden Delicious	(1, 2, 3)
	Greensleeves	(2)
	Idared	(2)
	Pilot	(2)
	Pinova	(1, 2, 3)
	Realka	(1, 2, 3)
	Reanda	(2)
	Reka	(1, 2, 3)
	Releta	(1, 3)
	Remo	(1, 2, 3)
	Rewena	(1, 3)
Wild type:	Malus robusta	(1, 3)

(1) axillary shoot formation.
(2) adventitious shoot formation.
(3) protoplast cultures.

Adventitious shoot formation from leaf explants

14 apple genotypes (Table 1) were chosen. The donor shoots were grown on a Murashige and Skoog proliferation medium supplemented with BA 0.5; IBA 0.2; GA_3 0.5 mg/l. The explants were taken from the four youngest, expanded leaves of non-rooted shoots. The leaf cultures were maintained in the dark for two weeks and then transferred to 16 h light at 24 °C. The selection of nutrient media for leaf cultures, the procedures for explant plating and physical culture conditions have been previously described (Hanke et al. 1991).

Protoplast isolation from mesophyll tissue

Protoplast isolation from mesophyll tissue in three apple genotypes was investigated. The protoplast sources were the most fully-expanded leaves from axenic shoots, that were monthly subcultured on seven different multiplication media (MS-media supplemented with 1.0 or 0.5 mg/l BA; 0.1, 0.2 or 1.0 mg/l IBA and 0, 0.1 or 0.5 mg/l GA_3). The donor shoots were grown at 26 °C under a 16 h light photoperiod at 2000 lux, 4000 lux or in the dark during the last subculture before protoplast isolation. For the pretreatment of tissues and for the conditions of the protoplast isolation the method of Ochatt & Power (1990) and Patat-Ochatt et al. (1988, 1990, 1993) was used. The results were obtained on the base of 194 isolations.

Results and discussion

Axillary shoot formation

These experiments were designed to test media with different hormone compositions for axillary shoot formation in dependence on the genotype (Table 2). The standard nutrient medium which was used in our laboratory for fruit tree propagation was less efficient in shoot production than other media tested. Beside the auxin-cytokinin-composition of the nutrient media the carbohydrate source was a very important component. Comparing media with the same hormone balance and different sugar components better results could be obtained with sucrose than with sorbitol. On media with a higher content of BA a higher number of axillary shoots were produced. The responses of the genotypes are positive effected by an exclusion of IBA. In general, rootstock-genotypes regenerated axillary shoots without any callus production in an higher amount than scion-genotypes (Table 3). The formation of axillary shoots was less in dwarf rootstocks (Pi-AU 7–33; Pi-AU 9–82; Pi-AU 9–16) than in semidwarf rootstocks, and less in the old German varieties than in the new varieties. An influence of the parents used in crosses for the scab resistant varieties on the shoot formation ability was found. To assess the regeneration ability of a cultivar which response in tissue culture is not investigated before it is useful to compare with the ability of other cultivars having the same cultivar as parent, like it can be seen from the results with the scab resistant cultivars.

The axillary shoot system forms the basis for the induction of genetic variation via regeneration of plants from explants taken from these micropropagated shoot cultures. Micropropagation is efficient for a lot of apple genotypes with most studies showing a large influence of the genotype (Skirvin et al. 1986). Nevertheless, it is necessary to understand the response of genotypes which are of greatest interest for breeders. Multiple

Table 2. Effect of nutrient medium on axillary shoot formation from shoot cultures in apple genotypes.

Medium: MS + 30 g/l Sucrose + Hormones (mg/l)			Number of Axillary Shoots/Explant	Number of Variants Tested; n/variant = 15
BA	IBA	GA_3		
0.5	0.5[1]	5.0	1.31 ± 0.80 d	43
2.0	0.1	0.1	1.38 ± 1.34 d	53
0.5	3.0	–	1.57 ± 0.97 d	52
0.2	0.1	0.1	1.61 ± 0.55 cd	21
1.0	1.0	0.1[2]	1.90 ± 0.98 cd	45
0.5	0.2	0.5	2.40 ± 1.09 c	91
1.0	1.0	–[3]	2.73 ± 1.35 bc	53
1.0	0.1	0.1[4]	3.19 ± 1.56 b	53
1.0	1.0	–	4.77 ± 1.96 a	39
1.0	0.1	1.0	4.81 ± 1.69 a	39
1.0	0.1	0.1	4.96 ± 1.46 a	39
1.0	–	0.5	5.73 ± 1.78 a	39

Dates with different letters differ significantly at $P < 0.05$.

[1] Instead of IBA: NAA.

[2] + phloroglucinol.

[3] Sorbitol instead of sucrose.

[4] Sucrose + sorbitol.

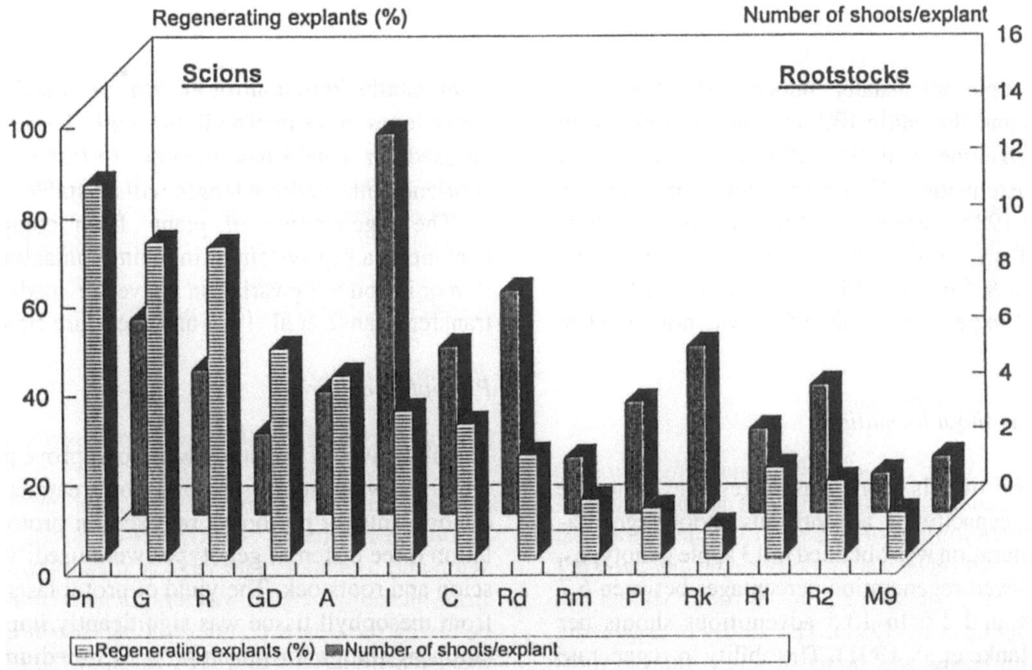

Fig. 1. Effect of the genotype on adventitious shoot formation from leaf explants in apple. Scions: Pn – Pinova; G – Greensleeves; R – Realka; GD – Golden Delicious; A – Alkmene; I – Idared; C – Clivia; Rd – Reanda; Rm – Remo; Pl –Pilot; Rk – Reka; Rootstocks: R1 – Pi-AU 36–2; R2 – Pi-AU 9–82; M9. (modified and supplemented from Hanke et al. 1991.)

Table 3. Effect of the genotype on axillary shoot formation from shoot cultures in apple genotypes.

Genotype		Number of Axillary Shoots/Explant	Number of Variants Tested; n/variant = 15
Rootstocks:	Pi-AU 9–16	2.82 ± 2.16 c	43
	Pi-AU 9–82	3.53 ± 2.32 bc	36
	Pi-AU 7–33	3.73 ± 2.00 bc	41
	Pi-AU 56–83	3.80 ± 2.00 bc	43
	Pi-AU 51–11	4.67 ± 2.26 ab	41
	Pi-AU 36–2	6.30 ± 1.41 a	15
Scions:	Spätblühender Taffettapfel	1.19 ± 0.66 d	26
	Ribston Pepping	1.40 ± 0.93 cd	26
	Kaiser Wilhelm	1.49 ± 1.20 cd	26
	Bohnapfel	1.56 ± 0.78 cd	25
	Realka	2.08 ± 0.98 cd	38
	Releta	2.16 ± 1.21 bcd	38
	Rewena	2.75 ± 1.49 abc	44
	Golden Delicious	2.97 ± 0.93 abc	18
	Pinova	3.35 ± 1.55 ab	45
	Reka	4.30 ± 1.83 a	18
Wildtype:	Malus robusta	2.14 ± 1.30	44
Rootstocks		3.88 ± 2.26	219
Scions		2.33 ± 1.49	304

Dates with different letters differ significantly at $P < 0.05$.

shoots in apple are usually induced by inclusion of cytokinins and the auxin IBA in concentrations from 0.1–1.0 mg/l (Jones et al. 1977, Zimmerman & Broome 1980). An exclusion of IBA seems to be more effective (Welander 1985, Scheewe 1990). The positive effect of sorbitol on shoot multiplication has been shown earlier (Pua & Chong 1985, Welander et al. 1989). In the present experiments this effect was not found in general.

Adventitious shoot formation

The purpose of this work was to test the genotype dependent capacity of adventitious shoot regeneration. Regeneration was obtained in 13 apple genotypes, which achieved regeneration percentages between 6.7 and 83.3% and 1.6 to 13.5 adventitious shoots per explant (Hanke et al. 1991). The ability to regenerate shoots from leaves was better in scions than in rootstocks, but less in the new bred resistant scions with the exception of 'Realka' (Fig. 1). The regenerated shoots were propagated *in vitro* as clones and transferred to soil. For 'Realka' two clones were found which differ significantly from control *ex vitro* trees and the other somaclones in the internodium length. These trees will be used for a field trial in order to test whether the shortened internodium length will be stable.

The regeneration of plants from complex leaf explants is a key system of inducing somaclonal variation or introducing variation via vector-mediated gene transfer (Hanke et al. 1991 and literature cited there).

Protoplast culture

The objectives of this study were to improve protoplast yield and viability by precondition treatments of the donor plants for protoplast release. For protoplast isolation three different genotypes were used: wild type, scion and rootstock. The yield of protoplasts obtained from mesophyll tissue was significantly improved by culturing the donor plants on a MS-medium supplemented with BA 1.0 and IBA 1.0 mg/l, 20 g/l sucrose in comparison with all other media tested. The size of protoplasts also varied depending on pretreatments of the donor shoots. Even shoots on media with lower sugar concentration released significantly larger protoplasts. The ability to release protoplasts was significantly lower in the rootstock (1.47 + 1.81 $\times$ 10^6 protoplasts/g fr.wt. of digested tissue) than in the scion (2.80 + 2.62 $\times$ 10^6 protoplasts/g fr.wt. of digested tissue) and in the wild type (2.49 + 1.81 $\times$ 10^6 protoplasts/g fr.wt. of digested tissue). As in other species the genotype and the growth conditions of the donor plant have proved to be of great importance in protoplast isolations (Wallin & Welander 1985, Nyman & Wallin 1988, 1992, Patat-Ochatt et al. 1988, Wallin & Johansson 1989).

Acknowledgements

Prof Dr M. Fischer and Prof Dr C. Fischer are gratefully acknowledged for scientific advice and provision of the genotypes. I wish to thank I. Schau and G. Klotzsche for valuable technical assistance.

References

Doughty, S., J.B. Power, 1988. Callus formation from leaf mesophyll protoplasts of *Malus* × *domestica* Borkh., cv. Greenleeves. Plant Cell Reports 7: 200–201.

Hanke, V., A. Rohde & C. Grafe, 1991. Untersuchungen zur Regeneration an somatischem Gewebe *in vitro*. I. Zur Adventivsproßbildung an Blattexplantaten bei Apfel (*Malus domestica* Borckh.). Gartenbauwissenschaft 56: 214–220.

James, D.J., 1987. Cell and tissue culture technology for the genetic manipulation of temperature fruit trees. In: G.E. Russell (Ed) Biotechnology and genetic engineering reviews 5, pp. 33–79. Intercept Ltd., Dorset.

Janick, J., 1992. Introduction. In: F.A. Hammerschlag & R.E. Litz (Eds) Biotechnology of perennial fruit crops, pp. 1–2. C.A.B. International, Oxford.

Jones, O.P., M.E. Hopgood & D.O'Farrell, 1977. Propagation *in vitro* of M 26 apple rootstocks. J. Hortic. Sci. 52: 235–238.

Laimer, M., A.D.A.C. Machado & V.V.A. Hanzer, 1988: *In vitro* Vermehrung der alten Lokalsorte 'Graf Uhlhorns August-Kalvill' (*Malus domestica*). Mitt. Klosterneuburg 38: 105–107.

Nyman, M. & A. Wallin, 1988. Plant regeneration from strawberry (*Fragaria* × *ananassa*) mesophyll protoplasts. J. Plant Physiol. 133: 375–377.

Nyman, M. & A. Wallin, 1992. Improved culture technique for strawberry (*Fragaria* × *ananassa* Duch.) protoplasts derived plants. Plant Cell Tissue Organ Culture 30: 127–133.

Ochatt, S.J. & J.B. Power, 1992. Plant regeneration from cultured protoplasts of higher plants. In: M. Moo-Young, G.S. Warren & M.W. Fowler (Eds) Comprehensive biotechnology, pp. 100–127, Pergamon Press, New York.

Patat-Ochatt, E.M. & J.B. Power, 1990. Advances in plant regeneration from apple protoplasts. Acta Hortic. 280: 285–288.

Patat-Ochatt, E.M., S.J. Ochatt & J.B.Power, 1988. Plant regeneration from protoplasts of apple rootstocks and scion varieties (*Malus* × *domestica* Borkh.). J. Plant Physiol. 133: 460–465.

Patat-Ochatt, E.M., J. Boccon-Gibod, M. Duron & S.J. Ochatt, 1993. Organogenesis of stem and leaf protoplasts of a haploid Golden delicious apple clone (*Malus* × *domestica* Borkh.). Plant Cell Reports 12: 118–120.

Pua, E.C. & C. Chong, 1985. Regulation of in vitro shoot and root regeneration in cultivar Macspur apple *Malus domestica* by sorbitol (D-glucitol) and related carbon sources. J. Amer. Soc. Hortic. Sci. 110: 705–709.

Pua, E.C., C. Chong & G.L. Rousselle, 1983. *In vitro* propagation of Ottawa 3 apple rootstock. Can. J. Plant Sci. 63: 183–188.

Scheewe, P., 1990. Entwicklung von Methoden zur *In vitro*-Mutationsinduktion und -selektion auf Resistenz gegen Apfelmehltau (*Podosphaera leucotricha* [Ell. et Ev.] Salm). Diss., Universität Hamburg.

Skirvin, R.M., M. Kouider, H. Joung & S.S. Korban, 1986. Apple (*Malus* × *domestica* Borkh.). In: Y.P.S. Bajaj (Ed) Biotechnology in agriculture and forestry, Trees I, pp. 183–197. Springer-Verlag, Berlin/Heidelberg/New York/Tokyo.

Wallin, A. & L. Johansson, 1989. Plant regeneration from leaf mesophyll protoplasts of *in vitro* cultured shoots of a columnar apple. J. Plant Physiol. 135: 565–570.

Wallin, A. & M. Welander, 1985. Improved yield of apple leaf protoplasts from in vitro cultured shoots by using very young leaves and adding L-methionine to the shoot medium. Plant Cell Tissue Organ Culture 5: 69–72.

Welander, M., 1985. *In vitro* shoot and root formation in the apple cultivar Akerö. Ann. of Bot. 55: 249–261.

Welander, M., N.T. Welander & A. Brackman, 1989. Regulation of *in vitro* shoot multi plication in *Syringa*, *Alnus* and *Malus* by different carbon sources. J. Hort. Sci. 64: 361–366.

Zimmerman, R.H. & O.C. Broome, 1980. Apple cultivar micropropagation. Proc. Conf. on Nursery Prod. of Fruit Plants through Tissue Culture-Application and Feasibility, April 21–22, Beltsville, Maryland, pp. 54–58.

H. Schmidt and M. Kellerhals (Eds.), Progress in Temperate Fruit Breeding, 371–375.

Plant regeneration from cotyledons and embryonic axes in apple: Sites of reaction and effect of pre-culture in the light

J. Keulemans & K. de Witte
Fruitteeltcentrum K.U. Leuven, Willem de Croylaan 42, B-3001 Leuven-Heverlee, Belgium

Key words: Apple, cotyledon, embryonic axis, *Malus domestica*, pre-culture, shoot regeneration, explant variation

Summary

Mature embryos of the cultivar Gloster were isolated from seeds of fruits stored at 1 °C. Embryonic axes and wounded cotyledons were pre-cultured in the light for 0 to 6 days on embryo growth medium (experiment 1) or for 0 to 24 hours on shoot induction medium (experiment 2). After a subsequent dark period of 3 weeks for induction of organogenesis, explants were cultured in the light on shoot elongation medium. Up to 95% of the embryos formed shoots, while the mean shoot number per embryo reached 6.4. Regeneration frequency was influenced by the type of explant: embryonic axes reacted better (75%) than the tops of the cotyledons (29%) or wounds on the cotyledons (9%). The mean total shoot number per embryo comprised means of 2.3 for embryonic axes, 1.8 for tops of cotyledons and 2.3 for wounds on cotyledons (experiment 1). Wounds on the proximal part of the cotyledon reacted much better than those on other parts. Shoot differentiation was highest when cotyledons were cultured with the adaxial side on the medium, suggesting that there exist gradations within the cotyledon for regeneration capacity (experiment 2). The number of shoots per reactive site was not normally distributed, while a high frequency of 6 or more shoots per reactive site was found (experiment 1). A pre-culture of more than 24 hours in the light seemed to have negative effect, but there was no clear evidence whether this was a light and/or a medium effect.

Abbreviations: BA: 6 benzylaminopurine; IAA: indole-3-acetic acid; IBA: indole-3-butyric acid; 2iP: 6-(gamma, gamma- dimethylallylamino)purine; GA_3: gibberellic acid

Introduction

Cotyledon regeneration can be used for several purposes such as *in vitro* multiplication, rescue of immature embryos, regeneration of difficult to germinate embryos (including haploid embryos) and organogenesis from mutation-induced or genetic transformed seeds.

Shoot differentiation from cotyledons or embryonic axes has been described for several fruit species, including apple (Kouider *et al.*, 1984a, James *et al.*, 1984), pear (Turner & Singha, 1988, Browning *et al.*, 1987), Japanese pear (Hiratsuka & Katagiri, 1988) and stone fruit (Bhansali *et al.*, 1991; Mante *et al.*, 1989, Lane & Cossio, 1986). To optimize plant regeneration most studies were concentrated on medium composition but little attention was paid to the influence of the explant material itself on plant regeneration efficiency.

The effects of seed maturity and dormancy in apple were studied by Kouider *et al.* (1985). The same authors reported also on the influence of embryonic dominance and polarity on adventitious shoot formation from cotyledons (Kouider *et al.*, 1984b).

The aim of our study was to investigate whether different parts of the embryo and/or different sites of the cotyledons had the same shoot regeneration capacity. The effect of pre-culturing the explants in the light was also studied.

Materials and methods

Preparation of the embryos

Mature embryos were isolated from seeds of stored fruits of the cultivar Gloster. Fruits were stored for 5 months (first experiment) or 2 months (second experiment). After desinfection of seeds (95% ethanol for 30 sec and 1,25% NaOCl for 15 min), the cotyledons were removed from the embryonic axis, wounded as described below and cultured together with the corresponding embryonic axis in petri-dishes on a solid medium.

Media

Pre-culture medium (ESM13): macro and micro salts: Lepoivre/2; Fe-Na EDTA (41.21 mg/l); mixture of vitamins: myo-inositol (100 mg/l), glycine (2 mg/l), nicotine acid (0,5 mg/l), pyridoxine-KCl (0,5 mg/l) and thiamine (0,1 mg/l); saccharose (20 g/l); agar (5 g/l) and plant growth regulators (PGR's) IAA (0.5 mg/l), BA (0.2 mg/l) and GA_3 (1 mg/l).

Shoot induction medium (ASIM): as ESM13 with macro salts: Murashige & Skoog; and PGR's: IBA (0.3 mg/l) and 2iP (8 mg/l).

Shoot elongation medium (ASEM): as ESM13 with BA (1 mg/l) and GA_3 (0.1 mg/l).

Culture conditions

Explants were cultured in the light (1600 Lux) at 23 °C.

Experiment 1

Each cotyledon was wounded 3 times transversely (proximal, midsection and distal) or 3 times longitudinally (2 outer wounds and 1 in the midsection area). Cotyledons, together with the corresponding embryonic axes, were cultured in the light with the adaxial side on the medium for 0, 2, 4 or 6 days on ESM13, then for 3 weeks on ASIM in the dark and for 3 weeks on ASEM in the light before evaluation. Each treatment contained 20 embryos, divided over 5 petri-dishes.

Experiment 2

Cotyledons were wounded transversely as in experiment 1. They were cultered with the adaxial or the abaxial side on the medium (ASIM). After a light period of 0, 6, 12 or 24 hours, petri-dishes were transferred to the dark for 3 weeks. After another 3 weeks on ASEM in the light explants were evaluated. For each treatment we used 40 embryos, divided between 10 petri-dishes.

Table 1. Influence of pre-culture in light on response frequency and shoot number per embryo

experiment 1			experiment 2		
duration pre-cult.	% response embryos	shoots/ embryo	duration pre-cult.	% response embryos	shoots/ embryo
0 days	95	8.9	0 hours	94	5.4
2 days	88	4.2	6 hours	96	5.7
4 days	96	3.9	12 hours	94	5.6
6 days	93	3.1	24 hours	98	6.0

Observations

For each embryo the number of shoots per potential reacting site (embryonic axis, top of cotyledon and wounds on cotyledon) was counted.

Results

Effect of pre-culture in the light

There was no clear influence of the pre-culture treatment on the responsiveness of the embryos: the percentage of responsive embryos ranged from 88 to 98% (table 1).

A pre-culture of 2 days or more on ESM13 had a negative effect on the regeneration efficiency, reducing it from 8.9 (0 days) to 4.2 (2 days) and 3.1 (6 days) shoots per embryo. A pre-culture treatment of 24 hours or less on ASIM apparently did not effect shoot regeneration.

Sites of reaction

Experiment 1 (table 2)

Embryonic axes were more responsive (75%) than the tops of cotyledons (29%) or the wounds on cotelydons (9%). Proximal wounds on cotyledons gave larger responses than distal wounds. These findings were confirmed by the second experiment (data not shown).

Table 2. Shoot regeneration among different embryo structures (ea, top, cot), type of wounds (cl, ct) and location of wounds on cotyledon (clo, clm, ctd, ctm, ctp)

sites:	A	B	C
ea	75	3.0	2.3
top	29	3.1	1.8
cot (cl and ct)	9	4.5	2.3
cl (clo and clm)	7	4.4	1.8
ct (ctd, ctm and ctp)	9	4.6	2.4
clo	4	7.0	1.0
clm	14	3.1	0.8
ctd	3	5.8	0.4
ctm	7	3.8	0.5
ctp	17	4.8	1.6

A: responsiveness of embryo (number of responsive sites/total number of sites x 100)
B: shoot number per responsive site and per embryo
C: total shoot number per site and per embryo
sites:
ea: embryonic axis
top: top of cotyledon
cot: wound on cotyledon
cl: longitudinal wound on cotyledons
ct: transverse wound on cotyledons
clo: longitudinal wound on the outer area of the cotyledon
ctd: transverse distal wound on the cotyledon
clm: longitudinal wound on the midsection area of the cotyledon
ctm: transverse midsection wound on the cotyledon
ctp: transverse proximal wound on the cotyledon

The number of shoots per responsive site was higher on wounds on cotyledons (4.5 shoots) than on embryonic axes (3 shoots) or tops of cotyledons (3.1 shoots). In the case of longitudinal wounds the shoot number was much higher on outer situated wounds than on wounds on the midsection of the cotyledon, but for the latter the response frequency was higher.

The total shoot number among embryonic axes, top or wounds on cotyledons was comparable and ranged from 1.8 to 2.3 shoots per embryo. Due to their higher response rates, transverse wounds near the top of the cotyledon bore more shoots than other wounds.

Experiment 2 (table 3)

In contrast to experiment 1 the number of shoots on the tops of cotyledons was much higher than on wounds on cotyledons or on embryonic axes. Cotyledons induced more shoots when placed with the adaxial side on the medium. This difference was expressed on cotyledon tops (4.5 versus 2.9 shoots per embryo) and on proximal wounds (0.5 versus 0.3 shoots). Midsection and distally situated wounds gave little or no reaction.

Table 3. Total shoot number per embryo and partition among different sites or structures of the embryo; embryos cultured with adaxial or abaxial side on the medium (experiment 2)

location of the shoots	number of shoots	
	adaxial	abaxial
embryonic axis	1.6	1.4
top of cotyledon	4.5	2.9
proximal wound on cotyledon	0.5	0.3
midsection wound on cotyledon	0.1	0.1
distal wound on cotyledon	0.0	0.0
total	6.6	4.7

Table 4. Frequency (%) of the shoot number per responsive site among different embryo structures (experiment 1)

embryo structure	frequency of shoot number per responsive site					
	1	2	3	4	5	> 5
embryonic axis	41	18	15	4	2	22
top of cotyledon	36	16	19	5	0	24
wound on cotyledon	35	5	2	0	0	58

Variability of reaction

Although most embryos responded, several potential response sites gave no reaction at all. The shoot number per embryo was variable among embryos and among petri-dishes. One shoot per responsive site was very common, but a frequency of 5 or more shoots per site was unexpectedly high, especially on cotyledon wounds (58%, table 4). The distribution of shoot frequency on embryonic axes and tops of cotyledons was comparable.

Discussion

A pre-culture of 2 days or more under light had a negative effect on shoot formation from excised embryonic axes or cotyledons. The negative influence of light on shoot formation in apple was also reported by Liu *et al.*, (1983), who suggested that light inhibited callus formation on cotyledon explants. However, our results showed that the pre-culture of embryonic structures for 24 hours or less on an induction medium had no effect on shoot regeneration.

Kouider et al (1985) mentioned that regeneration from mature embryos isolated from stored fruits decreased with storage time. This could not be confirmed by our results since in the first experiment (storage period of 5 months) embryos produced 8.9 shoots against 5.4 shoots per embryo in the second experiment (2 months storage period, treatments without pre-culture). Moreover it seemed that regeneration capacity on midsection and distal parts of the cotyledons was highest in the first experiment compared to the second experiment. In this context, Sinska (1988) mentioned that regeneration from callus, obtained from cotyledons, decreased with stratification time of the embryos.

The presented data suggest a pronounced basipetally orientated influence on shoot regeneration from cotyledons. Such gradation of shoot regeneration on cotyledons was also reported for sweet cherry (Schmidt & Kardel-Meisner, 1992), apple (Rubos & Pryke, 1984, Kouider *et al.*, 1984b), peach (Vujanic-Varga *et al.*, 1990) and other stone fruit species (Mante *et al.*, 1989), where most shoots were regenerated from the proximal part of the embryo. This polarity (negative or positive) was not found in pear (Browning *et al.*, 1987).

The increased shoot differentiation in apple when cotyledons were cultured with the adaxial side on the medium, is probably related to the position of the embryonic axis in intact seeds. A translocation of GA from axis to cotyledons, as postulated by Bogatek *et al.*, (1989), could play a role.

Several authors mentioned the influence of the physiological stage of the embryo on shoot differentation. This influence seems to be correlated with embryo development (Lane & Cossio, 1986) or dormancy (Kouider *et al.*, 1985). The great differences in regeneration efficiency between embryos in our experiments resulted from the embryos themselves since both cotyledons of the same embryo gave comparable shoot differentiation. From this observations we presume that differentiation potential is determined by internal (physiological) factors of the embryo.

The high frequency of 5 or more shoots per responsive site, especially on cotelydons, was remarkable. This high regeneration frequency suggests a stimulatory effect of a differentiating cell or tissue, probably through the gass atmosphere (Kumar *et al.*, 1987).

Acknowledgements

Part of this work was supported by 'Instituut voor aanmoediging van Wetenschappelijk Onderzoek in Nijverheid en Landbouw' (I.W.O.N.L.) – Brussels. We wish to thank Lia Verdoodt for her technical assistance.

References

Bhansali, R.R. Driver J.A. & D.J. Durzan, 1991. Adventitious embryogenesis and plant regeneration from rescued embryos of peach *Prunus persica* L. Indian. J. Exp. Biol. 29: 334–337.

Bogatek, R. Zarska-Maciejewska, B. Sinska, I. & S. Lewak, 1989. The embryonic axis controls lipid catabolism in cotyledons of apple seeds during germination. Physiol. Plant. 76: 557–562.

Browning, G. Ognjanov, V. Passey, A.J. & D.J. James, 1987. Multiple shoot and root regeneration from pear embryo cotyledon explants in-vitro. J. Hortic. Sci. 62: 305–312.

Hiratsuka, S. & M. Katagiri, 1988. Shoot and callus formation in explants from immature seeds of japanese pear. Sci. Hortic. (Amst) 34: 193–200.

James, D.J. Passey, A.J. & D.C. Deeming, 1984. Adventitious embryogenesis and the in-vitro culture of apple *Malus pumila* seed parts. J Plant Physiol 115: 217–230.

Lane, W.D. & F Cossio, 1986. Adventitious shoots from cotyledons of immature cherry and apricot embryos. Can. J. Plant. Sci. 66: 953–959.

Kouider, M. Skirvin, R.M. Korban, S.S. Widholm, J.M. & R. Hauptmann, 1984a. Adventitious shoot formation from cultivar Red Delicious *Malus domestica* apple cotyledons in-vitro. J. Hortic. Sci. 59: 295–302.

Kouider, M. Korban, S.S. Skirvin, R.M. & M.C. Chu, 1984b. Influence of embryonic dominance and polarity on adventitious shoot formation from apple *Malus domestica* cultivar Delicious cotyledons in-vitro. J. Am. Soc. Hortic. Sci. 109: 381–385.

Kouider, M. Korban, S.S. Skirvin, R.M. & H. Joung, 1985. The relationship of apple embryos and their cotyledons to maturity, dormancy, and the potential to form shoots in-vitro. J. Am. Soc. Hortic. Sci. 110: 93–96.

Kumar, P.P. Reid, D.M. & T.A. Thorpe, 1987. The role of ethylene and carbon dioxide in differentiation of shoot buds in excised cotyledons of Pinus radiata in vitro. Physiol. Plant. 69: 244–252.

Liu, J.R. Sink, K.C. & F.G. Dennis, 1983. Plant regeneration from apple seedling explants and callus culture. Plant Cell Tissue Organ Cult 2: 293–304

Mante, S. Scorza, R. & J.M. Cordts, 1989. Plant regeneration from cotyledons of *Prunus persica*, *Prunus domestica* and *Prunus cerasus*. Plant Cell Tissue Organ Cult. 19: 1–11.

Rubos, A.C. & J.A. Pryke, 1984. Morphogenesis in embryonic tissue cultures of apple *Malus pumila*. J. Hortic. Sci. 59: 469–476.

Schmidt, H. & U. Kardel-Meisner, 1992. Regeneration of adventitious shoots in-vitro in cherries III. Formation of adventitious shoots from cotyledons of sweet cherry cultivars. Gartenbauwissenschaft 57: 267–270.

Sinska, I., 1988. Callus formation and plant regeneration capacity of apple embryonic axes and cotyledons in relation to seed dormancy. Plant Sci. (Shannon) 54: 147–152.

Turner, S.R. & S. Singha, 1988. Shoot regeneration from pear cotyledons in-vitro. Hort. Science 23: 755.

Vujanic-Varga, D. Ognjanov, V. & V. Torbica, 1990. Plant regeneration from peach cotyledons. Jugoslovensko Vocarstvo 24: 37–41.

H. Schmidt and M. Kellerhals (Eds.), Progress in Temperate Fruit Breeding, 377–379.

In vitro induction of adventitious shoot and embryo formation in somatic tissues of apple (*Malus pumila* Mill.) zygotic embryos and stem segments

Lalichka Gavrailova Kolova & Nikola Atkov Stoyanov
Plant Biotechnology Department, Fruit Growing Research Institute, 12 Ostromila str., 4004 Plovdiv, Bulgaria

Key words: Adventitious embryos, adventitious shoots, embryo culture, *Malus pumila* Mill.

Summary

Somatic tissue of embryo origin represents a good source for adventitious regeneration to study somaclonal variation, mutation breeding, gene transfer, etc. Seed sections (cotyledon and hypocotyl segments) of *in vitro* stratified immature zygotic embryos and internodal stem segments of four-week-old *in vitro* grown seedlings of open pollinated apple cultivar Vista Bella and of controlled pollinated Vista Bella × Prima, Vista Bella × Liberty were used. The effects of type and concentration of mineral, organic and growth regulator components of culture media and its pH value on regenerative capacity of apple embryonic tissues were studied. The highest rate of shoot formation was obtained by hypocotyl segments cultured on MS (Murashige & Skoog 1962) medium containing benzylaminopurine (BA) 2 mg/l and naphtaleneacetic acid (NAA) 1mg/l at pH 5.2 prior to autoclaving. Up to 48% of cotyledon sections and 63% of internodal stem segments produced adventitious somatic embryos when cultured on SC (Skirvin & Chu 1977) medium supplemented with BA 2mg/l, NAA 0.5 mg/l and GA_3 0.1 mg/l filter sterilized at pH 5.8. Somatic embryos developed into normal plants only after being stored for 5 to 7 weeks at 4 °C.

Introduction

Morphogenesis in apple tissues of zygotic origin was first reported by Abbot & Whiteley (1976). The morphogenic capacity of apple embryonic tissue cultures was studied by Rubos & Pryke (1984). Apple cotyledon cultures were used to regenerate adventitious shoots and study the relationship of maturity and dormancy to regenerative potential *in vitro* (Koudier et al. 1984 1985). Regenerative potential of intact and sectioned apple embryo axes were compared (Korban & Skirvin 1985).

Somatic embryogenesis in apple was originally reported by Eichholtz et al. (1979). Leaf explants from apple seedlings (Liu et al. 1983), nucellus, endoosperm and intact zygotic embryos (James et al. 1984) were used in studies related to adventitious embryogenesis.

In this paper we present the results of our studies on adventitious morphogenesis of somatic tissues of immature apple zygotic embryos as a part of our research related to the genetic improvement of early ripening apples.

Materials and methods

Seeds of open-pollinated Vista Bella cultivar and controlled pollinated Vista Bella × Prima and Vista Bella × Liberty were excised 5–6 weeks post anthesis. Seeds were disinfected in 10% active Cl solution, rinsed three times in cold sterile water and embryos excised under aseptic conditions. *In vitro* stratification took place on SC (Skirvin & Chu 1977) medium containing 1 mg/l benzylaminopurine (BA), 0.25 mg/l naphtaleneacetic acid (NAA) and 0.01 mg/l gibberillic acid (GA_3) all filter sterilized. Cotyledon and hypocotyl segments were prepared for regeneration 10 days after stratification. Internodal segments were prepared 4 weeks after stratification. All types of explants studied were placed on either MS (Murashige & Skoog 1962) or SC medium both containing 0–3 mg/l BA, 0–2 mg/l NAA with or without the addition of 0–0.1 mg/l filter sterilized GA_3 and at a range of pH values from 5.0 to 6.0. Cultures were grown in darkness for 30–40 days before scoring for adventitious structures.

Table 1.

BA NAA mg/l^{-1}	0	0.5	1	2	3
0	C1 c=0% h=0% i=0%	C2 c= 0% h= 0% i= 0%	C3 c= 0% h= 0% i= 0%	C4 c= 0% h= 0% i= 0%	C5 c= 0% h= 0% i= 0%
0.5	C6 c=0% h=0% i=0%	C7 c= 0% h= 2% i= 10%	C8 c= 3% h= 2% i=22%	C9 c= 8% h=10% i=36%	C10 c=12% h=23% i=43%
1	C11 c=0% h=0% i=0%	C12 c= 9% h=10% i= 0%	C13 c=26% h=46% i= 0%	C14 c=39% h=78% i=12%	C15 c=41% h=63% i=51%
1.5	C16 c=0% h=0% i=0%	C17 c= 6% h= 0% i= 0%	C18 c=20% h=16% i=19%	C19 c=29% h=19% i=33%	C20 c=42% h=29% i=39%
2	C21 c=0% h=0% i=0%	C22 c= 0% h=12% i= 0%	C23 c= 0% h=21% i=0%	C24 c=11% h=36% i=0%	C25 c=16% h=22% i=0%

c – cotyledon segments
h – hypocotil segments
i – internodal segments

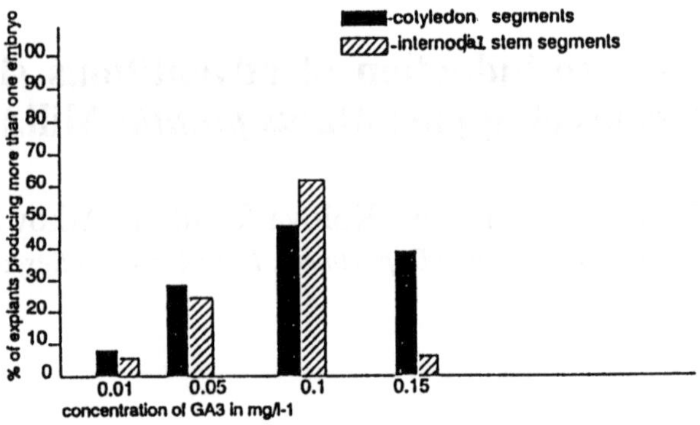

Fig. 1. Effect of the addition of GA_3 to culture media on the embryogenic response of cotyledon and internodal stem segments originating from immature zygotic embryos of open-pollinated Vista Bella seeds grown on SC medium supplemented with 2 mg/l BA and 0.5 mg/l NAA at pH 5.8.

Results and discussion

The best shoot regeneration was obtained from hypocotyl segments (Table 1). No shoot formation was detected on media lacking BA. At the highest BA concentrations, shoots appeared vitrified. These results are similar to those of James et al. (1984) and Korban & Skirvin (1985) who also found the presence of BA very important for multiple shoot formation by the embryonic axis. The fact that we obtained satisfactory shoot regeneration at half the BA concentration used by previous workers may be explained by enhancement of morphogenic activity through embryo axis sectioning (Korban & Skirvin 1985). The pH value of culture media strongly affected the regeneration capacity of all types of explants studied, the optimal being 5.2. In contrast to Koudier et al. (1985) we found no loss of regenerative potential of embryonic tissues after stratification.

Cultivation on SC medium containing 2 mg/l BA, 0.5 mg/l NAA and 0.1 mg/l GA_3 induced embyo production in cotyledon and internode stem segments at 48% and 63% respectively. Basal media containing high vitamin concentrations have been used with success in studies of embryogenesis of *Citrus* (Bitters et al. 1970) and *Malus* (Liu et al. 1983). The addition of GA_3 to regeneration media obviously diverted morphogenic activity to embryogenesis; optimal results were obtained at pH 5.8 (Fig. 1). Cotyledonary stage embryos developed into normal plants only when stored at 4 °C for 5 to 7 weeks on embryo stratification medium. This indicates that somatic embryos, produced by zygotic tissues can also be dormant. Further studies are needed to establish the physiological nature of this phenomenon.

References

Abbot, A.J. & E. Whiteley, 1976. Culture of Malus tissues *in vitro*. I. Multiplication of apple plants from isolated shoot apices. Sci. Hortic. 4: 183–189.

Bitters, W.P., T. Murashige, T.S. Rangan & E. Nauer, 1970. Investigations on established virus-free plants through tissue culture. Calif. Citrus Nurserymen's Soc. 9: 27–30.

Eichholtz, D.A., H.A. Robitaille & P.M. Hasegawa, 1979. Adventive embryony in Apple. HortScience 14(6): 699–700.

James, D.J., A.J. Passey & D.C. Deeming, 1984. Adventitious embryogenesis and the *in vitro* culture of apple seed parts. J. Plant Physiol. 115: 217–229.

Korban, S.S. & R.M. Skirvin, 1985. *In vitro* shoot regeneration from an intact and sectioned embryo-axis of apple seed. Plant Sci. 39: 61–66.

Koudier, M., R.M. Skirvin, S.S.Korban, J.M. Widholm & R. Hauptman, 1984. Adventitious shoot formation from Red Delicious apple cotyledons in vitro. J. Hort. Sci. 59(3): 295–302.

Koudier, M., S.S. Korban, R.M. Skirvin & H. Jong, 1985. The relationship of apple embryos and their cotyledons to maturity, dormancy and the potential to form shoots *in vitro*. J. Amer. Soc. Hort. Sci. 110(1): 93–96.

Liu, J.R., C. Sink & F.G. Dennis, 1983. Plant reneration from apple sedling explants and callus cultures. Plant Cell Tiss. Org. Cult. 2: 293–304.

Murashige, T. & F. Skoog, 1962. A revised medium for rapid growth and bioassays with tobacco tissue cultures. Phys. Plant 15: 473–497.

Rubos, A.C. & J.A. Pryke, 1984. Morphogenesis in embryonic tissue cultures of apple. J. Hort. Sci. 59(4): 469–475.

Skirvin, R.M. & M.C. Chu, 1977. Tissue culture may revolutionize the production of peach shoots. III. Res. 19: 18–19.

H. Schmidt and M. Kellerhals (Eds.), Progress in Temperate Fruit Breeding, 381–383.

Raising sweet cherry seedlings by using *in vitro* techniques

Hanna Schmidt & Andrea Ketzel
Institut für Zierpflanzenzüchtung, Bornkampsweg 31, D-22926 Ahrensburg, Germany

Key words: Breeding, grafting microshoots, *in vitro* adventitious shoots, seedlings, sweet cherry

Summary

Raising large progenies of sweet cherry seedlings is hampered by the low efficiency of crosses. Usually for every 100 pollinated flowers, no more than 1–5 seedlings can be raised. The proportion of seeds, from which plants can be produced is increased by using *in vitro* culture techniques. Ripe hybrid fruits were harvested, the seeds cracked, and the cotyledons excised. They were cultured on MS medium supplemented with 2 mg/L BA + 1 mg/L IAA, elongated on MS + 0.5 mg/L BA + 0.1 mg/L NAA and either rooted on 1/2 MS + 0.5 mg/L NAA, both media supplemented with additional $Ca(NO_3)_2$, or directly grafted onto rootstocks in the greenhouse. First shoots appeared after only 10–14 days. Vigorous shoots could be transferred to elongation medium after 3–4 weeks. Shoots rooted within one week and could be transplanted to the greenhouse a week later. *In vitro* propagated dwarfing rootstocks '473/10' were grafted with microshoots. Under optimal temperature, light and humidity conditions the grafted shoots resumed growth after 10 days. Embryos from prematurely dropping fruit could be regenerated to a certain extent. The younger the fruits were the higher was the percentage of 'empty' seed. Some 20% of the seeds containing embryos could be regenerated, which after stratification would not have germinated.

Introduction

Raising sufficiently large progenies of sweet cherry seedlings for breeding purposes is hampered by the low efficiency of crosses. Whereas apples may yield one plant per pollinated flower no more than 1–5 seedlings may be obtained from 100 pollinated flowers in cherries. This proportion can be raised by using *in vitro* culture techniques.

Experiments

Crosses were made on potted trees grafted onto dwarfing rootstocks, moved into the greenhouse before flowering and out again when the danger of spring frosts was over.

Cotyledon culture

Hybrid fruits, having reached at least 75% of their fruit developmental period, were harvested, the seeds cracked, and the cotyledons excised. In sectioned cotyledons the distal parts occasionally form shoots, however, only when the proximal part does so too. Thus, only the two proximal ends of the cotyledons were cultured on MS medium (Murashige & Skoog 1962), supplemented with 2 mg/L BA (6-benzylaminopurine) + 1 mg/L IAA (α-indole acetic acid). Adventitious shoots were harvested and cultured on MS medium, containing 1 g/L $Ca(NO_3)_2$ + 0.5 mg/L BA + 0.1 mg/L NAA (α-naphthyl acetic acid). Vigorous shoots were elongated under low light intensities, and rooted on half strength MS medium + 0.5 mg/L $Ca(NO_3)_2$ + 0.5 mg/L NAA (Yang 1992).

First shoots appeared on the cotyledons after only 10–14 days and were transferred to elongation medium after 3–4 weeks. Elongated shoots were rooted within one week and transferred to the greenhouse one week later. However, this procedure showed extreme variation, with the first plants rooted in August and the last ones 4–5 months later, mainly due to lack of elongation of the shoots. There was considerable variation in the percentage of cotyledons forming adventitious

Table 1. Shoot regeneration from cotyledons – genotypic and year to year variations.

Mother	% with Adventitious Shoots			
	1988	1991	1992	1993
Alma	68	44	39	53+[1]
Büttners	64	79	64	50+
Erika		59	53	42+
Hedelfinger		41		18+
Regina	50	90	12	25+
Valeska		57	8	52+
Van	40		50	35+

[1] Data up to 14/08/93.

Table 2. Embryo content and adventitious shoot regeneration in relation to the age of fruits at harvest in 1991.

Age of fruits (days)	% empty	No. of embryos	% Forming shoots
25–30	40	171	15
31–40	53	149	21
41–50	33	112	33
51–60	4	103	71
> 60	0	270	71

Table 3. Underdeveloped embryos/cotyledons with regeneration.

Mother	% with Regeneration[1]	
	1990	1991
Büttners	10	6
Mermat	31	
Regina		35
Stella	4	
Valeska	38	24
Van		33

[1] No. cultured in 1990: 185, 1991: 361.

Table 4. Age of fruit until excision of seed in relation to regeneration in 3 cultivars.

Mother	*'Flower → Excision'*	
	% of Ripe Fruit	% Regeneration of Seeds
Burlat	70	25
	77	41
	90–100	77
Regina	50	25
	57	45
	100	90
Sam	up to 60	22
	80	71
	100	91

shoots from year to year and between parental genotypes (Table 1). This was mainly due to differences in the quality of the embryo at harvest and to varying percentages of contamination in explants (Schmidt & Ketzel 1992).

Segregation for the 'albino' character was observed in many progenies, indicating heterozygosity for this character in cvs as Alma, Büttners Rote Knorpel, Erika, Hedelfinger, Lapins, Regina, Stella, Sunburst et al. Although some progenies clearly segregate 3:1 the majority had less than 10% albinos, suggesting that the mode of inheritance is not as simple as one recessive gene.

Of the hybrid seedlings raised from adventitious shoots plus the seedlings from normal germination in the greenhouse directly after harvest in 1990, 85% flowered in 1993 and 63% fruited.

Embryo rescue

In some years, especially under low light intensities in the greenhouse, there was a high percentage of prematurely dropping fruit. The younger these fruits were the higher was the percentage of 'empty' seeds (Table 2). Embryo-containing seeds could be induced either to germinate or to form adventitious shoots *in vitro*. Very young embryos, still transparent, were cultured on hormone free MS medium. For shoot formation cotyledons were transferred to induction medium when they became white. Altogether some 20% of the embryo containing-seeds could be induced to regenerate and produce plants (Tables 3 and 4). These seeds would not have germinated under normal conditions.

So-called 'genetic dwarfs' (Matthews 1970) are frequent in many sweet cherry progenies. It was obvious that they occurred at a much higher frequency among

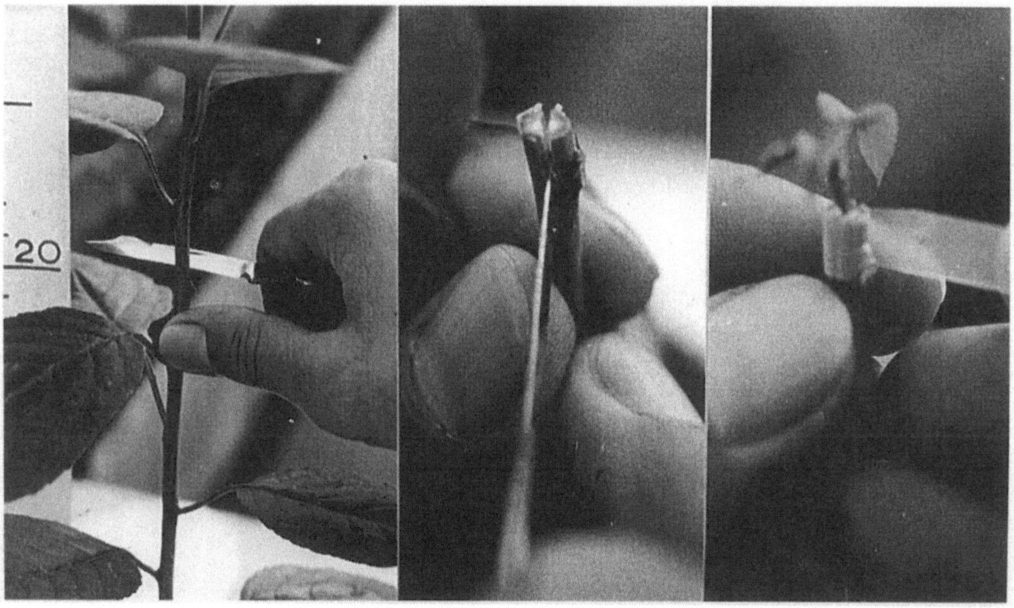

Fig. 1. Grafting procedure

plants regenerated from underdeveloped embryos than from normal seed, 45.0% versus 12.5% in 1991.

Grafting of microshoots

Sweet cherry cvs are usually grown on rootstocks. Dwarfing rootstocks, selected at Ahrensburg, induce early flowering. Some of the new generation of dwarfing rootstocks under test show incompatibility with specific cvs. A technique was developed to graft microshoots from *in vitro* culture directly onto rootstocks (Özzambak & Schmidt 1991). This has since been optimised and is now employed for hybrid seedling shoots so as to use the influence of a dwarfing rootstock from the very early life of the plant, inducing it to fruit early and to obtain some indication of compatibility relations between *Prunus avium* and a representative of the interspecific rootstock generation (473/10). The technique is demonstrated in Fig. 1.

From the seedlings grafted in 1992, 90% could be transferred to the field in spring 1993, altogether some 500 plants.

References

Matthews, P., 1970. The genetics and exploitation of dwarf seedlings in the sweet cherry. Proc. Eucarpia Fruit Breeding Symp., Angers, pp. 319–337.

Murashige, T. & F. Skoog, 1962. A revised medium for rapid growth and bioassays with tobacco tissue cultures. Physiol. Plant 15: 473–497.

Özzambak, E. & H. Schmidt, 1991. *In vitro* and *in vivo* micrografting of cherry (*Prunus avium* L.). Gartenbauwiss. 56: 221–223.

Schmidt, H. & A. Ketzel, 1992. Adventivsprossregeneration *in vitro* bei Kirschen. IV. Einsatz der Adventivsprossregeneration an Kotyledonen und 'embryo rescue' in der Kirschzüchtung. Gartenbauwiss. 58: 64–67.

Yang, H.Y., 1992. Untersuchungen zur *in vitro*-Regeneration, *in vitro*-Mutationsinduktion und Selektion bei Suesskirschen (*Prunus avium* L.). Thesis, Univ. Hannover.

H. Schmidt and M. Kellerhals (Eds.), Progress in Temperate Fruit Breeding, 385–388.

Rapid introduction of early season sweet cherry cultivars through *in vitro* micropropagation and virus elimination

Shmuel Zilkah[1], Evgenia Faingersh[1], Arie Rotbaum[1] & Adina Stein[2]
[1] *Institute of Horticulture, and* [2] *Institute of Plant Protection, Agricultural Research Organization, The Volcani Center, Bet Dagan 50250, Israel*

Key words: Prunus avium, quarantine, shoot culture, tissue culture

Summary

In vitro propagation techniques were used for the rapid introduction of new imported sweet cherry (*Prunus avium* L.) cultivars, 17 H 580, 40 E 50 and 106 EB 325. Media for bud establishment, bud shoot proliferation, shoot elongation and rooting were developed for each of them. The cultivars differed in their response to different media compositions. Indexing on several plant indicators under quarantine conditions found the cultivars 40 E 50 and 106 EB 325 to be virus infected. *In vitro* thermotherapy in combination with shoot-tip culturing was successfully applied. Virus-free cultures were rooted, acclimated and grown under greenhouse and field conditions.

Abbreviations: BA – 6-benzylaminopurine; GA_3 – gibberellic acid; IBA – indole-3-butyric acid; NAA – naphthaleneacetic acid

Introduction

One of the most important objectives for improvement of the sweet cherry industry in Israel is the extension of the marketing season. Early season cultivars have an economic importance for the development of new domestic and export markets. Sweet cherry (*Prunus avium* L.) cultivars 17 H 580, 40 E 50 and 106 EB 325 were selected by Zaiger's Inc. Genetics, USA. They were imported to Israel for potential use as early season cultivars.

Sweet cherry rootstocks and scion cultivars have been found difficult to propagate using hardwood cuttings (Feucht & Dausent 1976, Hartman & Brooks 1976). *In vitro* techniques have the potential to shorten significantly the time necessary for introduction of new imported plant varieties. Micropropagation could be performed concomitant to the indexing procedures of the quarantine station. Adequate proliferating material could be prepared *in vitro* and be readily used after certification of the cultivar for distribution. When virus infected material is identified by the quarantine procedures, *in vitro* heat therapy (Snir & Stein 1985, Stein et al. 1991) and shoot-tip culture could be used for virus elimination. In previous studies, cherry rootstocks (Jones & Hopgood 1979, Jones et al. 1984, Zilkah et al. 1992) as well as sweet cherry scion cultivars (Snir 1982a,b, Zilkah et al. 1993) were successfully micropropagated. The cultivars in the present study were new and had to be evaluated.

The present paper deals with the multiplication by tissue culture of three new sweet cherry cultivars to enable their rapid introduction.

Materials and methods

Cuttings with dormant buds were collected in December from one-year- old twigs of sweet cherry cultivars 17 H 570, 106 EB 325 and 40 E 50. The cuttings were incubated at 4 °C for 2–4 months until bud swelling. Short segments containing 1–2 buds were surface-sterilized by stirring in 80% isopropanol for 30 min. followed by rinsing with sterile distilled water. Each bud was dissected separately and transferred immediately to reducing solution containing 100 ppm dithiothreitol. Shoot-tips, about 1 mm length, including meristem-tip and several leaf premordia, were taken for culture initiation in 10 × 2.5 cm test-tubes, containing 10 ml of culture medium each and closed with plastic caps (Bellco-Kaput). The cultures were placed in a controlled environment room, with 16 h photoperiod at an irradiance intensity of 3 W/m^2 in the visible range and at a relative humidity of 50–60% at 25 °C.

Immersion of cv. 40 E 50 culture in a solution containing 50 mg/l rifampicin for 24 h significantly suppressed the endophytic bacterial contamination. When temporary treatments were not effective enough, some clones of this cultivar were continuously cultured on medium supplemented with 250 mg/l carbenicillin.

Virus elimination has been carried out through heat therapy (Stein et al. 1991) followed by shoot-tip culturing.

Results

Initiation and establishment

Dissected buds were transferred to different media already used for initiation of sweet cherry rootstocks (Zilkah et al. 1992). Tabachnic & Kester medium (Tabachnic & Kester 1977) supplemented with the growth regulators 1 mg/l IBA and 1 mg/l BA gave the best results for all three cultivars. Of the buds cultured for cvs 17 H 580, 106 EB 325 and 40 E 50, 14, 60 and 46%, respectively, developed to form a rosette with several leaves during the first 4–6 weeks in culture.

Proliferation, elongation and rooting

Established cultures were transferred to shoot proliferation media. Media were based on Murashige & Skoog (MS) medium (Murashige & Skoog 1962) and Almehdi & Parfitt (AP) medium (Almehdi & Parfitt 1986) with various combinations of IBA, BA and GA_3concentrations. The three cultivars responded differently to medium conditions. Cv. 17 H 570 proliferated at a rate of 3 shoots per subculture on MS medium supplemented with 1 mg/l IBA, 1 mg/l BA and 0.5 mg/l GA_3. On this medium, relatively elongated shoots were developed. No other medium for elongation was necessary and the cultures could be transferred directly to the rooting medium. The best rooting rate (> 90%) was obtained when the shoots were separately cultured on half strength MS medium containing 0.5 mg/l NAA.

In response to the same proliferating and elongating medium, cv. 106 EB 325 developed relatively elongated shoots but multiplied at a lower rate (2–2.5 per subculture at 4–5 weeks intervals). Shoots from this stage could also be transferred directly to rooting medium. For mass multiplication, however, shoots of cv. 106 EB 325 were transferred to AP medium containing 3.5 g/l agar, 1 g/l Gelrite, 0.01 mg/l IBA, 0.5 mg/l BA and 0.2 mg/l GA_3. The multiplication rate in this medium was > 3.5 shoots per subculture. However, the shoots did not elongate satisfactory (Fig. 1) and single shoots had to be transferred to MS elongating medium containing 1 mg/l IBA, 1 mg/l BA and 0.5 mg/l GA_3. High rooting rate (> 90%) was obtained by using half strength MS medium containing 0.5 mg/l NAA.

In contrast to the other two cultivars, shoots of cv. 40 E 50 could not be multiplied or elongated on MS medium. AP medium containing 3.5 g/l agar, 1 g/l Gelrite, 0.01 mg/l IBA, 0.5 mg/l BA and 0.2 mg/l GA_3 was a suitable proliferating medium (2.5–3 per subculture), but only about 20% of the developed shoots could be rooted. For elongation, the shoots were transferred to AP medium containing 0.01 mg/l IBA, 0.2 mg/l BA, 3.5 g/l agar and 1 g/l Gelrite. Over 90% of the shoots then developed sufficiently for root induction.

Cv. 40 E 50 was difficult to root. Various combinations of MS medium and the conventional rooting hormones NAA, IAA and IBA were tested. Fewer than 10% of shoots produced roots. IAA conjugates, which have been shown to act as rooting compounds (Wiesman et al. 1989), were tested too but with no significant success. After > 20 subcultures 40 E 50 started to form roots at an increasing rate (up to > 85%) on half strength of MS medium supplemented with 1 mg/l IBA. This medium gave better results than the rooting medium used for the other cultivars tested.

Virus elimination

During the development of the micropropagation methodology, the cultivars were indexed under the official quarantine procedure. Cv. 106 EB 325 was found virus-infected on *Prunus serrulata* cvs. Shirofugen and Kwanzan indicators. Cv. 40 E 50 were found infected on the latter indicators and also on *P. avium* cv. Bing and *P. persica* cv. Elberta. Cv. 17 H 580 was found infected with prunus necrotic ring spot virus detected by enzyme-linked immunosorbent assay (ELISA) as previously described (Snir & Stein 1985). *In vitro* thermotherapy in combination with shoot-tip culturing was applied. Plants were regenerated and indexed, and

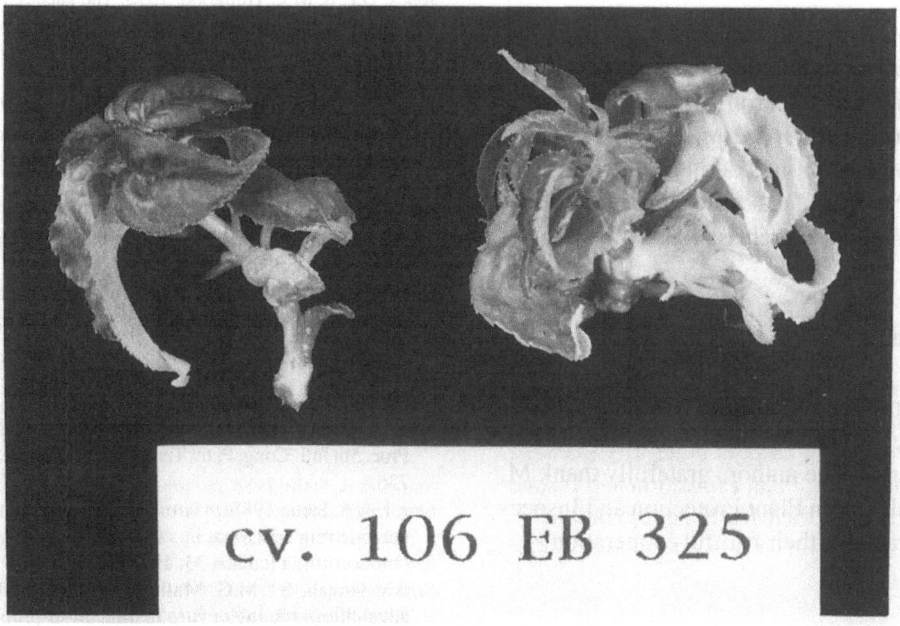

Fig. 1. Shoot culture of sweet cherry cv. 106 EB 325 grown on a single stage multiplication/elongating MS medium containing 1 mg/l IBA, 0.5 mg/l BA and 0.5 mg/l GA_3 (left). Same culture grown on multiplication AP medium containing 0.01 mg/l IBA, 0.5 mg/l BA and 0.2 mg/l GA_3 (right).

those found to be virus-free were clonally propagated. They were distributed as mother plants to nurseries and used in trial plots as a part of the introduction procedure.

Discussion

Although sweet cherry cultivars are difficult to propagate from hardwood cuttings (Feucht & Dausent 1976, Hartman & Brooks 1976), tissue culture methodology has opened new possibilities for proliferation of rootstock (Jones & Hopgood 1979, Jones et al. 1984, Zilkah et al. 1992) and scion (Snir 1982a,b, Zilkah et al. 1993) cultivars. Tabachnik & Kester medium gave the most satisfactory results for shoot establishment from buds for all the cultivars used in this study and also for other *P. avium* (Snir 1982a,b) or *P. avium* × *P. mahaleb* (Zilkah et al. 1992) cultivars. Endophytic bacterial contamination was a severe problem in establishing shoots from cv. 40 E 50 buds. No phenological variants were observed amongst plants derived from cv. 40 E 50 cultures grown on antibiotics.

Cultivar 40 E 50 was difficult to root. Only after > 20 subcultures was significant rooting obtained on a standard medium composition of 1/2 MS strength + 1 mg/l IBA. This phenomenon has been observed in other difficult to root species such as almond (Rugini & Verma 1982/83), apple (Sriskandarajah et al. 1982) and another early season sweet cherry cultivar, Burlat (unpublished data). We recently reported that difficult to root 40 E 50 cultivar was significantly different in IBA metabolism than easy to root cultivar (Epstein et al. 1993).

It is important to emphasize that the micropropagation procedures and thermotherapy treatments could be done routinely concomitant to the indexing procedures taken in the quarantine station. Certified cultivars officially released for distribution, could be immediately available for large-scale preparation of mother-plants for experimental and semi-commercial trials. These tissue culture approach is most advantageous for rapid introduction of difficult to propagate rootstocks such as sweet cherry (Zilkah et al. 1992). If the indexing and micropropagation were reliably done by the exporting country, then the movement among countries of new germplasm as pathogen-free tissue cultures would be much easier and faster.

Conclusions

Micropropagation methodologies have been developed for three potential early season sweet cherry cultivars concomitantly to the indexing procedures taken in quarantine. The *in vitro* methodology enabled rapid production of virus-free plants for testing in introductory experimental plots and release to nurseries.

Acknowledgements

Contribution from the Agricultural Research Organization, The Volcani Center, Bet Dagan, Israel. No. 1429–E, 1994 series. The authors gratefully thank M. Weiss and N. Malca from Plant Protection and Inspection Services, Israel, for their fruitful cooperation.

References

Almehdi, A.A. & D.E. Parfitt, 1986. *In vitro* micropropagation of peach: Propagation of Lovell and Nemaguard peach rootstocks. Fruit Var. J. 40: 12–17.

Epstein, E., S. Zilkah, E. Faingersh & A. Rotbaum, 1993. Transport and metabolism of indole-3-butyric acid in sterile easy- and difficult-to-root cuttings of sweet cherry (*Prunus avium* L.). Acta Horticulturae 329: 292–295.

Feucht, W. & B. Dausent, 1976. Root induction *in vitro* of easy-to-root *Prunus pseudocerasus* and difficult-to-root *Prunus avium*. Scientia Hortic. 4: 439–454.

Hartman, H.T. & R.M. Brooks, 1976. Propagation of Stockton Morello cherry rootstock by softwood cuttings under mist sprays. Proc. Amer. Soc. Hort. Sci. 71: 127–134.

Jones, O.P. & M.E. Hopgood, 1979. The successful propagation *in vitro* of two rootstocks of *Prunus*: The plum rootstock Pixy (*P. instititia*) and the cherry rootstock F12/1 (*P. avium*). J. Hort. Sci. 54: 63–66.

Jones, O.P., G.A. Gayner & R. Watkins, 1984. Plant regeneration from callus tissue cultures of the cherry rootstock Colt (*Prunus avium* × *P. pseudocerasus*) and the apple rootstock M. 25 (*Malus pumila*). J. Hortic. Sci. 59: 463–467.

Murashige, T. & F. Skoog, 1962. A revised medium for rapid growth and bioassay with tobacco tissue cultures. Physiol. Plant 15: 473–497.

Rugini, E. & D. Verma, 1982/83. Micropropagation of difficult-to-propagate almond (*Prunus amygdalus*, Batsch) cultivar. Plant Sci. Lett. 28: 273–281.

Snir, I., 1982a. *In vitro* propagation of sweet cherry cultivars. HortScience 17: 192–193.

Snir, I., 1982b. *In vitro* micropropagation of sweet cherry cultivars. Proc. 5th Intl. Cong. Plant Tissue & Cell Culture, Tokyo, pp. 735–736.

Snir, I. & A. Stein, 1985. *In vitro* detection and elimination of Prunus necrotic ring spot virus in sweet cherry (*Prunus avium*). Rivista Ortoflorofrutta Italiana 33: 191–194.

Sriskandarajah, S., M.G. Mullins & Y. Nair, 1982. Induction of adventitious rooting *in vitro* in difficult-to-propagate cultivars of apple. Plant Sci. Lett. 24: 1–9.

Stein, A., S. Spiegel, G. Faingersh & S. Levy, 1991. Responses of micropropagated peach cultivars to thermotherapy for the elimination of prunus necrotic ringspot virus. Ann. App. Biol. 119: 265–271.

Tabachnick, L. & D.E. Kester, 1977. Shoot culture for almond and almond-peach hybrid clones *in vitro*. HortScience 12: 545–547.

Wiesman, Z., J. Riov & E. Epstein, 1989. Characterization and rooting ability of indole-3-butyric acid conjugates formed during rooting of mung bean cuttings. Plant Physiol. 91: 1080–1084.

Zilkah, S., E. Faingersh & A. Rotbaum, 1992. *In vitro* propagation of three M × M cherry rootstocks. Acta Horticulturae 314: 93–97.

Zilkah, S., E. Faingersh, A. Rotbaum & A. Stein, 1993. *In vitro* micropropagation of indicator plants for indexing prunus necrotic ring spot virus. Acta Horticulturae 336: 121–125.

H. Schmidt and M. Kellerhals (Eds.), Progress in Temperate Fruit Breeding, 389–393.

Tissue culture approaches to peach improvement

V. Ognjanov, Dinka Vujanić-Varga & Ksenija Macet
Faculty of Agriculture, 21000 Novi Sad, Yugoslavia

Key words: Peach, embryo culture, micropropagation, genetic transformation

Summary

In-ovule embryo culture was used for obtaining seedlings from immature peach embryos 1–4 mm in length. Rapid micropropagation of Redhaven cultivar and B 52/2 vineyard peach selection was achieved by shoot tip proliferation. The two-stage rooting system, with a short auxin exposure followed by transfer to growth regulator-free half-strength basal media, proved to work much better than any other system published so far. Cotyledon regeneration was combined with *Agrobacterium* mediated genetic transformation. A clear difference in fluorescence between negative controls and the regenerants obtained after cotyledon inoculation with *A. tumefaciens* carrying the GUS gene was detected after 24 h indicating the production of chimeric peach plants.

Abbreviations: BA – benzyladenine; Cb – carbenicillin; GA_3 – gibberellin; GUS – β-glucuronidase; IAA – indole-3-acetic acid; IBA – indole-3-butyric acid; Kan – kanamycin; KIN – kinetin; NAA – 1-naphthaleneacetic acid; TDZ – N-phenyl-N′-1,2,3-thiadiazol-5-ylurea (Thidiazuron)

Introduction

One of the earliest uses for tissue cultures was embryo culture of *Prunus* immature embryos. The use of this techniques has subsequently been exploited by many stone fruit breeders to rescue embryos from early-ripening genotypes because the flesh ripens before the seed matures (Hesse 1975). Current procedures for peach embryo culture work well only with embryos with greater than 60% fill (Ramming 1985). In-ovule embryo culture allows embryo enlargement and accumulation of reserve materials *in vitro*.

Meristems and shoot tips have been the primary explants for peach micropropagation. There are several reports of *in vitro* propagation of peach cultivars and rootstocks but only in a few cases more than 50% rooting have been achieved (Hammerschlag et al. 1987, Vertesy 1980).

Several new methods and techniques have been developed in recent years which, in addition to their importance for fundamental research in the fields of genetics, biochemistry, physiology etc., may become exceedingly important in plant breeding. Conventional peach breeding is long-term commitment and is restricted by a very narrow genetic base. New approaches are needed to speed up the production of new agriculturally useful genotypes, especially for disease and virus resistance.

In peach, adventitious shoots have been regenerated from immature embryos (Hammerschlag et al. 1985) and cotyledons (Mante et al. 1989). By exposing callus initiated from immature embryos to bacterial toxins, peach plants resistant to *Xanthomonas campestris* pv. *pruni* and *Pseudomonas syringae* pv. *syringae* were obtained by Hammerschlag & Ognjanov (1990).

Reports of *Agrobacterium*-mediated transformation of mature and immature peach tissues have been published recently but transgenic peach plants have not so far been obtained (Hammerschlag et al. 1989, Martin et al. 1990, Scorza et al. 1990, Smigocki & Hammerschlag 1991).

In this study some aspects of in-ovule embryo culture, micropropagation and *Agrobacterium*-mediated transformation are discussed. The long-term objective is the incorporation of tissue culture approaches into conventional breeding programmes for disease and virus resistance.

Methods and materials

In-ovule embryo culture

Ovules from open pollinated cultivar Goldcrest were used. Ovules were removed from the fruit when embryos were less than 1 mm, 1–2 mm and 4 mm in length and cultivated on Stewart & Hsu (1977), Murashige & Skoog (MS) and White basal media in both liquid and solid form. The effect of sucrose concentration, NH_4^+ ions in form of $(NH_4)_2SO_4$, IAA and GA_3 treatment were examine. After 3 weeks the embryos were dissected, measured and transferred to test tubes containing Brooks & Hough (1958) media and stratified at 3 °C for 90 days to induce germination.

Micropropagation

Shoot cultures of cv. Redhaven and B 52/2 vineyard peach had been maintained on standard proliferation and elongation media described by Hammerschlag et al. (1987). Shoots longer than 10mm were transferred onto three rooting media consisting of 1/2-strength MS basal media supplemented with:

- R_1 – 0.8 μM NAA,
- R_2 – 5 μM IBA and
- R_3 – 15 μM IBA, and subjected to the three different treatments:
- T_1 – Shoots were three weeks on initial rooting media.
- T_2 – Shoots were three days on initial rooting media and than transferred to a 1/2 strength MS growth regulator-free medium.
- T_3 – Shoots were six days on initial rooting media and than transfered to a 1/2 strength growth regulator-free media.

Genetic transformation

Immature cotyledons used for all transformation studies were from open-pollinated Redhaven seeds harvested 102 and 109 days after anthesis. The procedure for regeneration of peach plants from cotyledons was as described by Mante et al. (1989). Explants were inoculated with *Agrobacterium tumefaciens* strains LBA4404 and BIN 19 containing the GUS and 35S GUS INT genes respectively. Culture media, growth of bacterial strains and cocultivation with *Agrobacterium* were carried out as described by James at al. (1989). Three different strategies for kanamycin selection were used – infection without selection, limited selection (10 and 20 days) and delayed selection (7, 14 and 21 days). The fluorogenic assay of GUS activity was performed as described by Jefferson (1987).

Results

In-ovule embryo culture

Successful rescue of fertilized ovules was dependent on the length of time following fertilization before *in vitro* culture. Only embryos 1–2 mm and greater in length were successfully cultured to maturity (Table 1). The increase in size of < 1 mm, 1–2 mm and 4 mm stage embryos after 3 weeks in culture was nearly the same on both solid and liquid media, about 32, 512 and 82%, respectively. Embryos less than 1 mm did not increase in size even with IAA and GA_3 treatment. The addition of 6% sucrose and ammonium ions to the culture medium were very effective for rapid growth and enhanced viability of the embryos. White basal medium was most effective. Satisfactory germination (up to 100%) and balanced seedling shoot and root growth were obtained only from the embryos cultured on liquid media.

Rooting of micropropagated shoots

Attempts to induce direct rooting by exposing shoots continuously to media containing NAA or IBA were not satisfactory (Table 2). Usually it resulted in abundant callusing with or without small stocky root formation.

Three days exposure of shoots to the NAA rooting media gave the best results. This treatment induced the highest rooting percentage (up to 100%) and the best balanced root-shoot growth. After 4 weeks no sign of callusing was observed. Roots were thin and long. The acclimatization of these plantlets was most successful because the shoot apex was always actively growing at the end of the rooting phase. Six-day treatment reduced the rooting to 46 and 49% respectively.

Addition of 15 μM IBA to the basal rooting media at all treatments induced the formation of abundant callus with or without numerous short, stocky roots. When IBA concentration was reduced to 5 μM only

Table 1. Seedlings produced from cultured ovules of Goldcrest peach.

Media	Sucrose (mg/l)	Growth regulators (mg/l)			Embryo Length After 3 Weeks in culture (mm)						Germination (%)					
					Liquid Media			Solid Media			Liquid Media			Solid Media		
		IAA	GA_3	KIN	< 1*	1–2	> 4	< 1	1–2	> 4	< 1	1–2	> 4	< 1	1–2	> 4
Murashige & Skoog	60	–	–	–	0.9	7.1	11.0	0.8	7.5	10.8	0	38	98	0	20	79
Murashige & Skoog	60	0.9	0.2	0.1	0.8	9.0	12.6	0.8	6.1	11.2	0	80	100	0	26	84
White	60	–	–	–	0.9	9.6	11.1	0.9	5.9	11.6	0	100	96	0	17	83
White**	60	–	–	–	0.7	9.5	11.4	0.8	6.0	10.4	0	58	89	0	22	79
Stewart & Hsu	20	0.9	0.2	0.1	0.8	6.4	10.3	0.8	5.9	10.7	0	33	91	0	12	87
Stewart & Hsu	60	0.9	0.2	0.1	1.0	8.1	12.8	0.8	7.4	11.5	0	65	100	0	15	91

[1] < 1 mm, 1–2 mm, > 4 mm – Initial embryo length.

[2] Addition of 100 mg/l NH_4^+.

Table 2. Rooting of Redhaven and vineyard peach B 52/2 shoots *in vitro*.

Treatment		Redhaven					Vineyard Peach B52/2				
		No. explants	Without change %	Callus + Roots %	Callus %	Roots %	No. explants	Without change %	Callus %	Callus = Roots %	Roots %
T_1	R_1	24	21	39	9	31	31	24	27	16	33
	R_2	22	0	49	51	0	22	11	35	54	0
	R_3	22	0	0	100	0	28	2	21	77	0
T_2	R_1	22	19	0	0	81	26	0	0	0	100
	R_2	21	0	36	33	31	28	6	61	0	33
	R_3	28	0	84	16	0	32	0	85	15	0
T_3	R_1	26	0	0	54	46	32	0	0	51	49
	R_2	22	2	38	32	28	28	0	34	46	20
	R_3	24	0	64	31	5	32	0	28	72	0

T_1 – Shoots were three weeks on initial rooting medium;

T_2 – Shoots were three days on initial rooting medium and transferred to 1/2 – strength MS growth regulator – free medium;

T_3 – Shoots were six days on initial rooting medium and then transferred to a 1/2 – strength MS growth regulator – free medium.

R_1 – 1/2 – strength MS supplemented with 0.8 μM NAA;

R_2 – 1/2 – strength MS supplemented with 5 μM IBA;

R_3 – 1/2 – strength MS supplemented with 15 μM IBA.

three days exposure to the rooting media gave up to 28% of rooted plants.

Well-rooted plantlets were successfully transferred to soil and had normal field performance in the nursery.

Genetic transformation

Cotyledon regeneration system in peach was combined with *Agrobacterium tumefaciens* mediated gene transfer techniques. Over 1000 peach cotyledons were infected with two strains of *Agrobacterium* carrying GUS and GUS intron genes using three different strategies for Kanamycin selection; infection without selection, limited selection and delayed selection (Table 3).

Infection with *Agrobacterium* reduced the percentage of cotyledons regenerating compared to uninfected controls (data not shown). Regeneration frequency was from 10% to 20% in treatments without kanamycin as selective agent. Exposure to kanamycin longer than 10 days and in concentrations over 20 μg/l reduced

Table 3. Regeneration of shoots from immature peach cotyledons and β -glucuronidase-specific activity in cotyledon callus 45 days after inoculation, infected with two strains of *Agrobacterium tumefaciens* carrying the marker gene β-glucuronidase.

Treatments		MS +		*Agrobacterium Tumefaciens*							
				Shoot Regeneration				GUS Activity in Callus			
		+		%				%			
		Kan	Cb	GUS		GUS-intron		GUS		GUS-intron	
		mg/l		I	II	I	II	I	II	I	II
Without selection	–	–	–	10.0	20.0	16.0	16.6	0.0	5.3	2.8	0.0
Limited	10 days	10	250	13.3	12.5	0.0	20.0	14.9	16.5	12.3	13.3
selection	20 days	10	250	11.7	0.0	12.5	8.7	16.3	16.2	15.3	14.9
Limited	10 days	20	250	12.3	11.1	14.3	16.7	27.3	33.2	26.8	29.1
selection	20 days	20	250	10.0	0.0	0.0	0.0	35.6	43.6	28.3	39.1
Limited	10 days	30	250	10.0	20.0	0.0	11.7	0.0	0.0	0.0	0.0
selection	20 days	30	250	0.0	0.0	0.0	0.0	0.0	0.0	0.0	0.0
Delayed	7 days	10	0	0.0	17.5	0.0	17.1	12.4	14.1	9.5	10.8
selection	14 days	20	0	11.1	11.0	15.5	15.8	5.6	6.5	4.4	2.9
	21 days	30	0	0.0	9.0	0.0	0.0	0.0	1.9	0.0	3.1

I – Fruit collected 102 days after anthesis, cultured on basal media supplemented with 1 mg/l TDZ.
II – Fruit collected 109 days after anthesis, cultured on basal media supplemented with 2 mg/l TDZ.

or inhibited the regeneration. The frequency of positive fluorogenic GUS assay on callus was from 2.9% to 35.6% and in most cases changed from non-fluorescent purple to intensive blue colour under u.v. within 15 min (Table 3). From 143 shoots we isolated 5 candidates expected to carry the GUS gene. In all five cases only a slow increase in fluorescence was observed, reaching bright blue colour after 24 h.

Discussion

Embryo culture is necessary to produce seedlings from early-ripening peach genotypes because the flesh ripens before the embryo accumulates enough oil as reserve materials for germination. When extremely early genotypes are used as female parents in crosses, like Goldcrest, Early Crest and Mayfire, in-ovule embryo culture has to be used. A total of 750 seedlings was raised, out of which 35 were propagated for further testing. Transgressive segregation of genotypes was observed in the progenies, advancing maturity dates for up to 10 days.

The delay in utilizing micropropagation for peach rootstocks and scion cultivars is due to the scarcity of information on *in vitro* rooting and phenotypic and genotypic stabilities of these genotypes after culture. Peach is considered to be a difficult-to root species. Similar positive effects of a two-stage process on rooting were noted in apple by James & Thurbon (1979). Successful clonal propagation shows that orchards on genetically uniform rootstocks can also be established using vineyard peach selections. In Yugoslavia, vineyard peach can be found as a seedling population. Seeds are used as the main source for rootstocks and are exported to other European countries for the same purpose.

Our preliminary results on peach transformation showed that completely transgenic peach plants have thus far not been obtained. We have demonstrated regeneration of peach plants in which probably some of the cells are expressing a foreign gene introduced by *Agrobacterium tumefaciens*. The regenerated plants may be chimeras composed of transformed and untransformed cells.

References

Brooks, H.J. & F.L. Hough, 1958. Vernalization studies with peach embryos. Proc. Amer. Soc. Hort. Sci. 71: 95–102.

Hammerschlag, A.F., B. Bauchan & R. Scorza, 1985. Regeneration of peach plants from callus derived from immature embryos. Theor. Applied Genet. 70: 248–251.

Hammerschlag, A.F., G. Buchan & R. Scorza, 1987. Factors Influencing In vitro multiplication and rooting of peach cultivars. Plant Cell Tissue Organ Culture 8: 235–242.

Hammerschlag, A.F., L.D. Owens & C.A. Smigocki, 1989. *Agrobacterium*-mediated transformation of peach cells derived from mature plants that were propagated *in vitro*. J. Amer. Soc. Hort. Sci. 114: 508–510.

Hamerschlag, A.F. & V. Ognjanov, 1990. Somaclonal variation in peach: Screening for resistance to *Xanthomonas campestris* pv. *pruni* and *Pseudomonas syringae* pv. *syringae*. Acta Horticulturae 280: 403–408.

Hesse, C.O., 1975. (The breeding of) peaches. In: J. Janick & J.N. Moore (Eds) Advances in fruit breeding, pp. 285–335. Purdue Univ. Press, West Lafayette 4, Indiana.

James, O.J. & I.J. Thurbon, 1979. Rapid *in vitro* rooting of the apple rootstocks. J. Hort. Sci. 54: 309–311.

James, J.D., J.A. Passey, J.D. Barbara & M. Bevan, 1989. Genetic transformation of apple (*Malus pumila* Mill.) using a disarmed Ti-binary vector. Plant Cell Rep. 7: 658–661.

Jefferson, R.A., 1987. Assaying chimeric genes in plants: the GUS gene fusion system. Plant Mol. Biol. Rep. 5: 387–405.

Mante, S., R. Scorza & J. Cordts, 1989. Plant regeneration from cotyledons of *Prunus persica*, *Prunus domestica* and *Prunus cerasus*. Plant Cell Tissue Organ Culture 19: 1–11.

Martin, G.C., A.N. Miller, L.A. Castle, J.W. Morris, R.O. Morris & A.M. Dandekar, 1990. Feasibility studies using β-glucoronidase as a gene fusion marker in apple, peach and radish. J. Amer. Soc. Hort. Sci. 115: 686–691.

Ramming, D.W., 1985. In-ovule embryo culture of early-maturing *Prunus*. HortScience 20: 419–420.

Scorza, R., P.H. Morgens, J.M. Cords, S. Mante & A.M. Callahan, 1990. *Agrobacterium*-mediated transformation of peach (*Prunus persica* L. Batsch) leaf segments, Immature embryos and long-term embryogenic callus. In Vitro Cell Dev. Biol 26: 829–834.

Smigocki, C.A. & A.F. Hammerschlag, 1991. Regeneration of plants from peach embryo cells infected with a shooty mutant strain of *Agrobacterium*. J. Amer. Soc. Hort. Sci. 116: 1092–1097.

Stewart, J.M. & C.L. Hsu, 1977. In-ovulo embryo culture and seedling development of cotton (*Gossypium hirsutum* L.). Planta 137: 113–117.

Vertesy, J., 1980. *In vitro* propagation of *Prunus persica* and *P. persica-davidiana* shoot tips in order to get virus-free plants. Acta Phytopath. Acad. Sci. Hung. 15: 261–264.

H. Schmidt and M. Kellerhals (Eds.), Progress in Temperate Fruit Breeding, 395–398.

Studies on plant regeneration from tissue-derived callus in *Fragaria* × *ananassa* Duch.

Valentina Isac, Aurel N. Popescu & Mihail Coman
Fruit Research Institute, 0300 Pitesti-Maracineni, Romania

Key words: Fragaria × *ananassa*, organogenesis, plant regeneration, tissue culture

Summary

The ability of the strawberry cvs. Aiko, Dana, Gorella and Premial to regenerate plants by organogenesis from leaf blade and petiole explants obtained from *in vitro* micropropagated plantlets was investigated. After an initial dark treatment of two weeks, both types of explants cultured on Murashige-Skoog medium supplemented with 2.4-dichlorophenoxyacetic acid and 6-benzyladenine were transferred to light. Maximum callus formation was achieved in a range from 94.2 % (Dana) to 100% (Premial). The differentiation of plants was induced on the same medium in eight to sixteen weeks. With all tested cultivars the regeneration process started earlier with leaf derived callus. Shoot organogenesis from both types of explants was strongly influenced by cytokinin concentration. The highest number of regenerated shoots per explant was obtained in the treatments with low auxin and high cytokinin concentration. This study revealed that the organogenetic potential in *Fragaria* × *ananassa* Duch. is highly dependent on genotype and to a certain extent on explant type.

Introduction

Strawberry (*Fragaria* × *ananassa* Duch.) is an important commercial fruit crop in most temperate regions of the world. Yield, fruit size and quality have been improved through conventional breeding. The prospects of somaclonal variation or transfer of genes coding for desired agronomic traits while conserving the other characteristics of an accepted cultivar has attracted strawberry breeders in recent years. Plant regeneration from somatic tissue has become an essential requirement for strawberry improvement by *in vitro* techniques and genetic engineering. Plant regeneration from various somatic tissues by organogenesis or by somatic embryogenesis has been reported in strawberry (Nishi & Ohsawa 1973, Lee & De Fossard 1975, Asahira & Kano 1977, Lee & Kim 1979, Wang et al. 1984, Yurgalevich et al. 1985, Foucault & Letouze 1987, Simon et al. 1987, Jones et al. 1988, Liu & Sanford 1988, Nehra et al. 1989, 1991, Miller & Chandler 1990). From most of these reports there is clear evidence of major influences of genotype and type of explant on regeneration ability.

This study was designed to investigate the organogenetic potential of leaf and petiole explants of the strawberry cultivars Aiko, Dana, Gorella and Premial.

Material and methods

Leaf (0.6 cm^2) and petiole (1.0 cm in lengh) explants excised from *in vitro* grown shoots of the cultivars Aiko, Dana, Gorella and Premial were cultured on Murashige-Skoog basal medium supplemented with 3.0% sucrose, 7.0% Noble-Difco agar, 4.5 μM of 2,4-dichlorophenoxyacetic acid (2,4-D) and 4.4 μM, 13.2 μM and 22.0 μM of 6-benzyladenine (BA), respectively. The pH of the culture medium was adjusted to 5.8 before adding agar. For each treatment there were 20 replications with 6 explants in each replicate. Leaf explants were placed with their adaxial surface in contact with the culture medium, according to Nehra et al. 1989. After two weeks at 20 °C in darkness, cultures were transferred to light (50 $\mu E\ m^{-2}\ s^{-1}$) at 26 °C under a 16/8 hours photoperiod to promote callus formation.

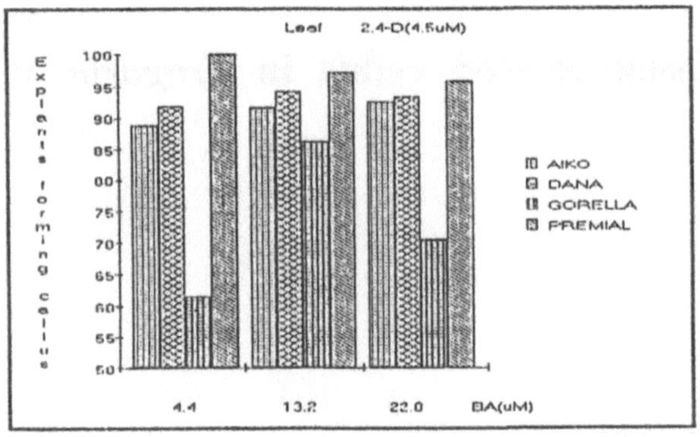

Fig. 1. Callus formation from leaf explants in response to 4.5 UM 2,4-D and different concentrations of BA.

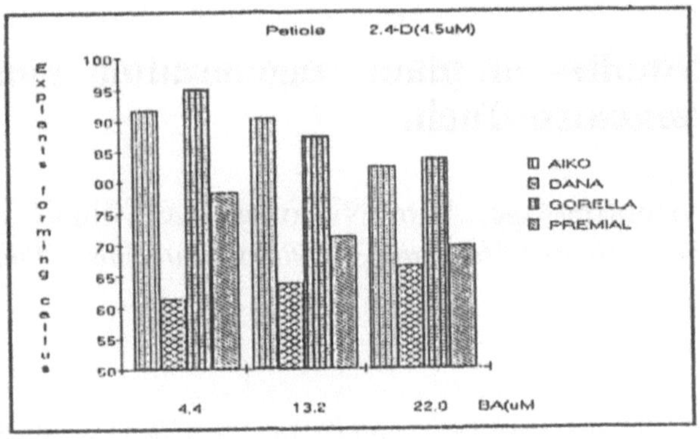

Fig. 2. Callus formation from petiole explants in response to 4.5 UM 2,4-D and different concentrations of BA.

In a second step leaf and petiole explants from Aiko and Premial were subcultured on media supplemented with 9.0 μM of 2,4-D while maintaining the range of BA concentrations. The aim was to investigate the influence of an increased auxin level on callus formation and plant regeneration. Callus initiation and the rate of proliferation in the leaf and petiole cultures were scored weekly. Shoot regeneration was also scored weekly, starting in the 6th week from the establishment of the cultures.

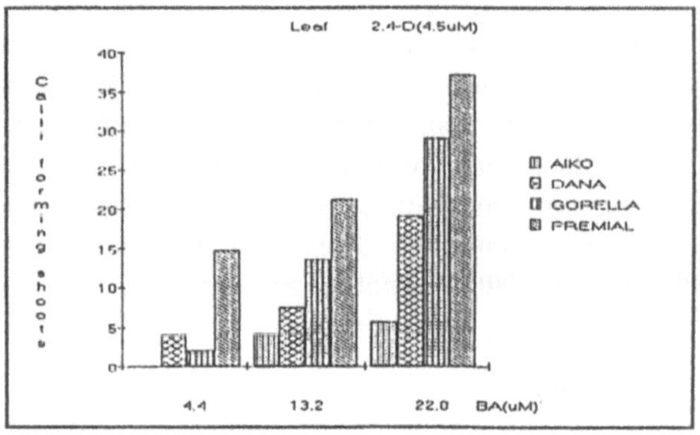

Fig. 3. Effect of 4.5 UM 2,4-D and different concentrations of BA on shoot regeneration from leaf-derived callus.

Results and discussion

None of the strawberry cultivars tested exhibited callus initiation within the two weeks of dark treatment. The callus formation occured three weeks after establishment of cultures in the cut edge of the leaves and petioles and gradually covered the entire explant.

The ability to produce callus was significantly influenced by genotype and to a certain extent by explant type. In treatments with 4.5 μM 2,4-D cv. Aiko exhibited a high potential of callus formation for both explant types tested whereas cvs. Dana and Premial showed a high percentage of callus only for leaf explants (Figs. 1 and 2). With cv. Gorella the best response for callus induction was obtained in cultures with petiole explants (Fig. 2).

In general, a significantly lower percentage of explants forming callus was recorded with increased BA concentration for leaf, but especially for petiole cultures (Figs. 1 and 2).

In the second step of this experiment, an increased concentration of 2,4-D (9.0 μM) resulted in a higher

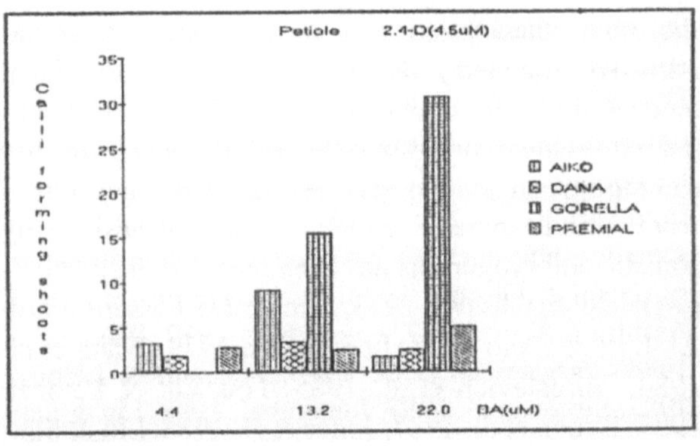

Fig. 4. Effect of 4.5 UM 2,4-D and different concentrations of BA on shoot regeneration from petiole-derived callus.

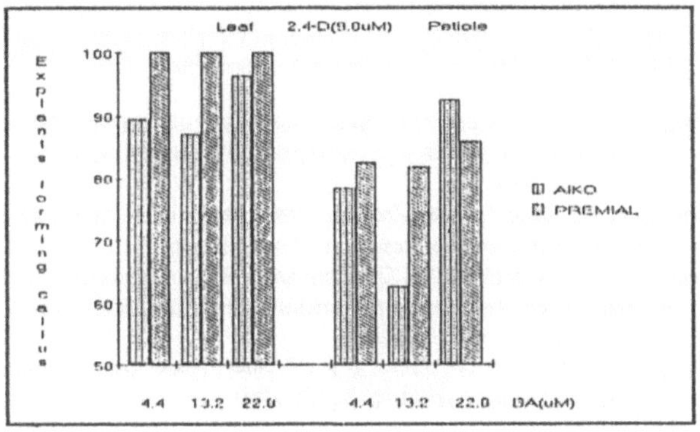

Fig. 5. Callus formation from Aiko and Premial leaf and petiole explants in response to 9.0 UM 2,4-D and different concentrations of BA.

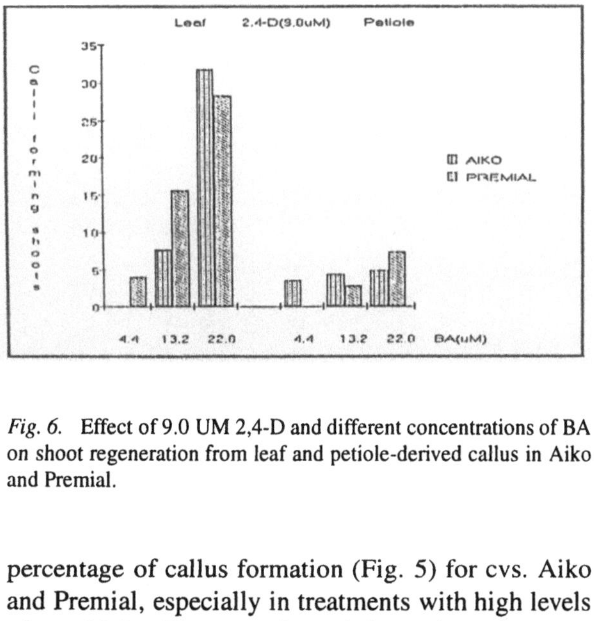

Fig. 6. Effect of 9.0 UM 2,4-D and different concentrations of BA on shoot regeneration from leaf and petiole-derived callus in Aiko and Premial.

percentage of callus formation (Fig. 5) for cvs. Aiko and Premial, especially in treatments with high levels of cytokinin. However, in petiole explants from cv. Premial, a decrease in callus formation was observed when cultured on medium with 9.0 μM of 2,4-D. A combination of 9.0 μM 2,4-D and 22.0 μM BA proved to be the most favourable for obtaining high percentage of callus formation, with a maximum which ranged from 96.4% in cv. Aiko to 100% in cv. Premial (Fig. 5).

The four strawberry varieties differed in response for shoot regeneration. The differentiation of shoots was induced on the original medium, without subculturing of leaf and petiole derived callus. The regeneration started earlier in compact yellow-green leaf-derived callus for all the tested combinations of growth regulators and the ability to regenerate shoots from both explant types was significantly influenced by the cytokinin (BA) level in the medium. A combination of 4.5 μM 2,4-D and 22.0 μM BA proved to be the best treatment for shoot regeneration, but in some calli regeneration also occured with low frequencies at 4.4 and 13.2 μM BA. The cv. Gorella exhibited a high regeneration potential in all treatments, with a maximum of 29.1% for leaf-derived callus, and 30.6% for petiole-derived callus (Figs. 3 and 4). In cv. Premial shoots were formed from leaf-derived callus in all hormone combinations tested, with the highest percent of regeneration (37.2%) associated with the highest BA concentration (Fig. 3).

In the second step of the experiment, an increased 2,4-D concentration (9.0 μM) resulted in a lower ability for shoot regeneration in cv. Premial but a significantly higher percent of regeneration (31.8%) was obtained in cv. Aiko (Fig. 6).

Generally, petiole-derived calli showed a lower organogenetic potential as compared to leaf-derived calli. In most of the auxin-cytokinin combinations tested, shoot regeneration from petiole derived calli occured later in comparison with leaf-derived calli. However, it is important to mention that, without subculturing, petiole-derived callus retained its ability to regenerate shoots for a longer period as compared with leaf derived callus. On culture medium with 4.5 μM 2,4-D and 4.4 μM BA, shoot formation from this explant type was recoreded even after 6 months.

It was noted that a high level of 2,4-D (0.9 μM) associated with low concentration of BA (4.4 μM) proved to be a poor hormonal combination for shoot regeneration from both leaf and petiole derived callus. Thus, only a high number of roots were regenerated. Furthermore, the callus obtained on medium with such a combination of growth regulators, gradually turned brown and stopped its growth in a short time.

All regenerated shoots from strawberry leaf and petiole derived calli had normal growth. They were successfully rooted and subsequently transferred into the greenhouse. This regeneration procedure may be useful for strawberry improvement through somaclonal variation.

References

Asahira, T. & Y. Kano, 1977. Shoot formation from cultured tissue strawberry fruits. J. Japan. Soc. Hort. Sci. 46: 317–324.

Foucault, C. & R. Letouze, 1987. *In vitro*: Régénération de plants de fraisier a partir de fragments de pétioles et de bourgeons floraux. Biologia Plantarum 29: 409–414.

Jones, O.P., B.J. Waller & M.G. Beech, 1988. The production of strawberry plants from callus cultures. Plant Cell Tissue Organ Culture 12: 235–241.

Lee, E.C.M. & R.A. de Fossard, 1975. Regeneration of strawberry plants from tissue cultures. Proc. Int. Plant Propag. Soc. 25: 277–285.

Liu, Z.R. & Y.R. Kim, 1979. Shoot formation in tissue culture of petiole and leaf blade segments of strawberry. Stud. Inst. Hort. Kyoto Univ. 9: 78–82.

Lee, Y.B. & J.C. Sanford, 1988. Plant regeneration by organogenesis from strawberry leaf and runner tissue. HortScience 23: 1057–1059.

Miller, R.A. & C.K. Chandler, 1990. Plant regeneration from excised cotyledons of mature strawberry achenes. HortScience 25: 569–571.

Nehra, N.S., C. Stushnoff & K.K. Kartha, 1989. Direct shoot regeneration from strawberry leaf disks. J. Amer. Soc. Hort. Sci. 114: 1014–1018.

Nehra, N.S., K.K. Kartha & C. Stushnoff, 1991. Nuclear DNA content and isozyme variation in relation to morphogenic potential of strawberry (*Fragaria* × *ananassa*) callus cultures. Can. J. Bot. 69: 239–244.

Nishi, S. & K. Ohsawa, 1973. Mass production method of virus-free strawberry plants through meristem callus. Japan. Agr. Res. Quart. 7: 189–194.

Simon, I., E. Racz & J.M. Zatyko, 1987. Preliminary notes on somaclonal variations of strawberry. Fruit Sci. Rep. 14: 151–154.

Wang, D., W.P. Wergin & R.H. Zimmerman, 1984. Somatic embryogenesis and plant regeneration from immature embryos of strawberry. HortScience 19: 71–72.

Yurgalevich, C.M., H.W. James & K.C. Chee, 1985. Somaclonal variation in *Fragaria*. HortScience 20: 450.

H. Schmidt and M. Kellerhals (Eds.), Progress in Temperate Fruit Breeding, 399–402.

In vitro androgenesis in apple: Induction, regeneration and ploidy level

Monika Höfer
Federal Centre of Breeding Research on Cultivated Plants, Institute of Fruit Breeding, Pillnitzer Platz 2, D-01326 Dresden, Germany

Key words: Anther culture, apple, embryogenesis, haploid, *Malus* × *domestica*

Summary

Status of haploid research in apple via anther culture and problems involved are briefly discussed. The anther culture response of 13 *Malus* genotypes has been analysed. The rate of embryoid induction was very low between 0.4 and 0.05% and occured only in 7 genotypes. Regeneration of shoots was obtained via formation of adventitious shoots from complexes of secondary embryoids on a medium supplemented with 0.1 mg/l TDZ. The assessed ploidy level of the embryoids and shootlets showed 24% haploid and diploid structures.

Introduction

In comparison with other plant species the possibilities for the induction of haploids in apple by *in vitro* androgenesis have been less studied and the success has been rather limited.

In 1981 and 1984 Chinese scientists (Fei & Xue 1981, Xue & Niu 1984) reported, for the first time, regeneration of androgenic plants of 6 apple cultivars. Johansson et al. (1987), Zhang et al. (1987) and our own report (Höfer, unpublished) list the details for the induction of androgenic embryoids of different apple cultivars, but only a few plants were produced from any of these embryoids.

The main object of experiments in producing haploids via anther culture in apple reported here is to increase the frequency of embryoid induction and to establish an embryo-to-tree-system.

Material and methods

From the apple breeding program at Dresden-Pillntiz 13 genotypes were tested and their anther culture response in respect of callus or embryoid induction was analysed.

Anthers with microspores at the uninucleate stage of the pollen development were cultivated at 27 °C in the dark. The MS-basal medium contained 5% sucrose and was supplemented with 2,4-D, NAA or IBA and Kinetin or BA in varying concentrations (0.2, 0.5 and 1 mg/l).

The cultures were examined during a culture period of 6 months. Counts were made of the number of anthers producing embryoids or callus. The embryoids were removed from the anthers and cultured at 23 °C with a photoperiod of 16 hours on MS-medium containing 3% sucrose, 0.1 mg/l IBA, 0.1 mg/l TDZ and 0.1 mg/l GA_3. The selection of these hormonal treatments was based on our earlier results. Regenerated shoots were transferred to MS-medium supplemented with 0.2 mg/l IBA, 0.5 mg/l BA and 0.5 mg/l GA_3.

For the determination of the ploidy level carried out on different stages of embryoids and shoots a Partec cytometer was used.

Results and discussion

Induction phase

About four to six weeks after putting the anthers in culture an obvious increase in their volume, and a beginning of callus formation could be observed. Callus was induced nearly in all cases from the connective tissue. During further cultivation the anther wall was locally

Table 1. In vitro androgenesis in apple – effect of genotype on the rate of callogenesis, embryogenesis and regeneration of plants. (For the last parameter exclusively embryoids cultured on regeneration medium with 0,1 mg/l TDZ were considered.)

Genotype	Anthers with Callus Formation %	Number of Embryoids	Embryoid Induction Rate %	Plant Regeneration Rate %
Alkmene	15.70	69	0.40	52.9
Remo	1.57	24	0.15	60.0
Malus zumi	0.52	10	0.24	–
Releika	8.90	4	0.05	50.0
Clivia	28.56	2	0.06	–
Reanda	9.40	2	0.10	–
Pinova	8.10	1	0.05	–
Average	10.39		0.21	54.2

– not investigated.

burst and the callus formation occurred progressively, mainly from inside the anthers.

After three to six months of anther incubation white embryoids appeared and developed directly from microspores, i.e., without callus formation. The rate of callogenesis and embryogenesis are genotype- (Table 1) and culture medium-dependent, especially with respect to the type of auxins and cytokinins used (Höfer, unpublished). The embryoid induction was very low between 0.4 and 0.05% and occurred only in seven genotypes. These observations emphasize the great importance of the genotype.

Similar observations were reported in previous studies, where the overall plating efficiency for embryogenesis in apple is generally very low (Xue & Niu 1984, Johannsson et al. 1987, Zhang et al. 1990).

Regeneration phase

After transfer to regeneration medium with 0.1 mg/l TDZ some embryoids increased in size or changed the colour from white to green and developed cotyledons and secondary embryogenesis could be found. After a number of subculturing operations the formation of adventitious shoots from the induced complexes of secondary embryoids was obtained. The adventitious shoots showed a different phenotype with respect to shape and colour. The plant regeneration rate in the means of the genotypes reached 54.2% (Table 1). Thereby this method represents a reproducible protocol for plant regeneration from androgenic embryoids. At present most of the shoots are in the phase of micropropagation and rooting.

Androgenic plants via direct germination have only been successfully regenerated by Fei & Xue (1981) and Xue & Niu (1984). On the other hand, the results of Zhang et al. (1990) and the present own experience demonstated the way of the adventitiuos shoot formation.

Ploidy level

24% of the structures showed haploid or diploid peaks in the histogramms of the cytometer, 28% are triploid and 48% are tetraploids. A comparison between our results and those obtained in other plant species summarized in a review by D'Amato (1985) showed similarities of the distribution of the ploidy level. Endoreduplication and nuclear fusion are mechanisms probably prevalent for the chromosomal variation in the anther culture (Sunderland et al. 1974).

A frequency analysis of obtained ploidy levels was carried out (Fig. 1). There is a clear relationship between the ploidy level of the embryiods or shoots and the type of cytokinin in the medium for embryoid induction. Haploid and diploid structures were nearly

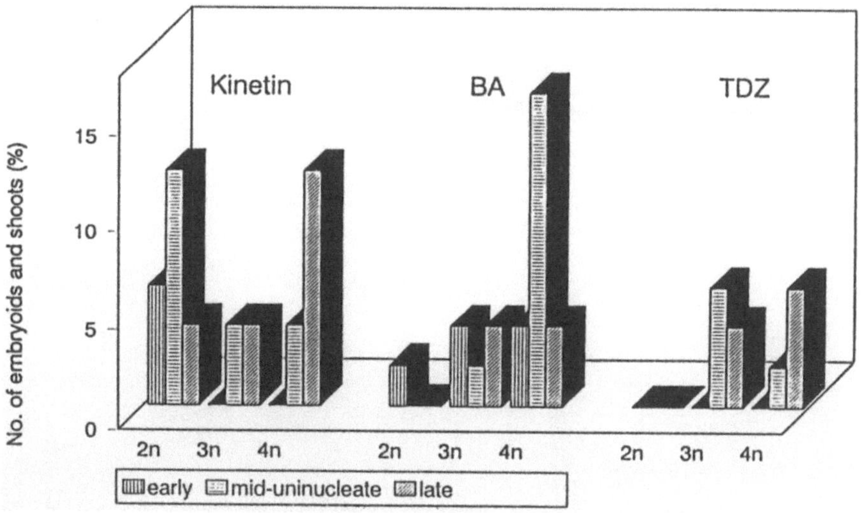

Fig. 1. Effect of cytokinins and stage of pollen development on the ploidy level of embryoids and shoots.

exclusively induced on media containing kinetin. 48% of all embryoids and shoots, induced on kinetin media, were haploids or diploids. A first three dimensional analysis for cytokinin and the stage of pollen development showed that triploid and tetraploid structures, which could be found on kinetin media, occurred from anthers, characterized mainly by the late-uninucleate pollen stage. Explaining these observations a very early determination of the ploidy level in the phase of androgenesis induction, which depends especially on the stage of the pollen development and the cytokinin, could be assumed. Following further cultivation, a spontaneous diploidization could be found.

Kasha et al. (1992) have indicated that the auxin-kinetin balance in the callus induction media can influence the occurrence of endoreduplication and nuclear fragmentation. Other experiments demonstrated the effect of kinetin (Guo 1972) and auxins (Ziauddin & Kasha 1990) on chromosomal status of callus cultures of anthers or immature embryos.

In conclusion, we have been able to establish a regeneration system of androgenesis in apple. On the base of a frequency analysis of obtained ploidy level a first relationship between factors affecting the induction phase (type of cytokinin and stage of pollen development) and the ploidy level of embryoids and shoots was derived. In order to realize the full potential of anther culture technology for apple, a greater understanding is still needed of factors that are involved in the morphogenetic competence.

Acknowledgements

I am grateful to Prof Dr Ch. Fischer for her contributions towards the selection and the provision of the genotypes; and S. Ernst, University of Halle, for determination of the ploidy level. I also wish to thank S. Schöber and R. Gläß for the technical assistance.

References

D'Amato, F., 1985. Cytogenetics of plant cell and tissue cultures and their regenerates. CRC Crit. Rev. Plant Sci. 3: 73–112.

Fei, K.W. & R.G. Xue, 1981. The induction of haploid plantlets by anther culture *in vitro* for the apple cultivar Delicious. Sci. Agric. Sin. 4: 41–44.

Guo, C.-Le., 1972. Effects of chemical and physical factors on the chromosome number in *Nicotiana* anther callus cultures. In Vitro 7: 381–386.

Höfer, M. & V. Hanke, 1990. Induction of androgenesis *in vitro* in apple and sweet cherry. Acta Horticulturae 280: 333–336.

Johansson, L., A. Wallin, A. Gedin, M. Nyman, E. Pettersson & M. Svensson, 1987. Anther and protoplast culture in apple and strawberry. Biennial Report 1986–1987. Division of Fruit Breeding – Balsgard, Swedish Univ. Agr. Sci., pp. 83–92.

Kasha, K.J., L. Kott & G. Seguin-Swartz, 1982. Genetic stability of haploid cell cultures. In: A. Fujiwara (Ed) Plant tissue culture, pp. 405–408. Maruzen, Tokyo.

Sunderland, N., G.B. Collins & J.M. Dunwell, 1974. The role of nuclear fusion in pollen embryogenesis of *Datura innoxia* Mill. Planta 117: 227–241.

Xue, G.R. & J.Z. Niu, 1984. A study on the induction of apple pollen plants. Acta Hort. Sin. 11: 161–164.

Zhang, Y.X., J. Boccon-Gibod & Y. Lespinasse, 1987. Obtention d'embryons de pommier (*Malus* × *domestica* Borkh.) après culture d'anthers. C.R. Acad. Sci. Paris, Ser. III. 305: 443–448.

Zhang, Y.X., Y. Lespinasse & E. Chevreau, 1990. Induction of haploidy in fruit trees. Acta Horticulturae 280: 293–306.

Ziauddiu, A. & K.J. Kasha, 1990. Long-term callus cultures of diploid barley (*Hordeum vulgare*). II. Effect of auxins on chromosomal status of cultures and regeneration of plants. Euphytica, 48: 279–286.

H. Schmidt and M. Kellerhals (Eds.), Progress in Temperate Fruit Breeding, 403–408.

Restrictions of the efficiency of haploid plant production in apple cultivar Idared, through parthenogenesis *in situ*

Krista De Witte & Johan Keulemans
Fruitteeltcentrum K.U. Leuven, Willem de Croylaan 42, B-3001 Leuven-Heverlee, Belgium

Key words: Apple, haploid induction, parthenogenesis *in situ*, *Malus domestica*, picking time, pollen irradiation, pollen quality, seed selection

Summary

For haploid plant production in Idared apple, parthenogenesis *in situ* was induced by irradiated pollen. The most important restrictions were seed set, embryo germination and green plant recovery from germinated embryos. The efficiency of green plant production was strongly influenced by irradiation dose, picking time and quality of the irradiated pollen. 250 Gy pollen was more efficient for green plant production than 500 Gy pollen. Using 250 Gy pollen with reduced germination capacity resulted in a five-fold lower green plant yield. Delaying picking of the fruits (140 instead of 70 days after pollination) had a significant positive effect. Up to 20 plants per 1000 pollinated flowers, showing no expression of a marker gene present in the pollen parent, were obtained in some treatments.

Induction efficiency was high when seed weight was high. This offers the possibility of selecting seeds with the best chances for haploid plant production. Heavy seeds generally contained visual normal developed embryos, which germinated better than those from lighter seeds and gave normal plant development. Delayed picking increased the proportion of large seeds, suggesting that pollination with irradiated pollen retards embryo development. From treatments with irradiated pollen of poor quality more small seeds were harvested.

Abbreviations: Gy – Gray; BA – 6 benzylaminopurine, IAA – indole-3-acetic acid; GA_3 – gibberellic acid

Introduction

As mentioned by Zhang & Lespinasse (1992), haploids have many applications in fruit crop improvement. Haploid induction in apple was assayed by androgenesis (Hoefer & Hanke, 1990; Zhang et al., 1987), gynogenesis (Zhang & Lespinasse, 1988) and parthenogenesis *in situ* (Zhang et al., 1988). Of these methods, parthenogenesis *in situ* seems the most promising at present (Zhang & Lespinasse, 1992).

Spontaneous parthenogenesis is rather exceptional in plant species but can be induced by several techniques. Colchicine treatments, high temperatures and pollination with foreign pollen were used for induction in cotton (Zhou et al., 1991). The latter method was successful in many other species. The stimulation of parthenogenesis *in situ* can also be achieved by treating pollinated flowers with toluidine blue (Illies, 1975) or by delayed pollination (Asker et al., 1983). Induction by pollination with irradiated pollen is commonly used for several crops, including kiwifruit (Pandey et al., 1990) and apple (Zhang & Lespinasse, 1992). According to Zhang et al. (1991), irradiation doses of 250 to 500 Gy are most suitable in apple although both seed and embryo set were reduced, but seed germination was good. Above 200 Gy, no plants showed expression of a marker gene, present in the pollen-parent.

The aim of our study was to investigate in Idared apple to what extent the restrictions in haploid production (fruit set, seed set, embryo set, embryo germination and green (maternal) plant production) could be manipulated by irradiation dose, pollen quality and picking time. For embryo set, embryo germination and green plant production, we studied the correlation with seed weight. The latter is much more variable after pollination with irradiated pollen compared to 'normal'

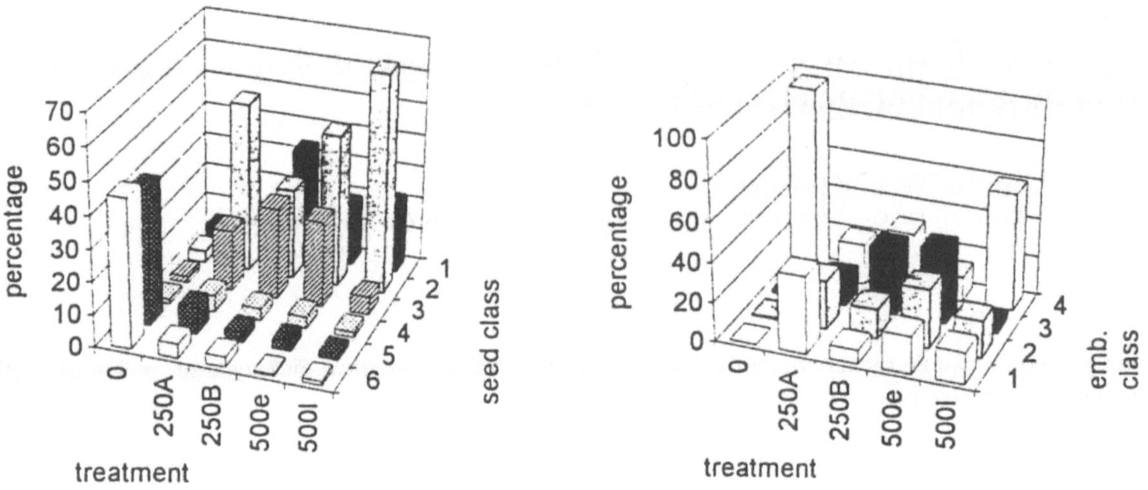

Fig. 1. Influence of irradiation dose, pollen quality (250A vs 250B) and picking time (500e vs 500l) on the distribution of seeds (left) and embryos (right) in seed- and embryoclasses.

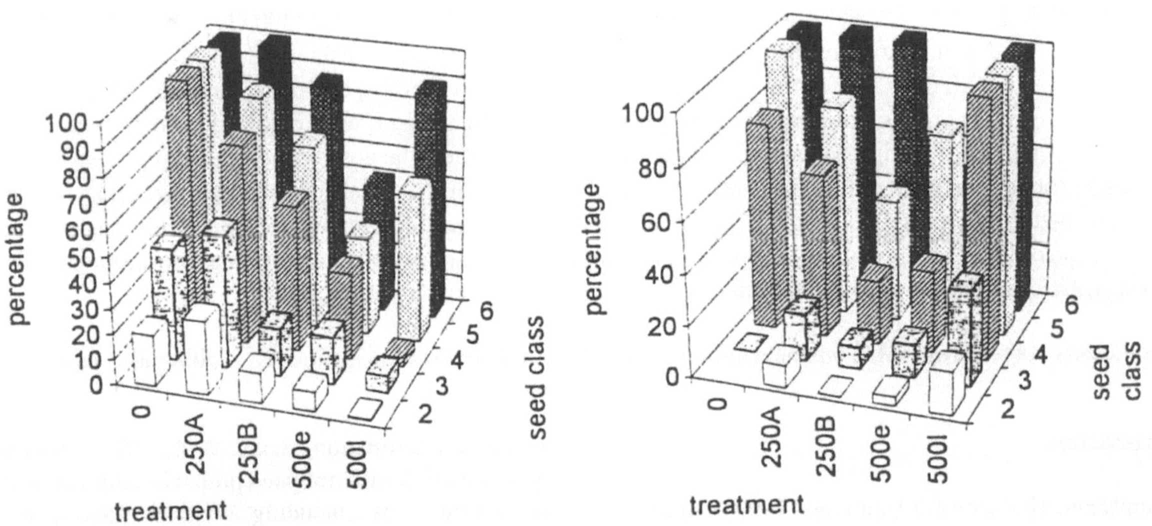

Fig. 2. Influence of irradiation dose, pollen quality (250A vs 250B) and picking time (500e vs 500l) on embryo set (left) (embryos per 100 seeds) and embryo germination (right) (%) among seed classes.

pollen and can probably be used for selecting the most promising seeds.

Material and methods

Plant material. Adult trees of the female parent Idared were isolated by a cage before flowering. The male parent was a hybrid which carries a homozygous dominant marker gene R, encoding for anthocyanin production in stem, leaves and fruits (Lespinasse & Chevreau, 1984).

Pollen and pollination. Pollen was irradiated at 250 and 500 Gy by γ rays from a Cs source (600 Gy per hour). Pollen germination was up to 70% after irradiation, except for a part of the 250 Gy irradiated pollen, where germination was less than 10%.

Flowers were emasculated at the ballon stage and pollinated with irradiated or control pollen. Flowers

Table 1. Influence of irradiation dose, picking time and pollen quality on fruit set (fruits/100 flowers), seed number per fruit, embryo set (embryos per 100 seeds), embryo germination (%), green structures/100 germinated embryos, green plants/100 green structures and green plants per 1000 pollinated flowers

	Treatments				
	0	250A	250B	500e	500l
fruit set	39.1	14.2	12.4	22.8	23.6
seeds/fruit	6.5	5.5	2.8	4.9	4.3
embryo set	93.0	45.3	18.2	12.3	4.2
germination	97.6	41.9	26.4	18.8	83.3
% green	–	38.8	59.4	71.4	86.0
% green plants	–	51.1	57.9	33.3	72.1
green plants/ 1000 flowers	–	27.8	5.7	6.1	22.0

were thinned to 3 flowers per cluster. Each treatment was distributed over several trees.

Fruit and seed manipulation. Fruits were harvested about 70 days after pollination. Half of the fruits from the 500 Gy treatment were harvested 140 days after pollination. Fruits were stored for 1 to 3 months at 2° C before seed collection. The mean storage time was the same for all treatments. Embryos were isolated from sterilised seeds (95% ethanol for 30 sec, 1.25% NaClO for 15 min) and stratified at 3° C on the culture medium for another 3 months.

Embryo culture. Embryos were cultured at 23° C in the light (16 h, 2800 Lux) on a medium containing: macro salts: modified Lepoivre/2 and micro salts: Lepoivre/2; Fe-Na EDTA (MS); mixture of vitamins: myoinositol (100 mg/l), glycine (4 mg/l), thiamine (30 mg/l): nicotinic acid (10 mg/l), pyridoxine-HCl (1 mg/l), chlorure cholin (0.4 mg/l), folic acid (0.4 mg/l), riboflavine (0.4 mg/l), biotin (0.04 mg/l); saccharose (3%); agar (0.65%); IAA (0.5 mg/l), BA (2 mg/l) and GA3 (1 mg/l). Before evaluation of the expression of the marker gene, plants were cultured for 5 days at 15° C.

Observations and treatments. The following treatments were considered:
0: control, normal marker pollen, non-irradiated (> 70% germination)
250A: 250 Gy pollen with good germination (> 70%)
250B: 250 Gy pollen with poor germination (< 10%)
500e: 500 Gy pollen with good germination (> 70%); picking time 70 days after pollination
500l: 500 Gy pollen with good germination (> 70%); picking time 140 days after pollination
For all treatments fruit set was evaluated. The seed number was counted for individual fruits.

Seeds were divided into classes according to their weight: class 1: 0 to 7.5 mg; class 2: 7.6 to 15.0 mg; class 3: 15.1 to 22.5 mg; class 4: 22.6 to 30.0 mg; class 5: 30.1 to 50.0 mg and class 6 (normal seed weight): more than 50.0 mg.

From each seed of known weight the embryo was isolated, if present. Embryos were evaluated for size and shape and divided into classes: class 1: very small and round-shaped; class 2: very small but normal shape or larger, but abnormal shape; class 3: normal shaped, small embryos; class 4: size and shape are comparable to control. Embryo germination was monitored at regular intervals; as germination we considered any development of a new structure from the embryo. The type of structure was evaluated as plant or 'non-plant'.

The expression of the marker gene, red colouration, was assessed 1 to 3 months after germination. This period was rather long since the red colouration was not always clearly expressed.

Results

Influence of dose, picking time and pollen quality on the restrictions of the green plant production process

As illustrated in Table 1, fruit set, seed number per fruit, embryo set and embryo germination were lower when flowers were pollinated with irradiated pollen compared to non-treated pollen. The last 3 characteristics decreased with increasing irradiation dose. The percentage of green structures reached 38.8% and 71.4% for the 250 dose and 500 dose respectively. Also after pollination with normal pollen some green plants were observed (data not recorded), suggesting that the expression of the marker gene was not absolute. The percentage of plants from the green structures was lower at the high irradiation dose. The efficiency of green plant recovery per pollinated flower was much higher from 250 Gy flowers (27.8 per thousand flowers) than from 500 Gy flowers (6.1), when fruits were harvested 70 days after pollination.

Pollination with pollen of low germination capacity (250B) resulted in a remarkable decrease in green plant production, due to a reduction of seed set, embryo set

Table 2. Embryo set (number of embryos per 100 seeds), embryo germination (%), green structures per 100 germinated embryos, green plants per 100 green structures and number of green plants per 1000 seeds (all treatments except control)

	Seed class					
	1	2	3	4	5	6
embryo set	0	8	7	29	64	89
germination	0	6.1	14.3	48.5	77.8	97.3
% green	0	53.8	56.7	54.5	60.4	54.8
% plant	0	0	23.5	50.0	54.5	80.0
plants/1000 seeds	0	0	1.3	38.3	163.9	379.6

Table 3. Total number of germinated embryos, number of green structures per 100 germinated embryos and number of green plants per 100 green structures among treatments

		Seed class				
		2	3	4	5	6
250A	total	9	10	19	46	32
	% green	56	30	42	44	28
	% plant	0	33	50	55	78
250B	total	0	5	3	8	16
	% green	0	60	100	75	44
	% plant	0	33	67	50	71
500e	total	3	12	7	18	2
	% green	67	75	43	78	100
	% plant	0	11	33	43	100
500l	total	1	3	4	19	23
	% green	0	68	100	79	96
	% plant	0	50	50	67	82
overall	total	13	30	33	91	73
	% green	54	57	55	60	55
	% plant	0	24	50	55	80

and embryo germination, while among the germinated embryos about 20% more green structures were found.

When 500 Gy fruits were harvested later (a few weeks before the commercial picking time), seed set was somewhat lower, but embryo set was dramatically reduced. Because of the better germination and the higher frequency of green structures, of which 72.1% were plants, an increase of green plant production from 6 (500e) to 22 plants (500l) per 1000 pollinated flowers was obtained.

Seed and embryo classes

Table 2 summarizes the effects of some factors on green plant production among the seed classes. The chance to obtain green plants from seeds of classes 1, 2 or 3 was less than 0.5% and increased with seed weight. The most restricting factors for these small seeds were low embryo yield, low germination capacity and low green structure frequencies when germinating. Germination and green plant development seems to be related. When only seeds of class 5 and 6 were maintained, the green plant recovery was 16 and 38%. The latter percentage was strongly determined by the seeds from late harvested fruits (500l). In this treatment a high frequency of class 6 seeds were present (if we ignore class 1 seeds, which never contained embryos). As shown in Table 3, these seeds gave more green structures and plants than class 6 seeds from the 250 Gy treatments. The distribution of seeds among weight classes was strongly influenced by irradiation dose, pollen quality and picking time (Fig. 1). As irradiation dose increased, seed weight decreased. There were more seeds found in the lower seed classes when flowers were pollinated with pollen of poor germination capacity (250B). The 500 Gy fruits had a higher frequency of class 1 and 2 seeds.

The distribution of the embryo classes (Fig. 1) showed another pattern, since seeds of lower seed classes rarely contained an embryo. The best developed embryos (class 4) were mostly found in control seeds and 500 Gy seeds of late harvested fruits (90% resp. 60% of the embryos). Pollination with 250 Gy pollen of poor quality resulted in a higher presence of embryos in classes 3 and 4, compared to pollination with good quality 250 Gy pollen.

Embryo set increased with seed class (Fig. 2) and for comparable seed classes decreased with increasing irradiation dose and declining pollen quality. When fruits were harvested late, seeds of classes 5 and 6 contained more embryos.

Most of the embryos of seed class 6 germinated. For seed classes 5 and lower, embryo germination was negatively influenced by higher irradiation and poor pollen quality, while a later harvest time was positive for embryo germination. When embryos germinated, green structures were found in all seed classes and for the overall mean of all treatments this frequency reached about 55%, irrespective of the seed class (Table 3). The presence of green structures was less pronounced in the 250 Gy than in the 500 Gy treatments. Most of the germinated embryos of 500 Gy late harvested fruits were green.

The percentage of green plants was highest in seed class 6 and decreased with the class number. This was the case for all treatments.

Discussion and conclusions

Fruit and seed set were reduced after pollination with irradiated pollen. The reduction in our experiments reached the same level as reported by Zhang & Lespinasse (1991). Visser & Verhaegh (1987) mentioned the positive relation of seed number and fruit set in some apple cultivars, while in others no relation was found. The lower fruit set after pollination with irradiated pollen could also be due to the lower sink activity of induced seeds, compared to fertilised seeds. This lower seed set in some cases could indicate that the stimulus effect of pollen on embryo development depends on irradiation dose and pollen quality or that the stimulus effect is insufficient, resulting in early abortion of induced embryos. The latter seems probable since a large number of seeds with low weight and containing no embryo were found. The presumed insufficient induction stimulus when pollen of poor quality is used, suggests a close contact/signal (fusion?) of the pollen tube and the cell which will develop into an embryo.

The negative influence of irradiation dose on seed set was also reported for soybean (Banga et al., 1984) and apple (Marcucci et al., 1984). In the latter study no seed set was obtained when 500 Gy irradiated pollen was used, while seed set was only 2/3 of the control for 1–150 Gy treated pollen. This is in contrast to our findings and those from Zhang & Lespinasse (1991). These authors reported also that irradiation had little effect on pollen germination, which is in accordance with our results, but Lecuyer et al. (1991) stated that irradiated pollen germinates slower.

The lower embryo germination in the case of 250 Gy irradiated pollen of poor quality compared to that of good quality should be considered together with the green structure percentage: the lower germination is completely compensated by the higher frequency of green structures, indicating that in the case of good pollen quality more fertilisation (red coloration) occurred. For the 500 Gy treatment the higher green plant production when fruits were harvested later can be explained by the better development of the remaining seeds, resulting in more seeds of higher seed weight. Such seeds germinated better and showed more green plant production. It cannot be excluded that the two picking times corresponded to differences in 'dormancy status' of the seeds, which would account for the differences in embryo germination and plant development.

We have not yet proved that the green plants are of haploid origin, but the absence of the expression of the marker gene suggests they are. However we also have indications that this expression is not absolute, because after pollination with control pollen, some green plants were found, but at low frequency. This was also the case in other comparable experiments (Lespinasse & Chevreau, 1984). Investigations are in progress to confirm the haploid origin of the green plants.

Acknowledgements

This work was supported by 'Vlaams Actieprogramma Biotechnologie' (VLAB) of the Flemish government and by N.V. Jo Nicolaï nurseries. We wish to thank B. Wertheim and H. Kemp from the Fruit Research Station Wilhelminadorp-Netherlands for delivering the pollen and Lia Verdoodt, Piet Wouters, Joeri Van Aerschot and Lieve Van Diest for their technical assistance.

References

Asker, S., A. Hagberg & G. Hagberg, 1983. Apomixis in barley. Sveriges Utsadesforenings Tidskrift 93: 75–76.

Banga, S.K., S.S. Banga & M. Srivastava, 1984. Induced parthenogenesis in soybean. Soybean Genetics Newsletter 11: 45.

Hoefer, M. & V. Hanke, 1990. Induction of androgenesis *in vitro* in apple and sweet cherry. Acta Horticulturae 280: 333–336.

Illies, Z.M., 1975. Induction of haploid parthenogenesis is aspen by postpollinisation treatment with Toluidine-Blue. Silvae Genetica 23: 221–226.

Lecuyer, M.P., Y.X. Zhang, M. Tellier & Y. Lespinasse, 1991. *In vitro* pollen tube division of irradiated an non-irradiated apple pollen. Agronomie 11: 483–489.

Lespinasse, Y. & E. Chevreau, 1984. Utilisation d'un gène marqueur pour la recherche de plantes haploides de pommier: comparaison entre pollinisation manuelle et pollinisation par les abeilles. In: Compte rendu de V^{e} symposium sur la pollinisation, Versailles-France, 27–30 septembre 1983: 391–396.

Marcucci, M.C., D. Ragazinni & S. Sansavini, 1984. The effects of Gamma-rays and laser rays on the functioning of apple pollen in pollination and mentor experiments. J. Hortic. Sci. 59: 57–62.

Pandey, K.K., L. Przywara & P.M. Sanders, 1990. Induced parthenogenesis in kiwifruit (*Actinidia deliciosa*) through the use of lethally irradiated pollen. Euphytica 51: 1–9.

Visser, T. & J.J. Verhaegh, 1987. The dependency of fruit and seed set of pear and apple on the number of styles pollinated. Gartenbauwissenschaft 52: 13–16.

Zhang, Y.X., J. Boccon-Gibod & Y. Lespinasse, 1987. Obtention d'embryons de pommier (*Malus* × *domestica* Borkh.) après cultures d'anthères. C.R. Acad. Sci. Paris, Série III 305: 443–448.

Zhang, Y.X. & Y. Lespinasse, 1988. Culture *in vitro* d'ovules non fécondés et d'embryons prélevés 8 jours après pollinisation chez le pommier cultivé (*Malus* × *domestica* Borkh) issues de parthenogénèse induite *in situ* par du pollen irradié et culture *in vitro* des pépins immatures. C.R. Acad. Sci Paris, Série III 307: 451–457.

Zhang, Y.X. & Y. Lespinasse, 1991. Pollination with gamma-irradiated pollen and development of fruits, seeds and parthenogenic plants in apple. Euphytica 54: 101–109.

Zhang, Y.X. & Y. Lespinasse, 1992. Haploidy. In: F.A. Hammerschlag & R.E. Litz (Eds) Biotechnology of perennial fruit crops, pp. 57–75. C.A.B. International, Wallingford.

Zhou, S.Q., D.Q. Qian & X.Y. Cao, 1991. Induction of parthenogenesis, and chromosome behavior in plants of parthenogenetic origin in cotton (Gossypium hirsutum). Genome 34(2): 255–260.

H. Schmidt and M. Kellerhals (Eds.), Progress in Temperate Fruit Breeding, 409.

Isolation and culture of apple microspores: Some preliminary results

Lalichka Kolova[1], Joke Janse[2] & Roeland Voorrips[2]
[1] *Genetic Manipulation Laboratory, Breeding Department, Fruit-Growing Research Institute, 12 Ostromila, 4004 Plovdiv, Bulgaria;* [2] *Department of Vegetable and Fruit Crops, Centre for Plant Breeding and Reproduction Research, CPRO-DLO, 6700 AA Wageningen, The Netherlands*

Key words: Apple, haploid cultures, isolated microspores

Summary

The establishment of true breeding lines is considered a critical step for cultivated variety development. Particularly in heterozygotic allogamous plants, such as apple (*Malus pumila* Mill), the establishment of homozygous lines offers new prospects for breeding. Isolated microspores may also offer an effective system for *in vitro* selection of haploid cells. Lespinasse et al. (1983) have discussed the potential of apple haploids in detail. The purpose of the present work was to study the potential of isolated apple microspores to develop *in vitro*.

Two apple varieties, Golden Delicious and Prima, and a *Malus pumila* × *M. Bascatong* hybrid heterozygous for anthocyanin colouration of vegetative tissues, were used as flower-bud donors. Three groups of flower buds were studied with respect to the importance of developmental stage for pollen plant induction, containing mainly early uninucleate, late uninucleate or late binucleate microspores. Full, 1/2 and 1/4 strength MS (Murashige & Skoog 1962), NLN 13 (Lichter 1985) and N_6 (Chu et al. 1975) basal culture media were used. Sucrose (130 g/l) was used as an osmoprotectant and a carbohydrate source. 2,4-D at 0–2 mg/l, BAP at 0–1 mg/l and GA_3 0–1 mg/l alone or in combinations were supplemented to filter sterilized culture media.

Apple flower-bud shape was established as the best morphological marker for contained microspores developmental stage. Multicellular structures of up to 18–36 cells were induced *in vitro*. Obviously depending on the physiological state of the donor plants and the stage of pollen the best response was obtained by early binucleate microspores. As a result of our studies we found that N_6 and 1/4 MS basic media was best for the promotion of multicellular structure formation by apple microspores. No obvious enhancing effect of growth regulators, added alone or in combinations to culture media, was detected.

The preliminary results of our studies on the potential of apple microspore culture show that it is possible to induce multicellular structure formation *in vitro*. We consider further detailed studies on the physiological condition of the microspore donor as the most important for successful production of microspore-derived apple plants.

References

Chu, C.C., C.C. Wang & C.S. Sun, 1975. Establishment of an efficient medium for anther culture of rice through comparative experiments on the nitrogen sources, Sci. Sin. 18(5): 484-488.

Lespinasse, Y., M. Godicheau & M. Duron, 1983. Potential value and method of production of haploids in apple tree, *Malus pumila* (Mill.). Acta Hortic. 131: 223–230.

Lichter, R., 1985. From microspore to rape plants: A temptative way to low glucosinolate strains, In: H. Sorensen (Ed), World crops: production, utilization, description, Vol. 11, pp. 268–277, Martinus Nijhoff Publ., Dordrecht/Boston/Langcaster.

Murashige, T. & F. Skoog, 1992. A revised medium for rapid growth and bioassays with tobacco tissue culture. Phys. Plant. 15: 473–497.

H. Schmidt and M. Kellerhals (Eds.), Progress in Temperate Fruit Breeding, 411–413.

Prospects for application of anther culture in selection of fruit plants

O.S. Zhukov, N.I. Savelyev & O.Ja. Oleynikova
All Russian Research Institute of Fruit Crop Genetics & Selection Michurinsk, Russia

Key words: Callus, gametes, morphogenesis, plant regenerants, haploidy

Summary

In order to provide future diploidization and production of homozygotic forms for their breeding usage, we studied the opportunity of haploid plant production. Thus, anther culture on nutritional medium was used. The methods developed enabled us to produce heterogenous callus tissues with a frequency up to 60%. Cytological investigations showed that pollen calli had both haploid (up to 30%) and diploid cells, and also tetraploid and aneuploid ones. Techniques and conditions of callus culture have been developed which enable us to obtain plant regenerants for the first time in our experience in Liberty, and Wijcik apple cultivars, in Vladimirskaya and Pamyati Vavilova cherry cvs, in *Prunus cerasus* × *Pr. Maackii* and *Pr. cerasus* × *Pr. avium* hybrids. We managed to develop special methods for successful adaptation of plant regenerants to natural conditions.

Introduction

Nowadays *in vitro* culture of isolated anthers and microspores by androgenesis is an actively developed method of great potential. The employment of this method promoted haploid production in more than 240 species of plants (Zhang, 1990). But as for perennial fruit plants, this method is still being studied. Anther culture helped to produce plants in certain forms of strawberry, apple, grapes and citrus (Lin Shunguan, 1987). The aim of our investigations is to establish methods for developing haploid production in apple and cherry cultivars which possess resistance to the most dangerous diseases, and show other valuable traits (O.S. Zhukov et al., 1993).

Material and methods

Anthers of the following cherry cultivars and hybrids were considered to be suitable material for investigations: Almaz, Malyutka, Padocerus – M, I/I hybrid, Lyubskaya, Novosyolka, Kharitonovskaya, Zhukovskaya, Pamyati Vavilova, Nadyozhnaya, Vladimirskaya & Komsomolskaya. Thus, Almaz, Padocerus M and Kharitonovskaya are donors of 'A' gene (*Coccomyces* resistance); Malyutka is a source of virus resistance (leaf spot group); and Zhukovskaya, Novosyolka and Pamyati Vavilova possess fruit of superior quality. The remainder of the cherry cultivars and hybrids is especially outstanding in their winter resistance and yielding capacity. Much attention was also paid to the investigations of Wijcik, Florina L. Prima, Liberty, 12-73 (Korichnoye novoye × Medium) apple cultivars and their anthers. These represent a source of 'Vf' and 'Co' genes (scab resistance and columnar growth habit). The methods of heterogenous callus tissue production, developed by us for cherry and apple anthers, include isolation of anthers on mononuclear and early binuclear stages of microspore development – taking into account their treatment with low temperature (3–5° C) for 24–48 hours. Anthers were planted on Heller or MS media supplemented by plant growth substances. After 7–10 days these calli were transplanted to MS medium for subsequent morphogenesis. In order to carry out cytological trials, e.g. calculation of chromosome number and investigations in microsporogenesis, we tried to use pre-set temporary preparations. The material was registered according to Carnua. The staining of preparations was performed with the use of acetocarmine and acetorsein.

Table 1. Ploidy level of plants and types of cherry gametes

Name of plant	Origin: maternal plant	Origin: paternal plant	Chromosome number: ♀	Chromosome number: ♂	Chromosome number: investigated plant	Types of gametes (x)
	Cherry cultivars					
Vladimirskaya	cv of unknown origin		–	–	32	16
Zhukovskaya	seedl. of I. Michurin cvs		–	–	32	16, 32
Malyutka	Lyubskaya × Zakharovskaya	Vladimirskaya	32	32	32	16
Nadyozhnaya	Lyubskaya	-"-	32	32	32	16
Pamyati Vavilova	seedl. of I: Michurin cvs		–	–	32	16, 32
Lyubskaya	cv of unknown origin		–	–	32	16
	Sour × sweet cherry hybrids					
No I/I	Prunus fruticosa	Zarya Vostoka	32	16	24	8, 16
Komsomolskaya	Ideal	Cherny Orel	32	16	32	16
Novosyolka	Lyubskaya	May-Duke	32	32	32	8, 16
	Cerasus × Maackii hybrids					
Almaz	Padocerus 15	Pamyati Varilova	32	32	32	8, 16
Padocerus M	X-ray mutant of Padocerus		–	–	32	16
Kharitonovskaya	Zhukovskaya	Almaz	32	32	32	16

Table 2. Comparative efficiency of processes, forming and regenerating callus in anthers of cherry *in vitro* culture

Cultivar, hybrid	Number of anthers: explanted on the whole	Formed callus tissues: number	Formed callus tissues: - %	Number of plant regenerants: number	Number of plant regenerants: - % concerning the number of of explanted anthers	Group for callus formation intensity
Almaz	734	316	43.1	–	–	
Malyutka	2378	1007	42.4	–	–	
No I/I (Prunus fruticosa × Prunus avium Zarya Vostoka)	1859	540	29.1	67	3.6	I. High
Pamyati Vavilova	2209	533	24.1	10	0.5	
Kharitonovskaya	543	118	21.7	–	–	
Padocerus M	1199	255	21.3	15	1.3	
Lyubskaya	1710	270	15.8	–	–	II. Medium
Nadyozhnaya	2554	392	15.4	–	–	
Novosyolka	2021	306	15.1	–	–	
Zhukovskaya	1548	208	13.4	–	–	
Vladimirskaya	2309	245	10.6	11	0.5	
Komsomolskaya	1923	74	3.9	11	–	III. Low

Results

We have calculated the number of chromosomes in somatic cells of the cherry and apple plants studied. The types of gametes appearing in them have also been studied (Table 1). The presence of different types of gametes promotes high level callus formation of certain genetical heterogenity and stimulates strain production with the levels of different ploidy. The apple cvs studied were diploid (2n = 34) and formed 17 chromosome gametes. It was found that the anthers of all 12 cherry cvs and hybrids, and 5 apple ones have formed callus tissues after *in vitro* culture. The cultivars and hybrids investigated can be subdivided into 3 groups according to callus formation intensity (Table 2). It is clear that all apple cvs belong to the group of high callus formation intensity. Our efforts to induce morphogenesis and to obtain plant regenerants of 4 cherry forms and 2 apple cultivars were successful (Table 2). Morphological traits, relative to haploids were revealed in 30–40% of plant regenerants. In order to guarantee successful adaptation of regenerants to soil conditions, techniques were developed which should create for them almost natural conditions via gradual exchange of agar medium for soil substrate under *in vitro* conditions.

Conclusion

After cytological investigation, the induced haploid regenerants are used for chromosome doubling tests and consequently for diploid homozygotic form production. There is no doubt that homozygotic plants (dihaploids) can also appear by the way of pollen callus growth on natural medium.

References

Lin Shunguan, 1987. Survey for application of *in vitro* culture methods in fruit breeding. J. Fujian Agr. Coll. 16, No 1, 73: 82.

Zhang Yongxiang & Yves Lespinasse, 1990. *In vitro* androgenesis in cultivated apple (*Malus* × *domestica* Borkh)/Bull. Soc. bot. Fr. Actual. bot. 137, No. 3–4, p. 144.

Zhukov, O.S., O.Ja. Oleynikova & N.I. Savelyev, 1993. Production of plant regenerants by anther cultivation of the most important fruit trees. Second symposium trends in plant biotechnology, Russia, Pushchino, May 18–20, 1993, p. 416.

H. Schmidt and M. Kellerhals (Eds.), Progress in Temperate Fruit Breeding, 415–420.

Somatic embryogenesis in grapevine *Vitis vinifera* L.*

R. Vallania[1], M.L. Miaja[2], G. Vergano[2], R. Botta[2], G. Me[2], R. Zanetti[1] & I. Gribaudo[1]
[1] *Centro di Studio per il Miglioramento Genetico della Vite, CNR, Torino, Italy;* [2] *Dipartimento di Colture Arboree, Frutticoltura, Università di Torino, Italy*

Key words: Callus, somatic embryos, leaf, ovule, petiole, *Vitis vinifera*

Summary

Somatic embryogenesis in grapes has been described by several authors, but till now a regeneration procedure applicable to all major cultivars has not been found. Young leaves and petioles of 3 grapevine cultivars (Barbera, Dolcetto, Nebbiolo) and of tetraploid plants of the cv Barbera were cultured *in vitro* to obtain somatic embryogenesis. In addition, endosperm was taken from mature seeds of 2n and 4n Barbera. Leaf discs and petioles were also excised from plantlets grown *in vitro*. Basal media used were Murashige & Skoog (MS) and Nitsch & Nitsch (NN) with added BA and 2,4-D or NOA at different concentrations and combinations. Ovules collected 45 days after anthesis from 14 stenospermocarpic cultivars were cultured on NN medium added with IAA, GA_3 and activated charcoal at different concentrations. Callus was easily obtained from all the explants with all the media and hormone combinations used. Root differentiation was observed on NN and MS media with added BA. Some ovules of stenospermocarpic cultivars Perlon, Superior S., Imperatrice, and Carina developed somatic embryos on the two media tested. Embryos were transferred to NN medium with BA; the plantlets that developed were transferred to NN without growth regulators.

abbreviations: BA – 6-benzyladenine; 2,4-D – dichlorophenoxyacetic acid; GA_3 – gibberellic acid; IAA – indolylacetic acid; NOA – 2-naphthoxyacetic acid

Introduction

Somatic embryogenesis in *Vitis* spp. has been obtained by several authors using different explants: Stamp & Meredith (1988a), Stamp et al. (1990) used leaves and anthers, Reisch & Roberts (1985), Colby et al. (1991), Robacker (1993) obtained regeneration from leaves and petioles, Srinivasan & Mullins (1980), Rajasekaran & Mullins (1985), Stamp & Meredith (1988b) used zygotic embryos and unfertilized ovules.

Since a regeneration procedure suitable for all the major cultivars of *Vitis vinifera* has not been found, *in vitro* culture of leaves, petioles, endosperm and ovules was carried out on a total of 17 cultivars.

Materials and methods

Three wine cultivars (Barbera, Nebbiolo, Dolcetto) widely cultivated in Piemonte, and 14 table stenospermocarpic grapes, were used for the experiments. Young leaves and petioles of cvs Barbera, Dolcetto, Nebbiolo and of tetraploid plants of cv Barbera (obtained by γ irradiation) were collected from shoots grown from cuttings kept at 22 °C with 16 h photoperiod and from plantlets cultured *in vitro* on Murashige & Skoog (MS) medium kept at 24 °C with 16 h photoperiod. Endosperm tissue was taken from mature seeds of 2n and 4n Barbera.

About 100 fertilized ovules from 14 stenospermocarpic cultivars were collected in different Italian regions 45 days after anthesis and were measured (Table 1).

* contribution no. 276 of Centro Miglioramento Genetico Vite

Table 1. Composition of the culture media used in the trial.

MS_1 = Murashige & Skoog (1962)	+ 1 mg/l 2,4-D + 0.1 mg/l BA + 30 g/l sucrose + 7 g/l Difco Bacto agar
MS_2 = Murashige & Skoog (1962)	+ 1 mg/l NOA + 1 mg/l BA + 30 g/l sucrose + 7 g/l Difco Bacto agar
NN_1 = Nitsch & Nitsch (1969)	+ 1 mg/l 2,4-D + 0.1 mg/l BA + 20 g/l sucrose + 7 g/l Difco Bacto agar
NN_2 = Nitsch & Nitsch (1969)	+ 1 mg/l NOA + 1 mg/l BA + 20 g/l sucrose + 7 g/l Difco Bacto agar
NN_3 = Nitsch & Nitsch (1969)	+ 1.75 mg/l IAA + 0.32 mg/l GA_3 + 20 g/l sucrose + 7 g/l Difco Bacto agar
NN_4 = Nitsch & Nitsch (1969)	+ 3.50 mg/l IAA + 0.32 mg/l GA_3 + 2 g/l activated charcoal + 20 g/l sucrose + 7 g/l Difco Bacto agar
NN_5 = Nitsch & Nitsch (1969)	+ 1.1 mg/l BA + 20 g/l sucrose + 7 g/l Difco Bacto agar
NN_6 = Nitsch & Nitsch (1969)	+ 20 g/l sucrose + 7 g/l Difco Bacto agar

Table 2. Ovule length of 14 cultivars at picking date (mean of 25 ovules ± standard error).

Cultivar	Picking place and date	Length (mm)
Perlon	Bari 16/7/90	5.2 ± 1.30
Flame S.	"	4.6 ± 0.08
Pasiga	"	4.1 ± 0.09
Nerona	"	2.6 ± 0.08
Imperatrice	"	5.6 ± 0.13
Carina	"	3.8 ± 0.09
Argentina	"	3.4 ± 0.11
Perlette	Torino 23/7/90	3.5 ± 0.09
Superior S.	Roma 23/7/90	3.4 ± 0.12
Centennial S.	"	3.7 ± 0.20
Giada	"	4.3 ± 0.13
Dawn S.	"	3.7 ± 0.08
Ruby S.	"	2.7 ± 0.11
Sultanina	Torino 27/7/90	2.5 ± 0.11

Leaves and petioles were surface sterilized by washing with agitation in 1% sodium hypochlorite solution (15% v/v commercial bleach) with a few drops of Tween 20 for 8 min, and rinsed three times in sterile distilled water.

Seeds and berries for ovule culture were surface sterilized by immersion for 10 min in a 2.25% sodium hypochlorite solution (33.75% v/v commercial bleach) containing surfactant, then rinsed three times in sterile distilled water.

The culture media and the different growth regulator combinations used in the trial are reported in Table 2.

The petioles, cut into 2 mm long segments, the leaves, cut into 3–6 mm discs, and the endosperm were cultured on MS_1, MS_2, NN_1, NN_2.

At least 100 leaf discs, petiole segments, and endosperms were cultured on each medium.

Every two months the callus obtained from the explants was transferred on fresh medium without growth regulators.

Ovules were cultured on NN_3 and NN_4. After four months of culture differentiated somatic embryos were transferred to NN_5; after a further two months plantlets that developed from somatic embryos were transferred to NN without growth regulators (NN_6).

All the explants were kept in a climatic chamber at 24 °C with 16 h photoperiod and 50 μmol s^{-1} m^{-2} PPF (photosynthetic photon flux).

Results

Callus was obtained from leaves, petioles and endosperm within 40 days with all the media and hormone combinations used (Table 3). Green, white, yellow or light brown and watery, soft or hard callus was observed.

Fig. 1. Root differentiation from callus of cv. Nebbiolo (leaves on MS_2).

Table 3. Percentage of explants developing callus after 6 weeks of culture on different media.

Explant	Barbera 2n			Barbera 4n			Dolcetto		Nebbiolo	
Medium	Leaves	Petioles	Endosperm	Leaves	Petioles	Endosperm	Leaves	Petioles	Leaves	Petioles
MS_1	42	35	24	44	45	26	40	40	41	42
MS_2	38	41	30	7	40	27	39	42	15	45
NN_1	42	45	25	26	29	20	37	4	28	32
NN_2	29	32	18	33	32	28	21	28	15	16

The explants excised from *in vitro* cultured plantlets produced 2–4% more callus than explants from shoots. Petioles callused more than leaves and endosperm.

In some cases callus that developed from leaves and petioles on MS_2 and NN_2 differentiated adventitious roots (Fig. 1). Endosperm produced white friable callus and differentiated adventitious roots only in Barbera 4n on MS_2 (Table 4).

On both media tested (NN_3, NN_4), the ovules of the stenospermocarpic cultivars developed callus at different rates (Table 5); the highest percentages of callusing ovules, observed after 3 months, were obtained in Imperatrice, Carina and Perlette.

After four months of culture some callused ovules of the cultivars Perlon, Superior S., Imperatrice and Carina developed somatic embryos on both of the tested media (Table 6, Fig. 2). The colour of the callus developing embryos was light green to light brown.

After transfer of embryos to NN_5 further somatic embryogenesis was observed in Perlon, Superior S. and Imperatrice. On this medium anomalous plantlet development occurred with high frequency; the percentages of somatic embryos that developed into whole plantlets were: Perlon 10.5%, Superior S. 8.8% (Fig. 3), Imperatrice 5.9%, Carina 8.8%. The plantlets were transferred to NN_6 to allow growth before acclimation.

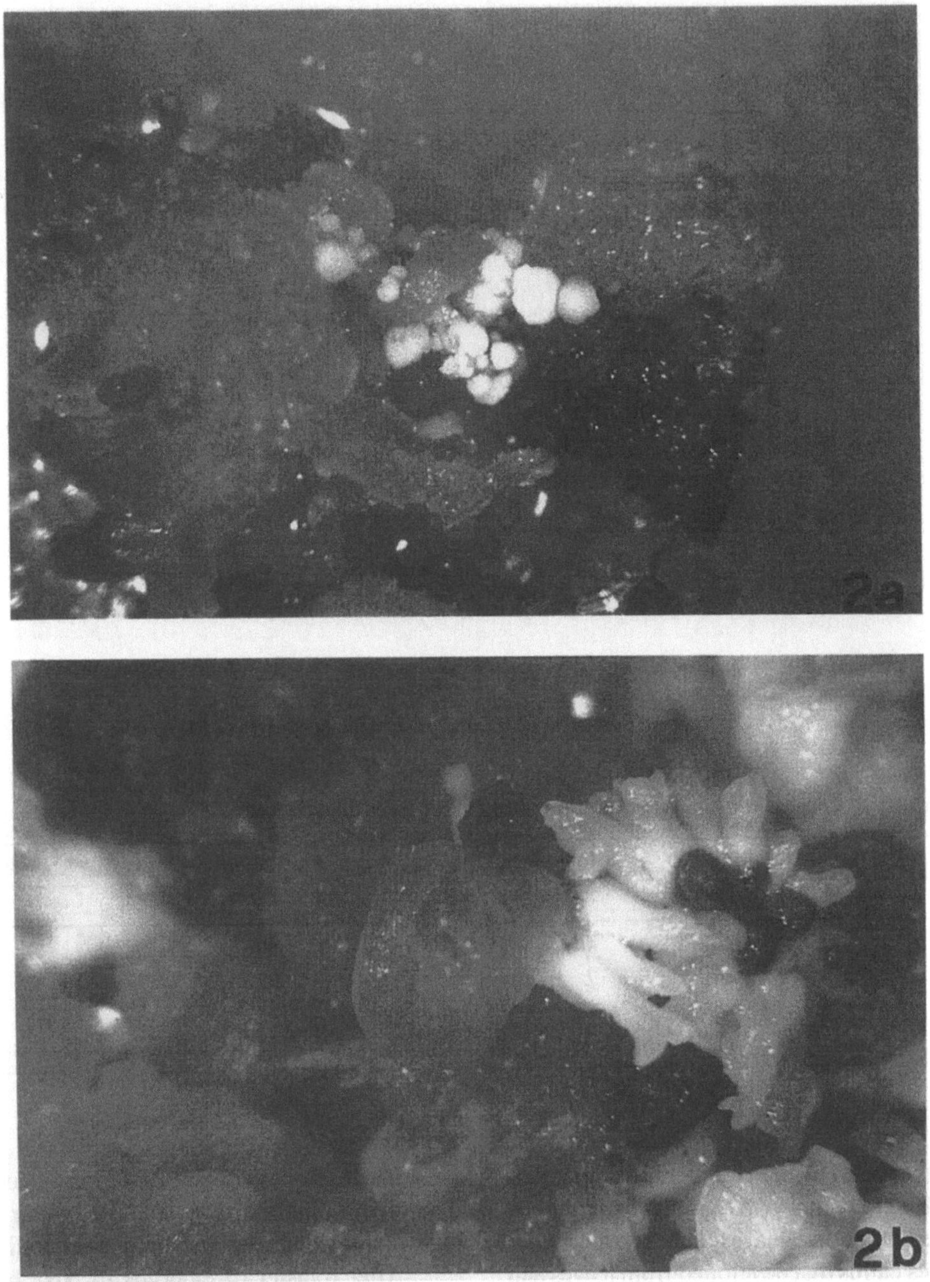

Fig. 2. Proliferating embryogenic callus obtained from ovule culture; somatic embryos at different stages of development in Superior S. on NN_4 (a) and Imperatrice on NN_3 (b).

Table 4. Percentage of callus differentiating roots after 2 months of culture on different media.

Explant	Barbera 2n			Barbera 4n			Dolcetto		Nebbiolo	
Medium	Leaves	Petioles	Endosperm	Leaves	Petioles	Endosperm	Leaves	Petioles	Leaves	Petioles
MS_2	35	12	0	32	2	15	30	28	34	27
NN_2	22	3	0	28	25	0	5	8	0	0

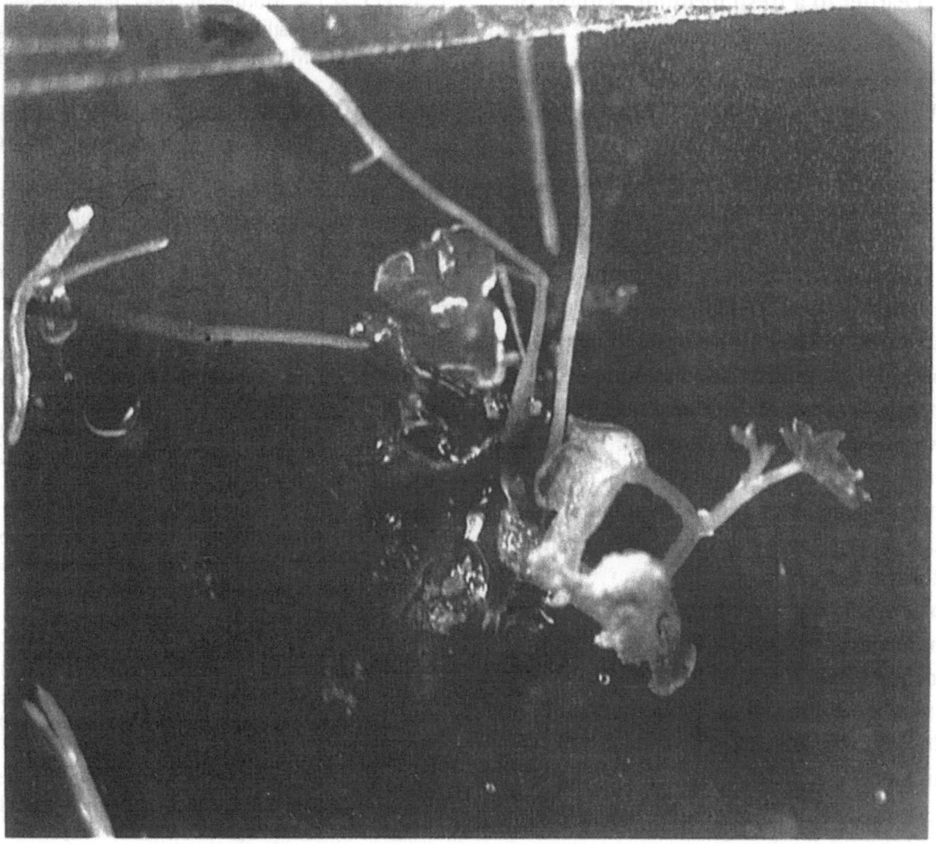

Fig. 3. Plantlet developed from somatic embryo in Superior S. on NN_5.

Table 5. Percentage of fertilized ovules developing callus (Gribaudo et al. 1993).

Cultivar	NN_3	NN_4
Pasiga	49.0	76.1
Flame S.	9.6	34.0
Perlon	51.5	69.1
Nerona	4.3	n.t.
Argentina	23.4	26.9
Imperatrice	81.2	87.5
Carina	81.6	87.6
Perlette	80.0	77.8
Superior S.	67.0	41.1
Centennial S.	1.1	20.2
Dawn S.	19.2	33.0
Ruby S.	10.0	43.7
Giada	34.4	32.6
Sultanina	26.6	n.t.

n.t. = not tested.

Table 6. Percentage of callused ovules developing somatic embryos on the two tested media.

Cultivar	NN_3	NN_4
Perlon	3.2	4.4
Imperatrice	6.0	6.6
Carina	8.2	8.6
Superior S.	10.6	9.0

Discussion

Leaf, petiole and endosperm culture did not successfully differentiate somatic embryos, although different types of callus developed abundantly. All the tested media were effective in promoting callusing but those supplemented with NOA (MS_2 and NN_2) induced root differentiation, indicating either that NOA concentration was too high or that callusing explants should have been transferred earlier to a fresh medium, without or with reduced levels of growth regulators.

Somatic embryogenesis from ovule culture of stenospermocarpic grapes was obtained in 4 out of 14 cultivars.

No relation could be found between ovule size and callus or somatic embryo development.

Although some of the cultivars easily developing callus (Pasiga and Perlette) did not give embryogenesis, all those which differentiated embryos showed over 65% callusing on at least 1 of the tested media.

In the second culture phase, BA was added to the medium according to Gray & Mortensen (1987), in order to overcome dormancy and stimulate germination. In fact, although somatic embryos can undergo secondary embryogenesis, they generally remain white and dormant and do not grow without a pre-treatment such as cold storage, GA_3 or BA.

In these experiments the medium NN_5 used for embryo development into plantlets induced a high frequency of abnormal growth, probably because the BA concentration was too high: only a low percentage of embryos developing plantlets could be transferred to a pre-acclimation medium.

References

Colby, S.M., A.M. Juncosa, J.A. Stamp & C.P. Meredith, 1991. Developmental anatomy of direct shoot organogenesis from leaf petioles of *Vitis vinifera* (*Vitaceae*). Amer. J. Bot. 78(2): 260–269.

Gray, D.J. & J.A. Mortensen, 1987. Initiation and maintenance of long term somatic embryogenesis from anthers and ovaries of *Vitis longii* 'Microsperma'. Plant Cell Tissue Organ Culture 9: 73–80.

Gribaudo, I., R. Zanetti, R. Botta, R. Vallania & I. Eynard, 1993. *In ovulo* embryo culture of stenospermocarpic grapes. Vitis 32: 9–14.

Murashige, T. & F. Skoog, 1962. A revised medioum for rapid growth and bioassays with tobacco tissue cultures. Physiol. Plant 15: 473–497.

Nitsch, J.P. & C. Nitsch, 1969. Haploid plants from pollen grains. Science 163: 85–87.

Rajasekaran, K. & M.G. Mullins, 1985. Somatic embryo formation by cultured ovules of Cabernet Sauvignon grape: effects of fertilization and of the male gameticide toluidine blue. Vitis 24: 151–157.

Reisch, B.I. & M.H. Roberts, 1985. Embryogenesis from petiole cultures of 'Horizon' grapes. 4th Int. Symp. on Grape-Vine Breeding, Verona, Italy, 13–18 April: 32–33.

Robacker, C., 1993. Somatic embryogenesis and plant regeneration from Muscadine grape leaf explants. HortScience 28(1): 53–55.

Srinivasan, C. & M.G. Mullins, 1980. High-frequency somatic embryo production from unfertilized ovules of grapes. Scientia Hort. 13: 245–252.

Stamp, J.A. & C.P. Meredith, 1988a. Somatic embryogenesis from leaves and anthers of grapevine. Scientia Hort. 35: 235–250.

Stamp, J.A. & C.P. Meredith, 1988b. Proliferative somatic embryogenesis from zygotic embryos of grapevine. J. Amer. Soc. Hort. Sci. 113(6): 941–945.

Stamp, J.A., S.M. Colby & C.P. Meredith, 1990. Direct shoot organogenesis and plant regeneration from leaves of grape (*Vitis* spp.). Plant Cell Tissue Organ Culture 22: 127–133.

CHAPTER 7

Genetic resources

H. Schmidt and M. Kellerhals (Eds.), Progress in Temperate Fruit Breeding, 423–424.

The Dresden-Pillnitz fruit tree genebank and its use

Short communication

Manfred Fischer & Rolf Büttner
Genbank Obst Dresden-Pillnitz am Institut für Pflanzengenetik und Kulturpflanzenforschung Gatersleben, D-01326 Dresden, Germany

The aim of the Pillnitz genetic resources programme for fruit is to conserve, evaluate and make available national and international cultivated and local varieties and wild species of apples, cherries, plums, pears, strawberries for genetics, fruit breeding, landscape conservation, pomological, taxonomical, phytopathological and pollination studies and for teaching.

At present the Pillnitz fruit tree genebank comprises 892 apple varieties and hybrids, 247 sweet cherry varieties and hybrids, 115 sour cherry varieties, 195 plum varieties, 135 pear varieties and hybrids, 244 strawberry varieties, 345 *Malus* species, 74 *Pyrus* species, 36 *Prunus* species and hybrids. In the next years we shall complete the genebank in strawberries, rootstocks and wild fruit cultivars.

In the *cultivar genebank* there are collected and documentated

- old German cultivars;
- resistant cultivars;
- new selections from abroad;
- new Pillnitz selections and cultivars;
- interesting hybrids.

In the *wild species genebank* there are collected and documentated

- different *Malus* and *Pyrus* species grown in Germany;
- resistant species;
- apomictic species;
- species used in the fruit breeding;
- ornamental shrubs and trees;
- interesting hybrids.

A lot of evaluation results of the *Malus* genebank is documentated in the IBPGR – 'European Apple Inventory', Part 11, German Democratic Republic, Rome 1988.

Results from using of *wild species in different breeding programmes* in Pillnitz in

- *Apple breeding* (Fischer, C.): Multiple resistant cultivars for fresh market and for processing (Remo, Reglindis, Retina, Rewena, Relinda, Reanda, Rene, Reka and others);
- *Apple rootstock breeding* (Fischer, M.): Dwarf to semi dwarf partial resistant rootstocks (Pillnitzer Supporter rootstocks);
- *Cherry rootstock breeding* (Wolfram): Dwarf to semi dwarf compatible rootstocks (Pi-KU – rootstocks);
- *Pear rootstock breeding* (Mildenberger, Fischer, M.): Semi dwarf frost resistant compatible rootstocks (Pi-BU – rootstocks).

By the use of different *cultivars in classical combination breeding programmes* there has been obtained in the

- *Apple breeding* (Murawski, Fischer, C., Schmadlak, Fischer, M.): High quality summer and winter cultivars (Piros, Pilot, Pinova, Pikant, Pikkolo, Havelgold and others);
- *Sweet cherry breeding* (Mihatsch): Very early to very late ripening cultivars (Nalina, Namosa, Nadino, Nanni, Namare and others);
- *Sour cherry breeding* (Wolfram): Early ripening sweet-sour cultivars (Korund, Karneol, Topas, Morina, Safir and others);
- *Plum breeding* (Mittelstädt, Schmadlak, Fischer, M.): Early ripening PPV tolerant cultivars (Gräfin Cosel, Graf Brühl and others);

– *Strawberry breeding* (Murawski, Ulrich): Cultivars for mechanical harvesting (Fratina, Fracunda and others).

References

Fischer, M., (Ed), 1990. Fortschritte in der Obstzüchtung. Tagungsber. der Akademie der Landwirtschaftswissenschaften der DDR, Berlin, Nr. 292.

Fischer, M., 1994. Langjähriger Aufbau und umfassende Evaluierung der Obstsortimente - Grundlage für die Pillnitzer Züchtungserfolge der Gegenwart. Vortr. Pflanzenzüchtung 27: 6–10.

Fischer, M. & R. Büttner, 1986. Die Bedeutung der Pillnitzer Malusarten-Kollektion für die Apfelzüchtung und als internationaler Genfonds. Archiv Gartenbau 34: 137–145.

Fischer, M. & R. Büttner, 1993. Pillnitzer Genbank für Obst – Bestand, Nutzung, Ergebnisse. Gartenbaumagazin 7/93: 52–53.

Fischer, M., G. Mildenberger, R. Büttner, K. Hammer & J. Schmidt, 1984. Der Genfonds an Malusarten in der DDR und seine Nutzung. Kulturpflanze 32: 123–142.

Haasemann, W., 1989. Ertragsanalyse des Pillnitzer Apfelsorten-Genfonds an zwei Standorten. Archiv Gartenbau 37: 3–22.

Watkins, R., M. Fischer & H. Kleinhempel, 1988. European Apple Inventory, Part 11, German Democratic Republic. IBPGR Rome.

H. Schmidt and M. Kellerhals (Eds.), Progress in Temperate Fruit Breeding, 425–431.

Screening fruit tree genetic resources in Belgium for disease resistance and other desirable characters

M. Lateur & C. Populer
State Plant Pathology Station, Chemin de Liroux, B-5030 Gembloux, Belgium

Key words: Breeding, disease resistance, fruit genetic resources, *Malus domestica, Nectria galligena, Podosphaera leucotricha*, nectria canker, powdery mildew, scab, *Venturia inaequalis*

Summary

The wide diversity of old fruit-tree cultivars originating or introduced into Belgium during the 18th and 19th centuries was collected as far as feasible over the last fifteen years at the State Plant Pathology Station in Gembloux. Out of the 2400 accessions now collected, one quarter was recovered from old public collections, and three quarters came from farms or gardens. The initial intention was to screen the material for disease resistance and other characters of agronomic interest with a view to using the best cultivars as breeding parents. However, as the collection developed, genetic resources conservation also became an objective *per se*. The collection presently contains 1150 apple, 850 pear and 300 plum accessions, and smaller numbers of other fruit species. Each accession is evaluated in an experimental orchard for at least ten years. In view of the growing public interest in old fruit-tree cultivars, the Plant Pathology Station has for several years been releasing to the nursery trade the better cultivars emerging from the evaluation, namely nine apple and four plum cultivars, and one peach cultivar. The principal features of the apple cultivars are presented in this paper. Since 1988, old apple and plum cultivars have been being used at the Station as parents in a breeding programme, with both controlled and open pollination. In some instances, old apple cultivars have also been crossed with a modern parent carrying the Vf gene for scab resistance. The preliminary observations on some of these seedlings are presented.

Introduction

Over the past fifteen years, the Plant Pathology Station in Gembloux has been collecting a wide range of old fruit-tree cultivars, originating in Belgium or introduced into the country during the last two hundred years and still existent in old orchards or school gardens. The collecting work is not yet finished, but temporarily proceeds at a slower pace (Populer & Lateur 1993).

The collection presently contains approximately 2400 accessions, with 1,150 apple, 850 pear and 300 plum accessions, and smaller numbers of other fruit species. The first 1,650 accessions to be collected are being evaluated in an experimental, unsprayed, low-stemmed orchard with the aim of screening for disease resistance and other characters of agronomic interest. The remaining accessions are still in the nursery stage. Two cordon trees of each apple or pear accession are also planted as duplicates in a separate repository at a density of 8,000 trees per ha.

After an evaluation period of at least ten years from budding, the better performing cultivars are presented to the Belgian nursery trade for propagation. This paper presents the principal features of nine old apple cultivars recommended for the nursery trade by the Station, in view of the growing public demand for hardy cultivars. Professional apple growers are also showing some interest, especially those involved in integrated pest management (I.P.M.), who still have to resort to 18–24 fungicide applications yearly for controlling the main fungal diseases on susceptible commercial cultivars such as Jonagold, Golden Delicious or Belle de Boskoop (Van Melckebeke 1990). Four original old plum cultivars and one peach cultivar were also released by the Station but they will not be dealt with here.

Since 1988, some of the better performing old apple cultivars have been used as parents in a disease resistance breeding programme, with both controlled and open pollination. Preliminary observations on some of these seedlings are also presented.

Material

Out of the 2,400 accessions now collected, one quarter was recovered from old public collections, mostly in horticultural schools, and the other three quarters were collected from some 500 farms or gardens, which we were invited to visit by their owners. The Belgian public has been showing a sustained interest over the years in our fruit genetic resources research project. So far, over 1,800 persons have reported ownership or knowledge of original old fruit cultivars. The total number of accessions represents approximately 1,800 cultivars. Of these, more than 600 originated in Belgium, either as landraces or produced by known breeders of past times. The other cultivars are either of foreign origin or unnamed material. In 1874, the Belgian pomologist Gilbert recorded 1,100 pear cultivars and 60 apple cultivars, produced by Belgian breeders during the 18th and 19th century but he made no reference to fruit landraces. More detailed information about the collected material was published previously (Populer 1979, 1980, 1982; Anonymous 1987).

Methods

Evaluation orchard

The accessions are planted in an experimental orchard at a density of 1200 trees per ha. The rootstock used for apple is M9. The orchard is managed under normal professional conditions, but with no fungicide or insecticide protection so that infection can occur freely, an abundant inoculum being provided by the more susceptible cultivars. Soil and climate conditions in the area are particularly favourable for the development of scab and Nectria canker. Well known commercial cultivars are planted as controls. All the trees are trained in the Oeschberg system, which can be adapted easily to any type of plant habit.

Assessment of disease susceptibility

Various keys have been described for assessing the intensity of apple scab in the field (Croxall *et al.* 1952 a & b; Jeger 1981) but they are too complex and time-consuming to be of practical value for evaluating a great number of accessions in the field.

A 1–9 descriptor state scale for the assessment of apple pests and diseases was proposed by the International Board for Plant Genetic Resources (IBPGR) and the EC Committee on Disease Resistance Breeding and the Use of Gene Banks (Watkins & Smith 1982). On this scale, class 1 indicates zero susceptibility, classes 3, 5 and 7 respectively low, medium and high susceptibility, and class 9 maximum susceptibility.

For the type of large-scale field evaluation we need to carry out, we developed a 0–5 assessment key, which is easier to envisage, can be adapted to different diseases and translated into the 1–9 IBPGR scale (Populer *et al.*, 1985). The assessment key for apple scab on fruit and leaves as used at the Plant Pathology Station is detailed in table 1. A 0–5 key built on the same principle is also used in the field for assessing susceptibility to mildew and canker, with a similar convertibility to the 1–9 IBPGR scale (Bauvin *et al.*, 1987; Populer & Lateur 1990).

The IBPGR scale, as we interpret it, is of an exponential type with a slow progression from class 1 to class 3, the "resistance zone" and with more widely spaced classes in the "susceptible zone" where it is unimportant to measure susceptibility precisely. Cultivars which are scored persistently over the years in classes 1 to 3, and exceptionally in class 4, are considered as sufficiently resistant for home-growing or for use by professional growers applying integrated pest management.

Plant data

Season of flowering

In spring of each year, the flowering stage is observed three times a week. The flowering season is divided into seven periods and the flowering intensity is scored on a 0–5 scale (Populer 1986).

Pollen quality

Pollen quality is evaluated over two years by *in vivo* hand pollination on different cultivars. At least 100 flowers per cultivar are emasculated, pollinated and protected with gauze or paper bags. The percentage of

Table 1. Assessment key used at the Plant Pathology Station for apple scab (*Venturia inaequalis*) on fruit and leaves; (X) = intermediate ratings

Field assessment scale	IBPGR scale	Field observations
0	1	No visible macroscopic symptoms
1	2	A few small scab spots detectable on close scrutiny of the tree
2	3	Scab immediately apparent, with lesions very thinly scattered, mostly in the lower or more shaded parts of the tree
2.5	4	(X)
3	5	Infection widespread over the tree; most fruits and leaves with at least one lesion; no defoliation or fruit drop
3.5	6	(X)
4	7	Heavy infection; multiple lesions on most leaves and fruits; defoliation up to 50%; fruit with skin cracks on scabbed lesions, often colonized by Monilia fruit rot
4.5	8	(X)
5	9	Maximum infection; fruits and leaves black with scab; eventually, total leaf fall and fruit drop

fruit set is indicative of pollen quality and/or compatibility.

Eating maturity
After picking, a 5 kg sample of each accession is stored in a ventilated fruit shed where temperature varies from 2 to 9 °C during most of the winter period. Eating maturity is rated on a 1–5 scale as proposed by Smith (1971).

Eating quality, dessert or cooked; fruit size and productivity
Dessert quality is evaluated at picking time for early and mid-season cultivars. Late keepers are tested at picking time if they have already reached eating maturity and 1–3 times during storage. Cooking quality is also evaluated for some of the cultivars. The total production per tree is weighed and the average fruit weight is computed on a sample. The ratings are made on assesment keys developed at the Station and are converted afterwards to the standard IBPGR descriptor states.

Results

Old apple cultivars released to the nursery trade

The nine old apple cultivars released to the Belgian nursery trade by the Plant Pathology Station were initially recommended for home-growing. Trials are now being undertaken with some of these cultivars by a cider-making concern and by a number of professional dessert fruit growers and grower associations.

The cultivars released to the nurseries are propagated as "FGR" cultivars; the acronym refers to the Fruit Genetic Resources project at the Plant Pathology Station.

Fruit susceptibility to scab
The FGR cultivars mostly fall within the 1 to 3 range of the IBPGR scab assessment scale, with an occasional score of 4 in years with high infection pressure, as in 1991 (Tab. 2). The less susceptible cultivars are Grenadier, Gris Braibant, La Paix and Président Roulin, with average ratings of 2 or less. The other cultivars are slightly more susceptible with an average rating around the "low susceptibility" level of class 3.

The assessment of leaf susceptibility to scab shows a good agreement with fruit ratings in most cultivars. The cultivar Gris Braibant falls out of line, however, as it consistently exhibits more susceptibility in leaves

Table 2. Susceptibility to fruit scab (*Venturia inaequalis*) of FGR apple cultivars recommended to the nursery trade by the Plant Pathology Station; ratings on the IBPGR scale (Watkins & Smith, 1982): 0 = no observation (no fruit); 1 = no infection; 3 = low susceptibility; 5 = medium susceptibility; 7 = high susceptibility; 9 = maximum susceptibility

Cultivars	1986	1987	1988	1989	1990	1991	1992	Average
1. Grenadier	1	1	1	1	2	0	1	1.2
2. Gris Braibant	2–3	2	2	2	2	2	1	1.9
3. La Paix	1	2–3	2	2	2	2	2–3	2.0
4. Président Roulin	2	2–3	2–3	2	2	1	2–3	2.1
5. Radoux	2–3	2	2	3–4	2	4	2–3	2.6
6. Blenheim Orange	2	2–3	2–3	2	3–4	4	2–3	2.7
7. Godivert	1	4–5	2–3	3	2	3–4	3	2.8
8. Joseph Musch	2	2	3–4	2–3	2–3	4	3–4	2.9
9. Cwastresse Double	1	3	3	4	3–4	4	3	3.1
Controls								
Priam (Vf)	1	1	1	1	1	1	1	1.0
Belle de Boskoop	5	6	5	5	5	5	5	5.1
Jonagold	4	5	5	6	6	6–7	6–7	5.6
Golden Delicious	4	7–8	7	7	7	7–8	0	6.7

than on fruit; Joseph Musch follows a similar pattern in occasional years.

Susceptibility to powdery mildew and Nectria canker

Table 3 shows that three FGR cultivars – Joseph Musch, Cwastresse Double and Président Roulin – have low susceptibility to mildew. The other cultivars range between class 3 and 4 on an average.

The same table summarizes the observations of the last seven years on canker susceptibility. Président Roulin and Grenadier perform best, with ratings lower than 3.

Plant data

The most important features are presented in table 4. Among the cultivars originating in Belgium, Cwastresse Double and Gris Braibant are Walloon landraces of unknown age, Joseph Musch and Radoux were raised by farmers in the Province of Liège around the middle of the 19th century and Président Roulin was bred by an amateur in Gembloux in 1912. La Paix is reportedly also of Belgian origin. The name Godivert was given at the Plant Pathology Station to an unidentified cultivar.

The cultivars are classified in table 4 according to their eating maturity. The eating season of Grenadier is second-early, Gris Braibant and Godivert are very late keepers and the other cultivars all range from mid to late season.

Flowering season and pollen quality are the main factors determining the possibility of pollination between cultivars. The risk of frost damage to flower buds is also partly dependent on the season of flowering. Grenadier, La Paix, Président Roulin and Radoux are good pollinators for cultivars within the same flowering group, or in the preceding or following group.

La Paix and Cwastresse Double are exceptionally good dessert apples. The quality of Blenheim Orange, Radoux, Joseph Musch and Gris Braibant varies from good to very good according to circumstances.

Président Roulin is the best cropper and an excellent cooking variety but the fruit lacks resistance to transport and is somewhat infected by *Monilinia fructigena*. Most of the other cultivars are in the intermediate productivity class. Joseph Musch is particular in having a compact growth habit and a high productivity/tree volume ratio.

Fruit size and productivity are evaluated without any thinning or fungicide protection. Table 4 also indicates the ratings of commercial cultivars with or without fungicidal protection as recorded in our experimental orchard. Other information about these cultivars was published previously (Populer 1986, Populer *et al.* 1987, Populer *et al.* 1989, Populer & Lateur 1990).

Table 3. Susceptibility to powdery mildew (*Podosphaera leucotricha*) and canker (*Nectria galligena*) of the FGR apple cultivars recommended to the nursery trade by the Plant Pathology Station; for ratings, see table 2

Cultivars	Powdery mildew 1989	1990	1991	1992	Average	Canker
1. Joseph Musch	2	2	3	3	2.5	4
2. Cwastresse Double	2	2	3	3	2.5	4
3. Président Roulin	3	3	3–4	3	3.1	2
4. Grenadier	3	3	3–4	3–4	3.3	2–3
5. Gris Braibant	3–4	3	4	3–4	3.5	4
6. Blenheim Orange	3	4–5	3–4	3	3.5	2–3
7. Godivert	2	4	4–5	3–4	3.5	4
8. La Paix	4	3	4	3–4	3.6	3–4
9. Radoux	3–4	4	4	4–5	4	3–4
Controls						
Golden Delicious	2	2	5	4	3.3	4
Belle de Boskoop	2–3	4	4	4	3.6	3–4
Jonagold	3	4–5	5	5	4.4	5
Priam	3	5	5	4–5	4.4	5

The cultivars Joseph Musch, Gris Braibant and to a minor extent Blenheim Orange and Cwastresse Double were chosen recently by an I.P.M fruit growers' group for evaluation on a commercial scale (Marc 1993). All the FGR apple cultivars, plus some other old cultivars from the Plant Pathology Station, are already on trial in an experimental orchard set up by an important cider and juice-making industrial concern.

The nine FGR apple cultivars released up to now emerged from a six to ten years evaluation process in an orchard which was planted from 1977 onwards and contains 300 apple accessions. The evaluation process is continuing in a new orchard planted from 1983 to 1992 and now containing 600 apple accessions.

Preliminary results of breeding work with old cultivars

Since 1988, a range of old apple cultivars has been used as parents in a breeding programme, with both controlled and open pollination; some of the crosses have also involved a cultivar with the Vf scab resistance as one of the parents (Fig.1). The seedlings are tested for scab resistance by artificial inoculation in the greenhouse (3×10^5 conidia/ml) and for mildew resistance by natural infection after planting out in the field.

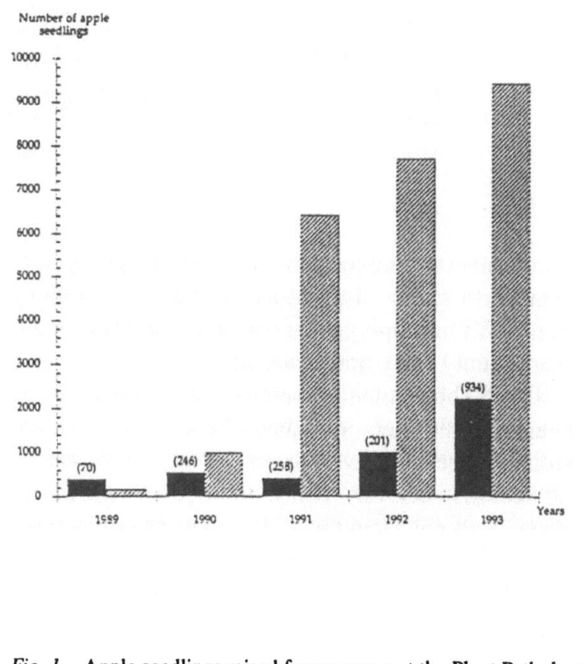

Fig. 1. Apple seedlings raised from crosses at the Plant Pathology Station from 1989 onwards; ■ apple seedlings raised from seed from cross pollination; ▨ seedlings raised from seed from open pollination; ( ) = number of seedlings with a Vf cultivar as one of the parents.

Table 4. Plant data for the FGR apple cultivars recommended to the nursery trade by the Plant Pathology Station

Cultivars	Provenance	Eating maturity	Season of flowering	Pollen quality	Eating quality Dessert	Cooked	Fruit size	Productivity
1. Grenadier	UK	2	2–3	++	2	8	6	4
2. La Paix	B?	3	3–4	++	8		5	4
3. Président Roulin	B	3–4	2–3	++	5–6	8	7–8	6
4. Blenheim Orange	UK	3–4	2–3	–	6–7	7	7	5/b
5. Cwastresse Double	B	3–4	3–4	–	8–9		7	5
6. Radoux	B	3–4	3–4	++	7		4–5	5/b
7. Joseph Musch	B	3–4	2	–	7		7	4–5
8. Gris Braibant	B	5	2–(3)	?	7–8		4–5	4–5
9. Godivert	?	5	3–4	–	3	7–8	7	5
Controls								
Belle de Boskoop	NL	4–5	2	–	5	6	(7)5	6/b
Jonagold	USA	4–5	3–4	–	8		(7)5	(7)4–5
Golden Delicious	USA	4–5	4	++	6		(5)2–3	(7)2–3
Priam	F/USA	3–4	2	++	3		3	5

Eating maturity: evaluation in a ventilated fruit shed, at 2–9 °C; 1 = mid-July - mid-August; 2 = mid-August - September; 3 = October - November; 4 = December - January; 5 = February and later. *Season of flowering*: 1 = extremely early; 7 = extremely late. *Pollen quality*: tested *in vivo* by controlled hand pollination: ++ = very good; – = poor (usually connected to triploidy). *Eating quality*: 1 = extremely poor; 9 = extremely good. *Fruit size*: scale: 1 = extremely small; 9 = extremely large; ( ) = rating when cultivar kept free from diseases. *Productivity* (*cropping efficiency*): 1 = extremely low; 9 = extremely high; ( ) = rating when cultivar kept free from diseases. /b = biennal bearing.

The aim is to produce new cultivars with polygenic resistance to scab and powdery mildew or a combination of Vf and polygenic resistance and with good dessert quality and storage capacity.

Table 5 shows that the percentage of seedlings in progenies from open pollinated trees of La Paix and Radoux in the first two classes of the scab assessment scale is very low. If it is thought acceptable to include seedlings from the third class, the percentage of seedlings passing the scab test rises to 16%, which makes the subsequent screening procedure for other characters a workable proposal.

Table 5. Scab resistance in progenies of open pollinated trees of La Paix and Radoux evaluated after inoculation in the greenhouse and a first year in the field; for ratings see IBPGR 1–9 scale, table 1

	Number of seedlings	% seedlings in scab infection classes 1–2	% seedlings in scab infection classes 1 to 3
La Paix	269	1.2	16.8
Radoux	264	3.4	17.8

Conclusion

The systematic evaluation of fruit tree genetic resources in Belgium has led to the introduction in the Belgian nursery trade of a number of old apple cultivars with low to moderate susceptibility to diseases. Some of these cultivars may be valuable for professional fruit growers interested in integrated pest management or for producing fruit for the industry. They also deserve attention as parents in breeding programmes.

References

Anonymous. 1987. Les anciennes variétés fruitières en Wallonie. Station de Phytopathologie (Ed.), Centre de Vulgarisation Agricole, Gembloux.

Bauvin, J.P., M. Lateur & C. Populer. 1987. La résistance aux maladies chez les anciennes variétés d'arbres fruitiers : évaluation et valorisation commerciale. Meded. Fac. Landbouwwet. Rijksuniv. Gent. 52 : 763–769.

Croxall, H.E., D.C. Gwyne & J.E.E. Jenkins. 1952a. The rapid assessment of apple scab on fruit. Plant Pathol. 1 : 89–92.

Croxall, H.E., D.C. Gwyne & J.E.E. Jenkins. 1952b. The rapid assessment of apple scab on leaves. Plant Pathol. 2 : 39–41.

Gilbert, C. 1874. Les Fruits Belges. Imprimerie Callewaert, Bruxelles.

Jeger, M.J. 1981. Disease measurement in a study of apple scab epidemics. Ann. appl. Biol. 99 : 43–51.

Marc, P. 1993. Développement de la lutte intégrée en verger de pommiers résistants aux maladies cryptogamiques - Coopération Transfrontalière Wallonie/Nord-Pas de Calais. Le Fruit Belge, 443 : 82–83.

Populer, C. 1979. Liste des anciennes variétés belges de poiriers et de pommiers réunies à la Station de Phytopathologie à Gembloux. Note Tech. 3/23, Cent. Rech. Agron. Etat, Gembloux.

Populer, C. 1980. Old apple and pear varieties - What for? UPOV Newsletter 23 : 9–33.

Populer, C. 1982. Why old kind of apples, pears, and plums? Acta Hortic. 142 : 45–53.

Populer, C., C. Delmotte & J.P. Bauvin. 1985. Evaluation de la résistance aux maladies chez les anciennes variétés fruitières en Belgique. A.C.F.E.V. & B.R.G. Journal d'Agriculture Traditionnelle et de Botanique Appliquée, pp. 131–137. Paris.

Populer, C. 1986. Variétés anciennes d'arbres fruitiers recommandées pour la culture d'amateurs. Fiches techniques 1 à 3. Publ. Cent. Rech. Agron. Etat, Gembloux.

Populer, C., J.P. Bauvin & M. Lateur. 1987. Variétés anciennes d'arbres fruitiers recommandées pour la culture d'amateurs. Fiches techniques 4 à 6. Publ. Cent. Rech. Agron. Etat, Gembloux.

Populer, C., J.P. Bauvin & M. Lateur. 1989. La réintroduction de variétés anciennes d'arbres fruitiers dans le commerce des pépinières. Rev. Agric. 42 : 113–124.

Populer, C. & M. Lateur. 1990. Variétés fruitières anciennes et résistantes aux maladies : comportement des variétés de pommier et de prunier RGF recommandées aux pépiniéristes. Meded. Fac. Landbouwwet. Rijksuniv. Gent 55 (3a) : 851–858.

Populer, C. & M. Lateur. 1993. Sauvegarde et valorisation des ressources génétiques fruitières. Annales de Gembloux. In press.

Smith, M.W.G. 1971. National Apple Register of the United Kingdom. Ministry of Agriculture, Fisheries and Food, London.

Van Melckebeke, J. 1990. La lutte intégrée permet-elle de limiter l'utilisation de produits phytopharmaceutiques? Agricontact 23 : (IIa) 1–5.

Watkins, R. & R.A. Smith (Eds) 1982. Descriptor list for apple (Malus). CEC Secretariat, Brussels & IBPGR Secretariat, Rome.

H. Schmidt and M. Kellerhals (Eds.), Progress in Temperate Fruit Breeding, 433–437.

Genetic resources in apple, pear and vineyard peach populations in former Yugoslavia

Dinka Vujanič-Varga, V. Ognjanov, Jelica Balaž, Ksenija Macet & Marija Krstič
Faculty of Agriculture, Trg Dositeja Obradoviča 8, 21000 Novi Sad, Yugoslavia

Key words: Apple, pear, peach, germplasm evaluation, *Malus × domestica, Pyrus communis, Prunus persica*

Summary

During the period 1988–90, several germplasm collecting trips were made to all republics of former Yugoslavia. A total of 56 old apple cultivars, many of which are represented in up to 5 types, 38 old pear cultivars and 367 genotypes of vineyard peaches were collected. The availability of so much genetic and genotypic wealth made it possible to start apple disease resistance breeding and peach cultivar and rootstock breeding programmes.

Introduction

The Balkan peninsula is located within or near the borders of the Mediterranean centre of cultivated plant origin. Most of the land in the fruit-growing areas is hilly, between 200 and 500 m in altitude, although there are smaller areas of lowlands in addition to the Panonian Plain in the north. Inland, the annual rainfall is between 600 and 800 mm, minimum winter temperatures between - 20° C and - 30° C and average temperatures during the growing season around 17° C.

The Balkan peninsula is an extremely rich source of genetic variation in apple, pear and peach which may contribute greatly to the improvement of economically important traits. Although it still contains areas which until recently were effectively isolated, numerous local populations have either disappeared or reduced to a limited number of trees during the modernization of plant production and agriculture.

The main objective of collecting trips made between 1988 and 1990 was to collect, evaluate and conserve apple (*Malus × domestica*), pear (*Pyrus communis* L.) and peach (*Prunus persica* (L.) Batsch) germplasm in former Yugoslavia and make it available to fruit breeding programmes.

Material and methods

Accessions were collected over wide areas of the continental and Mediterranean Yugoslavia. They were sampled as budwood, fruit and seed. The methodological approach was a compromise between a search for local varieties, specific characters and random sampling of populations.

The collection for *in vivo* conservation of collected material is located in the Fruška Gora mountain near Novi Sad. The accessions were described using IBPGR descriptor list for apple (1982). Harvest maturity was defined by iodine – starch and pressure tests. Storage life was assessed under refrigerated storage at 1° C.

Evaluation of disease and pest resistance was based on a 1–9 scale of general field susceptibility where 3 was low, 5 medium and 7 high susceptibility. The 1–9 scale corresponds to the van der Zwet scale (1970) and represents the portion of leaves infected by mildew (*Podosphaera leucotricha* (Ell. et Ev.) Salm) or the severity of aphid (*Aphis pomi* de Geer) attack. Symptoms were observed three times a year at month interval beginning from the end of June. Data are mean ratings for three years.

A computerized database system was established using a programme for record-keeping and variety and seedling evaluation.

Results

Apple

Out of over 1000 apple trees evaluated *in situ*, 56 local cultivars, 26 unnamed accessions and 14 spontaneous seedlings were collected. Genetic erosion has been detected. Many unnamed vegetatively propagated local types of apples have been observed. Today, even well known old cultivars are very difficult to find and are threatened with extinction.

In many cases collected materials clearly reflect preferences for a specific type of apple in a particular area. Most of them have large fruit, good skin finish and attractive colour. Storage life of very late maturing varieties is often over a year. Details on local cultivar origin, harvest maturity, taste, mildew and aphid resistance are given in Table 1. Mildew and aphid resistance evaluation *in vivo* revealed a few resistant, late maturing genotypes like Krstovača, Popadinka and Prancija. A strong hypersensitivity reaction to mildew was observed in Šumatovka and spontaneous seedling HE 1. Since 1992 our resistance breeding programme has been based on Krstovača, Popadinka and Šumatovka as donors of genes for field resistance to apple mildew and scab, as they are also scab resistant. Our observations for field resistance to scab (*Venturia inaequalis* (Cooke) Aderh.) were not reliable because there was no outbreak of scab for three years even in unsprayed conditions.

Pear

Systematically exploration of pear germplasm resources in Bosnia, Macedonia and Serbia, revealed considerable genetic variation. Out of over 2000 pear trees evaluated *in situ*, 38 local cultivars and 32 unnamed genotypes were collected. Collected local cultivars listed in alphabetical order are:

Arapka	Evropejsko Avče	Lojevača	Susakuša
Argandče	Grofica	Lubeničarka	Takuša
Badnička kruška	Ilinjača	Medunac	Tropuša
Bela Arapka	Ječmenka	Miolika	Trupnjak
Belaja	Jeribasma	Mioljak	Turšijara
Bostanka 1	Kajkuša Zimska	Misirka	Vašerka
Bostanka 2	Kaličanka	Panadurka	Zimnjača
Brašanka	Karamanka	Petrovača	Žutica
Carigradsko avče	Lisica	Svrljiški medunac	
Crvena Stolovača	Litrenjača	Sinka	

Additionally in Bijelo Polje, Montenegro, 60 small fruited native pear germplasm were collected as random sample of very rich genetically diverse population.

All accessions belong to *Pyrus communis* L. Only characterization work was done on collected material. Season of ripening stretches from late June to late October, with a peak from late August to mid September. Fruit size varied widely from small to large, oblong to pyriform in shape. In general, accessions had poor to intermediate eating quality with high amount of grit cells. Very late accessions had firm and extremely coarse texture which could be consumed only after flesh breakdown. Fruits are used mainly for drying or for the production of pear brandy.

Disease and pest evaluations are available only from *in situ* observations. Occurrence of scab (*Venturia pirina* Adelh.) and pear psylla (*Psylla piricola* Foerster) was low although entirely clean accessions like Žutica were rare. Surrounded by heavily scabbed pear trees, Žutica looked completely resistant.

Peach

The collection contains 602 accessions, of which 235 are varieties originating from Italy, USA, China, Japan and Yugoslavia and 367 are vineyard peach accessions. Within vineyard peaches as regards flesh colour six groups were distinguished: white flesh without red colour at the pit cavity (25.9%), white flesh with red colour at the pit cavity (50.1%), yellow flesh without red colour at the pit cavity (8.9%), yellow flesh with red colour at the pit cavity (12.4%), mostly red flesh with white stripes (2.4%) and entirely purple flesh with white colour at the pit cavity (0.3%). Size of the fruit varied from 43 g up to 214 g. Most fruits have very soft flesh with the special flavour, aroma and fragrance of dessert fruits. A few large-fruited, late-maturing, white fleshed, clingstone peaches with firm, non-melting flesh texture suitable for processing were also observed. The ripening period stretches over more than four months, from August to December. Present research work is conducted in two main areas:

a Selection of vineyard peach genotypes suitable as seedling rootstocks,

b Selection of vineyard peach genotypes suitable as dessert fruit.

Evaluation for peach leaf curl, caused by *Taphrina deformans* (Berk) Tul., and mildew, caused by *Sphaerotheca pannosa* (Wallr.) Lew. were done in unsprayed conditions in years with outbreaks of diseases in orchards. Completely clean accessions, like

Table 1. Origin, harvest maturity, taste, mildew and aphid ratings of apple accessions collected from 1988 to 1990

Local cultivar	Origin	Harvest maturity	Taste	Mildew rating	Aphid rating
Avajlija	BiH	7			3
Akpetrič	SLO			3	3
Alamanka	Mac	8	sweet	2	2
Babovača	MN	5	sweet/acid		
Bedrika Sitna	Sr	8	sweet	3	5
Bedrika Krupna	Sr	8	sweet	2	2
Bela Čejna	Sr			4	1
Bela jabuka	Sr	4	sweet		
Bobovec	SLO	8	acid	2	5
Budimka	Sr	9	sweet/acid	4	5
Gospoinjača	Sr			2	4
Grekinja	Sr				
Dulabija	BiH	5	sweet	4	
Ilinjača	Sr			4	7
Jardolika	Sr			2	5
Karapaša šarena	Mac	7	sweet		
Kiseljača	BiH	5	acid	1	1
Kožara	SR	6	sweet	3	3
Kolačara	Sr, BiH	7	sweet/acid	1	5
Kraljica	Sr				7
Krivopecelj	SLO	8	acid	3	3
Krstovača	MN	8	sweet	1	1
Krstorača Bela	MN	8	sweet	1	1
Krstorača Crvena	MN	8	sweet	1	1
Krupna divljaka	BiH	6		1	1
Kumovača	Sr			1	7
Lipovača	BiH	3	sweet	2	5
Mirišljavka	Sr				
Mušanika	Sr				
Muškinja	BiH	8		2	3
Orahovača	Sr				
Pamuklija	Sr			4	7
Pašinka	MN	7	sweet		
Pepeljača	Sr				
Petrovka	BiH	1	sweet		
Pljosna široka	Sr	8			
Popadinka	Sr	7		1	3
Pramenka	Sr			1	5
Prancija	BiH	5	sweet	1	2
Prijedorska zelenika	BiH	7	acid	3	4
Prijedorska šarenika	BiH	5	acid	3	1
Senabija	BiH	8	sweet	4	4
Sitna strekača	Sr	9	sweet	4	5
Slavonska Srčika	CRO	7	acid	3	2

Table 1. continued

Local cultivar	Origin	Harvest maturity	Taste	Mildew rating	Aphid rating
Tetanka	Mac	7	sweet	2	7
Tetovka	Mac	8	sweet	2	3
Tvrdokorka	Sr	9	sweet		6
Zejtinka	Sr	6	sweet	1	
Zelenika	Sr	8	acid/sweet	4	3
Zlatna Kraljica	Sr	2	sweet	6	7
Šarunka	Sr	8	sweet		
Šečerka	Sr	6	sweet	1	
Šogorka	Sr				
Šumatovka	Sr.MN	9	acid	2	3
Ciganka	Mac	8	acid	3	1
Crvena jabuka	Sr	4	sweet	4	4
Idared		7	acid/sweet	8	4
Jonagold		7	sweet	5	5
Golden Delicious		6	sweet	7	5
Granny Smith		9	acid	4	3

SLO - Slovenia, Sr - Serbia,
CRO - Croatia, MN - Monte Negro,
Mac - Macedonia, BiH - Bosnia and Hercegovina.

JM 1, EA 1, AV2 and B 35/4 for mildew and JJ5 and NJ5 for leaf curl, were very rare.

Discussion

Although, many names of provinces and settlements originate from the words jabuka (apple) and kruška (pear), there is no written document that Slovenians, moving to the Balkan peninsula, found apple and pear as domesticated species. Probably, the first cultivated apples and pears were introduced from Greece and Italy to Macedonia and Slovenia at the beginning of the new era, spreading further on to the north and west. During the 500 years of Turkish reign the predominant taste was sweet which explains why nearly all local apple varieties found south from river Danube are without acid. Fruit growing in Slovenia and North Croatia was more influenced by Italy, Austro-Hungary and Germany, where acid apples are preferred. This distribution of genotypes has not changed much to this day.

The peach was introduced to the Balkan peninsula between 400 and 300 B.C., probably in the form of vineyard peach. In Yugoslavia today, vineyard peach can be found as a generative population. It is propagated by seeds and grown as semi-cultivated single trees in vineyards or on hillsides. Fruits are eaten fresh and used in home canning. Seedlings are used as the main source for rootstocks in Yugoslavia and are exported to other European countries for the same purpose.

The only successful conservation programme on pear germplasm in the Balkan peninsula region was led by van der Zwet (1970) when over 400 accessions were collected.

Apple, pear and peach germplasm had never benefited from nation-wide efforts to conserve it. It is threatened with extinction as a result of the modernization of agriculture. Introduction of more advanced, large fruited cultivars has been so successful that they have almost entirely replaced local types. Many valuable local types have already disappeared or will be lost in the next few years. Fortunately genetic resources in sparsely populated and less developed areas of Yugoslavia have been less eroded.

Conclusion

Apple, pear and peach germplasm in the Balkan peninsula is an extremely rich source of genetic variation which can contribute greatly to the improvement of economically important traits. Their dessert fruits very often have a special flavour, aroma and fragrance. Other useful traits include winter hardiness, drought and disease resistance or tolerance, which are rare and the result of centuries of natural selection. The germplasm has become adapted to many different climates and soil types. Because of the existing variability in the region, differing ecological conditions and human activities, the Balkan peninsula can be regarded as one of the secondary centers of genetic diversity of apple, pear and peach.

A long-term programme to conserve the existing germplasm is urgently needed, hopefully as an international project together with Bulgaria, Greece and Romania. Within former Yugoslavia, more thorough collecting should be carried out in South-west Serbia, Macedonia and Montenegro.

References

Stankovič, D. & P.D. Mišič, 1978. Yugoslavia – an important source of fruit germplasm. Chron. Hortic. 18(1): 3–4.

Van der Zwet, T., D. Stankovič & V. Cociu, 1983. Collecting *Pyrus* germplasm in Eastern Europe and its significance to the USDA pear breeding program. Acta Horticulturae 140: 43–45.

Vujanič-Varga, Dinka & V. Ognjanov, 1992. Conservation of vineyard peach population in Yugoslavia. FAO/IBPGR Plant Genetic Resources Newsletter 90: 28–30.

Watkins, R. & A.R. Smith, 1982. Descriptor list for apple (*Malus*). IBPGR Secretariat, Rome.

H. Schmidt and M. Kellerhals (Eds.), Progress in Temperate Fruit Breeding, 439–445.

Genetic resources in apricot and peach trees in Romania

Viorica Balan & Antonia Ivascu
Research Station for Fruit Tree Growing Baneasa, Bucharest, Bd. Ion Ionescu de la Brad nr. 4 sect. 1, cod 71592, Romania

Key words: Apricot, collections, genitors, heredity, peach, genetic resources, *Armeniaca vulgaris, Amygdalus persica*

Summary

By investigating the phenotype of 443 apricot and 609 peach and nectarine cultivars and selections in collections, as well as by individually selecting potential genitors and by using the results of genetic study we have been able to draw out strategies for the breeding of new genotypes.

Genetic transgression of late blossoming, late fruit ripening, higher frost resistance of flower buds, resistance against *Cytospora cincta, Stigmina carpophila, Sphaerotheca panossa* var. *persicae*, content of dry matter, overall sugar and vitamin C were noticed in those combinations where the cultivars below were used as sources for genes: apricot Don Gaetano, Comandor, Excelsior, Dacia, Goldrich, 77.3.52 BV, 83.4.12 BI, 83.12.17 BI, Patriarca temprano and peach and nectarine – Flacăra, Cresthaven, July Elberta, Albertina, Compact Red Haven, 77.43.40, Superba de toamnă, Hardyred, Nectared 10, NJN 58, ARK 125, ARK 134, Regina.

The heterosis phenomenon involved in physiological adaptability and resistance against *Sphaerotheca pannosa* var. *persicae* and *Cytospora cincta* was noticed in the combination between the following genitors: apricot – Dacia, 8K1739, De Kostiugeni, Murfatlar 3/28, Olimp, Stella, Sulina, Farmigdale, Monaco Bello, Maria Matilde, Erevani, D3R29T91 and peach and nectarine – Amsden, Cherrired, Harbrite, Redwing, Victoria, Cardinal, Redskin, Triumf, ARK 134, ARK 125, HNA, Crimsongold, NJN 58, Romamer I, Armking, ARK 141.

The dominance of fruit early ripening in the genitors of: apricot – Patriarca Temprano, 77.3.52 BV, 80.3.60 BV, 83.4.12 BI, 83.4.70 BI, 83.16.28 BI, NJA 53, Early Orange, Baneasa 4/73 and peach and nectarine: – Early Red, Collins, Favorita Morettini 1, Primerose, Crimsongold, Romamer 1, PSCR17P26, HSR1P11, has made it possible to obtain hybrids with this feature.

Segregation in F1 at a ratio of 3:1 or 1:1 made the selection of hybrids with round fruit possible by using the following genitors: apricot – Goldrich, Excelsior, Dacia, Harogem, Early Orange and peach – Collins, Congres, Cardinal, Harko, Morton.

Introduction

To breed new apricot, peach and nectarine genotypes that are better adapted to the local environmental conditions and have extra early or very early ripening fruits, implies the development of new genetic combinations with better disease resistance and high quality fruits.

Considering that the number and value of such combinations greatly depends on the diversity and value of the available genes the researches of the ‘Research Station for Fruit Tree Growing – ‘Baneasa’ have set up new apricot, peach and nectarine collections made up of species, cultivars and selections from various geographical areas. The specificity of the collections resides in their being part of an apricot, peach and nectarine breeding programme, the goal of it being both the preservation of the entire existing genetic variability, and the preservation and turning to account of the useful genetic variability, as well as the finding of the genetic resources of such species.

Materials and methods

The phenotypes of 443 apricot and 609 peach cultivars and selections were studied from the large apricot and peach growing zones of the world: Europe, America, Asia, as well as smaller ones: Australia, New Zealand.

290 apricot cultivars and selections and 251 peach cultivars and selections from the following European countries were examined: Romania: 195 apricot and 181 peach; France: 22 apricot and 23 peach; Hungary: 14 apricot and 4 peach; Holland: 2 apricot; Italy: 22 apricot and 14 peach; Czech and Slovac Republic: 15 apricot; Ucraina: 13 apricot; R. Moldova: 2 apricot; Spain: 1 apricot and 1 peach; Yugoslavia: 2 apricot; Bulgaria: 1 peach; Germany: 2 apricot.

From America there are 138 apricot cultivars and selections and 324 peach cultivars and selections from S.U.A. 117 apricot and 294 peach, from Canada 21 apricot and 24 peach, from Argentina 6 peach.

From Asia have been studied 14 apricot cultivars and 35 peach cultivars.

The countries of Asia from which we obtained the cultivars are: Iran: 6 apricot; Turkey: 2 apricot; China: 3 apricot; Armenia: 1 apricot; Afghanistan: 1 apricot; India: 1 apricot and 1 peach; CIS: 25 peach and Japan: 4 peach.

Australia is represented in the collection with 1 apricot cultivar and New Zealand with 2 peach cultivars. The morphological and physiological variation and fruit quality characteristics have been examined. Potential genitors of apricot and peach which correspond to breeding strategies have the following characteristics: early and late fruit maturity; low vigour stable resistance to diseases of apricot and peach, and high quality of fruits.

According to their features and characteristics, the genitors were used in various hybridizing schemes, obtaining hybrid populations which were used for the study of genetics and of heredity. The inheritance to the descendents of blossoming period, maturity fruit period, fruit quality (shape, size, dry substance content and C vitamin, disease resistance and frost resistance) was studied.

Resistance or sensitivity reaction of genitors and hybrid descendents to the action of pathogenous agents were registered with artificial infection conditions applying *Cytospora cincta* and under natural infection conditions without applying fungicides with *Stigmina carpophila* and *Sphaerotheca pannosa* var. *persicae*.

Results and discussion

Finding potential apricot and peach genitors

57% of the apricot and 22% of peach phenotypes were selected as potential genitors of fruit early maturity, while 6% of the apricot and 19% of the peach phenotype of late maturity. 7.5% apricot and 5% peach phenotypes were proper for low vigour. 51% apricot and 31% peach phenotypes featured good fruit quality, while 5% apricot and 2% peach phenotypes out of the whole number of trees existent in our collections proved very high fruit quality.

10% of the apricot and 15% of the peach genitors proved frost resistance, under the conditions of a temperature drop down to - 25° C in January or of a temperature variation of almost one week from + 16° C to - 16° C during February.

A 1.9% of potential disease resistant genitors could be found in apricots, these standing the cumulative attack of four pathogens: *Monilinia laxa, Stigmina carpophila, Cytospora cincta* and Plum-pox, 11% were resistant to 1 or 2 diseases.

7.2% of peach phenotypes proved resistant against *Cytospora cincta* and *Sphaerotheca pannosa* var. persicae.

Heredity of certain features and characteristics of potential genitors and the development of breeding strategies

Types of heredity on to F1 hybrid descendent

We found out heredity transgression on to the descendents of some hybrid families, where the genitors were:

- for apricot: Mektep, Don Gaetano, Early Orange, Skaha, Comandor, Dacia, 80.3.60 BV, 77.3.52 BV, Mr. 69/5, Excelsior, 6630052, Goldrich, HW 412,83.4.12 BV, 83.12.17 BI, Patriarca temprano.
- for peach: Flacăra, Roubidoux, Superba de toamnă, Albertina, Madison, Triumf, Harbrite, July Elberta, Cardinal, Splendid, Redwing.
- for nectarine: Romamer 1, Nectared 10, NJN 21, NJN 58, ARK 125, Crimsongold, Regina, ARK 141.

The transgressions that were noticed in hybrid descendance stand for:

Blossoming stage start up: 2–5 days later in apricot

Table 1. Apricot and peach genetic resources in which the dominance of fruit early maturation is made use of

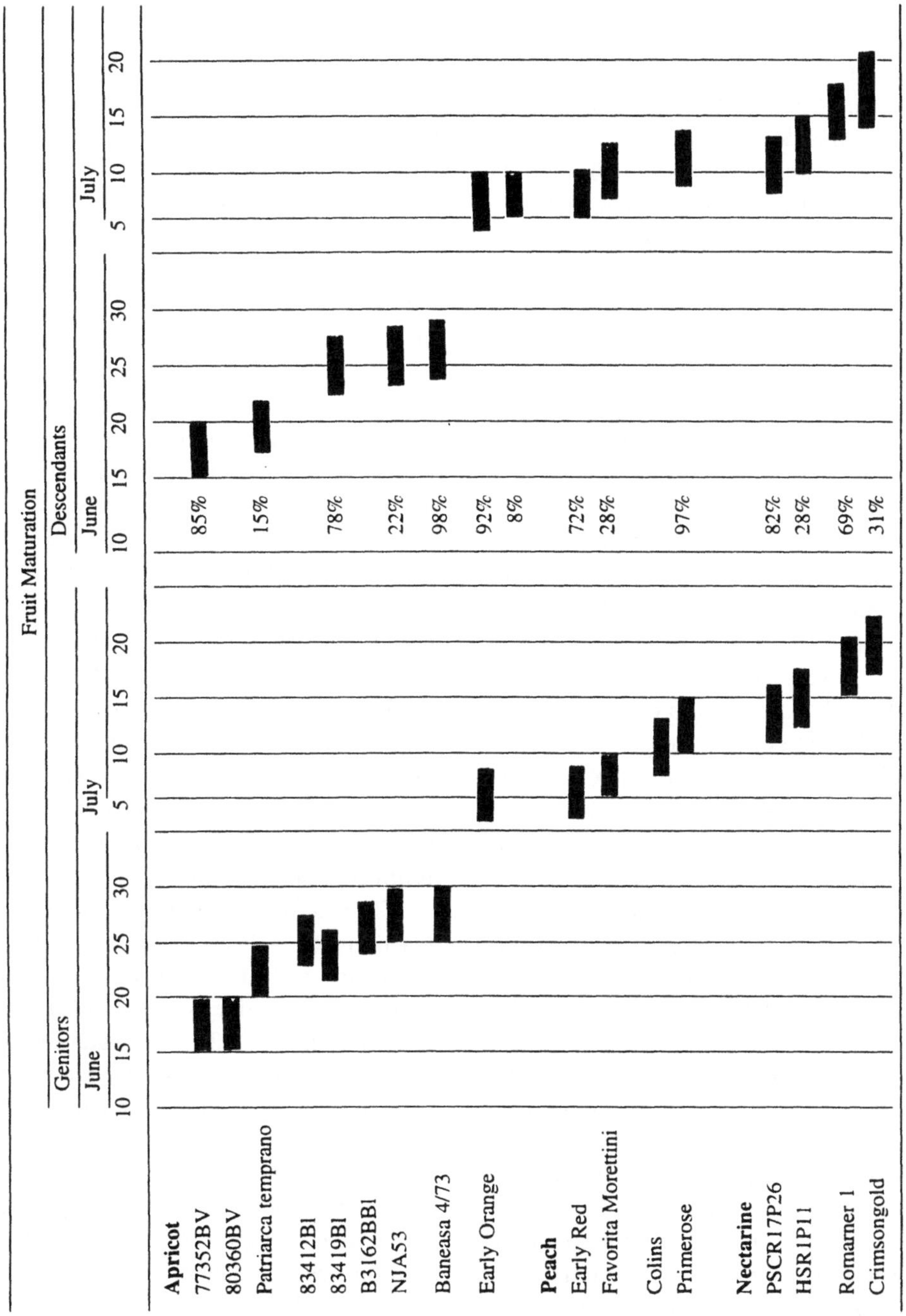

and 2–3 days earlier in peach. Maturation stage start up: 3–14 days earlier in apricot and 3–10 days in peach. High dry substance content: over 18% in 20–80% of apricot and 14% of the peach descendents.

Higher content of C vitamin; over 14% in apricot, peach and nectarine.

Stable-disease resistance: a-apricot descedents more resistant against Cytospora cincta in 31 families out of the 51 studied, against *Stigmina carpophila* in 23 hybrid families, against *Monilinia laxa* in 29 hybrid families, against *Pseudomonas syringae* in 30 hybrid families: b-more resistant peach descendent: against *Cytospora cincta* in 22 hybrid families out of 65 families, against *Sphaerotheca pannosa* var. *persicae* in 28 hybrid families.

Frost resistance: 5–15 of descendents are resistant. The study of fruit ripening period of fruits point out at the both species three situations: 1 transgressive-for very early in the following types of combinations: late maturity × medium Comandor (5VIII) × Dacia (10VII); Favorit (15VIII) × Dacia (10VII) at apricot and Roubidoux (25VIII) × Collins (10VII); Superba de toamna (10IX) × Favorita Morettini 1 (15VII) at peach; 2 dominant transmission of very early maturity feature (15VII) in combinations of type: late maturity × very early maturity – apricot; Comandor (5VIII) × 77.3.52 BV (15VI); Comandor (5VIII) × 80.3.60 BV (15VI) and at peach Elberta (25VIII) × HB 26-65 (18VI); Flacara (1IX) × HB 25-38 (20VI) in combinations of type: very early maturity × early maturity – 77.3.52 BV (15VI) × Patriarca temprano (20VI); 80.3.60 BV (15VI) × Patriarca temprano (20VI); 83.4.12 BI (20 VI) × NJA 53 (25VI); NJA 53 (25VI) × Baneasa 4/73 (25VI) in combinations of type: very early maturity × early maturity – 83.16.28 BI (25VI) × Early Orange (2–5VII) at apricot and in combinations of type: early maturity × early maturity – Early red (5VII) × Favorita Morettini 1 (5VII) or Collins (10VII) × Primerose (25VII) at peach and PSCR 17 P26 (15VII) × HSR1P11 (20VII) or Romamer 1 (15VII) × Crimsongold (15VII) at nectarine; 3. intermediate situation of fruit ripening period in comparison with both two parents generally nearby of genitor with medium maturity in combinations of type: medium maturity × late maturity-Comandor (5VIII) × Excelsior (25VII) and Excelsior × Comandor; Favorit (15VIII) × Excelsior (25VII); Olimp (10VII) × Harogem (15VII) at apricot, and Harbrite (5VII) × Flacara (1IX); Triumf (10VII) × HB 9-46 (5IX) at peach.

The important objectives of the breeding programme in apricot and peach constitute extra early maturity (15-25VI) and early maturity (25VI-5VII) of fruits, breeding strategy could be based on transgressive transmission or dominance of maturity period of fruits.

The parents of the apricot descendents were: 77.3.52 BV (15VI) × 80.3.60 BV (15VI) and Patriarca temprano (20VI) had ripening period 85% on 15 of June and 15% on 20 of June, and those with parents: 83.4.12 BI, 83.4.19 BI, 83.16.28 BI, and NJA 53 had ripening period 78% on 20–24 of June and 22% on 23–28 of June (Table 1).

98% from descendents in which Baneasa 4/73 (25VI) was one of parents and 92% from descendents in which E. Orange (5VII) was parent had ripening period of fruits in the same period with the parents.

At peach resulting from parents as Early red (3–8 July) and Favorita Morettini 1 (5–10 July) ripened in 78% on 5–8 July and 28% on 6–10 July, and those descendents have as from parents: Collins (8–14 July) and Primerose (12–15 July) ripened 97% in the same period with their parents (Table 1).

Similary situations registered also at the nectarine descendents resulted from parents: PSCR17P26, HSR1P11, Romamer 1 and Crimsongold (Table 1).

The interaction apricot host plant – *Cytospora cincta* and *Stigmina carpophila* and peach and nectarine host plant – Cytospora cincta and Sphaerotheca pannosa var. persicae show improved resistance at one of the apricot, peach and nectarine descendents and this is effect of heterosis or hybrid vigour phenomenon.

The heterosis phenomenon improved the resistance reaction (R) against *Cytospora cincta* artificial infection action, at the F1 hybrid descendent from the sensitive × sensitive type combination as, Mektep × Don Gaetano, B22/32 × Wenatchee, Don Gaetano × E. Orange, Mektep × Skaha, Excelsior × Goldrich, Comandor × Excelsior Comandor × 77.3.52 BV or from the sensitive × medium resistant type as, Comandor × Dacia (Table 2).

The resistance reaction (R) against *Stigmina carpophila* was improved in the F1 from: Excelsior × Goldrich, Comandor × Dacia, 8K1739 × Maria Matilde, Sulina × Erevani, Olimp × Monaco Bello sensitive × medium resistant type).

The heterosis phenomenon improved the resistance reaction (R) against *Cytospoara cincta* and *Sphaerotheca pannosa* var. *persicae* at the F1 peach and nectarine descendent of the following hybrid families: Cherrired × Redskin, ARK 125 × Crimsongold,

Table 2. Heterosis phenomenon of descendent resistance against the attack of certain steady diseases of apricot and peach

Hybrid combinations	*Cytospora cincta* (necrosis) cm				*Stigmina carpophila* GA%			
	$\bar{x}\frac{P_1+P_2}{2}$	$\bar{x}(F_1)$	$F_1-\frac{P_1+P_2}{2}$	$\frac{\text{Test'T'}}{P_1F_1P_2F_1}$	$\bar{x}\frac{P_1+P_2}{2})$	$\bar{x}(F_1)$	$F_1-\frac{P_1+P_2}{2}$	$\frac{\text{Test'T'}}{P_1F_1P_2F_1}$
a. apricot								
Mektep^x Don Gaetano	32.5	20.0	-12.5	*** ***	–	–	–	–
B22/32^x Wenatchee	24.0	13.5	-10.5	*** ***	–	–	–	–
Don Gaetano ^{x}E. Orange	27.5	15.0	-12.5	*** ***	–	–	–	–
Mektep^x Skaha	31.5	12.0	-19.5	*** ***	–	–	–	–
Excelsior^x Goldrich	15.0	4.8	-10.2	*** ***	17.5	15.0	- 2.5	** **
Comandor^x Excelsior	23.5	2.8	-20.7	*** ***	–	–	–	–
Comandor^x Dacia	24.5	3.9	-20.6	*** ***	15.0	10.0	- 5.0	** **
Comandor^x 77352BV	29.0	3.9	-24.1	*** ***	–	–	–	–
8K1739^x Maria Matilde	–	–	–	–	13.0	11.0	- 2.0	** **
Olimp^x Monaco Bello	–	–	–	–	16.5	6.0	-10.5	*** **
Sulina^x Erevani	–	–	–	–	19.5	13.0	- 6.5	*** ***
b. Peach and nectarine								
Cullinan^x Triumf	29.5	12.9	-16.6	*** ***	–	–	–	–
Victoria^x Cardinal	–	–	–	–	12.5	3.4	- 9.1	*** ***
Cherrired^x Redskin	27.6	14.3	-13.3	*** ***	18.2	7.9	-10.3	*** ***
Victoria^x Harbrite	34.2	19.9	-14.3	*** ***	–	–	–	–
Redwing^x HB9-10	–	–	–	–	11.8	9.2	- 2.6	** **
Ark125^x Crimson-gold	19.2	6.8	-12.4	*** ***	14.2	5.7	- 8.5	*** ***

Table 2. continued

Hybrid combinations	*Cytospora cincta* (necrosis) cm				*Stigmina carpophila* GA%			
	$\bar{x}\frac{P_1+P_2}{2}$	$\bar{x}(F_1)$	$F_1-\frac{P_1+P_2}{2}$	$\frac{\text{Test'T'}}{P_1F_1P_2F_1}$	$\bar{x}\frac{P_1+P_2}{2})$	$\bar{x}(F_1)$	$F_1-\frac{P_1+P_2}{2}$	$\frac{\text{Test'T'}}{P_1F_1P_2F_1}$
Romamer 1^x NJN 58	21.6	7.4	-14.2	*** ***	9.7	4.3	- 5.4	** **
HNA^x Armking	20.8	12.7	- 8.1	*** ***	–	–	–	–
ARK141^x Regina	16.5	9.2	- 7.3		8.2	3.6	- 4.6	** **

$\bar{x}\frac{P_1+P_2}{2}$ parents overage
$\bar{x}$ F_4 : overage

Romamer 1 × HJN 58 and ARK 141 × Regina sensitive × medium resistant).

At the F1 from Cullinan × Triumf, Victoria × Harbrite (peach combination) and HNA × Armking (nectarine combination) was improved the resistance only against *Cytospora cincta*, and at the F1 from Victoria × Cardinal and Redwing × HB 9-10 (peach combination) the resistance was improved only against *Sphaerotheca pannosa* var. *persicae* (Table 2).

The fruit shape characteristic segregated in certain hybrid families at a 3:1 ratio. Here are the apricot families: Comandor × Excelsior ($\chi^2 = 0.44$; P% = 70), Dacia × Comandor ($\chi^2 = 0.58$; P% = 70), E. Orange × Skaha ($\chi^2 = 0.44$; P% = 70), Dacia × Goldrich ($\chi^2 = 0.26$; P% = 70), Harogem × Comandor ($\chi^2 = 0.08$; P% = 80), and the peach families: Collins × Congres ($\chi^2 = 0.14$; P% = 70), Cardinal × Roubidoux ($\chi^2 = 0.21$; P% = 70), Redhaven × 437-45 ($\chi^2 = 0.22$; P% = 70), Harko × Roubidoux ($\chi^2 = 0.34$; P% = 70), Morton × Collins ($\chi^2 = 0.18$; P% = 70).

Genetic resources on apricot and peach trees and development of breeding strategies

The apricot and peach breeding goals concerning late and early maturity aim at extending the harvesting and consumption period, late blossoming, higher frost resistance, stable-disease resistance. The high content of dry substance, sugar and C vitamin can be obtained by making use of the transgressive-type heredity of certain genitors such as: Don Gaetano, Comandor, Excelsior, Dacia, Goldrich, 77.3.52 BV, 83.4.12 BV, Patriarca temprano, Mektep, Skaha, HW 412, 6630052, Mr. 69/5 (in apricot); Superba de toamna, Albertina, Madison, Triumf, Harbrite, July Elberta, Cardinal, Splendid, Redwing (in peach); Romamer 1, Nectared 10, NJN 21, NJN 58, ARK 125, Crimsongold Regina, ARK 141 (in nectarine).

Fruit early maturity can be obtained by using the dominant feature in the following genitors: apricot-77.3.52 BV, 80.3.60 BV, Patriarca temprano, 83.4.12 BV, 83.4.19 BV, 83.16.28 BI, NJA 53, Baneasa 4/73, Early Orange; peach: Early red, Favorita, Morettini 1, Collins, Primerose and nectarine: PSCR17P26, HSR1P11, Romamer 1, Crimsongold.

The resistance of F1 hybrid progenies where the parents were intermediary or even sensitive was possibly due to the heterosis phenomenon and is liable to back up the goal of creating a phenotype resistance against *Cytospora cincta* (in apricot), and *Cytospora cincta* and *Sphaerotheca pannosa* var. *persicae* (in peach).

Genetic resources used:

- apricot: Don Gaetano, B22/32, Wenatchee, E. Orange, Skaha, Comandor, Excelsior, Dacia, 77.3.52 BV, 8 K 1739, Maria Matilde, Olimp, Monaco Bello, Sulina, Erevani.
- peach and nectarine: Cullinan, Triumf, Victoria, Cardinal, Redskin, Cherrired, Romamer 1, NJN 58, HNA, Armking, ARK 141, Regina.

Spheric fruits can be obtained by separating F1 segregants developed in hybrid families where the partners were: apricot-Comandor, Excelsior, Dacia, E. Orange,

Skaha, Harogem and in peach: Collins, Congres, Cardinal, Roubidoux, Redhaven, HB7-45, Harko, Morton.

In some apricot, peach and nectarine genetic resources, we could explore simultaneously their transgressive-type early maturity, dominance, disease resistance, heterosis and the segregation of fruit shape characteristic.

They are: 77.3.52 BV, 80.3.60 BV, 83.4.12 BV and for peach and nectarine: Collins, Romamer 1, Crimsongold.

References

Bailey, C.H. & F. Hough, 1975. Apricots. In: J. Janick & J. More (Eds) Advances in fruit breeding. West Lafayette. Ind. Prude Univ. Press.

Bassi, D. & F. Bartolozzi, 1993. Il miglioramento genetico dell' albicocco: obiettivi e strategie. Rivista di Fruticoltura 5: 41–53.

Balan, V., 1992. Variability study of initial apricot breeding material to identify new genes sources for adaptability, productivity and quality. Lucr. Stiintifice I.C.P.P. Pitesti-Maracineni, vol. XV: 55–70.

Couranjou, J., 1989. Main genetic results from a study of an apricot semi-dialel. IX International Symposium on Apricot Culture.

Hartl, D., D. Freifelder & L. Snyder, 1988. Basic genetics. Jones and Bartlett Publishers, Boston Portola Valley, Chapter 1: 7–33, Chapter 8: 195–213.

Ivascu, A., 1992. Researches concerning the range of nectarine trees varieties grown in the are of Bucharest. Lucr. Stiintifice I.C.P.P. Pitesti-Maracineni, vol XV: 97–105.

Liverani, A. & C. Fideghelli, 1993. Il miglioramento genetico del pesco: risultati e prospettive. Rivista di Frutticoltura 5: 11–23.

H. Schmidt and M. Kellerhals (Eds.), Progress in Temperate Fruit Breeding, 447–449.

Genetic diversity and scope of walnut improvement in India

R.L. Sharma & K. Kumar
Department of Fruit Breeding, University of Horticulture and Forestry, Nauni-Solan (HP) 173 230, India

Key words: Explorations, nut characters, *Juglans regia*, persian walnuts, tree selections, walnut production in India

Summary

Persian walnut (*Juglans regia* L.) improvement is in a primitive stage despite the fact that there exists in it a great amount of genetic diversity. All walnut plantations are of seedling origin. These are on an area of 35,000 hectares and produce around 20,000 MT nuts annually. About half of the produce is exported. All the seedling populations exhibit tremendous genetic variation in tree and phenological traits, colour, shell sealing and in hardness of nuts as well as quality and percentage of kernel. Therefore, systematic evaluation of this genetically diverse germplasm needs to be taken up immediately for selecting superior genotypes to build gene repository. Fortunately the selection for most of the traits, being highly heritable, can be accomplished easily. The paper envisages the nature and extent of existing variation for various phenological prospects. The genetic improvement of this nut crop with high export potential is also suggested.

Present status

Persian walnut (*Juglans regia* L.) is the only walnut grown in India. Currently, its annual production is around 20,000 MT (Qureshi & Dalal 1985, FAO 1989) and half of the produce is exported. Almost the entire produce comes from seedling trees covering roughly 35,000 hectares in temperate region of the North-Western Himalayas 1200 m to 3500 m above mean sea level. Besides a few hundred thousand trees in Jammu & Kashmir province, thousands of walnut trees with nuts of good quality also occur further South-East in the interior of Himachal Pradesh.

Genetic diversity

Persian walnut is the only species under commercial cultivation. A mention in literature (Dode 1909) of another species *J. duclouxiana* Dode of thin-shelled walnuts being of commercial occurrence in northern India has not attracted any attention and taxonomists have preferred to include it under *J. regia* L. only (McGranahan & Leslie 1990). Some strains of persian walnuts with unusually thick shell and small kernel have been found to exist in forests near perpetual snowline far beyond human inhabitation in Chamba district of Himachal Pradesh (Sharma 1993). The variable thick-shelled nuts are mostly of very poor quality, hard enough to break small stones used for breaking the nuts. The kernel is only 10–12% and so difficult to extricate that it can be taken out only in small bits with the help of a needle. Existence of similar walnuts near Gangotri in Uttar Pradesh hills was also reported earlier by Hayes (1957). These come under *J. regia* L. A number of other walnut species viz., *J. nigra*, *J. sieboldiana*, *J. cordiformis*, *J. major*, *J. chenova*, *J. mandshurica* and *J. drianorski* were introduced from abroad but none of them has been utilized in cultivation or as rootstock or in variety improvement programmes in India (Sharma 1993).

The vast genetic variation within the *regia* species in India is because of: 1) the area being a part of the centre of origin of walnuts, 2) monoeciousness and dichogamy in the species, and 3) high heterozygosity due to continuing sexual propagation.

There are no standard walnut cultivars in cultivation in the country. The seeds of superior strains are used for raising planting material which give variable stand. The age of bearing tree extends to more than

Table 1. Range of variation in phenological traits of walnut seedling populations.

Trait	Range of Variation	Reference
Time of leafing out	2nd wk March to 4th wk April	15
Time of full bloom		
Male:	3rd wk March to 3rd wk April*	15
Female:	4th wk March to 4th wk April*	
Time of nut maturation	2nd wk August to 3rd wk October	2, 9, 14, & 15

* In late leafing types the male catkins emerged and matured earlier than or simultaneously with leaf emergence.

Table 2. Range of variation in nut and kernel traits of walnut seedling populations.

Trait	Range of Variation	Reference
Nut colour	Dark brown to straw yellow	14, 15
Nut shape	Round-broad ovate-narrow elliptic	6, 14, 15
Nut length (cm)	2.50–5.46	2, 9, 14, 15
Nut diameter (cm)	2.00–4.68	2, 8, 14, 15
Nut weight (g)	4.2–26.0	8, 9, 14, 15
Shell thickness (mm)	0.2–4.3	2, 7, 14, 15
Shell-out (%)	24.0–80.0	2, 7, 14, 15
Kernel colour	Dark brown to light cream	7, 8, 14,15
Kernel weight (g)	1.2–21.0	8, 9, 14, 15

100 years. Many of the trees are as good as of any standard cultivar. During the past 20–30 years, several workers viz., Motial & Gilkar (1970) and Qureshi & Dalal (1985) in Kashmir valley; Nauriyal et al. (1970) and Singh (1993) in Kullu valley; Chauhan & Sharma (1979) in Kinnaur and Pandey & Sinha (1984) in Jaunsar Bhabhar area of UP hills, have tried to survey superior seedling trees but ended up in describing the nature and extent of variation for various characteristics in the seedling populations. The outcome of these surveys has been confined to quantifying the variation in metric traits (Tables 1 and 2).

The range of variation in the nut characters (Table 2) is tremendous. 'Karan' walnut from Kashmir (Motial & Gilkar 1970) for instance is an extremely thin-shelled easy enough to be shelled with very negligible pressure of fingers. The yield is 100–125 kg nuts with more than 80% kernel. Likewise, 'Gobind' is a similar selection popularized in Himachal Pradesh. This is high yielding with semi-soft shelled nuts. Excessively thin-shelled character is considered by some as undesirable from in-shell transportation view point but there exist intermediate types with 60–65% kernel content (Qureshi & Dalal 1985). Singh (1993) observed the occurrence of two trees (out of 104) which since inception of their fruiting decades back have borne tri- and tetra locular nuts.

In addition to surveys of indigenous seedling populations, efforts have been made to introduce a few exotic varieties with improved characteristics (Rathore 1984, Rathore & Mittal 1986). The preliminary evaluation indicates that 'Lake English', 'Turtle 31', 'Payne' and 'Hansen' hold promise at least in early stages of trees. Evaluation work on 19 germplasm accessions is also in progress in the University (Anonymous 1992).

Future scope

In spite of extensive variation prevalent in the native walnut germplasm, there have been only sporadic attempts to evaluate it. There have been no genetic studies or hybridization work. The selected genotypes have not been multiplied for testing and commercial growing due to poor success in vegetative propagation. Sometimes even the rootstocks were not available with the explorers. Vegetative multiplication of seedling-tree selections would have made the process of selection easier in a very short period. It would be desirable to include in future surveys the following characters: precocity (lateral bearing) and higher yield, regularity in bearing, delayed leafing and flowering, insect-pest and disease resistance, large, round, thin shelled and well sealed nuts, better kernel quality covering the quality of proteins and fat, adaptation to environmental stresses especially extreme of cold and low chilling, synchronous maturity of male and female flowers, and short duration maturity.

Walnut provides a typical example of crop improvement through explorations and evaluation of individual trees in seedling plantations by utilizing its vast genotypic variation created through natural cross pollination and perpetuation through seed and harnessing the natural selection facilitated by the extreme climatic diversities.

References

Anonymous, 1992. Annual Progress Report (1991–92), Dept. of Fruit Breeding & Genetic Resources, Univ. of Hort. & Forestry, Nauni-Solan (HP).

Chauhan, J.S. & S.D. Sharma, 1979. Phenotypic variability in walnuts of Kinnaur district of Himachal Pradesh. Indian J. Agric. Sci. 49(6): 420–22.

Dode, L.A., 1909. Contribution to the study of the genus *Juglans*. Bul. Soc. Dendrologique de France 11(22–90) 12: 165–215.

F.A.O., 1989. FAO Production Yearbook. F.A.O., Rome.

Hayes, W.B., 1957. Fruit growing in India. Kitabistan, Allahabad.

McGranahan, G. & C. Leslie, 1990. Walnuts (*Juglans*). In: Genetic resources of temperate fruits and nut crops. Acta Hort. 290(2): 907–951.

Motial, V.S. & A.M. Gilkar, 1970. The exchange earner – Kashmir Walnuts. Indian Fmg. 20(6): 31–33.

Nauriyal, J.P., K.L. Chadha & Harish Kumar, 1970. Some promising seedling walnut trees in the Kullu Valley. PAU J. Res. 6(3) (Suppl.): 852–857.

Pandey, D. & M.M. Sinha, 1984. Some promising seedling selections of walnut from Jaunsar Bhabhar area of Garhwal hill: Morphological studies. Progr. Hort. 16(3–4): 183–187.

Qureshi, A.S. & M.A. Dalal, 1985. Status of nut crops in Jammu & Kashmir. Progr. Hort. 17: 197–205.

Rathore, D.S., 1984. Evaluation of genetic resources of walnut. In: T.R. Chadha, V.P. Bhutani & J.L. Kaul (Eds) Advances in research on temperate fruits, pp. 47–50. Dr. Y.S. Parmar Univ. of Hort. & Forestry, Nauni-Solan.

Rathore, D.S. & S.K. Mittal, 1986. Conservation and preliminary evaluation of genetic resources of temperate fruits. Proc. Silver Jubilee Seminar on role of plant genetic resources in the development of hill agriculture, NBPGR, Regional Station, Phagli, Shimla, pp. 181–190.

Sharma, R.L., 1993. Genetic resources of temperate fruits. In: K.L. Chadha (Ed) Advances in horticulture, Vol. 1, pp. 243–266. Malhotra Publishing House, New Delhi.

Sharma, S.D. & J.S. Chauhan, 1980. A note on variability in nut characters of walnut seedlings. Pb. Hort. J. 20(1–2): 78–79.

Singh, D., 1993. Genetic variability in bearing seedling walnut (*Juglans regia* L.) in Kullu valley. M.Sc. Thesis, Dr. Y. S. Parmar Univ. Hort. & Forestry, Solan, pp. 155 + iv.

H. Schmidt and M. Kellerhals (Eds.), Progress in Temperate Fruit Breeding, 451–454.

Resistance to a single UK isolate of *Colletotrichum acutatum* in strawberry germplasm from Northern Europe

D.W. Simpson, C.Q. Winterbottom,[1] J.A. Bell & M.L. Maltoni[2]

Horticulture Research International, East Malling, West Malling, Kent ME19 6BJ, UK; [1] present address: P.O. Box 73113, Davis, California 95617-3113, USA; [2] present address: Istituto Sperimentale per la Frutticoltura, Piazzale della Vittoria, 15. 47100 Forli, Italy

Key words: Anthracnose, *Colletotrichum acutatum, Fragaria* × *ananassa*, resistance, strawberry

Summary

Colletotrichum acutatum on strawberry was first introduced to the UK on runners of cultivar Brighton, which were imported from California in 1983. Recently further outbreaks have been reported following the importation of infected runner plants from mainland Europe.

Phytosanitary precautions have so far been successful in controlling the disease but the current trend in the UK is for increasing late season production. This increases the likelihood of there being fruiting plants when the environmental conditions are favourable to *C. acutatum*.

Fifteen cultivars and 19 HRI breeding lines were tested for susceptibility to *C. acutatum* using the most pathogenic isolate found on plants imported into the UK. It is likely this isolate was originally of Californian origin. Potted plants were spray inoculated and subsequently rated in the glasshouse using a disease severity rating (DSR) with the range from 0–4, where a rating of 4 indicated severe disease symptoms.

In contrast to results from California and Italy standard deviations were variable and often large and consequently it was considered unsafe to classify the susceptibility of the genotypes on the basis of mean DSR alone. Classification was thus based on a combination of the mean value and the distribution of the scores but in some cases the results were inconclusive.

It was possible to classify five lines as having a high level of resistance (Elvira, Honeoye, EM99, EM224, EM255); six lines had a useful level of resistance but did show symptoms on some plants (Cambridge Favourite, Gorella, Pandora, Pantagruella, EM17 and EM290); five lines were highly susceptible (Elsanta, Redgauntlet, Tamella, EM237 and EM319). The remaining 18 lines could not be classified with the same degree of certainty but 10 were probably resistant.

Introduction

Black Spot or anthracnose of strawberry, caused by the fungus *Colletotrichum acutatum*, was first introduced to the UK on runners of cv Brighton, which were imported from California in 1983. The first outbreak was effectively controlled by eradicatory measures but recently further outbreaks have been reported following the importation of infected runner plants from mainland Europe. It is likely that these outbreaks can also be traced back to California via European nurseries who had imported mother plants of Californian cultivars, which are grown in southern Europe.

In the past *C. acutatum* has been associated with warmer strawberry growing regions such as Florida and California, and recently the disease has become serious in southern Italy and spain. The optimum temperature for the fungus is 28° C and epidemics are usually associated with hot periods accompanied by high rainfall. It is splash dispersed, so frequent rainfall or overhead irrigation will exacerbate the problem. Madden et al. (1992) showed that substantial dispersal occurred within 30 seconds during a rain shower and maximum incidence of rot in test plots occurred after showers of only sixteen minutes. Only a short wetness duration (c 6 hrs) is required at or near the optimum temperature for infection of ripe fruit to occur (Ellis & Madden, 1993). The disease does not disappear at low temperatures and Ellis & Madden (1993) reported that it overwinters well on mummified fruit, although Eastburn & Gubler (1990) reported poor survival in infested petiole pieces under California conditions. Problems arise when *C. acutatum* becomes established in the nurseries, particularly if these are raising everbearing cultivars, as it is impossible to remove all the fruits from these plants in the nursery bed and the disease can rapidly spread from a few infected fruits to the leaves and stolons of other plants. Often infected plants will show no symptoms until the optimum conditions arise.

Increasing late season production in the UK means it is becoming more likely that plants will be fruiting at times when the environmental conditions are favourable to the disease. If, as a consequence, the disease becomes established then resistant cultivars could potentially offer the simplest solution to the problem.

Winterbottom (1988) studied 50 Californian cultivars and breeding lines and found that over half were resistant. Furthermore, he concluded that the resistance was conferred by a major dominant gene. This suggests that the resistance may be widespread and might also be present in European germplasm since the material tested in California had not been bred with any deliberate intention of incorporating resistance to *C. acutatum*. Tests on the susceptibility of European cultivars and HRI breeding lines are reported here.

Materials and methods

Cold stored (frigo) plants of 15 cultivars available in the UK were obtained from a commercial nursery and 19 HRI breeding lines were propagated from misted tips. Five isolates of *C. acutatum* were obtained, four of which had been isolated from plants imported into the UK and one from strawberry fruits imported from Costa Rica. On receipt single- spore cultures were derived from each isolate.

To produce inoculum the pathogen was cultured on potato dextrose agar (PDA) and grown under continuous fluorescent light at 25° C ± 1° C. Conidia were harvested from PDA plates by flooding with 10 ml of sterile distilled water and agitating the culture with a glass rod. Aliquots (0.5 ml) of the conidial suspension were spread on PDA plates, harvested after 8–10 days as previously described, and then filtered through four layers of muslin to remove mycelial fragments. Conidial concentrations were adjusted to 1.5×10^6 conidia ml^{-1} for plant inoculation.

Prior to inoculation all plants were grown in 11 cm pots in an unheated polyethylene house and given liquid feed on alternate days to encourage rapid growth of the foliage. Just prior to inoculation all but the youngest 3 to 4 leaves were removed from plants which were 6 to 10 weeks old. All plant parts were spray inoculated to the point of runoff with the conidial suspension and were immediately placed in a dark growth chamber for 48 hr at 28° C ± 1° C, and 100% humidity was maintained by enclosing each plant in a polyethylene bag. Plants were removed from the growth chamber after 2 days and were subsequently maintained in a glasshouse at 30° C ± 5° C day and 22° C ± 5° C night temperature. All experiments were completed in the period July to mid September.

Initially the pathogenicity of the four isolates derived from plants was tested by inoculating four Pandora plants (a known resistant – Faedi et al., 1991) and four Elsanta plants (a known susceptible) with each. The Costa Rican isolate did not sporulate well and therefore was not used. For subsequent testing the most pathogenic isolate (IMI351247-1) was used throughout and two to four plants of the test cultivar or selection were inoculated on each occasion. Four Elsanta plants were included as susceptible standards at each inoculation date; two plants were inoculated and two were left uninoculated. Inoculations were repeated at least once using different plants and on occasions when the standard cultivar did not behave as expected, or, where the results for the test plants were inconsistent, further test inoculations were done. Plants were rated daily for 15 days starting two days after inoculation. The disease severity rating (DSR) scale of Smith & Black (1987) was used on the runners and occasionally petioles: 0 = no lesions; 1 = lesion(s) < 3 mm; 2 = lesion(s) 3–10 mm, not girdling runner; 3 = lesion(s) 10–20 mm, girdling runner; 4 = lesion(s) > 20 mm, runner necrotic; 5 = youngest leaf wilted in absence of lesions, indicating crown infection; 6 = dead. This DSR scale was developed for use with the closely related fungus *Colletotrichum fragariae* and the two highest ratings on the scale (5 & 6) would not normally be expected when working with *C. acutatum* of presumed Californian origin. The mean DSR was calculated for runners for each genotype.

Table 1. Mean and maximum scores for *Colletotrichum acutatum* symptoms and susceptibility classification for 34 strawberry cultivars and breeding lines

Cultivar or line	Number of plants*	Mean score**	SD	Max score	Plants with max score	Class
Cambridge Favourite	24	0.38	0.97	4	1	RES
Cambridge Vigour	8	0.25	0.71	2	1	RES
Elsanta	30	2.77	1.52	4	16	SUS
Gorella	14	0.5	1.16	4	1	RES
Hapil	6	0.17	0.41	1	1	RES
Pandora	19	1.16	1.34	4	2	RES
Pantagruella	12	0.50	1.17	4	1	RES
Pegasus	25	1.20	1.38	4	3	RES?
Redgauntlet	36	2.56	1.68	4	17	SUS
Tamella	16	3.19	1.42	4	11	SUS
EM17	12	0.67	0.65	2	1	RES
EM20	6	0.67	1.03	2	1	RES
EM22	17	1.00	1.32	4	1	RES?
EM33	7	1.14	1.07	2	4	RES?
EM90	9	0.44	0.88	2	2	RES
EM237	15	3.93	0.26	4	14	SUS
EM290	15	0.73	1.22	4	1	RES
EM317	16	2.00	1.32	4	3	SUS?
EM319	13	3.08	1.04	4	7	SUS
EM404	10	1.3	1.49	4	1	SUS?
EM407	7	0.29	0.76	2	1	RES
LA934	5	0.80	1.10	2	2	RES?

* For two accessions (Gardena and EM320) less than five plants were tested and these are excluded from the table.

** Elvira, Honeoye, EM99, EM224, EM255 (12–16 plants) and Bogota, LA937, LA969, LA1043 (5–8 plants) all expressed no symptoms.

SD = standard deviation.

Results

In the initial pathogenicity tests no symptoms were seen on Pandora plants but two isolates, from Cornwall and Kent, produced good symptoms on Elsanta. The Cornish isolate (IMI351247), originally from cv Brighton, was the more pathogenic and was thus used in all subsequent tests.

For his original study in California Winterbottom had rated genotypes with a mean DSR of 3.0 or greater 20 days after inoculation (DAI) as susceptible whereas those with a DSR less than 3.0 were classified as resistant. However, in the experiments reported here the results were much less consistent than had been experienced when using the same technique in either California or Italy (Winterbottom, 1988; Faedi et al., 1991). For example the standard susceptible cultivar, Elsanta, had an overall mean of 2.77 with a standard deviation (SD) of 1.52 (Table 1) whereas in Italy the mean for Elsanta had been 3.99 and the SD 0.04 (Faedi et al., 1991). Overall the standard deviations were variable and often large (Table 1) and consequently it was considered unsafe to classify the susceptibility of the genotypes on the basis of the means alone. Classification was thus based on a combination of the mean value and the distribution of the scores, with genotypes having a mean ≥ 2.0, but where a proportion of the plants achieved the maximum DSR (4.0), being recorded as susceptible. Other genotypes are classified as resistant, although in some cases the results were inconclusive and classifications which are uncertain are indicated by a question mark (?). Additionally, any where the number of experimental plants was less than 10 should also be treated with some caution.

Despite the problems with symptom expression it was possible to classify five lines as having a high level of resistance: Elvira, Honeoye, EM99, EM224, EM255 (no symptoms, 12–16 plants); while six lines had a useful level of resistance but did show symptoms on some plants: Cambridge Favourite, Gorella, Pandora, Pantagruella, EM17 and EM290. Five lines were highly susceptible: Elsanta, Redgauntlet, Tamella, EM237 and EM319; and the remaining 18 lines could not be classified with the same degree of certainty but 10 were probably resistant: Bogota, Cambridge Vigour, Hapil, EM20, EM90, EM320, EM407, LA937, LA969 and LA1043 (4–9 plants).

Discussion

Overall the findings were less decisive than had been anticipated from earlier studies (Winterbottom, 1988; Faedi et al., 1991). Even Elsanta, the susceptible standard, which was found to be highly and consistently susceptible in Italy (Faedi et al., 1991), was recorded with a score of 4 on only 53% of the inoculated plants and five plants (17%) showed no symptoms. Another anomalous finding was for Bogota which was classified as susceptible by Faedi et al. (1991) but showed no symptoms here. Interestingly, Bogota was tested against a single isolate at two locations in Italy and symptoms were either absent or mild in Cesena but it was consistently highly susceptible when tested in Trento.

These inconsistences may have arisen because the isolates of *C. acutatum* available were less pathogenic than those used in California or Italy. Smith & Black (1990) showed that isolates can vary considerably in their pathogenicity and they also proposed the possible presence of a race structure based on a significant genotype × isolate interaction. However, their data were inconclusive in this respect since a high proportion of the genotype × isolate combinations gave an intermediate disease response and the LSD's were large. An alternative explanation for our inconsistent symptoms could be that the temperature and environmental conditions in the glasshouses where the testing was done were not always favourable to the disease. This second explanation seems more likely since this would also explain the inconsistency of the results during the course of the study. The British weather is less constant than that of California and Italy and temperature fluctuations were unavoidable.

This study has shown that useful levels of resistance to *C. acutatum* are present in both HRI breeding lines and northern European cultivars. If necessary resistance screening could be incorporated into the HRI breeding programme using the method described here but with closer control over environmental variation. Furthermore, two commercially acceptable resistant cultivars are already available to UK growers, namely Honeoye and Elvira. Pegasus may also have a useful level of resistance but requires further testing.

Acknowledgements

The work was funded by the Horticultural Development Council, Contract SF19b. We are grateful to Dr Roger Cook, of the Ministry of Agriculture, Fisheries and Food (MAFF) Central Science Laboratory, who kindly supplied the isolates of *C. acutatum*. The work was carried out under MAFF licence number PHF1300/63/62.

References

Eastburn, D.M. & W.D. Gubler, 1990. Strawberry anthracnose: Detection and survival of *Colletotrichum acutatum* in soil. Plant Disease 74: 161–163.

Ellis, M.A. & L.V. Madden, 1993. Studies and observations on the epidemiology and control of anthracnose fruit rot of strawberry in Ohio. Acta Horticulturae 348: 449–457.

Faedi, W., G.L. Bagnara, P. Turci, C.Q. Winterbottom & R. De Clauser, 1991. Valutazione della suscettibilità ad antracnosi di diversi cloni di fragola. In: P. Rosati, S. Fraccaroli & A. Febi (Eds) Nazionale sulla Fragola, Verona 8 November 1991, pp. 261–269. Società Orticola Italiana.

Madden, L.V., L.L. Wilson, X. Yang & M.A. Ellis, 1992. Splash dispersal of *Colletotrichum acutatum* and *Phytophtora cactorum* by short duration simulated rains. Plant Pathology 41: 427–436.

Smith, B.J. & L.L. Black, 1987. Resistance of strawberry plants to *Colletotrichum fragariae* affected by environmental conditions. Plant Disease 71: 834–837.

Smith, B.J. & L.L. Black, 1990. Morphological, cultural and pathogenic variation among *Colletotrichum* species isolated from strawberry. Plant Disease 74: 69–76.

Winterbottom, C.Q., 1988. Resistance of strawberry to *Colletotrichum acutatum*. MS Thesis of Science in Horticulture. University of California, Davis.

H. Schmidt and M. Kellerhals (Eds.), Progress in Temperate Fruit Breeding, 455–457.

Strawberry genetic resources in Romania

Mihail S. Coman, Aurel N. Popescu, Irina N. Popescu & Stefan N. Chiriac
Fruit Research Institute Pitesti-Maracineni, 0300 Romania

Key words: Fragaria, genetic resources, variation

Summary

Since 1970 variation in phenotypic traits in a field collection of strawberry cultivars, including foreign varieties and clones as well as Romanian clones, have been investigated for potential use in breeding. The strawberry germplasm collection has been enlarged with genotypes obtained by intra and interspecific crosses, androgenesis, irradiation and organogenesis. The ample variability among the 192 *Fragaria* and 3 *Potentilla* genotypes, including 2×, 4×, 5×, 6×, 8× and 10× ploidy level, covers a wide range of potentially useful traits such as fruit quality, yield and pollen fertility.

Introduction

Fragaria × *ananassa* Duch., the modern cultivated octoploid strawberry ($2n = 8\times = 56$), is the culmination of all strawberry species, resulting from human intervention in its evolutionary process (Bringhurst 1990). According to Staudt (1962) it originated from Europe. Although the cultivated strawberry is derived from two American octoploid compatible wild species, *F. virginiana L.* and *F. chiloensis* Duch., many breeders use *Fragaria spp.* of equal or lower ploidy level of different American, European or other origin to incorporate some useful traits. Most of the successful incorporation of wild derived traits have come through backcrossing, notably the incorporation of day-neutrality from *F. virginiana glauca* into the cultivated strawberry by Bringhurst (1990). The success in strawberry breeding is directly connected with the availability of germplasm resources. Proper and efficient usage of genetic resources in breeding is based on a good knowledge of behaviour under specific ecological conditions and also on analysis of genetic variation.

Table 1. Strawberry germplasm under assessment at Pitesti

Ploidy level	Species	Genotypes	Origin
	Fragaria		
2 ×	F. vesca	6	I,S,F, USA, Ro
	F. daltoniana	1	Asia
4 ×	F. orientalis	1	Asia
6 ×	F. moschata	3	Europe
8 ×	F. virginiana	1	USA
	F. chiloensis	3	USA, Au
	F. × ananassa	169	Europe, America, Asia
	Other polyploids		
2 ×	Dihaploid (F. × ananassa)	1	Ro
4 ×	Tetrahaploid (F. × ananassa)	1	Ro
5 ×	F. vesca × F. ananassa	5	Ro
10 ×	F. vescana	1	D
	Potentilla		
2 ×	P. fruticosa	1	Europe
	P. rupestris	1	Europe
6 ×	P. palustris	1	Europe

Material and methods

Plant material The strawberry germplasm under assessment at Pitesti is given in Table 1. It has been

Table 2. Genotypes under evaluation

	VE	A4 ×	Mo	CH	AN
a. interspecific crosses					
F. vesca (VE)	–	31	–	39	73
Aiko 4 × (A4 ×)	–	–	55	–	–
F. moschata (Mo)	–	–	61	–	–
F. chiloensis (CH)	4	–	8	–	92
F × ananassa (AN)	70	38	12	33	–
F. vescana (VN)	–	–	34	–	–
b. androgenesis	43	–	–	–	–
c. irradiation	–	–	–	–	12
d. in vitro organogenesis	–	–	–	–	3

divided in three groups: a) wild species and genotypes, except the octoploids, b) old commercial cultivars, c) new cultivars and clones (Cociu 1976). 134 European, 57 North American, 3 Asian and 1 Australian genotypes are included in the present collection. These have been obtained through exchange of biological material, intra and interspecific crosses, androgenesis, gamma irradiation and *in vitro* organogenesis.

Characters measured The interspecific and androgenetic genotypes were checked for their cytogenetical status by using the squash method with root tips stained by acetoorcein, before planting them in the germplasm collection. For wild species and other polyploids, microscopical observations on the size and density of stomata and on the pollen fertility (fertile or semifertile) were carried out. For octoploid genotypes, field and laboratory observations and measurements were made according to methods described by Coman (1992).

Data analysis The paper presents only two main quantitative traits: yield and fruit weight. For some promising genotypes general combining ability (GCA) has been estimated according to Griffing (1956).

Results

Characterization

Aiko 4× was obtained in 1987 by direct androgenesis from an octoploid Aiko cultivar. Phenotypically it is similar to the octoploid, but the pollen fertility is very uneven (18–84%).

Aiko 2× was obtained in 1989 by the same method from an Aiko 4× genotype. It is less vigorous and pollen fertility does not exeed 30%.

Both clones are continuously used in interspecific crosses.

Pentaploids, according to Bringhurst (1990) true polyploid, $2n = 5\times = 35$, are octoploid × diploid hybrids. At present 70 clones are being assessed but so far only five have been used in backcrosses with strawberries, or for chromosome doubling. These have shown semifertile pollen and a vigorous plant habit. Besides 558 clones originating from different interspecific hybridizations, a further 43 clones were obtained from *F. vesca* androgenesis, 12 through gamma irradiation and 3 by *in vitro* culture of somatic tissue. These are under evaluation (Table 2).

Among 169 investigated cultivars and clones (Table 3), only 8 (4%) cropped over 650 g per plant and 19 (10%) beween 500–650 g, such as

Table 3. Yield of strawberry varieties, clones and species

Yield (g./plant)	octoploid cvs. and clones			Species and other polyploids	
	indigenous	foreign			
	no.	no.	%	no.	%
< 100	–	3	12	10	5
100–250	11	28	20	8	4
250–350	21	36	30	3	2
350–500	12	31	22	2	1
500–650	8	11	10	–	–
> 650	3	5	4	–	–

variety/clone/species	Yield (g/plant)
79.17.1	820
Spadeka (10 ×)	380
S. Sengana	710
F. chiloensis B2 (8 ×)	250
78-62-1	620
Capron (6 ×)	240
Dukat	510
Aiko (4 ×)	210
Redgauntlet	510
F. vesca Ro (2 ×)	50
Benton	500
Addie	500

Table 4. Fruit weight of strawberry varieties, clones and species (1989–1992 period)

Mean weight (g)	octoploid cvs. and clones indigenous	foreign		Species and other polyploids	
	no.	no.	%	no.	%
< 8.5	9	23	17	21	11
8.5–10.5	19	18	19	2	1
10.5–15.0	16	41	29	–	–
15.0–20.0	11	20	16	–	–
> 20.0	–	12	7	–	–

variety/clone/species	Mean weight (g)
83–44–1	24.9
Spadeka (10 ×)	8.5
Linda	21.2
F. chiloensis B2 (8 ×)	7.6
Dana	21.2
79–244–2	20.1
Cesena	19.4
Gea	15.7
Brio	15.5
Premial	13.2
Redgauntlet	10.8

Table 5. Estimates of general combining ability (CGA) in some strawberry parents

Traits	G C A Females 83–44–1	79–244–2	Brio	78–62–1
Yield	- 0.65	- 0.79	- 0.30	1.73+
Fruit Weight	2.44+	0.17	- 1.26	- 1.34
	G C A Males			
	Premial	Redgauntlet	Benton	Gea
Yield	- 0.14	0.29	- 0.26	0.11
Fruit Weight	- 1.10	- 0.46	- 0.16	1.73

Explanation: + level of significance at P = 0.05

79.17.1, Senga Sengana, Dukat and Redgauntlet. The yield of most genotypes ranged between 250–300 g per plant. The yield potential of species or other polyploids is lower, but Spadeka (10×) and *F. chiloensis* B2 have cropped over 250 g per plant.

Large variation in fruit weight has been recorded among cultivars and species (Table 4). The data revealed only 12 cultivars (7%) with a mean berry weight over 20 g, and 31 (16%) between 15–20 g such as 83–44–1, Linda, Dana and 79–244–2. 17% of the cultivated strawberry varieties had very small fruits of less than 8.5 g.

Depending on the parental material selected to increase fruit size and yield, the general combining ability can be important in breeding. The estimates of GCA for those quantitative traits are listed in Table 5. Positive GCA significantly different from 0 were found. The female parent 78–62–1 had the highest GCA for yield and clone 83–44–1 for fruit size. As male parents, positive values for yield were shown by Redgauntlet and Gea and for fruit size only Gea. Consequently, the breeding strategy would be based on the parents with the highest GCA when the trait to be improved is quantitatively inherited.

Conclusion

To attain our breeding objectives the existing strawberry germplasm collection has sufficient genetic resources for the main traits and it is up to the plant breeder to take advantage of this. Continuous efforts are required to enlarge the germplasm collections, data collections and improve data processing.

References

Bringhurst, R.S., 1990. Cytogenetics and evolution in American *Fragaria*. Hort. Sci. 25(8): 879–881.

Cociu, V., 1976. Establishment, preservation and utilization of fruit germplasm in Romania. Fruit Var. J. 33(2): 45–48.

Coman, M., 1992. Fondul de germoplasma la capsun in Romania. Probl. Genet. Teor. Aplic. XXIV(1–2): 51–59.

Griffing, B., 1956. Concept of general and specific combining ability in relation to diallel crossing system. Aust. J. Biol. Sci. 9: 465–493.

Staudt, G., 1962. Taxonomic studies on the genus *Fragaria*. Can. J. Bot. 40: 869–886.

H. Schmidt and M. Kellerhals (Eds.), Progress in Temperate Fruit Breeding, 459–461.

Screening *Actinidia* species germplasm for frost tolerance

J. Chat
Station de Recherches Fruitières, I.N.R.A. – Centre de Bordeaux, BP 81, 33883 Villenave d'Ornon Cedex, France

Key words: Actinidia sp., frost tolerance, germplasm, screening method

Summary

A breeding programme was initiated in 1990 to assess the frost-tolerance of four *Actinidia* species, i.e. *A. deliciosa, A. arguta, A. kolomikta* and *A. polygama.* Two-year-old plants were raised at an outdoor site for natural cold acclimation. In February 1992, resistance to frost was assessed by the recovery of plants in a glasshouse after freezing at - 18° C in a controlled environment chamber. *A. arguta* clone 16 showed the highest level of frost-tolerance, while all the *A. deliciosa* vines examined suffered severe frost damage following freezing.

Introduction

Frost may cause serious yield losses in kiwifruit (*Actinidia deliciosa* (A. Chev.) C.F. Liang & A.R. Ferguson var. *deliciosa*), particularly in south-west France where severe frost damage was evident in many kiwifruit orchards in January 1985. However frost damage may occur throughout most of the year. Previous observations have indicated that plants may incur serious frost injury in spring (young shoots and flower buds), autumn (fruits and leaves), and winter (trunk, canes and dormant buds) (Blanchet, 1985). Several methods have been used for measuring cold hardiness. Exposure to natural frost conditions is the oldest and ultimate test of plant cold hardiness (Li, 1984). However the main pitfall of such a test is that the magnitude of the frost stress cannot be controlled with the consequence that experimental results are not reproducible and therefore difficult to interpret. Potted plants (or plant parts) may be subjected to artificially induced frost conditions. The use of a controlled temperature chamber is the most common and suitable method for assessing frost tolerance though its use in early generation selection is problematic where only single plants or unreplicated progeny are assessed (Blum, 1988).

Five species of *Actinidia* are reported to be hardier than *A. deliciosa*: *A. arguta, A. kolomikta, A. melanandra, A. polygama* and *A. purpurea* (Avery, 1991). The objective of our research was to screen germplasm for climatic adaptation to low-temperature stress in order to identify suitable parental genotypes for a breeding programme.

Materials and methods

Frost tolerance was assessed on 2-year-old plants under controlled freezing conditions. Several clones belonging to different *Actinidia* species were used. Each clone was propagated by cuttings in July 1990 and grown indoors until May 1991. Vines were transferred to an outdoor site for natural cold acclimation until tested.

A completely randomised experimental design was adopted consisting of six replications of each treatment (genotype). Pots were placed inside the climatic chamber in a Styrofoam box. A 20 cm thickness of glass wool, surrounded by polyethylene, covered the pots, thus reducing the incidence and duration of root system exposure to freezing injury. Chamber temperature was held at 6° C for 4 h, followed by a temperature drop of 2° C/h to - 18° C, a temperature reported to be damaging to kiwifruit plants (Monet & Bastard, 1985).

The most direct method of measuring freezing injury is to observe if the plants recover (Quamme, 1978). In the present paper, plant material was assessed for frost injury six weeks after transfer from freezing to warm (glasshouse) conditions. Vines were considered to be dead if no growth occurred, either above or

Table 1. Survival of kiwifruit cuttings exposed to - 18° C in February 1992

Species Number of clones/ cultivars	Number of plants in each class Dead No regrowth	Alive: Regrowth below 20 cm	 above 20 cm	Total number of plants
A. deliciosa (8 cultivars)	6	40	2	48
A. arguta (3 clones)	0	0	18	18
A. kolomikta (2 clones)	0	0	12	12
A. polygama (1 clone)	0	0	6	6

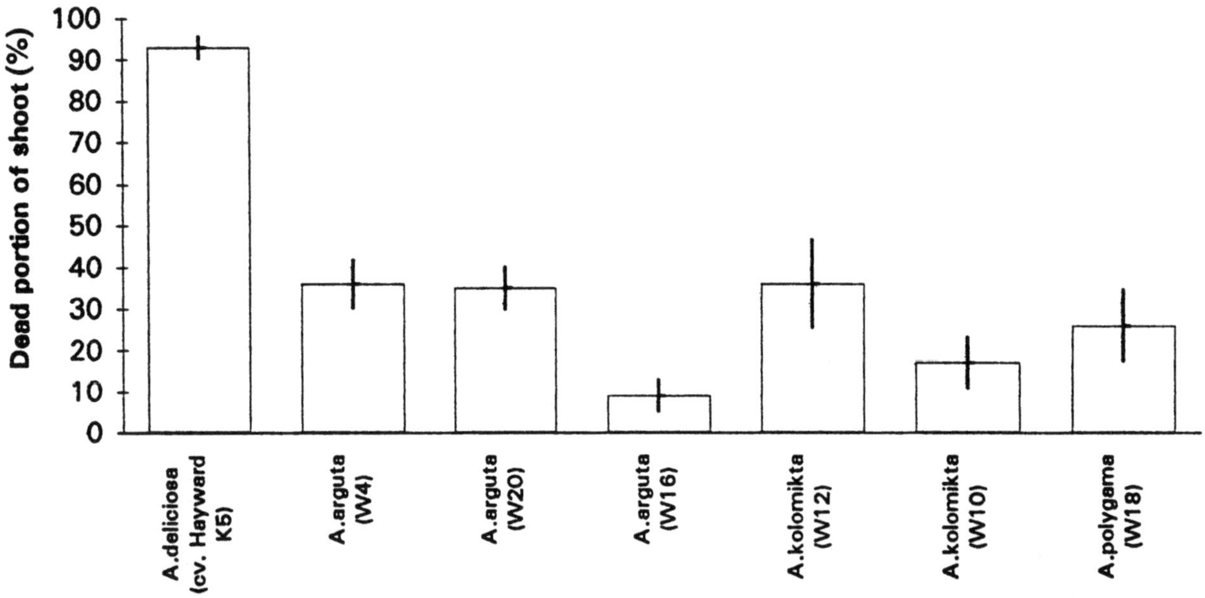

Fig. 1. Effect of controlled freezing on kiwifruit plants in February 1992. (Mean of 6 cuttings ± SE)

below the glass wool insulation layer covering the pots. In addition, direct frost damage was assessed from the degree of browning visible on the woody tissues.

Results

The *Actinidia* species examined differed significantly in frost hardiness. Six vines of *A. deliciosa* were entirely killed when frozen at - 18° C whereas all the vines of the botanical species survived at this temperature (Table 1). For the *A. deliciosa* cultivars, no growth occurred above the area previously protected by glass wool during the frost, except for two plants. Young growing shoots developed to greater heights for all botanical species than in cultivated species.

All clones had some tissue browning in February (Fig. 1) though *A. deliciosa* sustained more damage than the botanical species. The clone W16 showed least injury and was classified as tolerant to - 18° C in late winter.

Discussion

Table 1 shows that kiwifruit is a species in which the possibility of improvement in winter hardiness appears to be limited within the commercial cultivars. This is in accordance with results obtained previously with 'Hayward' and 'Bruno' cultivars by Pyke et al. (1986). As indicated by the data, the botanical species *A. arguta, A. kolomikta* and *A. polygama* are more frost tolerant in winter than 'Hayward', the only cultivar grown (Fig. 1). Our results, obtained under controlled freezing conditions are consistent with observations reported by Blanchet (1989) after natural freezes in 1985. In our study, *A. arguta* clone 16 would appear to have potential in the breeding of freezing tolerant cultivars, provided that frost-hardy genotypes appear within the segregating progenies.

Nevertheless, a single latewinter determination of frost tolerance is insufficient. The genotype ranking obtained for frost-tolerance in this study needs to be confirmed or clarified by an evaluation of frost hardiness during the fall, winter and spring.

Acknowledgements

This research was funded by a grant from the Commission of the European Communities (8001-CT91-0110).

References

Avery, J., 1991. *Actinidia*. Annual Report Northern Nut Growers Association 82: 175–179.

Blanchet, P., 1985. Les dégâts de gel sur kiwi (*Actinidia sinensis* Pl.), les risques du verger français. Arboric. fruit. 370: 43–48.

Blanchet, P., 1989. Description et comportement d'espèces d'*Actinidia* à fruits glabres dans le Sud-Ouest de la France. Fruits 44 (10): 543–552.

Blum, A., 1988. Plant breeding for stress environments. CCR Press (Ed.) Boca Raton, United States, 213 p.

Li, H., 1984. Subzero temperature stress physiology of herbaceous plants. Hort. Rev. 6: 373–416.

Monet, R. & Y. Bastard, 1985. Les kiwis et les gelées de janvier. Arboric. fruit. 377: 51–52.

Pyke, N.B., C.J. Stanley & I.J. Warrington, 1986. Kiwifruit: frost tolerance of plants in controlled frost conditions. N.Z.J. Agric. 14: 443–447.

Quamme, H.A., 1978. Breeding and selecting temperate fruit crops for cold hardiness. In: P.H. Li & A. Sakai (Eds) Plant cold hardiness and freezing stress mechanism and crop implications, pp. 313–332. Academic Press, New York.

H. Schmidt and M. Kellerhals (Eds.), Progress in Temperate Fruit Breeding, 463–467.

Genetic diversity in *Hippophae* L. and its use in plant breeding

Yingmou Yao & Peter M. A. Tigerstedt
Department of Plant Biology, University of Helsinki, P.O. Box 27 (Viikki D), 00014 Helsinki, Finland

Key words: Fruit, genetic diversity, growth rhythm, *hippophae*, isozyme, vitamin C

Summary

In the last decade, *Hippophae* species, particularly *H. rhamnoides* L., has received special attention in several countries for its multiple uses. The plant produces edible fruit with high nutrient and medicinal values and is characterized by its nitrogen-fixing symbiosis. It is widely distributed on the Eurasian continent (27°–69° N). Wide genetic variation is the basis of its distribution and provides good opportunities for selection and breeding. *Hippophae* is extremely variable in height, from a small bush less than 50 cm to a tree more than 20 m high. Phenological observations have shown a clear gradient of growth rhythm and plant size corresponding to the geographical distribution from north to south in *H. rhamnoides*. Studies on vitamin C concentration in the fruit have revealed significant differences between and within natural populations. The fruit size varies from 4 to 60 g/100 berries, the fruit color, from yellow, orange to red, and the shape, from flat, round, oval to cylindrical. Isozyme studies have shown that the mean number of alleles per locus per population is 2.1 and the percentage of polymorphic loci was 40.3% (at 0.95 standard). There is large genetic diversity residing within and between subspecies and species in the genus *Hippophae*.

Introduction

Sea buckthorn (*Hippophae* species) is dioecious and is widely distributed on the Eurasian continent (latitude 27°–69° N). It produces edible fruit with high nutrient and medicinal values and can fix nitrogen by its root nodules. Sea buckthorn has been used for producing fruit and firewood, for ornamental purpose, for preventing soil erosion, for improving agricultural and forest soil, and for reclaiming wasteland.

Since the early 1980s, *Hippophae*, particularly *H. rhamnoides*, has received special attention and become an important subject for domestication in many countries. On the other hand, the high nutrient and medicinal values of the fruit have led to uncontrolled exploitation and even destruction of sea buckthorn resources in some parts of its natural distribution. Thus, protection and preservation of the valuable germplasm have become urgent.

Genetic diversity is the basis for plant adaptation, selection and breeding. Knowledge of genetic variation is of importance for germplasm collection and conservation. This paper briefly summarizes some results of our observations in recent years. In addition some results from the literature concerning genetic variation in sea buckthorn are presented.

Material and methods

Since 1985, we have made several investigations on the growth, morphology and ecology in *Hippophae* natural populations and plantations in China, Finland, Sweden, Norway and Estonia. In the autumn of 1989 we made an extensive berry (and seed) collection and investigation of sea buckthorn from natural populations along the Finnish west coast and some islands. In 1989 and 1990, seeds of *H. rhamnoides* ssp. *sinensis*, ssp. *turkestanica*, ssp. *yunnanensis*, *H. gyantsensis*, *H. neurocarpa*, *H. salicifolia*, *H. tibetana* and the open pollinated seeds of a Russian cultivar as well as the berries/seeds from the F1 progenies of eight Russian cultivars were obtained from China, Russia and Finland, respectively. Hybrid seeds from our crosses among the Finnish, Danish, Chinese and Russian material were obtained in 1990.

Seedlings from above material have been grown in greenhouse and field. In the spring of 1991/1992, two multipurpose field trials were established in Helsinki and the bushes began to fruit in 1993. A modified complete block design with noncontiguous plots was used in these trials. The above material (berries and seeds) were also used for the isozyme and vitamin C studies of which parts of these results have been published (Yao et al., 1992; Yao & Tigerstedt, 1993).

On the basis of the above investigations of natural populations and the experiments in the nursery, greenhouse and field, this paper briefly describes the genetic diversity in *Hippophae* and its use in plant breeding.

Table 1. Variation of fruit size and vitamin C concentration in *Hippophae* (data based on our investigations and several reports)

	Weight g/100 berries	Vitamin C mg%
H. rhamnoides		
ssp. *sinensis*	4–40	307–2500
ssp. *yunnanensis*	10–20	958–1300
ssp. *rhamnoides*	20–60	28–360
ssp. *turkestanica*	6–21	98–1100
ssp. *mongolica*	28–60	350–400
H. tibetana	40–42	141–257
H. neurocarpa	4–5	3–15
H. salicifolia	19	1729
H. gyantsensis	6.5	23

Results and discussion

Genetic variation in plant height

According to recent taxonomical studies, the genus *Hippophae* includes five species, of which *H. rhamnoides* was further divided into eight subspecies. Rousi (1971) divided it into nine subspecies but Lian (1988) raised one subspecies to specific status. Except for *H. rhamnoides* which is widely distributed on the Eurasian continent, all other species in the genus are distributed in China and some neighboring countries along the Himalayan Mountains (Rousi, 1971; Liu & He, 1978; Tian, 1985; Lian, 1988; Yu et al., 1989).

One of the most variable characters in *Hippophae* is plant height. Of the five species in the genus, *H. salicifolia* Don is a tree of 3–11 m while *H. tibetana* Schlecht and *H. neurocarpa* Liu & He are bushes. However, *H. tibetana* is a small bush only 5–50 cm high, but *H. neurocarpa* is usually 1–3 m high. On the other hand *H. rhamnoides* and *H. gyantsensis* Lian may be bushes or trees depending on subspecies and individual genotypes. For example *H. rhamnoides* ssp. *sinensis* Rousi usually grows as a bush but trees more than 10 m high have been reported repeatedly. A tree of 16 m high and 320 years old (Lu, 1988) and a tree of 11.5 m high with 57 cm diameter at breast high (Xu, 1989) have been reported in ssp. *sinensis*. Lian (personal communication, 1993) found a virgin forest of *H. rhamnoides* ssp. *yunnanensis* Rousi with trees of more than 20 m high and with a diameter above 100 cm at breast high.

The great variation of plant height in *Hippophae* is an adaptation to the variable environmental conditions. *H. tibetana* is an alpine species which is distributed at

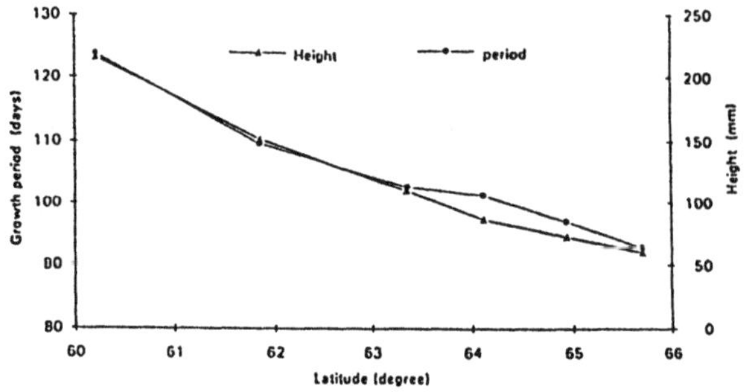

Fig. 1. Effects of latitude on growth period and height of 1-year-old sea buckthorn seedlings (n = 5 for each data point)

altitudes from 3300 to 5200 m. The low form is an adaptive response to cold and strong wind conditions. The same adaptation strategy is also found in other species and subspecies. For example, *H. rhamnoides* ssp. *rhamnoides* is normally a bush or a small tree 1–5 m high, but at latitude 65°40′ N it appears only 20–50 cm high on the Finnish sea shore where strong sea wind and harsh winter prevail. On the other hand, to compete with other plants, *Hippophae* may grow into a tall bush or tree. The variation of plant height can thus be considered as a balance between competition with other plants and the adaptation to its physical environmental conditions such as nutrients, drought, cold, and winds.

Genetic variation in growth rhythm

Table 2. Genetic variability at 6 loci in 25 populations of *Hippophae*. Populations 1–9 *ssp. rhamnoides*, 10–11 *ssp. turkestanica*, 12–22 *ssp. sinensis*, 23 open pollinated seeds from a Russian cultivar, 24 *H. neurocarpa*, 25 *H. tibetana*. (Yao & Tigerstedt, 1993)

Population	Mean sample size per Locus	Mean no. of alleles per locus	Percentage of loci polymorphic		Mean heterozygosity	
			0.95[a]	0.99[b]	Direct-count	HdyWbg expected[c]
1. Hailouto	152.3	1.7	0.0	33.3	0.020	0.022
2. Himanka	170.5	1.8	50.0	66.7	0.176	0.181
3. Björköby	128.2	2.2	50.0	66.7	0.169	0.194
4. Merikarvia1	230.2	2.0	50.0	50.0	0.233	0.237
5. Merikarvia2	192.8	1.7	50.0	50.0	0.247	0.238
6. Pori	307.8	1.8	66.7	66.7	0.215	0.226
7. Rauma	49.8	1.8	50.0	66.7	0.158	0.145
8. Aland	112.0	1.7	33.3	66.7	0.106	0.105
9. Helsinki	307.8	2.0	33.3	50.0	0.138	0.160
10. Chabuchar	187.3	3.0	33.3	83.3	0.086	0.093
11. Kongliu	66.8	2.3	50.0	100.0	0.118	0.151
12. Hualong	133.5	3.0	33.3	83.3	0.105	0.113
13. Tatong	53.5	2.2	33.3	50.0	0.156	0.142
14. Wushan	203.7	2.5	66.7	83.3	0.181	0.188
15. Taian	86.7	2.7	16.7	66.7	0.088	0.104
16. Tianshui	74.2	2.3	33.3	83.3	0.118	0.135
17. Yongshou	36.2	1.7	50.0	66.7	0.114	0.129
18. Linfen	122.7	2.0	50.0	50.0	0.115	0.137
19. Kelan	97.7	1.7	50.0	66.7	0.104	0.106
20. Youyu	86.8	2.2	33.3	50.0	0.081	0.080
21. Yuxian	191.2	2.7	16.7	66.7	0.085	0.093
22. Chifeng	184.7	1.5	33.3	50.0	0.064	0.061
23. Novosibirsk	19.0	1.5	50.0	50.0	0.184	0.148
24. Neurocarpa	10.0	1.7	33.3	33.3	0.150	0.167
25. Tibetana	2.0	1.0	0.0	0.0	0.000	0.000

a: A locus is considered polymorphic if the frequency of the most common allele does not exceed 0.95.
b: A locus is considered polymorphic if the frequency of the most common allele does not exceed 0.99.
c: Unbiased estimate of Hardy–Weinberg expectation (Nei, 1978).

Phenology, especially growth rhythm, is an important indicator of plant adaptation to local conditions. The results of our field experiment show a clear clinal variation in growth rhythm (growth period, growth cessation and leaf fall) and in plant height corresponding to the geographical distributions from north to south in *H. rhamnoides* (Fig. 1). The higher the latitude, the shorter the growth period and plant height. The correlation (R) between growth period and latitude is – 0.83 ($P < 0.0001$) and between height and latitude –0.88 ($P < 0.0001$). To adapt to climate towards the north, sea buckthorn reduces its growth period and height.

The clinal variation of growth rhythm and plant height provide good opportunity for plant breeding and selection, particularly for designing breeding programs. Breeders may be able to predict the performance of hybrids based on their parental populations, and design their breeding plans to obtain an "ideal" type with certain growth period, maturity time and plant height for a particular cultivation region.

Genetic variation in fruit morphology

Sea buckthorn is a small berry plant. Nonetheless large variation in fruit size exists in natural populations, both between species and within species (Table 1). Of the five species, *H. tibetana* has the largest berries while *H. neurocarpa* has the smallest. The largest variation is found in *H. rhamnoides* (4–60 g/100 berries). In some Russian cultivars, the berry size exceeds 60 g/100 berries. It is believed that there is much room for improvement in berry size by selection and crossing between different species, subspecies and populations.

Studies have shown that the vitamin C concentration varies from 3–2500 mg% (milligram per 100 g berries) in sea buckthorn fruit (Table 1). Asian subspecies of *H. rhamnoides* have higher vitamin C than the European subspecies. The results of our earlier studies showed that there were significant difference between and within natural populations (Yao et al., 1992).

Sea buckthorn fruit has beautiful colours, varying from yellow, orange to red. In autumn the fruit provides attractive scenery, particularly in late autumn after the leaves have been shed but the berries are still on the bushes. The berry flesh and juice have the same colours as the berry skin, and provide valuable natural colour for the food industry.

Sea buckthorn shows diverse fruit shapes, from flattened spherical, spherical, cylindrical, ovate, elliptic to many irregular shapes. The combination of fruit colour and fruit shape provides diverse choice and increases the ornamental value of the plant.

Genetic diversity based on isozymes

The results of our isozyme studies show high genetic variation in *Hippophae*. Measures of genetic variation at six loci (Adh, Sdh, Pgm, Pgi, Mdh1 and Mdh2) in 25 populations are summarized in Table 2. The mean number of alleles per locus, excluding population 25 (*H. tibetana*) which had only two seeds analyzed, varies from 1.3 to 3.0 with an average of 2.1 alleles per population per locus, higher than the means of 1.44 in dicots, 1.79 in woody plants, and 1.79 in wind-pollinated species quoted by Hamrick & Godt (1990). The percentage of polymorphic loci ranged from 0–66.7% (mean = 40.3%) at the 0.95 standard and 33.3–100% (mean = 62.5%) at the 0.99 standard among populations (see Table 2 for explanation). The mean heterozygosity per population (0.140) was also higher than average (0.113) given by Hamrick & Godt (1990).

Analysis of gene diversity shows that the total gene diversity averaged over all loci, was 0.461, much higher than the mean 0.310 for all plants (Hamrick & Godt, 1990). The mean gene diversity within populations was 0.135, accounting for 53.9% of the total gene diversity in *H. rhamnoides* (0.251). The gene diversity within species (*H. rhamnoides* and *H. neurocarpa*) accounted for 44.4% (0.205) and between species for 55.6% (0.257) of total gene diversity.

In addition, great variation is also found in crown forms, in leaves and branch shapes and colours, and in number of thorns on branches. Our trials show that *Hippophae* is a very good multipurpose plant having both nutritional and ornamental values. Thorn density directly affects the efficiency of management and harvest of plantation. The thornless type is found in natural populations and thornless or nearly thornless cultivars have been produced in the former USSR. The leaf colour is of high ornamental value.

In summary, there is much genetic variation residing in *Hippophae*, at species, subspecies, population and individual levels. Germplasm conservation should concern all levels of genetic diversity. Clinal variation in growth rhythm and plant height indicate that germplasm sampling should pay more attention to populations along latitude or altitude. The genetic variation in *Hippophae* provides plant breeders with good opportunities for breeding and selection, and helps them to design their breeding programmes.

References

Hamrick, J.L. & M.J.W. Godt, 1990. Allozyme diversity in plant species. In: A.H.D. Brown, M.T. Clegg, A.L. Kahler & B.S. Wier (Eds), Plant population genetics, breeding and genetic resources, pp. 43–63. Sinauer, Massachusetts.

Lian, Y., 1988. New discoveries of the genus *Hippophae* L. Acta Phytotaxonomica Sinica 26: 235–237 (in Chinese, abstract in English).

Liu, S.W. & T.N. He, 1978. The genus *Hippophae* from Qing–Zang plateau. Acta Phytotaxonomica Sinica 16: 106–108 (in Chinese).

Lu, R., 1988. A study of resource of *Hippophae* in Sichuan. Hippophae 1: 10–14 (in Chinese).

Nei, M., 1978. Estimation of average heterozygosity and genetic distance from a small number of individuals. Genetics 89: 583–590.

Rousi, A., 1971. The genus *Hippophae* L. A taxonomic study. Ann Bot. Fennici 8: 177–227.

Tian, H., 1985. *Hippophae*: Cultivation and Development. Shaanxi Agric. Info. Press (in Chinese).

Xu, Y., Y. Zhang, Y. Meng, C. Tong & Y.S. Zhang, 1989. The development and utilization of *Hippophae* resources in the Qilian Mountain areas. Hippophae 1: 11–13 (in Chinese).

Yu, Z., F. Ao & Y. Lian, 1989. Discussion on the problems of origin, classification, community and resources of sea buckthorn in China. Proc Intern Symp Sea Buckthorn (*H. rhamnoides* L). Xian, China. pp. 21–30.

Yao, Y., P.M.A. Tigerstedt & P. Joy, 1992. Variation of vitamin C concentration and character correlation between and within natural sea buckthorn (*Hippophae rhamnoides* L.) populations. Acta Agric. Scand. 42: 12–17.

Yao, Y. & P.M.A. Tigerstedt, 1993. Isozyme studies of genetic diversity and evolution in *Hippophae*. Genetic Resources and Crop Evolution 42: 153–164.

Author index

MIX
Papier aus verantwortungsvollen Quellen
Paper from responsible sources
FSC® C105338

If you have any concerns about our products,
you can contact us on
ProductSafety@springernature.com

In case Publisher is established outside the EU,
the EU authorized representative is:
Springer Nature Customer Service Center GmbH
Europaplatz 3, 69115 Heidelberg, Germany

Printed by Libri Plureos GmbH
in Hamburg, Germany